Securing 5G and Evolving Architectures

Pramod Nair

To, MARK

Looking forward to interesting conversations !!

♠ Addison-Wesley

Boston • Columbus • New York • San Francisco • Amsterdam • Cape Town • Dubai •
London • Madrid • Milan • Munich • Paris • Montreal • Toronto • Delhi • Mexico City •
São Paulo • Sydney • Hong Kong • Seoul • Singapore • Taipei • Tokyo

For information about buying this title in bulk quantities, or for special sales opportunities (which may include electronic versions; custom cover designs; and content particular to your business, training goals, marketing focus, or branding interests), please contact our corporate sales department at corpsales@pearsoned.com or (800) 382-3419.

For government sales inquiries, please contact governmentsales@pearsoned.com.

For questions about sales outside the U.S., please contact intlcs@pearson.com.

Visit us on the Web: informit.com/aw

Library of Congress Control Number: 2021917555

ISBN-13: 978-0-13-745793-9
ISBN-10: 0-13-745793-6

1 2021

Vice President, Editorial
Mark Taub

Director, ITP Product Management
Brett Bartow

Executive Editor
Nancy Davis

Development Editor
Christopher A. Cleveland

Managing Editor
Sandra Schroeder

Project Editor
Mandie Frank

Copy Editor
Bart Reed

Indexer
Erika Millen

Proofreader
Donna Mulder

Technical Reviewers
Dave Hucaby
Keith O'Brien

Editorial Assistant
Cindy Teeters

Designer
Chuti Prasertsith

Compositor
codeMantra

Graphics
codeMantra

Pearson's Commitment to Diversity, Equity, and Inclusion

Pearson is dedicated to creating bias-free content that reflects the diversity of all learners. We embrace the many dimensions of diversity, including but not limited to race, ethnicity, gender, socioeconomic status, ability, age, sexual orientation, and religious or political beliefs.

Education is a powerful force for equity and change in our world. It has the potential to deliver opportunities that improve lives and enable economic mobility. As we work with authors to create content for every product and service, we acknowledge our responsibility to demonstrate inclusivity and incorporate diverse scholarship so that everyone can achieve their potential through learning. As the world's leading learning company, we have a duty to help drive change and live up to our purpose to help more people create a better life for themselves and to create a better world.

Our ambition is to purposefully contribute to a world where:

- Everyone has an equitable and lifelong opportunity to succeed through learning.
- Our educational products and services are inclusive and represent the rich diversity of learners.
- Our educational content accurately reflects the histories and experiences of the learners we serve.
- Our educational content prompts deeper discussions with learners and motivates them to expand their own learning (and worldview).

While we work hard to present unbiased content, we want to hear from you about any concerns or needs with this Pearson product so that we can investigate and address them.

- Please contact us with concerns about any potential bias at https://www.pearson.com/report-bias.html.

Credits

FIGURE	CREDIT/ATTRIBUTION
Figure 4-24	Courtesy of O-RAN Alliance e.V.
Figure 5-51a	Courtesy of Google Cloud
Figure 5-51b	Courtesy of Amazon Web Services, Inc.
Figure 5-51c	Courtesy of Microsoft Corporation
Figure 3-1	Anand R. Prasad, Alf Zugenmaier, Adrian Escott and Mirko Cano Soveri, 3GPP 5G Security, 3GPP, August 6, 2018
Figure 3-2	Anand R. Prasad, Alf Zugenmaier, Adrian Escott and Mirko Cano Soveri, 3GPP 5G Security, 3GPP, August 6, 2018
Figure 3-13	D. Hardt, Ed., The OAuth 2.0 Authorization Framework, Internet Engineering Task Force, 2012
Figure 10-3	Courtesy of Cisco Systems
Figure 10-4	Courtesy of Cisco Systems
Figure 10-11	Courtesy of Cisco Systems
Cover	Yurchanka Siarhei/Shutterstock

Dedication

I would like to dedicate this book to my family both near and far. Thank you all for your unwavering support, motivation and patience throughout the development of this book.

Table of Contents

Part IV Emerging Discussions

Foreword

Society is about to embark on a digital upgrade—the next generation of the world's mobile communication infrastructure—5G. Along with new and innovative capabilities, 5G also introduces new security features, vulnerabilities, and risks. 5G does not just represent significantly increased bandwidth and lower latency, but it is expected to fundamentally change the mobile ecosystem with new partnership models, network slicing, massive deployment of Internet of Things (IoT) devices, and ultimately, an increasingly critical dependency on the technology for society to function. Due to this, our ability to secure 5G will directly affect the resilience of critical infrastructure and national security.

Some of the security key risks affecting 5G confidentiality, integrity, and availability are supply chain risks, increasing complexity leading to new vulnerabilities, and inherent weaknesses in the standards. The supply chain risks have reached the geopolitical center stage due to the high societal impact of 5G, and this has led to national and EU-level regulations, risks assessments, and GSMA's accreditation scheme Network Equipment Security Assurance Scheme (NESAS). The inherent increased complexity of 5G leads to a wide range of new potential vulnerabilities that will require increased vigilance from product vendors, service providers, and users alike.

In order to manage these risks, 5G is equipped with a broad range of security features and capabilities, and GSMA has outlined a list of critically sensitive functions—virtualization infrastructure, controller, orchestrators, Internet gateways, network slicing, mobile edge computing, routing and switching of IP traffic at the core, database functions, authentication, and access control. As always, a security by design approach following a zero-trust approach, with secure deployments and good operational hygiene, is key to securing the world's 5G deployments.

In this book, Pramod Nair guides us through the evolution of cellular technologies from a security perspective, the security architecture, deployment modes and use cases of 5G, as well as discusses end-to-end security architecture and prioritizing security investments. His unique outlook as the Lead Security Architect, head of 5G security architecture in Cisco Systems, and from more than 20 years in security allows him to combine a theoretical and applied perspective for the benefit of both business and technical readers.

André Årnes, PhD
Senior Vice President and Chief Security Officer at Telenor Group
Professor II at the Norwegian University of Science and Technology

Preface

5G technology will redefine the way we perceive cellular networks and will touch almost every aspect of our lives. 5G is not about just being faster, bigger, or better; it's about enabling multiple services that we'll all consume on an everyday basis. It will give rise to a new ecosystem of developers building applications that exploit the openness of 5G to help you develop new use cases for consumption by enterprises and subscribers alike. New features in 3GPP Releases 16 and 17 help further enable new use cases for non-public deployment of 5G by industry verticals and tighter convergence of 3GPP and non-3GPP technologies, bringing in multiple deployment methods—including on-premises, hybrid, and fully public cloud-based deployments. The 5G ecosystem will see a breakout from 3GPP-only based architecture to an open, multi-technology, multi-standard, polyglot ecosystem.

This evolution of the technology landscape also requires an evolution of the security mindset. We should start thinking of security as a foundational layer. It should be one of the primary foundations for any planned 5G use case implementation. This requires embracing multilayered security beyond the requirements in 3GPP specifications.

The business operational risk, legal risk, and reputational risk exist not only for the companies providing 5G software and hardware infrastructure, but for all companies, nation-states, and individuals who provide and consume 5G technology.

The time is now to evaluate the cyber risk posture and apply innovative thoughts to how we can approach these challenges today and build for what's to come tomorrow.

Motivation for Writing This Book

Security in evolving cellular technologies is not an easy concept to grasp, as the technologies have evolved rapidly and are becoming increasingly complex and nuanced as they become more open, especially when you add 5G to the mix.

5G will also enable enterprises and industry verticals to deploy private 5G/non-public 5G networks (5G NPN networks) on their own, without any integration with service providers. This necessitates private and government sectors to fully understand the 5G threat surfaces, develop methods to mitigate the threats, and prioritize the investments in security.

The existing material on security and cellular technologies is dispersed across many resources and does not cover the end-to-end 5G threat surfaces, threat mitigation, examples of real-life deployment scenarios, and prioritization of security controls based on use cases and deployment scenarios. The learning curve for a person trying to understand the evolution in cellular technologies, new architectures, multiple deployment methods, threat surfaces, and mitigation techniques is extremely steep and sometimes unnerving.

It is not surprising that the topic of securing cellular technologies tends to flummox newcomers and even seasoned network security engineers.

This book brings all the information together and arranges the key topics in such a way that they can be easily consumed and understood. The main purpose of this book is to enable any person to understand the key aspects of securing 5G and evolving technologies. This book covers a range of topics; it will take you through the evolution of technologies from 2G to 5G, with deep dives into specific topics, such as securing non-public 5G networks/private 5G deployments and prioritizing security investments.

The goal of this book is to provide pragmatic views on securing 5G and evolving networks. The knowledge and information gathered through numerous customer workshops, brainstorming sessions with service providers, industry verticals, industry experts from multiple vendors, proof of concepts, and lessons learned from actual security deployments for 5G networks are detailed in this book. Discussions with multiple CSOs and CTOs have enlightened me on the key data points required for prioritizing security, which you will see highlighted in this book. Apart from service providers, industry verticals are expected to adopt 5G technology, and this area has been expanded into specific use cases, threats, and mitigation techniques. This book closes with a chapter discussing the key areas of security evolution that will motivate you to investigate different aspects of security as the network evolves. It is aimed at helping you create a new mindset while securing your networks of the future.

Who Should Read This Book

I have designed this book so that you can begin without any prior knowledge about 5G or any preceding cellular technologies. This book is written to be suitable for multiple levels of technical expertise, including the following:

- Security experts looking to understand the history of cellular technology evolution to 5G, key 5G security enhancements, and security challenges

- Early-in-career telecom engineers, transport design engineers, and radio engineers looking to design and implement mobile networks

- Government departments looking at security impacts of 5G deployment for use cases such as smart city and looking at implementing security measures

- Management consultants advising governments and service providers on 5G security strategy

- CSO and CTO teams from service providers looking at securing 5G deployments

- CSO and CTO teams from enterprises deploying NPN/private 5G

- Enterprise network design and implementation teams deploying NPN/private 5G deployments

- Security architects responsible for securing the mobile infrastructure

- Enterprise solution architects and enterprise security architects working with enterprises integrated with service provider 5G networks

- Security strategy teams within service providers, enterprise and industry verticals deploying 5G

- Cloud computing and data center teams involved with 5G strategy and deployment

- Enterprise solution and security architects deploying standalone private/NPN 5G or utilizing service providers' 5G slice network

- Audiences of varying levels of expertise from the military and defense community

- Audiences from industry verticals such as smart manufacturing, critical infrastructure entities and vendors, and autonomous vehicle manufacturers

- Cybersecurity vendor product managers looking for use cases or features to enhance security products to cater for secure 5G deployments

- Students who would like to get a quick understanding of cellular technologies and a look at the new features in 5G

Throughout the book, you will see practical examples and real-life scenarios of how you might architect a solution to mitigate threats and improve the security posture of your network.

How This Book Is Organized

To allow technical and nontechnical audiences to consume the book in an effective and organized way, it is split into four parts. The parts and chapters cover specific topics.

Part I, "Evolution of Cellular Technologies to 5G, Security Enhancements, and Challenges," explains the evolution of cellular technologies toward 5G as well as new security enhancements and new security challenges brought in by 5G. It will also take the reader through different deployment modes, including private 5G / non-public networks (NPN). This part will mostly cater to the audience who wants a high-level view of 5G technology and its security aspects. It includes the following chapters:

- **Chapter 1**, "Evolution from 4G to 5G," covers the evolution of cellular technologies and will provide you with a basic understanding of the 5G technology features. It will also take you through some of the key enhancements in 3GPP Rel-16 and Rel-17.

- **Chapter 2**, "Deployment Modes in 5G," covers the different non-standalone and standalone deployment modes and use cases, which can be mapped to specific deployment modes.

- **Chapter 3**, "Securing 5G Infrastructure," covers new security enhancements and new security challenges brought in by 5G. It also discusses the reasons why you should have an external layer of security controls, even though 3GPP provides some enhancements in security.

Part II, "Securing 5G Architectures, Deployment Modes, and Use Cases," covers the security controls for 5G network components such as RAN, transport, 5GC, and devices. It then takes you through securing 5G enablers—such as multi-access edge compute (MEC), software-defined networks (SDNs), network slicing, orchestration, and automation—and protecting different deployment methods such as on-premises, private and public cloud based MEC, and hybrid cloud, including open RAN deployments. It finally covers securing key 5G use cases such as critical infrastructure, vehicle-to-everything (V2X), and smart factory. This part of the book will be of keen interest to readers who would like to deep-dive into the security aspects of 5G and its key use cases. It includes the following chapters:

- **Chapter 4**, "Securing RAN and Transport Deployments in 5G," covers the 5G RAN and transport threat surfaces and threat mitigation for the 5G public and non-public deployments, including open RAN. This chapter also takes you through some real-world attacks and mechanisms to mitigate them.

- **Chapter 5**, "Securing MEC Deployments in 5G," covers various MEC deployment models, network functions deployed in the private and public cloud based MEC, its threat surfaces, and methods to mitigate the threats. The chapter also provides some real-world risk and risk mitigation scenarios.

- **Chapter 6**, "Securing Virtualized 5G Core Deployments," covers the threats due to virtualized 5G Core deployments and new methods of software development and deployment. This chapter also provides some key recommendations to secure your virtualized 5GC deployments with vendor-agnostic approaches and includes some real-world scenarios.

- **Chapter 7**, "Securing Network Slice, SDN, and Orchestration in 5G," covers network slicing and enablers of network slicing such as software-defined networks (SDNs), orchestration, and automation. The chapter also explains the threat surfaces and threat mitigations specific to network slicing and its enablers. This chapter also delves into the network slice as a service (NSaaS) offering, its threat surface, and methods to mitigate the threats.

- **Chapter 8**, "Securing Massive IoT Deployments in 5G," covers the risks related to IoT devices and related connectivity and management. The chapter then goes on to explain different security mechanisms and best practices to secure your network from any IoT device-based attacks.

- **Chapter 9**, "Securing 5G Use Cases," covers critical infrastructure, V2X, and smart manufacturing use cases, which use different types of IoT devices—some smart, some semi-smart—as well as non-smart devices. The chapter takes you through the risks within these three use cases and methods to mitigate the risks.

Part III, "End-to-End 5G Security Architecture and Prioritizing Security Investments," provides an overview of the various security recommendations for end-to-end 5G security and discusses the factors based on which certain security controls can be prioritized among other security controls for 5G networks. This part will be of keen interest to an audience who would like to have an end-to-end view of security and understand the methods to prioritize investments in security. It includes following chapters:

- **Chapter 10**, "Building Pragmatic End-to-End Security 5G Architecture," covers the key building blocks for creating an end-to-end security layer for 5G deployments. This chapter also provides you with a checklist for each of the 5G domains and includes zero-trust design principles.

- **Chapter 11**, "Prioritizing 5G Security Investments," covers the considerations and recommendations for prioritizing investments to secure your 5G network. This chapter takes two primary scenarios—one related to a service provider providing mobile service, and the other related to the non-public deployment methods for industry verticals and enterprises.

Part IV, "Emerging Discussions," takes you through the topics aimed at new features being discussed for 5G and evolving architectures, security enhancements using machine learning (ML) and artificial intelligence (AI), and the method to make your network quantum safe. This part will be of keen interest to readers who would like to understand the key discussions in the security industry around 5G and evolving technologies. It includes following chapter:

- **Chapter 12,** "5G and Beyond," covers the adoption and adaptation of 5G standalone technology with new use cases, convergence of non-3GPP and 3GPP technologies, application of AI and ML in securing 5G and evolving technologies, and the importance of deploying crypto-agile mobile networks.

Due to ongoing developments, Chapter 12 will occasionally be updated with relevant new content and insights on the book's website at www.informit.com. Register your copy of *Securing 5G and Evolving Architectures* on the InformIT site for convenient access to these updates and/or corrections as they become available. To start the registration process, go to informit.com/register and log in or create an account. Enter the product ISBN (9780137457939) and click Submit. Look on the Registered Products tab for an Access Bonus Content link next to this product, and follow that link to access any available bonus materials. If you would like to be notified of exclusive offers on new editions and updates, please check the box to receive email from us.

Please note that this book is written with a vendor-neutral approach, and it does not give recommendations on what vendor should be deployed. Each service provider or industry vertical planning to deploy 5G can evaluate the security controls required and make decisions based on their own criteria, circumstances, and targeted use cases. This book covers the details of the security controls, required features, and functions required for securing 5G and evolving networks, allowing you to make better informed decisions.

Happy reading, and I hope you enjoy reading this book as much as I enjoyed writing it!

Acknowledgments

I would like to acknowledge the tremendous support I received from the Cisco staff, especially my management team and colleagues.

Similarly, I would like to thank the reviewers, Keith O'Brien and David Hucaby, for their comments, feedback, and insights that enriched the content of this book.

I would like to thank Dr. André Årnes, PhD, Senior Vice President and Chief Security Officer at Telenor Group and Professor at the Norwegian University of Science and Technology, for writing the foreword.

I would like to extend my appreciation to the people from multiple companies who provided constructive comments during those numerous 5G security customer workshops and brainstorming calls.

I would like to thank executive editor, Nancy Davis, for her guidance, feedback, and massive support. I would also like to extend my thanks to the Pearson/Addison-Wesley team, especially to Chris Cleveland, the development editor, for his robust guidance throughout the editing process, and to Mandie Frank, for her support through the production process.

Finally, I would like to thank the many standards organizations, technologists, security experts, and industry peers who continue to contribute to the fields of both mobile communications and security.

About the Author

Pramod Nair is a Lead Security Architect at Cisco Systems focusing on service providers. During his 20 years of experience in the industry, Pramod has worked in multiple areas, including research and development, designing end-to-end mobile networks, and technical consulting on military and defense projects.

Among other responsibilities in his current role within Cisco, Pramod leads 5G Security Architecture, driving its adoption globally, and has been instrumental in architecting secure next-generation networks for customers across the globe. He is a regular speaker on the subject at large conferences and industry events.

Pramod is an active member of the security community. His role is to help mobile network providers, service providers, industry verticals, the national security and defense sectors, and other agencies dedicated to securing critical infrastructures. He is also deeply involved with industry trade organizations, has co-chaired the 5G security white paper within the 5GAmericas work group, and works with the National Institute of Standards and Technology (NIST) on 5G security.

Pramod holds a patent in fraud detection and has published various white papers and articles covering security-related topics.

Chapter 1

Evolution from 4G to 5G

After reading this chapter, you should have a better understanding of the following topics:

- Evolution of cellular technologies
- Enhancements in 5G as compared to 4G
- Key 5G features and enhancements in 3GPP Releases 15, 16, and 17
- Key 5G Advanced features and enhancements under study in 3GPP Release 18

This chapter takes you through the evolution of cellular technologies and provides you with a basic understanding of 5G technology features.

This chapter will be of particular interest to the following individuals and teams from enterprise/5G service providers deploying 5GC and cybersecurity vendors planning product developments for 5G security use cases:

- Security leaders within government organizations who would like a quick look into the brief history of cellular communications
- Teams from enterprise and industry verticals who want an understanding of cellular technologies
- Enterprise cybersecurity vendor security architects looking to understand the evolution of cellular technologies
- Cybersecurity vendor security architects and product managers looking to understand what's new in 3GPP Releases 15 and 16 and what's planned in Release 17
- Cybersecurity students who would like to get a brief understanding of cellular technologies and new features in 5G

In mobile technologies, the "G" in 1G, 2G, 3G, 4G, and 5G refers to "Generation," where there has been a major shift in the capabilities offered in terms of spectrum efficiency, higher throughput, making the architecture more "flat," and enhanced algorithms in the air interface for providing better security. 1G was based on analog technology, and 2G was the first digital technology to be used in mobile cellular communication.

The foundation of 4G was set by International Telecommunications Union–Radio communications sector (ITU-R) around 1998. In 2009, 4G was established and called Long Term Evolution before eventually being called 4G (pure marketing), although it did not meet the criteria of 1Gbps for low-mobility use cases (pedestrian users). In 2011, LTE-Advanced (LTE-A) fulfilled the criteria set by ITU-R and specified in Releases 8 and 9 of Third-Generation Partnership Project (3GPP) and was actually "true 4G." The U.S. Defense Advanced Research Projects Agency (DARPA) had also envisioned 4G by using end-to-end IP networking and distributed architectures.

Figure 1-1 illustrates the evolution of cellular digital networks from 2G to 5G.

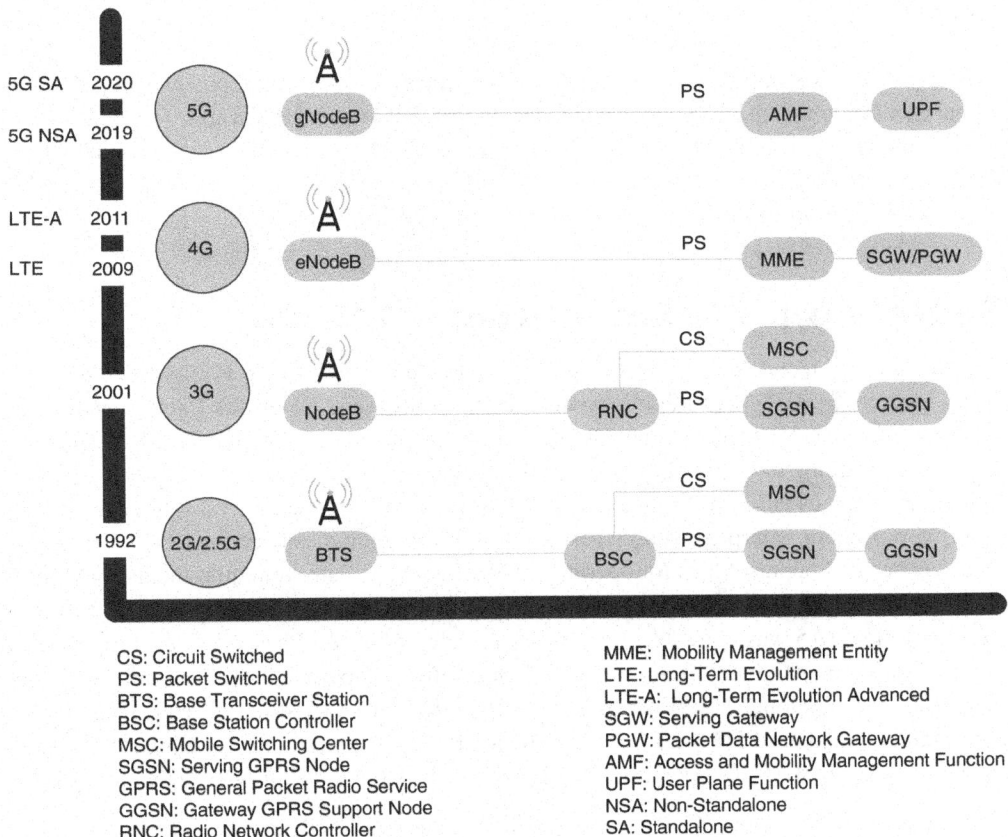

CS: Circuit Switched
PS: Packet Switched
BTS: Base Transceiver Station
BSC: Base Station Controller
MSC: Mobile Switching Center
SGSN: Serving GPRS Node
GPRS: General Packet Radio Service
GGSN: Gateway GPRS Support Node
RNC: Radio Network Controller

MME: Mobility Management Entity
LTE: Long-Term Evolution
LTE-A: Long-Term Evolution Advanced
SGW: Serving Gateway
PGW: Packet Data Network Gateway
AMF: Access and Mobility Management Function
UPF: User Plane Function
NSA: Non-Standalone
SA: Standalone

FIGURE 1-1   Evolution from 2G to 5G

Mobile technologies had come a long way from 1G in 1981, 2G in 1992, 3G in 2001, to 4G in 2009/2011(LTE-A). As shown in Figure 1-1, 4G brought a shift in designing the mobile network, as it was based on all Internet Protocol (IP) deployment, allowing better throughput due to reduced latency in the back haul, better modulation techniques, advanced dynamic channel allocation, and adaptive traffic scheduling. Advanced antenna systems such as Multiple Input Multiple Output (MIMO) with beamforming helped achieve better throughput, but its use was restricted due to cost and was aimed at specific dense clusters that required high throughput. 4G also brought in a major shift in the network design, as it moved away from the requirement of having both circuit-switched and packet-switched networks and enabled the simplification of network architecture to a purely packet switched (PS) all-IP flat network architecture with reduced latency. For example, the following systems used separate circuit-switched networks for voice and packet-switched networks for data:

- 2.5G, also called General Packet Radio Service (GPRS)
- 2.75G, also called Enhanced Data Rates for GSM Evolution (EDGE)
- 3G systems

For a service provider, 2G and 3G systems required two separate sets of equipment in parallel, as they had separate circuit switch (CS) traffic for voice and packet switch (PS) traffic for packet data. The IP-based flat network architecture in LTE helped to reduce costs and complexity by replacing the separate circuit- and packet-switched networks with a single Evolved Packet Core (EPC), which also had the capability to support handovers from legacy 2G and 3G technologies.

Mobile Network Evolution from 4G to 5G

5G is the successor of 4G technologies, and it defines a system that uses 5G New Radio (NR) software. The standards are set by 3GPP, and the minimum specifications were set by the International Telecommunications Union (ITU) International Mobile Telecommunications–2020 (IMT-2020) standard in 2015. 5G is not just going to be about high-speed data connections for enhanced mobile broadband; it will enable several new capabilities that can cater to several new enterprise use cases. 5G will not just be about serving consumer/enterprise subscribers with high-throughput connectivity. 5G will enable new revenue avenues and opportunities for operators by being able to cater to requirements for several new enterprise use cases. The first NR specification was made in 2015, whereas the phase of 3GPP 5G specifications in Release 15 was completed in 2019 (also referred to as "Phase 1" standardization of 5G). The initial Phase 1 deployment of 5G was based on a software upgrade of 4G LTE infrastructure and was called Non-Standalone (NSA) mode, which depended on an existing 4G LTE network for control plane signaling and used 5G NR for the user plane. Phase 2 is based on Standalone (SA) mode with 5G packet core network (5GC) providing the control plane and 5G NR providing the user plane.

Figure 1-2 illustrates the high-level 5G architecture with comparisons to 4G architecture.

Apart from higher data rate/throughput and utilizing a new spectrum, the key differentiators and technology evolutions in 5G as compared to 4G are discussed in the sections that follow.

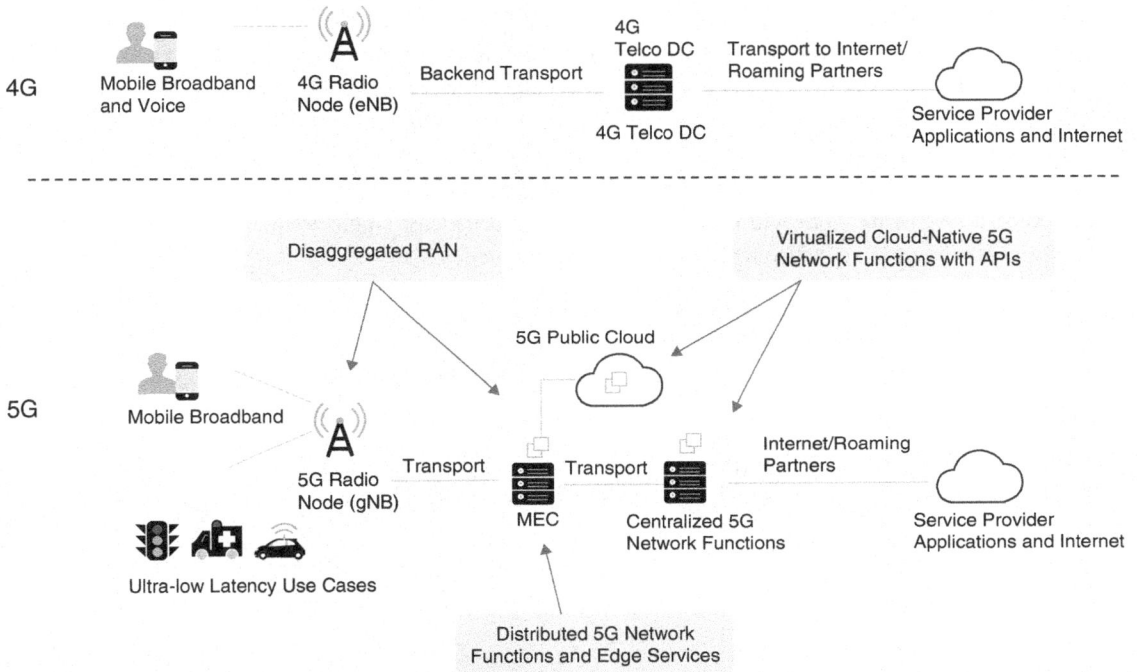

FIGURE 1-2 Comparison of 5G and 4G Architectures

5G New Radio Features

New Radio (NR) for 5G should support the following key requirements:

- The ability to handle high data rates and a wider bandwidth

- Support for ultra-reliable low-latency communications (URLLC) with fast short bursts of data

- Support for energy-efficient communications for low-powered devices catering for Internet of things (IoT)

The requirements are met by using cyclic prefix-based orthogonal frequency-division multiplexing (OFDM), which is the same waveform LTE uses for downlink. For downlink, only cyclic prefix-based OFDM (CP-OFDM) is used, while for uplink, both CP-OFDM and Discrete Fourier transform spread OFDM (DFT-S-OFDM) can be used. Compared to CP-OFDM, DFT-S-OFDM has the merit of a lower peak-to-average ratio.

Dynamic Spectrum Sharing (DSS) and Flexible NR Protocols

During the initial stages of 5G adoption, Non-Standalone (NSA) mode is the preferred mode of deployment for service providers, before transitioning to the full-fledged 5G Standalone mode of deployment. The 5G NR can adapt to existing 4G deployments and share the spectrum used exclusively by 4G today, using a mechanism called Dynamic Spectrum Sharing (DSS). DSS allows the service providers to share the spectrum between 4G and 5G without having to re-farm the 4G spectrum. This method is therefore spectrum efficient and facilitates a smooth transition to 5G SA. In DSS, the 5G NR device, such as user equipment (UE), will have two active radios—one for 4G and the other for 5G NR—that detect the synchronization signal blocks (SSB) to access the network. SSB is sent periodically by the network within the 4G multimedia broadcast single-frequency network (MBSFN) subframes to maintain synchronization in time and frequency.

Operating Spectrum and Bandwidth

The frequency range in which 5G NR can operate is split between FR1 and FR2, as documented in Table 1-1.

TABLE 1-1   5G Frequency Range

Frequency Range Designation	Frequency Range
FR1	410–7125 MHz
FR2	24250–52600 MHz

5G NR also supports hierarchical system bandwidth from 5 MHz to 100 MHz for FR1 and from 50 MHz to 400 MHz for FR2, respectively, as indicated in Table 1-2.

TABLE 1-2   5G Bandwidth

Frequency Range Designation	Bandwidth
FR1	5–100 MHz
FR2	50–400 MHz

Massive MIMO

One of the key enablers for 5G's higher data rates is Multiple Input Multiple Output (MIMO). MIMO as a technology is not new and has been around for quite some time. The key reason MIMO is important in 5G is due to the use of higher frequencies in 5G, which leads to higher losses and therefore lower coverage. Massive MIMO with its higher number of transceivers will help in improving capacity, coverage, and spectrum efficiency as well as the user experience. Apart from the higher number of transceivers, beamforming will also help in identifying coverage requirements for moving users and optimizing the offered coverage and capacity.

Disaggregated Architecture

5G introduces disaggregation of network functions in the Radio and Packet Core domains, which is a dramatic departure from proprietary closed hardware and software sourced from a single vendor toward completely decoupled, open source network functions that are then combined to provide the 5G-specified services.

Figure 1-3 illustrates the functional split of Next-Generation RAN (NG-RAN) and 5G Core, along with their main tasks.

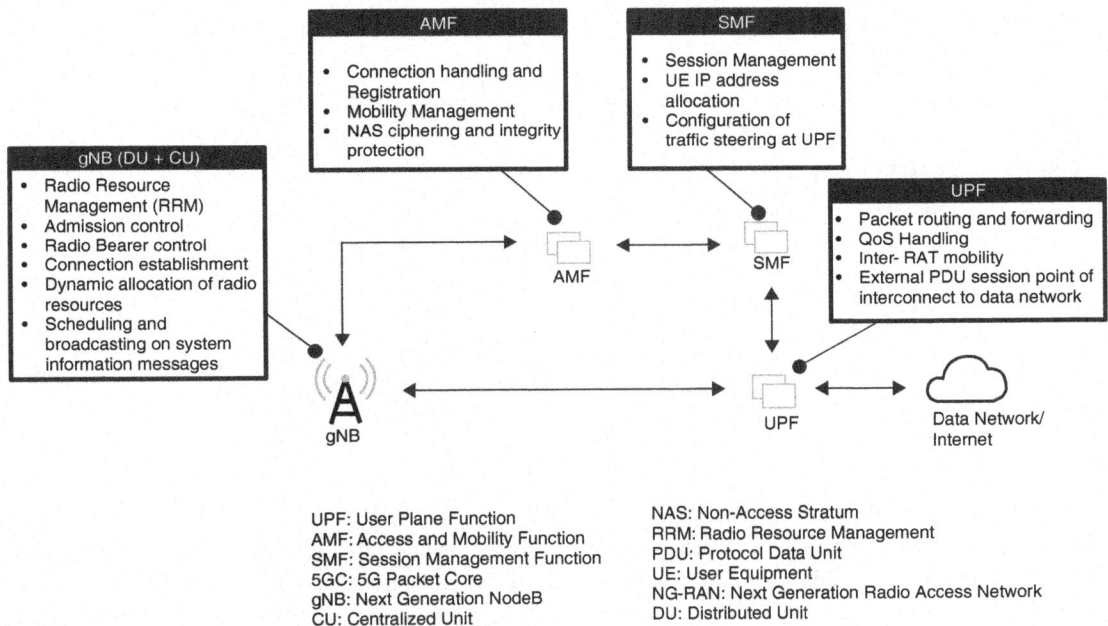

AMF
- Connection handling and Registration
- Mobility Management
- NAS ciphering and integrity protection

SMF
- Session Management
- UE IP address allocation
- Configuration of traffic steering at UPF

gNB (DU + CU)
- Radio Resource Management (RRM)
- Admission control
- Radio Bearer control
- Connection establishment
- Dynamic allocation of radio resources
- Scheduling and broadcasting on system information messages

UPF
- Packet routing and forwarding
- QoS Handling
- Inter- RAT mobility
- External PDU session point of interconnect to data network

AMF

SMF

gNB

UPF

Data Network/ Internet

UPF: User Plane Function
AMF: Access and Mobility Function
SMF: Session Management Function
5GC: 5G Packet Core
gNB: Next Generation NodeB
CU: Centralized Unit

NAS: Non-Access Stratum
RRM: Radio Resource Management
PDU: Protocol Data Unit
UE: User Equipment
NG-RAN: Next Generation Radio Access Network
DU: Distributed Unit

FIGURE 1-3 Functional Split of NG-RAN and 5GC

As shown in Figure 1-3, the next-generation NodeB (gNB), which is the next-generation base station supporting the new 5G New Radio (NR), performs the access node–related tasks, including radio resource management, radio bearer, connection, scheduling, and admission control for the UE. In the 5GC, Access and Mobility Function (AMF) caters for control plane, Session Management Function (SMF) caters for session establishment, and User Plane Function (UPF) handles the user plane.

Apart from the functional split between NG-RAN and 5GC, as illustrated in Figure 1-3, the gNB can be deployed as virtual centralized unit (VCU) and virtual distributed unit (VDU) instances on commercial off-the-shelf (COTS) hardware, due to the disaggregation and decomposition introduced by 5G, as shown in Figure 1-4, thus enabling flexible 5G deployment.

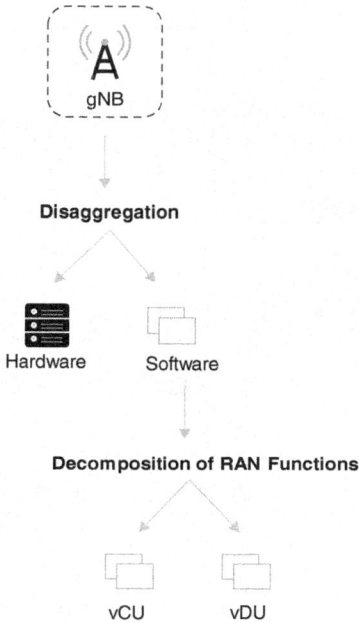

FIGURE 1-4   Disaggregation and Decomposition of gNB

The CU processes non-real-time protocols and services, and the DU processes PHY-level protocol and real-time services. This allows you to build an end-to-end public or non-public 5G network by employing virtualized network functions from multiple vendors and deploying it on COTS servers using merchant silicon.

Network disaggregation eliminates vendor lock-in and enables you to select the best combination of hardware and software vendors based on the offered strengths and features of their respective solutions. The decomposition and disaggregation of 5G enables a major architectural shift to an edge infrastructure that combines decomposed subscriber management with access functionality. This shift will also apply to wireline networks with increasing adoption of the 5G Fixed Access Networks. Figure 1-4 illustrates the disaggregation and decomposition of Radio Access Network (RAN) functions. Disaggregation in 5G brings you lots of flexibility because it fundamentally changes the way networks are procured, built, and operated. Here are some of the benefits:

- Efficient partitioning of network functions and resources to support the 5G use cases such as enhanced mobile broadband (eMBB), ultra-reliable low-latency communication (URLLC), and Internet of Things (IoT) by using technologies such as software-defined networking, edge computing, service function chaining, and service orchestration.

- In 5G deployment models such as Cloud RAN (C-RAN), pre-provisioning of baseband resources network isn't required for the maximum capacity of the site. You can use baseband resource pooling provisioned for the traffic profile of the entire network. These traffic profiles can be scaled up or down during peak-hour usage or in the case of high-traffic events.

- Scalability to quickly and cost effectively adapt to various network topologies and use cases.

- Enhanced network availability and reliability already built into NFV technologies such as OpenStack, the virtual infrastructure manager, and service orchestrator.

- Disaggregation of the hardware from the software using commercial off-the-shelf, high-processing density hardware and open interfaces.

- Enablement of advanced algorithms such as cooperative multipoint and inter-site carrier aggregation.

As mobile technologies evolved from one generation to the next, the functions performed by base stations and other devices have been distributed using cloud technologies. Cloud-RAN decouples the baseband processing from the radio units, allowing the processing power to be pooled at a central location, thus reducing the required redundancy. One of the most important aspects of C-RAN architecture is the splitting of base station functions at distributed locations and centralized C-RAN servers. There are eight standardized options for this split, with two (options 2 and 7) being seen most frequently. Figure 1-5 illustrates the functional split consideration in the 5G RAN.

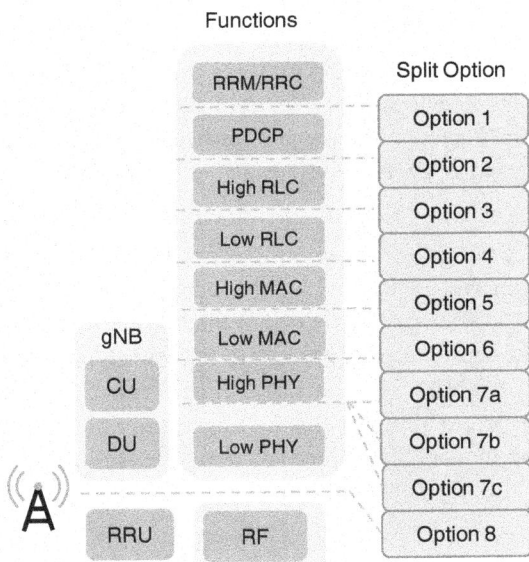

FIGURE 1-5   Functional Split Options in 5G RAN

The chosen radio split option will have a direct impact on your transport (xHaul) design and deployment because the transport requirements vary dramatically with radio split options; therefore, you should consider your existing transport design, its capabilities, and future enhancements in your transport layers before you decide on the radio split options. The 3GPP specifications 38.801 and 38.816 provide additional details about the radio split options.

Flexible Architecture

Traditional mobile networks are based on defined perimeters, where most of the mobile packet core functions are centralized. 4G did bring in Control Plane and User Plane Separation (CUPS), but it was not used much due to limited use cases and advantages. 5G SA Packet Core is inherently equipped with several new built-in capabilities so that you have flexibility and capability to face new challenges thrown open by the new set of requirements for varying new use cases in 5G. The network functions in the new 5G Core are broken down into smaller entities, such as the Session Management Function (SMF) and User Plane Function (UPF), which can be used on a per-service basis. Gone are the days of huge network boxes—welcome to services that can be deployed on private cloud and public cloud and automatically register and configure themselves over the service-based architecture (SBA), which is built with new functions like the Network Resource Function (NRF) that borrow their capabilities from cloud-native technologies. This flexible architecture will equip you with more capabilities to cater to enterprise customer needs to support their current use cases as well as new use cases. Enterprise and industry verticals can also deploy their own standalone non-public network (NPN) or public network integrated non-public network (PNI-NPN) to fulfill the use cases.

5G Non-Standalone (NSA) Solution

In Non-Standalone (NSA) architecture, the 5G RAN and its NR interface are used in conjunction with the existing LTE and Evolved Packet Core (EPC) infrastructure core network (respectively 4G Radio and 4G Core), as shown in Figure 1-6, thus making the NR technology available without network replacement.

In this configuration, only the 4G services are supported but enjoy the capacities offered by the 5G New Radio (lower latency and so on). The NSA is also known as E-UTRA-NR Dual Connectivity (EN-DC). This is where you will leverage your existing EPC (4G) Packet Core to anchor the 5G NR using the 3GPP Release 12 dual connectivity feature. This will help you with aggressive 5G launch needs to launch 5G in a shorter time and with less cost. The 5G NSA solution will suffice for some initial use cases. However, 5G NSA has limitations with regard to getting a much cleaner, truly 5G-native solution, so you will eventually be expected to migrate to a 5G Standalone solution. The section, "5G Non-Standalone (NSA) Deployments," in Chapter 2, "Deployment Modes in 5G," covers the different options of the NSA architecture in further detail.

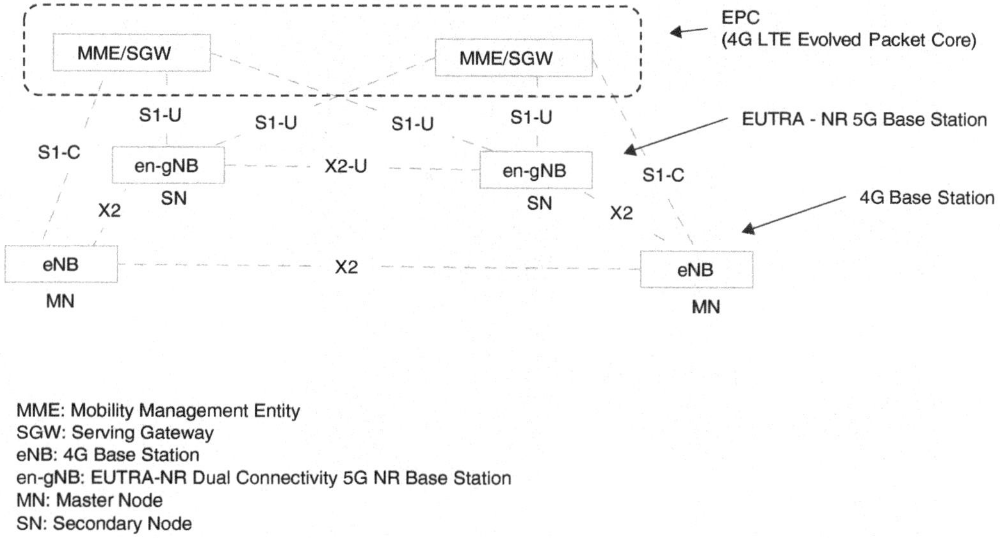
EPC
(4G LTE Evolved Packet Core)
MME/SGW
MME/SGW
S1-U
S1-U
S1-U
S1-U
S1-C
en-gNB
X2-U
en-gNB
S1-C
EUTRA - NR 5G Base Station
X2
SN
SN
X2
4G Base Station
eNB
X2
eNB
MN
MN
MME: Mobility Management Entity
SGW: Serving Gateway
eNB: 4G Base Station
en-gNB: EUTRA-NR Dual Connectivity 5G NR Base Station
MN: Master Node
SN: Secondary Node

FIGURE 1-6 Non-Standalone (NSA) 5G Architecture

5G Standalone (SA) Solution

In Standalone (SA) architecture, the 5G Next-Generation Radio Access Network (NG-RAN), in conjunction with the 5G Core (5GC), is used for both the control plane and user plane, thereby removing any dependency on the 4G LTE Radio and Core network, as shown in Figure 1-7. This flexibility in deployment without any dependency on the 4G network and the openness from 5G Radio and Core allow industry verticals such as smart factories and healthcare to deploy private 5G infrastructures, also known as non-public networks (NPNs).

Figure 1-7 illustrates the 5G SA architecture.

5G SA has several new capabilities, such as network slicing, CUPS, virtualization, automation, multi-Gbps support, ultra-low latency, and other such aspects that are natively built in to the 5G SA Packet Core architecture. Chapter 2 covers the SA architecture in more detail.

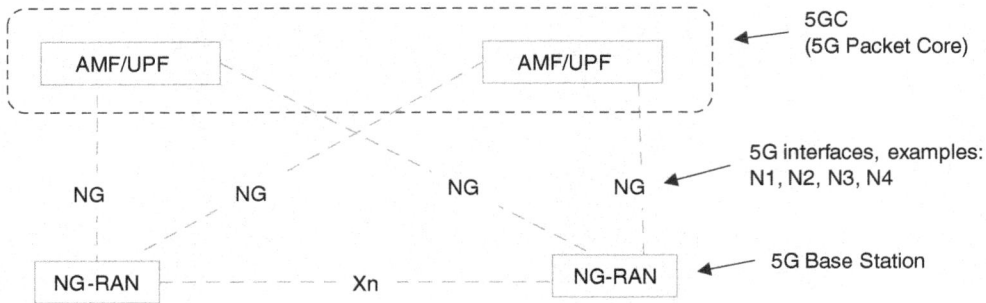

AMF: Access and Mobility Management Function
UPF: User Plane Function
NGRAN: Next-Generation Radio Access Network
NG: Next Generation

FIGURE 1-7   5G SA Architecture

Service-Based Architecture

5G SA introduces an all-new 5G Packet Core, as illustrated in Figure 1-8. 5G introduces service-based architecture (SBA), which is a framework defined by 3GPP that enables multivendor applications deployment and communications between them. In the SBA framework, the architecture elements are defined in terms of Network Functions (NFs) rather than by traditional Network Entities. Via interfaces of a common framework, any given NF offers its services to all the other authorized NFs and/or to any "consumers" that are permitted to make use of these provided services, thereby offering modularity and reusability.

As shown in Figure 1-8, SBA is a control plane functionality using the application programming interface (API) with REST interface using HTTP/2, enabling different 5G network functions to exchange information within the 5GC control plane network functions. The SBA approach enables a fully distributed, fully redundant, stateless, and/or fully scalable deployment of 5G NFs. The capability to use APIs for communication and having the NFs fully virtualized allows 5G services to be provided from several locations, be it on-premises private cloud, public cloud, or hybrid cloud.

Figure 1-9 illustrates the comparison of 4G and 5G control plane communications. API-based communication is now used for all 5G control plane communications instead of GTP-C and Diameter in 4G.

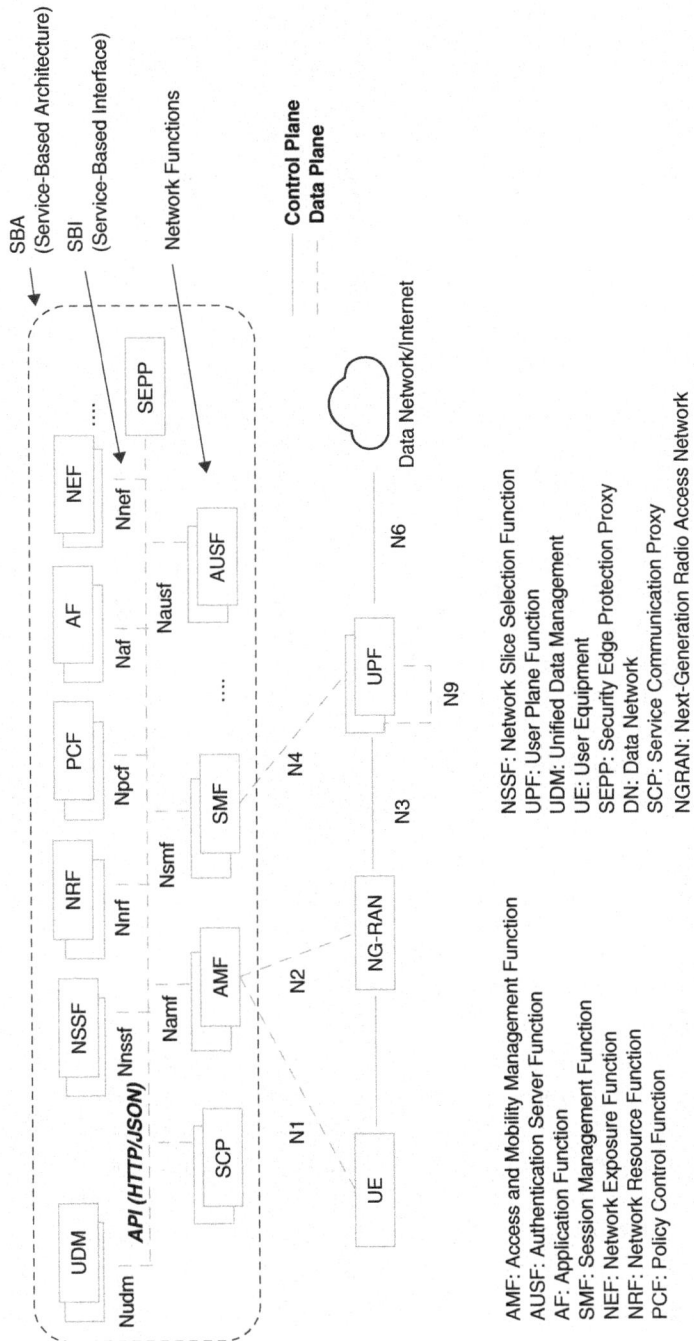

FIGURE 1-8   5G Service-Based Architecture

AMF: Access and Mobility Management Function
AUSF: Authentication Server Function
AF: Application Function
SMF: Session Management Function
NEF: Network Exposure Function
NRF: Network Resource Function
PCF: Policy Control Function

NSSF: Network Slice Selection Function
UPF: User Plane Function
UDM: Unified Data Management
UE: User Equipment
SEPP: Security Edge Protection Proxy
DN: Data Network
SCP: Service Communication Proxy
NGRAN: Next-Generation Radio Access Network

4G   5G

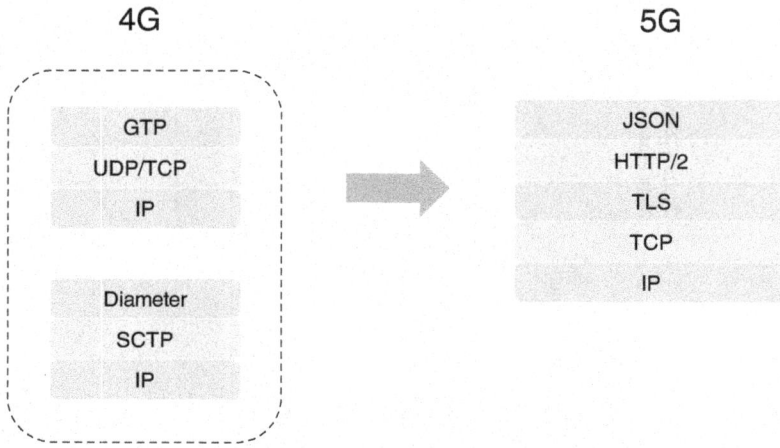

FIGURE 1-9 Comparison of 4G and 5G Control Plane Protocols

The API is also used for exposing the capabilities of 5GC NFs with external third-party applications using the Network Exposure Function (NEF). This allows streamlining all communication between the 5GC and external third-party applications through the NEF. APIs are also extensively used in the orchestration and SDN deployments in 5G. The network slice as a service (NSaaS) offering will make use of API-based calls in the service layer for management access between the communication service provider (CSP) and communication service customer (CSC). Seeing the importance of APIs in 5G, 3GPP has also introduced Common API Framework (CAPIF) in Release 15, with enhancements in Release 16. The main purpose of CAPIF is to have a unified northbound API framework across several 3GPP functions. There is a single and harmonized approach for API development, with a number of 3GPP specifications in the works—to specify a framework to host APIs of PLMN and also to allow third parties to leverage the CAPIF framework to host their APIs.

Adoption of Cloud-Native Technology

To achieve the economies of scale and enable new use cases that demand low latency and high scalability, 5G allows you to adopt cloud-native technologies to deploy 5G. Cloud-native is an approach to building and running applications that fully exploits the benefits of the cloud computing model. Adoption of cloud-native technology in 5G means that the network functions such as User Plane Function (UPF), Session Management Function (SMF), and so on can be built using open source software components and deployed on-prem or on a multistack public cloud as Cloud-Native Functions (CNFs), exploiting all the benefits of the cloud-native architecture. Here are some of the key benefits of cloud-native technology adoption in 5G:

- **Avoid vendor lock-in:** The basic tenet of cloud-native architecture is based on open source and cloud technologies. All the 5G CNFs communicate using the API in the service-based interface (SBI) based on the service-based architecture (SBA). This makes it easier for you to select any vendor based on your criteria and then use the API for integration, thereby avoiding any vendor lock-in, including infrastructure-independent deployment of 5G CNFs on any hypervisor, bare metal or any public cloud provider.

- **Scalability and resiliency:** In 5G deployments, specifically for user plane CNFs like the User Plane Function (UPF), there will be the need to scale up and scale down based on capacity requirements due to a surge of traffic in a particular location. There is also a requirement for high availability in 5G to ensure service availability and prevent any downtime, as there might be revenue impacts and deployments related to national security. Auto scalability, high resiliency, and auto-provisioning are supported by cloud-native applications and network functions. Vulnerabilities can also be fixed faster, as cloud-native applications support DevOps process and automation, allowing you to patch the risks by deploying patched images quickly.

- **Orchestration:** Virtualized 5G deployments in multistack public cloud/multi-cloud and on-prem, along with the implementation of use cases such as network slice as a service (NSaaS), require end-to-end orchestration to ensure that all the related 5G CNFs have the right configuration to enable smooth running of the use cases without any need for manual intervention. Cloud-native technology such as Kubernetes ensures smooth automation of deployment, scaling, and management of containerized deployment.

A majority of the cloud-native technology projects are being handled by the Cloud-Native Computing Foundation (CNCF), which incubates the projects and brings them into fruition, which implies a level of technology maturity. One such example is Open Policy Agent (OPA), which is an open source, general-purpose policy engine that enables unified, context-aware policy enforcement across the entire stack and is integrated with other CNCF projects like Kubernetes, Envoy, CoreDNS, Helm, SPIFFE/SPIRE, and more. Projects with such integrations will allow you to have a consistent policy across your 5G CNFs, which are deployed across on-premises and multistack public cloud.

Multi-access Edge Computing (MEC)

Multi-access edge computing (MEC) is acknowledged as one of the key pillars for meeting the demanding key performance indicators (KPIs) of 5G, especially as far as low latency and bandwidth efficiency are concerned. Edge computing as an evolution of cloud computing brings 5G network functions and applications from centralized data centers down to the network edge, closer to consumers and the data generated by devices and applications. However, not only is edge computing in telecommunications networks a technical enabler for the demanding KPIs, it also plays an essential role in the transformation of the telecommunications business, where telecommunications networks are turning into versatile service platforms for industry and other specific customer segments. This transformation is supported by edge computing, as it opens the network edge for applications and services, including those from third parties. MEC provides a new ecosystem and value chain by enabling third-party application developers and the cloud-computing capabilities of content providers, as well as an IT service environment at the edge of the network. Operators can open their Radio Access Network (RAN) edge to authorized third parties, allowing them to flexibly and rapidly deploy innovative applications and services toward mobile subscribers, enterprises, and vertical segments. Depending on your network design, MEC applications and user plane functions could be set up on private cloud, public cloud, or hybrid deployments, enabling ultra-low latency and high bandwidth as well as real-time access to radio network information that can be leveraged by applications.

The European Telecommunications Standards Institute (ETSI) MEC framework, as illustrated in Figure 1-10, provides an architecture reference to create a standardized, open environment that allows the efficient and seamless integration of applications from vendors, service providers, and third parties across multivendor MEC platforms.

The key components of the MEC architecture framework, as shown in Figure 1-10, are discussed in detail next.

The MEC host level provides compute, storage, and network resources for the MEC applications, which includes a data plane that executes the traffic rules received by the MEC platform, and routes the traffic among applications, services, DNS server/proxy, 3GPP network, other access networks, local networks, and external networks. The core entities of the MEC host level are composed of the MEC platform (MEP), applications, UPF data plane, MEC platform manager (MEPM), and virtualization infrastructure. The MEPM is responsible for managing the MEP.

MEC system-level core entities are composed of the MEC orchestrator, which helps in on-boarding of application packages, including checking the integrity and authenticity of the packages, validating application rules and requirements, and maintaining an overall view of the MEC system.

To avoid duplication and inconsistency of approach between different API specifications, 3GPP has considered the development of a common API framework for 3GPP northbound APIs (CAPIF) that includes common aspects applicable to any northbound service APIs.

Network Slicing

5G SA provides flexibility and openness for service providers to serve vertical use cases by peeling off a part of your network and offering it to the industry verticals with a network slicing feature that provides you with the right level of isolation, ease of management, and service level agreements (SLAs). This type of service offering is called network slice as a service (NSaaS). The type of private 5G network that integrates with service providers is called the public network integrated non-public network (PNI-NPN). Private networks can use the licensed, unlicensed, dedicated, and shared spectrum with other service providers. Deploying 5G in the industry verticals drives tremendous gains in productivity, economic growth, and other benefits.

A network slice can be seen as a logical network serving a defined business purpose or customer, consisting of all required network resources configured together. Network slicing divides your physical network into multiple logical networks or domains. These logical networks/domains would permit the implementation of tailor-made functionality and network operation specific to the needs of each slice customer, rather than a one-size-fits-all approach, as witnessed in the current and previous mobile generations.

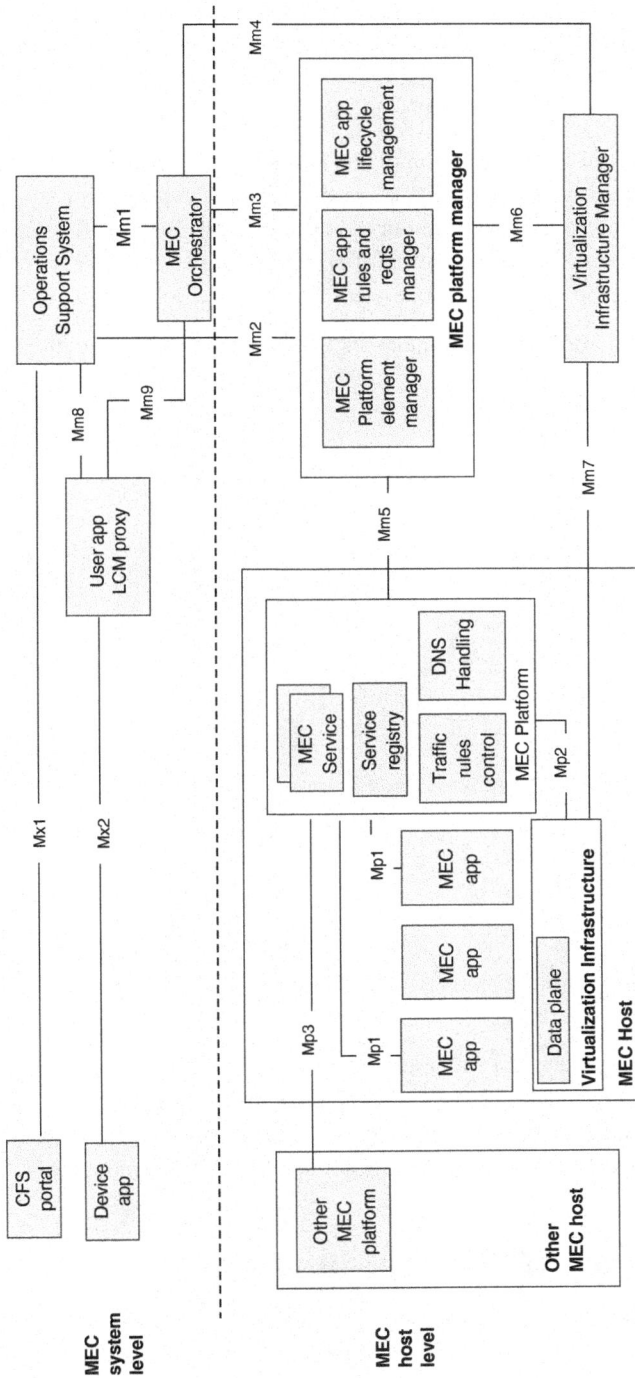

MEC system level

CFS portal

Device app

Operations Support System

User app LCM proxy

MEC Orchestrator

MEC platform manager

MEC Platform element manager

MEC app rules and reqts manager

MEC app lifecycle management

Virtualization Infrastructure Manager

MEC host level

MEC Host

MEC Platform

MEC Service

Service registry

Traffic rules control

DNS Handling

MEC app

MEC app

MEC app

Virtualization Infrastructure

Data plane

Other MEC host

Other MEC platform

Mx1, Mx2, Mm1, Mm2, Mm3, Mm4, Mm5, Mm6, Mm7, Mm8, Mm9, Mp1, Mp2, Mp3

MEC: Multi-Access Edge Compute
CFS: Customer-Facing Service
DNS: Domain Name System
LCM: Lifecycle Management

FIGURE 1-10 ETSI MEC Framework

The network slicing domains can consist of Radio Access Network (RAN), Transport Network, 5G Core (5GC) Network Functions, and also third-party applications. Some of the domains can overlap multiple slices; for example, certain 5GC network functions such as the Network Slice Selection Function (NSSF) and Policy Control Function (PCF) can be shared between multiple slices.

As shown in Figure 1-11, successful implementation of a network slice requires network slice member management, configuration of slice SLAs and quality of service (QoS), a multidomain software-defined network (SDN), and end-to-end orchestration. Without the capability of end-to-end slice management with end-to-end domain components, implementation of a successful network slice would be nearly impossible and extremely difficult to manage and operate.

Figure 1-11 illustrates the example of multiple domains in a network slice implementation.

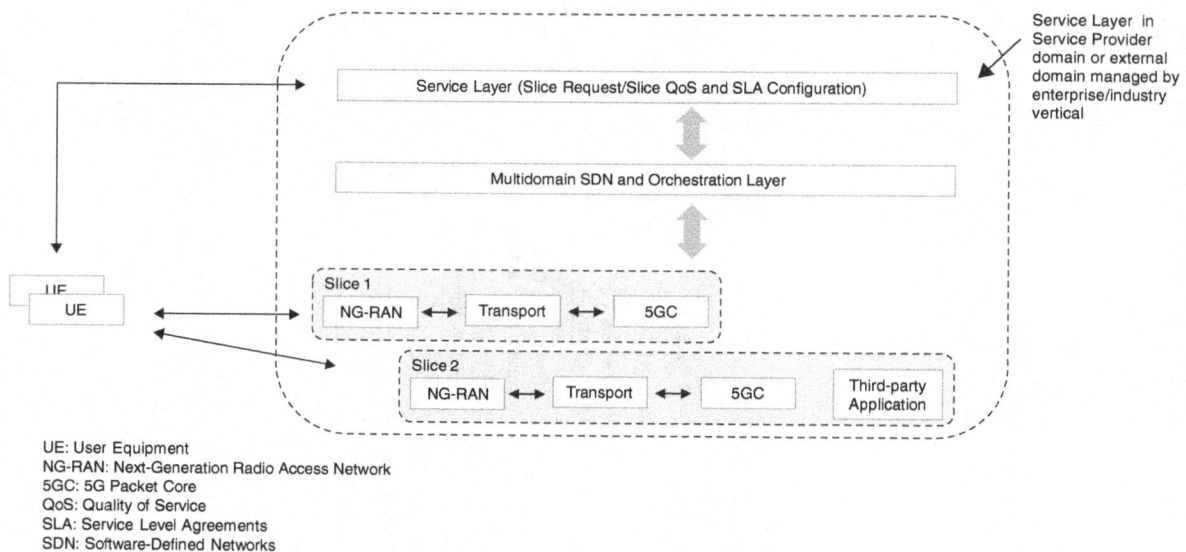

UE: User Equipment
NG-RAN: Next-Generation Radio Access Network
5GC: 5G Packet Core
QoS: Quality of Service
SLA: Service Level Agreements
SDN: Software-Defined Networks

FIGURE 1-11   Multiple Domains Within a Network Slice Deployment

Chapter 7, "Securing Network Slice, SDN, and Orchestration in 5G," covers the topic of network slicing, its threat surface, and mitigations in more detail. The section "Private 5G/Non-Public Networks" in Chapter 2, "Deployment Modes in 5G," covers the topic of network slicing deployment modes for the non-public network (NPN) and public network integrated non-public network (PNI-NPN) in more detail.

Key 5G Features in 3GPP Releases

This section discusses the key 5G features in Releases 15, 16, and 17. The overall trend in Release 16 onward is to make the 3GPP 5G System (5GS) more open and more of an enabling platform for industry verticals such as smart factories, transportation industry, and healthcare. The features in present and future releases, as illustrated in Figure 1-12, are geared to improving the adaptability of 5G toward non-3GPP industrial networks with enhancements in URLLC, network slicing, MEC, cellular IoT (C-IoT), secure communication based on APIs with third-party Application Functions (AFs), non-public networks, positioning services, and LAN-type services.

Figure 1-12 lists the key features for 5G in 3GPP Releases 15 and 16 and the key the features planned for Rel-17.

5G Phase 1
Release 15
(2017–2019)

5G System and Packet Core:
- SBA-based architecture
- Massive MTC and IoT
- API Exposure to third party
- Cloud friendly (SBI)
- Network Slice as a Service
- MEC

RAN:
- New Radio specification
- NR NSA (option 3x)
- NR SA (option 2/4/5/7)
- Wide bandwidth (~400 MHz)
- Wide Freq Range (up to 28GHz)

Industry verticals:
- V2X Phase 2
- Mission Critical Interworking
- WLAN and unlicensed spectrum

5G Phase 2
Release 16
(2018–2020)

5G System and Packet Core:
- Anchored NR-U
- Standalone NR-U
- Integrated Access and Backhaul (IAB)
- CAPIF enhanced

RAN:
- NR Positioning
- NR Coverage
- UE power saving techniques
- HARQ enhancement
- Enhanced Scheduling
- Coordinated multi-point (CoMP)

Industry verticals:
- Time-Sensitive Networks
- C-V2X enhancement
- IIoT and URLLC support
- NPN support

Phase 3
Release 17
(2020–2022 expected)

5G System and Packet Core:
- 5G LAN type services
- 5G Wireline and Wireless Convergence
- Edge Computing in 5G
- IAB enhancement
- Network Slicing - phase 2
- Multi-SIM
- Access Traffic Steering

RAN:
- NR Coverage enhancement
- NR Positioning enhancement
- NR Quality of Experience
- NR MIMO
- RAN Slicing
- Multi-Radio DCCA enhanced

Industry verticals:
- IIoT and URLLC enhanced
- NB-IoT and LTE-MTC enhancement
- Unmanned Aerial Systems
- Satellite components
- Enhanced V2X Services
- NPN enhanced

www.3GPP.org/specifications/work-plan

SBA: Service-Based Architecture
RAN: Radio Access Network
MTC: Machine Type Communications
IoT: Internet of Things
API: Applications Programming Interface
MEC: Multi-Access Edge Compute
NSA: Non-Standalone
SA: Standalone
V2X: Vehicle to Everything

WLAN: Wireless Local Area Network
NR: New Radio
NR-U: New Radio Unlicensed
HARQ: Hybrid Automatic Repeat Request
IIoT: Industrial Internet of Things
URLLC: Ultra-Reliable Low-Latency Communications
NPN: Non-Public Networks
LAN: Local Area Networks
MIMO: Multiple Input Multiple Output

FIGURE 1-12 Key 5G Features in 3GPP Releases 15, 16, and 17

In 3GPP Release 16 (Rel-16), the key areas of enhancements in APIs are toward aligning northbound APIs with the Common API Framework (CAPIF), as defined in 3GPP TS 29.222, where the following are introduced:

- Enhanced event report to support detailed event reporting of CAPIF events

- Network Exposure Function (NEF)/Services Capability Exposure Function (SCEF) Northbound (NB) API registration with CAPIF Core Function to enable discovery of the NB APIs by external Application Functions (AFs)

- Defining the Uniform Resource Identifier (URI) structure for APIs related to N33 and T8 interfaces

While a basic Common API Framework is made available in 3GPP Rel-15, there are several enhancements that are considered for eCAPIF in Rel-16 to enable the API exposing function, the API publishing function, and the API management function of the API provider domain within the third-party trust domain interaction with the CAPIF core function in the PLMN trust domain.

The main purpose of CAPIF is to have a unified northbound API framework across several 3GPP functions. There is a single and harmonized approach for API development, with a number of 3GPP specifications in the works—to specify a framework to host APIs of the PLMN and also to allow for third parties to leverage CAPIF to host their APIs.

Key 5G Advanced Features

This section discusses the key 5G features, which are under study in 3GPP Release 18. 5G naming conventions will evolve to 5G-Advanced with the forthcoming Release 18 from 3GPP. Figure 1-13 illustrates the key features under study for 5G Advanced.

In 3GPP Release 18 (Rel-18), the key areas of enhancements are related to immediate and longer-term commercial needs for the following:

- Evolved Mobile BroadBand (eMBB)

- Non-eMBB evolution

- Cross functionalities for both eMBB and non-eMBB driven evolution

Majority of the enhancements under study in Rel-18 is aimed at getting improved data throughput rates in uplink (UL), improvements in getting lower latency and higher bandwidth for gaming and eXtented Reality (XR), and enhancements to allow seamless deployments of Non-Public Network deployments. As per the 3GPP work plan proposal, Rel-18 is expected to be ready by December 2023.

5G Advanced
Release 18
(2022 - 2023 Dec expected)

5G System and Packet Core:
- Enhanced support for V2X
- High Altitude Platform System (HAPS)
- Dual Active Protocol Stack (DAPS) improvements
- Network slicing enhancements
- Private 5G/Non-Public Network (NPN) enhancements aimed at non-3GPP deployments
- Service Functionality Chaining (SFC) in 5G System
- 5G-enabled fused location service (LCS) capability
- Enhanced 5G Timing Resiliency
- Inclusion of MEC in roaming

RAN:
- Enhancements for eXtended Reality (XR)
- Sidelink Enhancements aimed at unlicensed, power saving enhancements, efficiency enhancements, and so on
- Evolution of Downlink MIMO with enhanced handling of multiple Transmission Reception Points (multi-TRP) and multibeam
- Evolution of Non-Terrestrial Network aimed at 5G NR and IoT
- Inclusion of traffic characteristics and KPIs for AI/ML
- Evolution of V2X with sidelink enhancements
- Enhancements for Unmanned Aerial Vehicle (UAV)
- Enhanced inter-gNB coordination
- Minimization of Drive Test (MDT) and Self Organizing Networks (SON)
- High Altitude Platform System (HAPS)
- UE Power Savings
- Extended support for frequency bands beyond 52.6GHz
- Carrier Aggregation (CA)/Dual Connectivity (DC) enhancements
- Flexible spectrum integration
- Multicast Broadcast Service (MBS)

Industry verticals:
- V2X
- Non Terrestrial Networks
- Railway
- Non-3GPP and unlicensed spectrum

SBA: Service-Based Architecture   WLAN: Wireless Local Area Network
HAPS: High Altitude Platform System   NR: New Radio
RAN: Radio Access Network   SFC: Service Functionality Chaining
MEC: Multi-Access Edge Compute   NTN: Non-Terrestrial Networks
NSA: Non-Standalone   LCS: Location Services
SA: Standalone   NPN: Non-Public Networks
V2X: Vehicle to Everything   LAN: Local Area Networks
                                        MIMO: Multiple Input Multiple Output

FIGURE 1-13   Key 5G Advanced Features under Study in 3GPP Release 18

Summary

As discussed in this chapter, the earlier generations of cellular technology such as 4G focused on ensuring connectivity, whereas 5G takes connectivity to the next level by delivering connected experiences from the cloud to clients. Even antenna technologies will see increased adoption of MIMO and algorithms to enhance beamforming based on the density of users and type of application. This is a major shift from passive antenna technologies such as panel antennas with just remote down-tilts and up-tilts. 5G technology uses a new spectrum but is not limited to it. 5G is designed to support a converged, heterogeneous network, combining licensed and unlicensed wireless technologies. This will add bandwidth available for users and devices being catered for by 5G.

5G networks are virtualized and software driven, with open, flexible deployment models, and they exploit cloud technologies and software-defined platforms, in which networking functionality is managed through software rather than hardware. Advancements in virtualization, cloud-based technologies, adoption of APIs, IT, and business process automation enable 5G architecture to be agile and flexible and to provide anytime, anywhere user access. 5G networks can create software-defined subnetwork constructs known as network slices, which enable you to create an isolated end-to-end network consisting of both virtualized and physical components tailored using automated provisioning and proactive management of traffic and services to fulfill diverse requirements requested for a particular application.

Although Rel-17 5G and Rel-18 5G are looking at enhanced capabilities and evolution of features for MIMO, NPN, and better security in API communications, it has to be noted that there is usually a 2 or 3 year delay in the features, which has impacts on radio access network (RAN) and user devices (UE). Being pragmatic, real-world deployments of 5G standalone (SA) will be around at the end of 2022 and the beginning of 2023. For Rel-17, the deployments are expected to be around 2024 and Rel-18 around 2025–2026.

Acronym Key

The following table expands the key acronyms used in this chapter.

Acronym	Expansion
5GC	5G Core network
AKA	Authentication and key agreement
API	Application programming interface
BH	Back haul
DAPS	Dual Active Protocol Stack
eMBB	Enhanced mobile broadband
eNB	E-UTRAN NodeB
EPC	Evolved Packet Core
FH	Front haul
gNB	New Radio NodeB
GTP	GPRS Tunneling Protocol
HAPS	High Altitude Platform System
HTTP	Hypertext Transfer Protocol
HW	Hardware
IAB	Integrated access and back haul
IIoT	Industry Internet of Things
IoT	Internet of Things
IP	Internet Protocol
IR	Incident response

Acronym	Expansion
IT	Information technology
JSON	JavaScript Object Notation
KPI	Key performance indicators
M2M	Machine-to-machine
MBB	Mobile broadband
MBS	Multicast Broadcast Service
MEC	Multi-access edge compute
MDT	Minimization of Drive Test
MH	Mid haul
MitM	Man in the middle
MIMO	Multiple Input Multiple Output
NGAP	NG Application Protocol
NID	Network identifier
NPN	Non-public network
NR	New Radio
PLMN	Public land mobile network
PNI-NPN	Public network integrated NPN
QAM	Quadrature amplitude modulation
RNC	Radio network controller
S-GW	Serving gateway
SCTP	Stream Control Transmission Protocol
SFC	Service Functionality Chaining
SNPN	Standalone Non-Public Network
SON	Self Organizing Networks
TLS	Transport Layer Security
UAV	Unmanned Aerial Vehicle
UE	User equipment
UPF	User Plane Function
URLLC	Ultra-reliable and low-latency communications
V2I	Vehicle-to-infrastructure
V2N	Vehicle-to-network
V2P	Vehicle-to-pedestrian
V2V	Vehicle-to-vehicle
V2X	Vehicle-to-everything
WLAN	Wireless local area network
X2 GW	X2 gateway
X2-C	X2–control plane

Acronym	Expansion
X2-U	X2–user plane
Xn-C	Xn–control plane
Xn-U	Xn–user plane
XnAP	Xn Application Protocol
XR	eXtended Reality

References

Pramod Nair, "Securing Your 5G Infrastructure to the Edge and Beyond," Cisco Public, April 1, 2021

Pramod Nair, "Securing 5G and Evolving Architectures," Cisco Knowledge Networks, Cisco Public, September 17, 2020

ETSI GS MEC 003, "Multi-Access Edge Computing (MEC): Framework and Reference Architecture"

3GPP work program, https://www.3gpp.org

3GPP TS 23.501, "System Architecture for the 5G System: Stage 2"

3GPP TS 38.401, "NG-RAN: Architecture Description"

3GPP TS 33.501, "Security Architecture and Procedures for 5G System"

3GPP TS 29.122, "T8 Reference Point for Northbound Application Programming Interfaces (APIs)"

3GPP TS 38.470, "NG-RAN: F1 Application Protocol (F1AP)"

3GPP TS 38.300, "NR: NR and NG-RAN Overall Description"

3GPP TR 21.915, "Release 15 Description: Summary of Rel-15 Work Items"

3GPP TR 21.916, "Release 16 Description: Summary of Rel-16 Work Items"

TS 29.222, "Common API Framework for 3GPP Northbound APIs"

TS 29.522, "5G System: Network Exposure Function Northbound APIs: Stage 3"

3GPP TS 36.101, "Evolved Universal Terrestrial Radio Access (E-UTRA): User Equipment (UE) Radio Transmission and Reception"

Cisco Systems, "Reimagining End-to-End Network Mobile Network in the 5G Era," Cisco Public, 2019

Chapter | 2

Deployment Modes in 5G

After reading this chapter, you should have a better understanding of the following topics:

- Multiple 5G SA and NSA deployment options
- Public and non-public network deployment models in 5G
- Interworking between 4G and 5G
- 5G LAN type deployment

The previous chapter took you through the evolution of cellular technologies—in particular, the evolution of cellular technology from 4G to 5G. This chapter takes you through the different modes of 5G deployment and the use cases that can be mapped to deployment modes.

This chapter should be of particular interest to the following individuals and teams from enterprise/5G service providers deploying 5GC and cybersecurity vendors planning product developments for 5G security use cases:

- Security leaders within government organizations who would like an understanding of 5G deployment modes

- Teams from enterprise and industry verticals that want an understanding of 5G deployment modes

- Enterprise cybersecurity vendor security architects looking to have an understanding of 5G deployment modes

- Cybersecurity students who would like to get a brief understanding of 5G deployment methods

5G brings in real openness and deployment flexibility, with many options to choose from. Service providers who have already invested billions in network infrastructure in 2G, 3G, and 4G technologies can reuse parts of the 4G infrastructure for initial deployment of 5G. Such deployments are called 5G Non-Standalone (NSA). 5G also introduces deployment options where all the components of 5G System (5GS) are purely 5G and have no dependencies on 4G and legacy cellular technologies. Such deployments are called 5G Standalone (SA).

Although 5G NSA provides a good step from 4G in your journey toward 5G, it's not a full-fledged 5G SA deployment. There are also multiple variants of 5G NSA, and the choice of your deployment will depend on multiple areas such as your existing network topology, the existing bandwidth between the eNB and enhanced LTE eNB (ng-eNB), and your investment appetite.

This chapter takes you through the different options of 5G NSA and SA, including the following:

- Evolution options from 5G NSA toward 5G SA as well as 4G and 5G interworking

- Different options of non-public networks (NPNs) and public network integrated non-public networks (PNI-NPNs)

- New 5G LAN type deployments

- Deployments enabling network slice as a service (NSaaS) offerings

5G NSA and SA Deployments

Two solutions are defined by 3GPP for 5G networks:

- **5G Non-Standalone (NSA):** The existing LTE radio access and core network (Evolved Packet Core, or EPC) is used as an anchor for mobility management and coverage to add the 5G carrier. This solution enables operators to provide 5G services within a shorter timeframe and for less cost.

- **5G Standalone (SA):** An all new 5G Packet Core will be introduced with several new capabilities built in. The SA architecture is composed of 5G New Radio (5G NR) and 5G Core Network (5GC). Network slicing, CUPS, virtualization, multi-Gbps support, ultra-low latency, and other aspects will be natively built in to the 5G SA Packet Core architecture.

Figure 2-1 illustrates the NSA and SA deployment options.

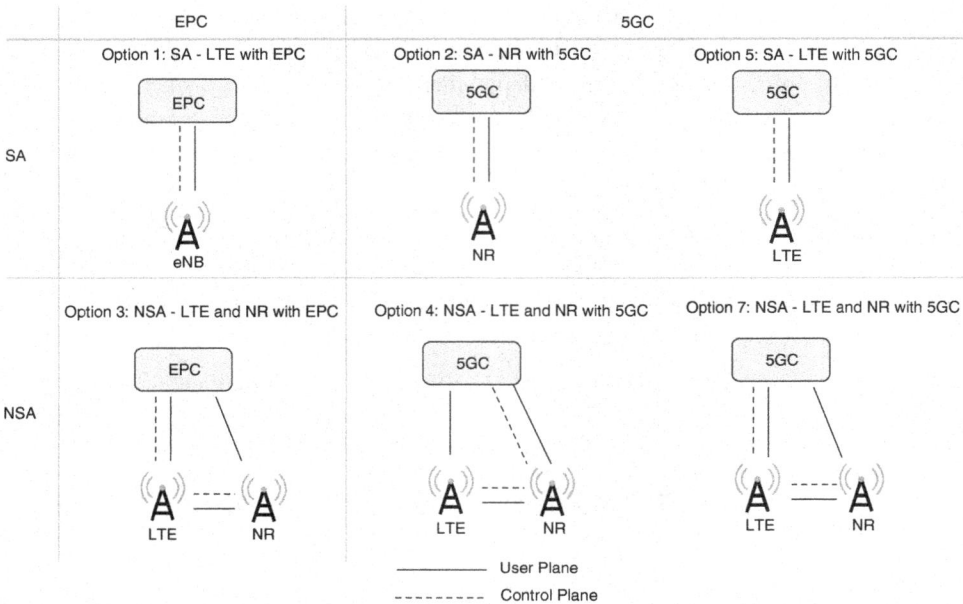

FIGURE 2-1   NSA and SA Deployment Options

5G Non-Standalone (NSA) Deployments

The 5G Non-Standalone (NSA) solution enables operators using 4G EPC Packet Core to launch 5G services within a shorter timeframe and leverage the existing infrastructure. NSA leverages the existing LTE radio access and core network (EPC) to anchor 5G NR using the dual connectivity feature. This solution provides a seamless option to deploy 5G services with very little disruption in the network. As shown in Figure 2-1, there are three primary options for NSA deployment: Option 3, Option 4, and Option 7. NSA Options 5 and 7 are based on using 5GC for supporting LTE deployments. NSA Option 3 is based on upgrading the existing LTE components, and the initial deployments of 5G are based primarily on this option.

Option 3 has three variants, as described in the sections that follow.

Option 3

In NSA Option 3, traffic is split across 4G and 5G at eNodeB, as illustrated in Figure 2-2.

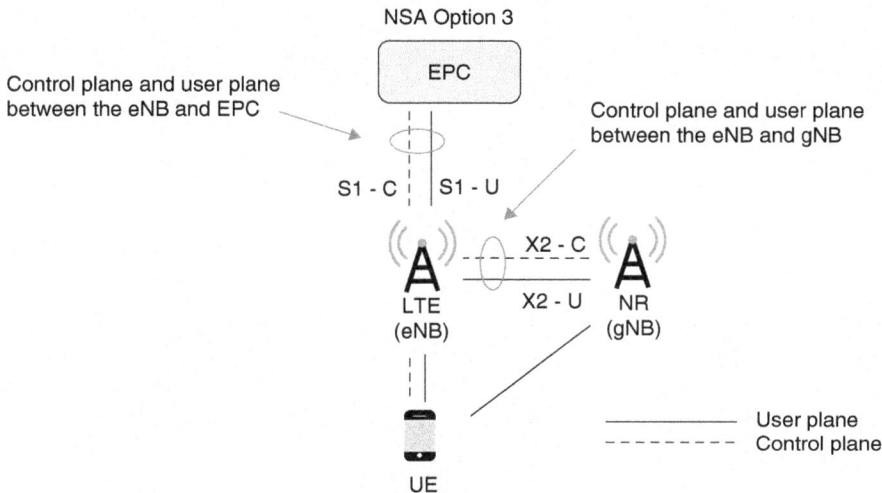

FIGURE 2-2   5G NSA Option 3

The X2 interface is the interface between eNBs used in 4G and 5G NSA deployments; it performs the following control plane and user plane functions:

- Context transfer from source eNB to target eNB/gNB

- Control of user plane transport bearers between source eNB and target eNB/gNB

- Handover cancellation

- UE context release in source eNB

- Load management

- Inter-cell interference coordination

- Uplink interference load management

- Downlink interference avoidance

- General X2 management and error-handling functions

- Application-level data exchange

As shown in Figure 2-2, in NSA Option 3 deployment mode, both the user plane (X2-U) and control plane (X2-C) are required to be implemented between the 4G eNB and 5G gNB. For this reason (as compared to the other Option 3 deployment variants), this deployment option has the highest X2 traffic, specifically due to the X2-U traffic.

Option 3a

In NSA Option 3a, the traffic is split across 4G and 5G at EPC, as illustrated in Figure 2-3.

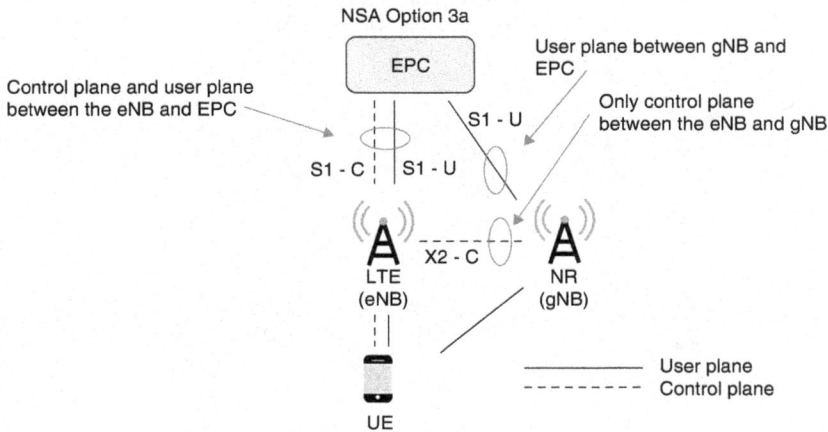

FIGURE 2-3   5G NSA Option 3a

As shown in Figure 2-3, in NSA Option 3a deployment mode, only the control plane is implemented between the 4G eNB and 5G gNB, and the user plane is implemented between the 5G gNB and the 4G EPC. As only the X2 control plane is implemented between the eNB and gNB, the traffic is quite sparse as compared to the 5G NSA Option 3 deployment mode.

Option 3x

In NSA Option 3x, the traffic is split across 4G and 5G at 5G gNB, as illustrated in Figure 2-4.

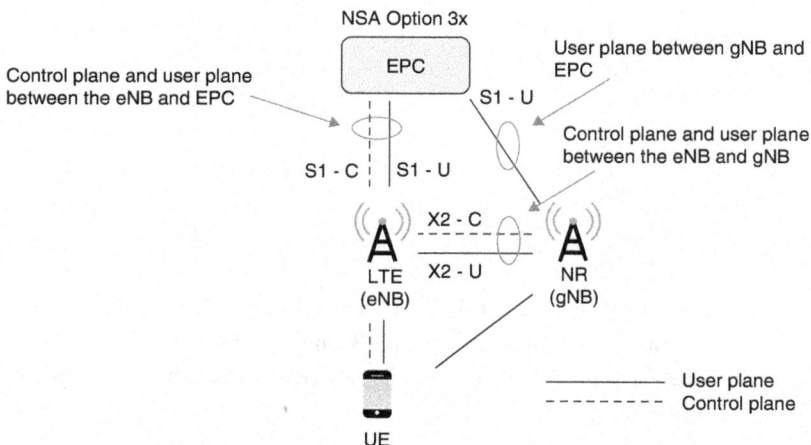

FIGURE 2-4   5G NSA Option 3x

As shown in Figure 2-4, in NSA Option 3x deployment mode, although the control plane and user plane are implemented between the 4G eNB and 5G gNB, the amount of user plane and control plane LTE traffic in these interfaces is quite negligible. The user plane is also implemented between the 5G gNB and the 4G EPC, which enables data to route directly to the NR gNB, thereby avoiding excessive user plane load (which was the case in Option 3) on the existing LTE eNB, which was designed for 4G LTE traffic load and not additional NR traffic load.

Among the three variants of 5G NSA Option 3, Option 3x is the widely deployed option. In this variant, the control plane is implemented in the 4G eNB, where the call can be anchored, thus providing good service continuity. Option 3x also has negligible impact on the transport catering for the X2 and S1 interfaces, which is one of the most expensive parts of the infrastructure and requires careful planning and configuration.

5G Standalone (SA) Deployments

For use cases such as enhanced mobile broadband (eMBB), the target key performance indicators (KPIs) are 4ms one-way latency, 10–20Gbps peak data rate, and 50–100Mbps user-experience data rate, which can be achieved by 5G SA Option 2 deployment.

Figure 2-5 illustrates one of the methods for evolving from 4G to 5G NSA and from 5G NSA to 5G SA.

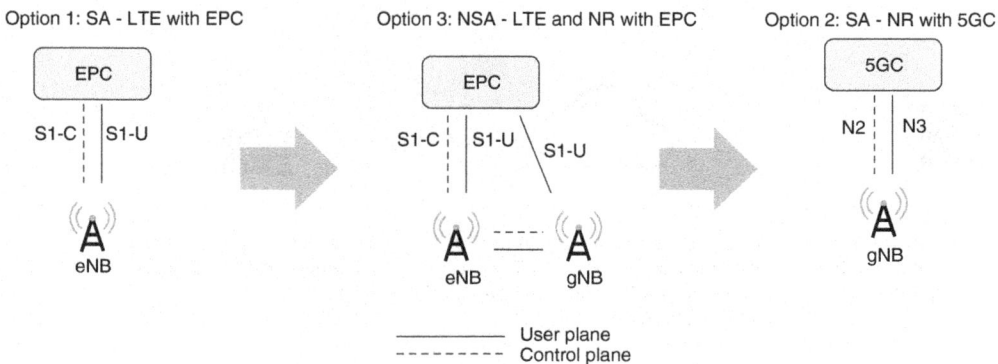

FIGURE 2-5   Standalone (SA) Options

As shown in Figure 2-5, the existing service providers can choose to migrate to 5G NSA mode of deployment using Option 3, which is the most common step used by service providers. In 5G SA Option 2, the New Radio (NR) access network consisting of gNBs is connected to 5G Core (5GC). The user plane (UP) and control plane (CP) of SA Option 2 use NR and are completely independent of Long-Term Evolution (LTE), and many deployments plan to cater for the distributed UP and CP implementations using multi-access edge compute (MEC).

5G SA Option 2 has many deployment options based on the use case, industry vertical, and service offering planned by the service providers. In many cases, related to industry verticals such as smart manufacturing that require ultra-reliable low-latency communications (URLLC), high bandwidth, and integration with third-party applications, the edge computing platform and applications will be co-located with the User Plane Function (UPF). In cases where latency is not the primary requirement, only the automation part will need co-location with the UPF in the MEC. In cases where MEC is deployed on the gNB location, it will usually have limited hardware resources and site conditions and therefore require virtualization infrastructure with a very low footprint (for example, the size of the room), cooling, limited power supply, and so on. This can be achieved by ensuring that only critical services with container-based technology are deployed in the edge.

With the 5G technology being used in diverse verticals, the requirements of building an open ecosystem and having interaction protocol refinement across multiple parties (including vendors and operators) are critical. This inherently necessitates flexibility in deployment of 5G-related network functions and applications. Integration with the public cloud and deployments of network functions on the public cloud are very much dependent on the regulations being stipulated per country, as some countries (specifically in Europe) have very strict regulations on public cloud deployment. Figure 2-6 illustrates the main locations where 5G network functions (NFs) could be deployed.

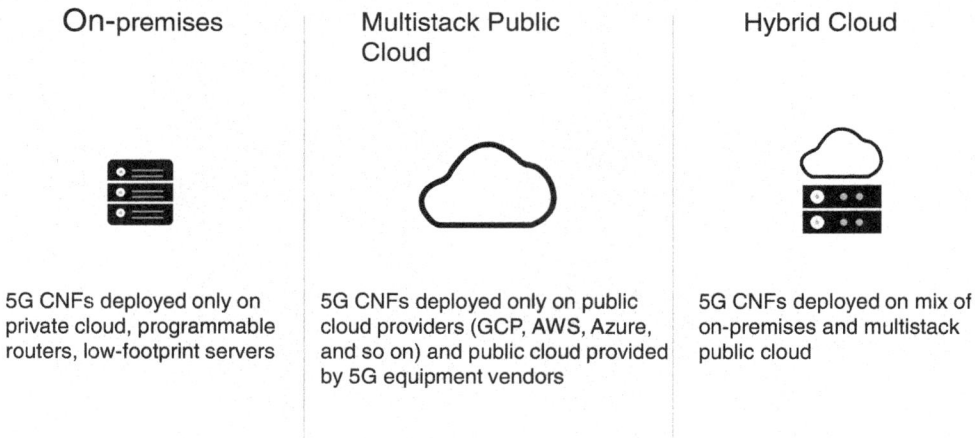

On-premises	Multistack Public Cloud	Hybrid Cloud
5G CNFs deployed only on private cloud, programmable routers, low-footprint servers	5G CNFs deployed only on public cloud providers (GCP, AWS, Azure, and so on) and public cloud provided by 5G equipment vendors	5G CNFs deployed on mix of on-premises and multistack public cloud

FIGURE 2-6   Deployment Location Options for 5G Network Functions

Direct and Indirect Communication for NF and NF Service

You can deploy 5G SA Option 2 in direct and indirect inter-NF communication models, as illustrated in Figure 2-7.

5G NF: 5G Network Function
API: Application Programming Interface
SCP: Service Communication Proxy

FIGURE 2-7 Direct and Indirect Communication for NF and NF Services

The direct and indirect communication models for NF and NF services as specified in 3GPP Rel-16 are further explained next.

The direct model consists of two options:

- **Model A:** Direct communication without Network Resource Function (NRF) interaction. In this model, the NFs do not interact with NRF and there is no service discovery.

- **Model B:** Direct communication with Network Resource Function (NRF) interaction. In this model, the NFs interact with NRF for service discovery.

The indirect communication model consists of the following two options:

- **Model C:** Indirect communication without delegated discovery. In this model, Service Communication Proxy (SCP) is introduced in the NF communication path, which aggregates Hypertext Transfer Protocol (HTTP) links and provides monitoring of centralized signaling.

- **Model D:** Indirect communication with delegated discovery. In this model, SCP takes over service discovery and selection for NF consumers; hence, NF consumers no longer need to perform discovery and select service providers.

You can deploy both the direct and indirect communication models at the same time in your network, depending on your architecture and the use case you are trying to achieve. The indirect communication model makes use of the Service Communication Proxy (SCP) network function, as shown in Figure 2-7. SCP was introduced in 3GPP Rel-16 and defined in 3GPP TS 23.501. Figure 2-8 illustrates the key functions of SCP, which are further explained in the list that follows:

1. **Centralized monitoring:** SCP provides centralized monitoring, which is very critical to identify threats in 5G multivendor deployments.

2. **Distributed SCP deployments:** SCP can be deployed in distributed mode, enabling you to design the network in a flexible manner. SCP also provides message forwarding and routing to other SCPs in your network. In order to enable an SCP to route messages through several SCPs, an SCP can register its profile in the NRF. If NRF integration is not preferred, then a local configuration can be used to determine the SCP in the next hop.

3. **Load balancing:** SCP provides real-time congestion control, handling of overload conditions and message forwarding and routing between NF/NF service.

4. **Resilient integration with third-party vendors and application developers:** SCP provides communication authorization and continuous service discovery and selection for the NF consumers by selecting a different instance of NF. This ensures continuous service to the NF consumer and third-party vendors and application developers.

5. **Encrypted communications:** Apart from preventing unauthorized communications, SCP can also implement cryptographic verification using techniques such as Transport Layer Security (TLS)/mutual Transport Layer Security (mTLS). You can maintain complete control over the distribution of keys and certificates, with the vendors, application developers, and NF consumers completely detached from the process.

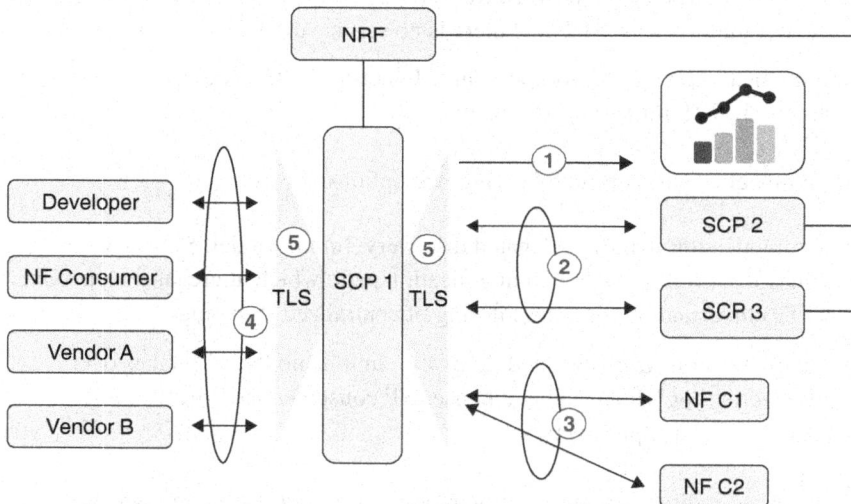

FIGURE 2-8   Key Functions of SCP

The aforementioned features and functionalities of SCP provide added security to your 5G deployments.

4G and 5G Interworking

4G and 5G interworking must have inter–radio access technology (inter-RAT) handovers (HOs). Inter-RAT HO is supported by functions such as combined 4G and 5G subscription data access function

(HSS and UDM), 4G and 5G control plane (PGW-C and SMF), and 4G and 5G user plane (PGW-U and UPF), as shown in Figure 2-9.

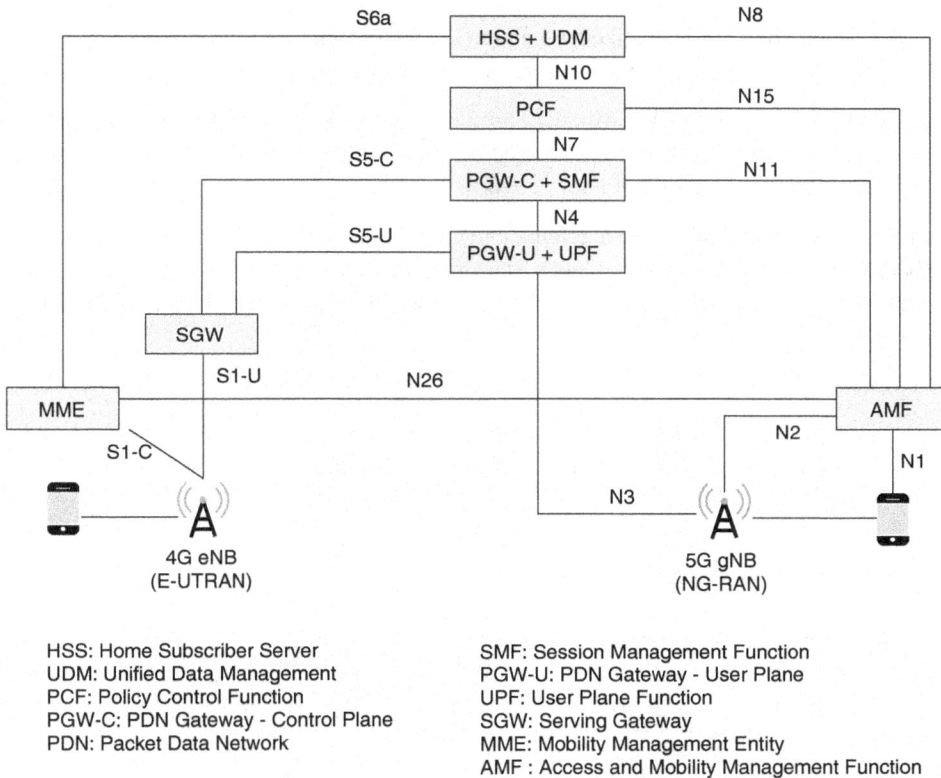

FIGURE 2-9   4G and 5G Interworking

As shown in Figure 2-9, the 4G and 5G Radio Access Networks (RANs) have their own control plane and user plane implementations and work independently of each other, unlike 5G NSA, where there are some dependencies between 4G and 5G, depending on the Option 3 variants.

The interworking is achieved by having combined 4G and 5G network functions in the packet core. The interworking includes idle mode and active mode procedures. Idle mode is when your phone is not on a call and is on a RRC_IDLE state. Cell selection and reselection modes are used for moving between cells. In an active mode, the RRC state is RRC_CONNECTED, and the same mechanism of moving between cells is called a handover.

VRAN/O-RAN/C-RAN and Related Deployment Models

Virtualized RAN (VRAN) and Open RAN (O-RAN) are terms used in conjunction with each other, and though they are closely inter-linked, there are some differences between them.

VRAN relates to providing you with virtualized 4G and 5G network functions that can be deployed in any commercial off-the-shelf (COTS) hardware server or multistack public cloud.

O-RAN, which is specified by the O-RAN Alliance, provides you with the ability to utilize machine learning systems and artificial intelligence backend modules to empower network intelligence through open and standardized interfaces in a multivendor network.

C-RAN has been known as Centralized, Clean, Cloud, and Collaborative RAN but is widely just called Centralized RAN or Cloud RAN. C-RAN uses open platforms and real-time virtualization technology rooted in cloud computing to achieve dynamic shared resource allocation of Centralized Unit (CU) and Distributed Unit (DU).

This section will discuss O-RAN architecture and deployment options and then delve into C-RAN.

Open RAN (O-RAN) deployments are driven by the requirements of open interface, flexible deployments and total cost of ownership (TCO) reduction. Figure 2-10 illustrates the O-RAN high-level architecture.

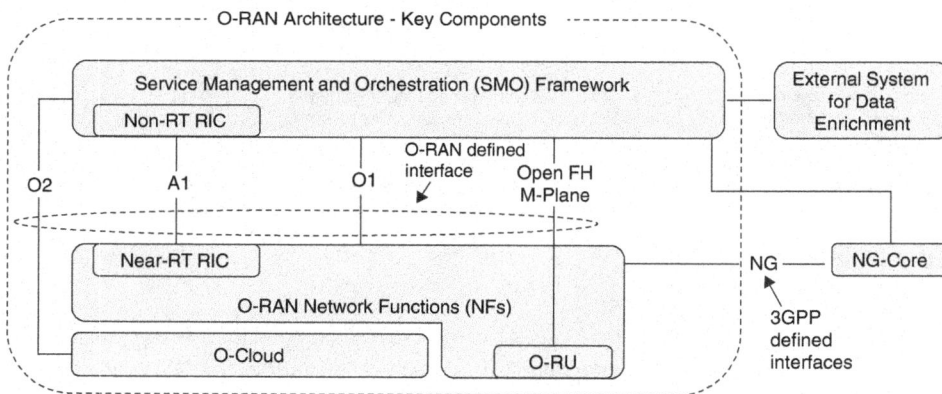

FIGURE 2-10   4G and 5G High-Level O-RAN Architecture

The key components of the O-RAN architecture are as follows:

- **Service Management and Orchestration framework:** The Service Management and Orchestration (SMO) framework provides an integration fabric that enables orchestration and management of multivendor O-RAN Network Functions. The SMO framework also contains the Non-Real Time (Non-RT) RAN Intelligence Controller (RIC) function, which has the goal of supporting intelligent RAN optimization in non-real time (that is, greater than one second) by providing policy-based guidance using data analytics and AI/ML training/inference.

■ **Non-RT RIC:** The Non-RT RIC is a central component of the SMO, which enables non-real-time control and optimization of RAN elements and resources and support for AI/ML workflows. The Non-RT RIC provides fine-grained policy-based guidance to the Near-RT RIC through the AI interface and is composed of the Non-RT RIC framework and Non-RT RIC applications (rApps).

■ **O-RAN Network Functions:** The Network Functions for the radio access side consist of Near-Real-Time RAN Intelligence Controller (Near-RT RIC), Open-RAN Compatible Centralized Unit Control Plane (O-CU-CP), Open-RAN Compatible Centralized Unit User Plane (O-CU-UP), Open-RAN Compatible Distributed Unit (O-DU), and Open-RAN Compatible Radio Unit (O-RU). Figure 2-11 illustrates the planes of the O-RAN front-haul interface and its protocol stack.

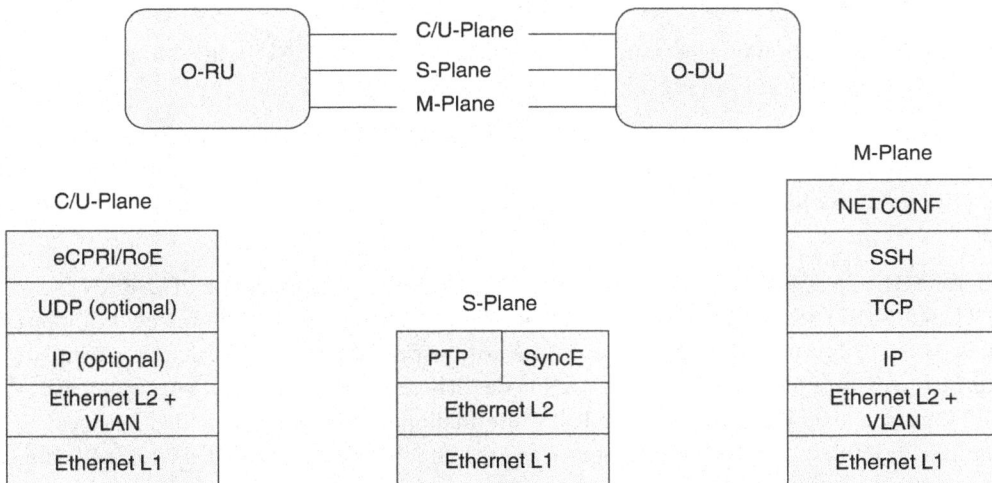

FIGURE 2-11 Details of Each of the Planes of the O-RAN Front-haul Interface

The key functionalities of the C/U, S, and U planes are as follows:

■ Control plane (C-Plane) messages provide:

  ■ Scheduling and beam-forming commands

  ■ Downlink (DL) precoding configuration

  ■ Physical random access channel (PRACH) handling

- User plane (U-Plane) messages provide:

  - Support for data compression

  - In-phase and Quadrature Phase (I/Q) data transfer

  - DL data precoding

- Sync plane (S-Plane) provides:

  - Timing and synchronization between O-RU and O-DU

  - PTP and SyncE profiles

  - Synchronization topologies

- Management plane (M-Plane) provides:

  - Support for hierarchical/hybrid model

  - C/U plane IP and delay management

  - Fault, configuration, accounting, performance, security (FCAPS), including sync configuration and status

SMO also utilizes the O1 interface for OAM toward the O-RAN Network Functions, except for the O-RU. For hybrid management, the Open FH M-plane is utilized between the SMO and the O-RU, while for hierarchical management the SMO manages the O-RU indirectly via the O-DU.

- **Near-RT RIC:** Near-RT RIC enables near real-time control and optimization of O-RAN (O-CU and O-DU) nodes and resources over the E2 interface with near-real-time control loops (that is, 10ms to 1s). The Near-RT RIC uses monitor, suspend/stop, override, and/or control primitives to control the behaviors of O-RAN nodes. The Near-RT RIC hosts xApps that use the E2 interface to collect near-real-time RAN information to provide value-added services using these primitives, guided by the policies and the enrichment data provided by the A1 interface from the Non-RT RIC.

- **O-Cloud:** O-Cloud facilitates flexible deployment options and service provisioning models of O-RAN virtualized network elements in telco clouds. The O-Cloud is an infrastructure element (or collection of elements) that is based on COTS servers, uses hardware accelerator add-ons as needed, and has a software stack that is decoupled from the hardware. O-Cloud will host the relevant O-RAN functions, the supporting software components, and the appropriate management and orchestration functions.

There are multiple modes of O-RAN deployment, as shown in Figure 2-12. In the initial stage, CU/DU co-located deployment is more suitable, which can reduce E2E latency, CAPEX, time to market, and complexity in network planning and operation. Many implementations are also aimed at CU/DU split deployment, where the DU is located closer to the gNB. The CU is deployed at the edge site farther to the DU and aggregates multiple DUs. Due the flexibility of 5G NF deployments, the DUs, CUs, UPFs, and all other virtualized components of 5G can be deployed on multistack public cloud.

Figure 2-12 illustrates the different deployment methods for O-RAN.

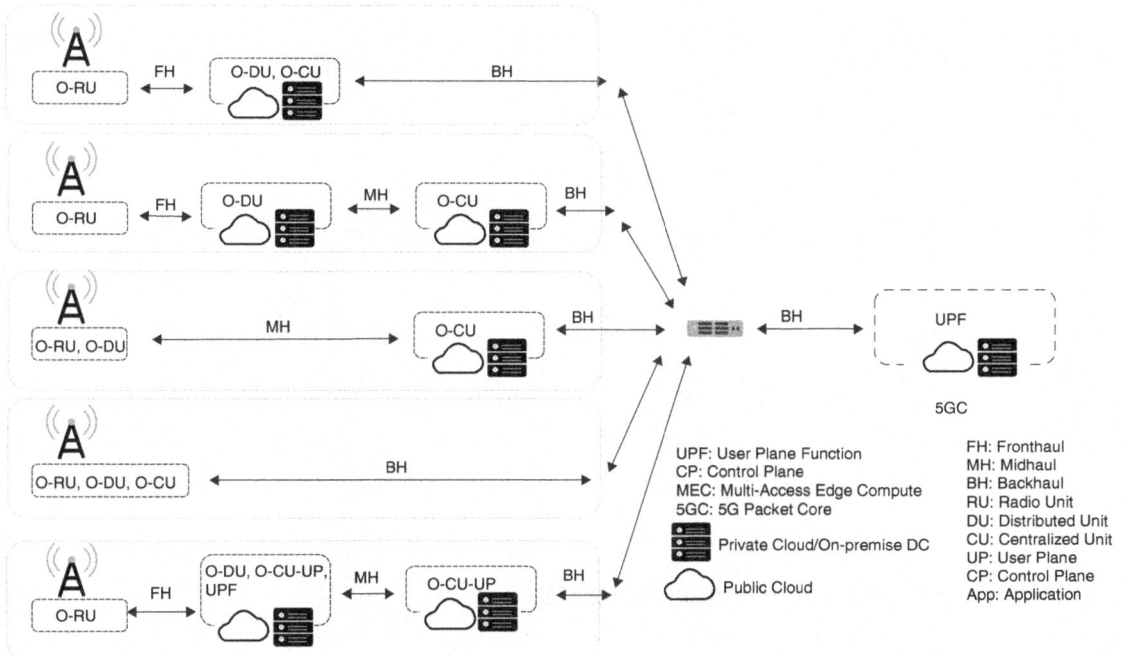

FIGURE 2-12   O-RAN Deployment Options

O-RAN Alliance and O-RAN-related communities are building partnerships and ecosystems to allow wider use of O-RAN. An example is the O-RAN Software Community (OSC), which is a collaboration between the O-RAN Alliance and Linux Foundation with the mission to support the creation of software for the RAN and Open RAN MoU Group of the Telecom Infra Project (TIP), which supports the advancement and adoption of standards-based Open RAN solutions. Furthermore, O-RAN Alliance Security Focus Group (SFG) is mitigating security risks to reduce the likelihood and impact of attacks on third-party RAN applications, open front-haul interface, network management, and UE identity privacy within the O-RAN.

Cloud RAN (C-RAN), which is also sometimes referred to as Centralized RAN, has the baseband unit (BBU) deployed at the centralized location, which enables you with a lightweight deployment of Radio Unit (RU), which is also sometimes referred to as Radio Resource Unit (RRU) and Antenna at the cell site. In 5G, the BBU is further split into Distributed Unit (DU) and Centralized Unit (CU).

C-RAN architectures use a front-haul interface between the BBU in the centralized core and the RU in the cell site, as shown in Figure 2-13. Although C-RAN provides several advantages, such as better resource virtualization, joint processing, and the possibility of cooperative radio sharing, it has yet to take off as a mainstream deployment option. As compared to C-RAN, O-RAN is seeing wider adoption due to the openness, cost effectiveness, and flexibility in deployment.

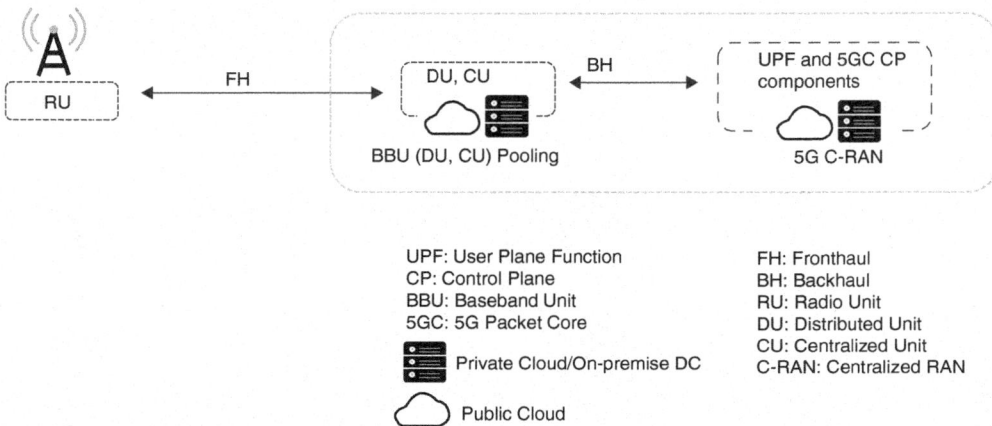

UPF: User Plane Function
CP: Control Plane
BBU: Baseband Unit
5GC: 5G Packet Core

Private Cloud/On-premise DC

Public Cloud

FH: Fronthaul
BH: Backhaul
RU: Radio Unit
DU: Distributed Unit
CU: Centralized Unit
C-RAN: Centralized RAN

FIGURE 2-13   C-RAN Deployment Example

Network Slice as a Service (NSaaS)

5G SA Option 2 also enables smart manufacturing/industry 4.0 by providing lower latency for URLLC by using edge computing and flexible deployment methods such as deploying 5GC Network Functions (NFs) in public cloud. These offerings can be made as a service by the service providers, as shown in Figure 2-9, using network slice as a service (NSaaS), thereby providing them with a new potential revenue source.

NSaaS can be offered by a communication service provider (CSP) to its customer service consumer (CSC) as a service. Note that some vendors refer to CSP as *network slice provider (NSP)* and CSC as *network slice consumer (NSC)*.

Business-to-Everything (B2X)

The network slice offered as a service to the CSC can be utilized to fulfill its requirements. This model is generally referred to as the B2X model, which consists of business-to-business (B2B) and business-to-consumer (B2C) models and is illustrated in Figure 2-14. B2B is generally used for providing services from service providers/CSPs to enterprise or industry vertical customers, and B2C is aimed at providing Internet-related services to mass consumers.

Figure 2-14 shows the NSaaS deployment method for the B2X offering.

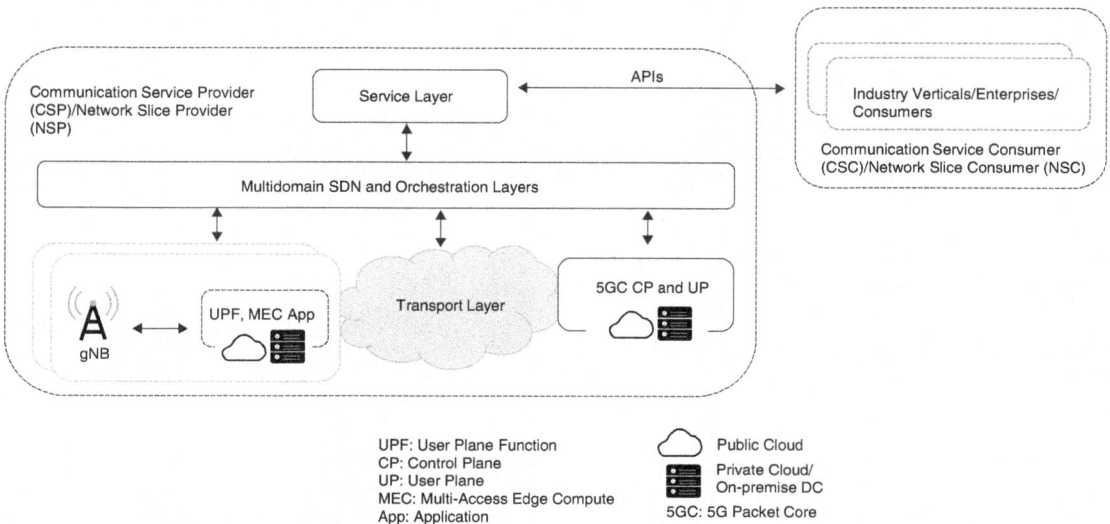

FIGURE 2-14   NSaaS Deployment Method for B2X Offering

Business-to-Business-to-Everything (B2B2X)

The network slice obtained by the CSC can also be utilized to play the role of CSP and offer its own services on top the slice services offered by the primary CSP. These types of services, illustrated in Figure 2-15, are called B2B2X services.

Depending on the service offering, the CSP offering NSaaS can impose certain limits on the NSaaS management capabilities exposure to the CSC, and the CSC can manage the network slice according to NSaaS management capabilities exposed and agreed upon in the SLAs between the CSP and CSC.

The NSaaS offered by the CSP could be characterized by certain KPIs and SLAs. Here's a list of some sample KPIs:

■ Type of Radio Access Technology (RAT)

■ Minimum Guaranteed Downlink (DL) Throughput

■ Minimum Guaranteed Uplink (UL) Throughput

■ Maximum Latency

■ Minimum Availability

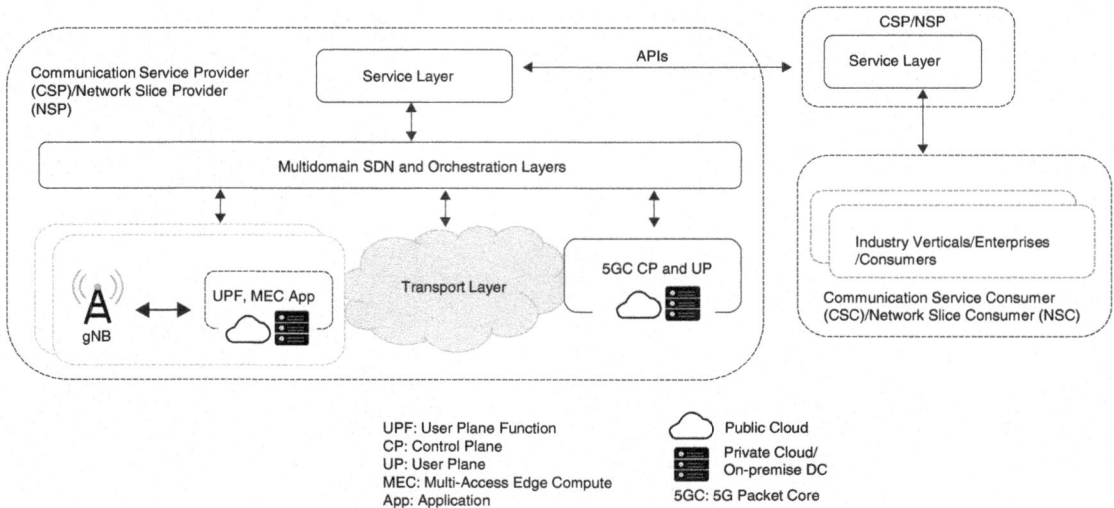

FIGURE 2-15   NSaaS Deployment Method for B2B2X Offering

5G Time-Sensitive Networks

The key advantages of deploying 5G technology are realized when 5G can be integrated and converged with existing and upcoming technologies being used in industry verticals and enterprises to enable new use cases such as smart industry 4.0, smart factory, and so on. For this to happen, 5G systems should work hand in hand and be integrated with the IEEE 802.1 working group covering time-sensitive networks (TSNs).

The IEEE TSN set of specifications is aimed at convergence and enabling low-latency communication in the factories of the future. 5G Time-Sensitive Communication is a service that supports deterministic and/or isochronous communication with high reliability and availability. It provides packet transport with quality of service (QoS) characteristics such as bounded latency and reliability, where end systems and relay/transmit nodes can be strictly synchronized.

When 5G is used to support TSN, the whole 5G system (5GS) is considered to be a TSN bridge. For 5GS to act as a TSN bridge, it needs to be time aware. This requires 5GS to be compliant with IEEE 802.1AS, which is an adaptation of Precision Time Protocol (PTP) for use with time-sensitive networks. To achieve this, TSN translators (TTs) are deployed at both the edge of the network—at the UE side and the UPF side—as shown in the Figure 2-16. The TT at the UE side is called the device-side TSN translator (DS-TT) and the TT at the UPF end is called network TSN translator (NW-TT). When 5GS needs to support a TSN flow, a PDU session will be initiated with QoS characteristics, as required by the TSN traffic flow.

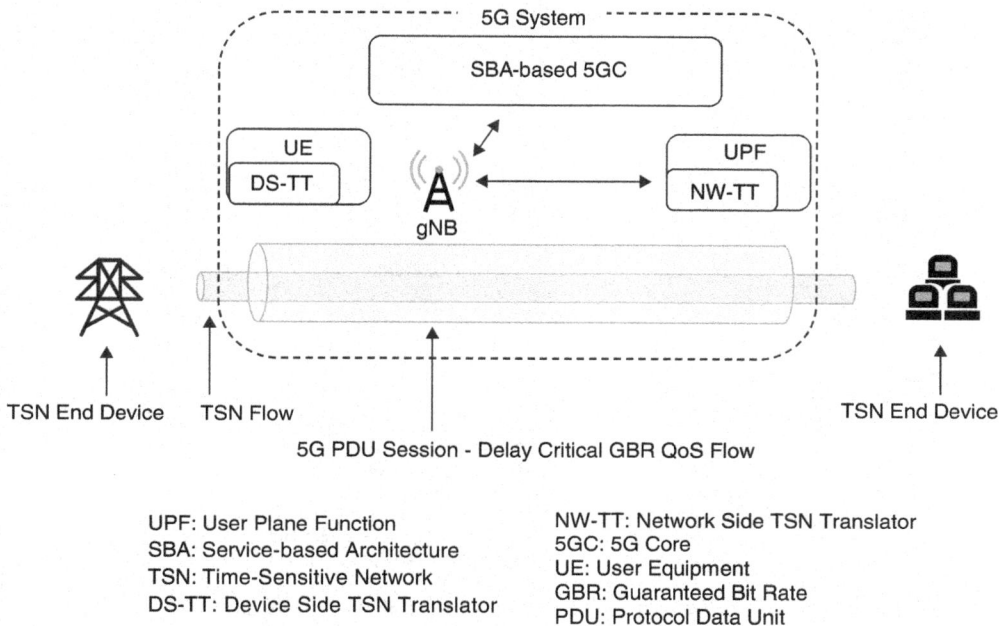

FIGURE 2-16 TSN Flow in 5G System

To maintain timing synchronization, UE, gNB, UPF, and TT will use the 5GS grandmaster (GM) clock. For the TSN domain synchronization, the GM in the TSN domain will be used. The PDU sessions with the specific QoS flows will be established based on the Application Function (AF) TSN translator integration with 5GC via the Network Exposure Function (NEF) or Policy Control Function (PCF), as shown in Figure 2-17. The AF TT is also integrated with the central network controller (CNC), which defines the schedule for transmitting TSN frames. To maintain time synchronization between the TSN and 5GS domain, the following procedure is followed.

Whenever the UPF receives the PTP message from the TSN domain, it will be sent to the DS-TT via the UE using the PDU session. The DS-TT will include the time taken for the packet to traverse from NW-TT to DS-TT in the egress PTP message before it sends it to the TSN domain. This time taken for the packet to traverse from the NW-TT to DS-TT is called the *calculated reference time*. By adding the calculated reference time to the egress PTP message, the accuracy of the PTP timing information is maintained.

Ongoing work to enhance TSN support for 5G-related use cases is prolific. The IEC/IEEE 60802 joint project is currently defining a TSN profile specification for the TSN features to be supported for industrial automation. This will enable interoperability, testing, and certifications. The Open Platform Communications (OPC) Foundation's Field Level Communications (FLC) initiative also aims to provide specifications based on IEC/ IEEE 60802 to achieve a single common multivendor converged TSN network infrastructure. 3GPP Release 17 (Rel-17) will bring in further enhancements to provide enablers for time-sensitive communications and time synchronization.

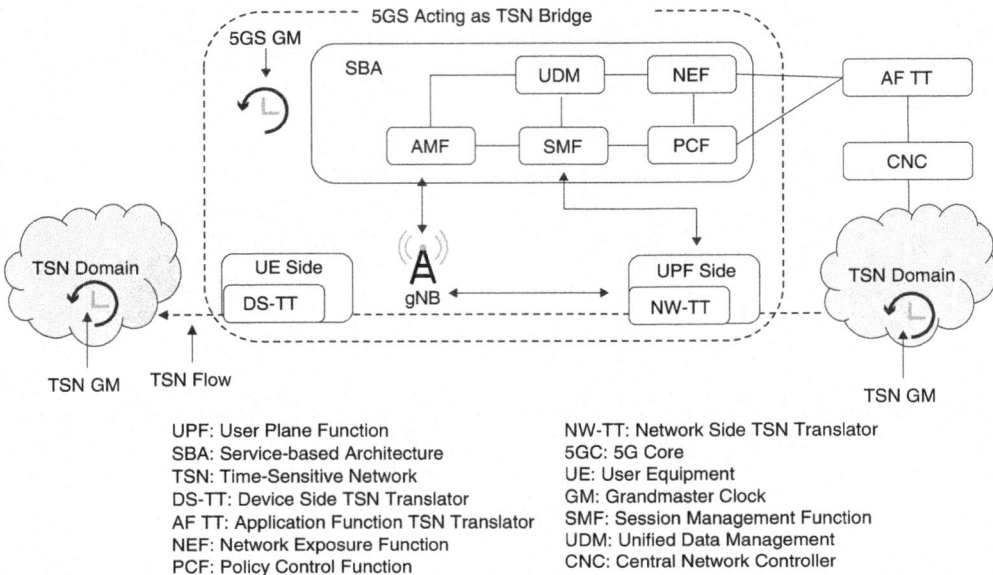

UPF: User Plane Function
SBA: Service-based Architecture
TSN: Time-Sensitive Network
DS-TT: Device Side TSN Translator
AF TT: Application Function TSN Translator
NEF: Network Exposure Function
PCF: Policy Control Function

NW-TT: Network Side TSN Translator
5GC: 5G Core
UE: User Equipment
GM: Grandmaster Clock
SMF: Session Management Function
UDM: Unified Data Management
CNC: Central Network Controller

FIGURE 2-17   Maintaining Timing Synchronization Between 5GS and TSN Domains

5G Local Area Network–Type Service

The real value of 5G is realized when we extend the scope of 3GPP-defined technologies to industry verticals and enterprises. Looking at the enterprise deployment scenarios, there are use cases such as local area networks (LANs) that can be fulfilled by 5G.

5G LAN-type services are being introduced by 3GPP, which provides LAN-type abstraction to 5G user equipment (UE) to send packets across the 5G network. The functionalities provided by 5G LAN-type service will be similar to your LAN, including support of virtual networks (VNs), which are termed 5G LAN–virtual networks (5G LAN-VNs). For example, in the enterprise environment, the UEs in the 5G VN group can communicate with each other within a 5G VN. The Network Exposure Function (NEF) also exposes services to dynamically manage the 5G VN group data. Furthermore, 5GS supports optimized routing by enabling support for local switching at the UPF without having to traverse the data network for UE-UE communication when the two UE(s) are served by the same user plane function.

Three types of traffic-forwarding methods are allowed for 5G VN communication, as listed next:

- **N6-based:** In this method, the UL/DL traffic for the 5G VN communication is forwarded to/from the DN; Figure 2-18 illustrates the non-roaming user plane architecture to support 5G LAN-type service using N6 tunnel.

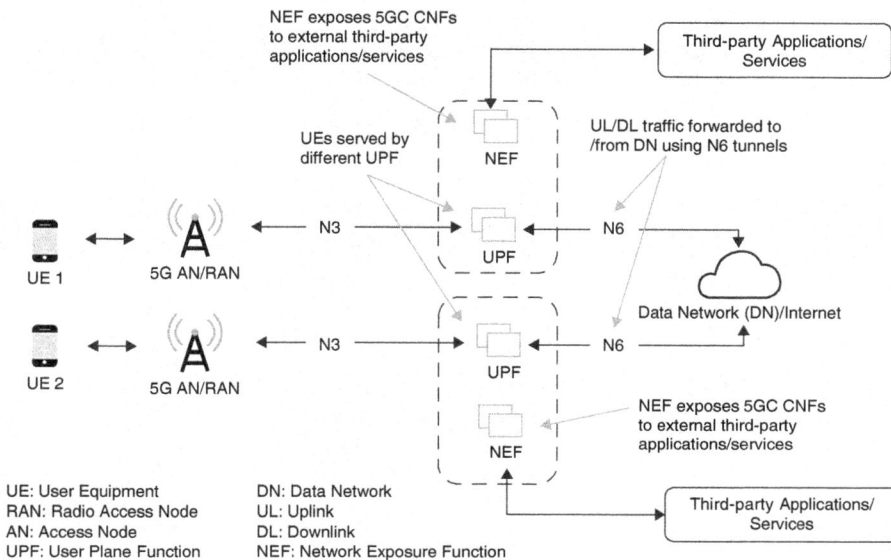

FIGURE 2-18   N6-Based Traffic Forwarding for 5G LAN-Type Service Deployments

- **Local switch based:** In this method, the traffic is locally forwarded by a single UPF if this UPF is the common PSA UPF of different PDU sessions for the same 5G VN group. Figure 2-19 illustrates the non-roaming user plane architecture to support 5G LAN-type service using a local switch.

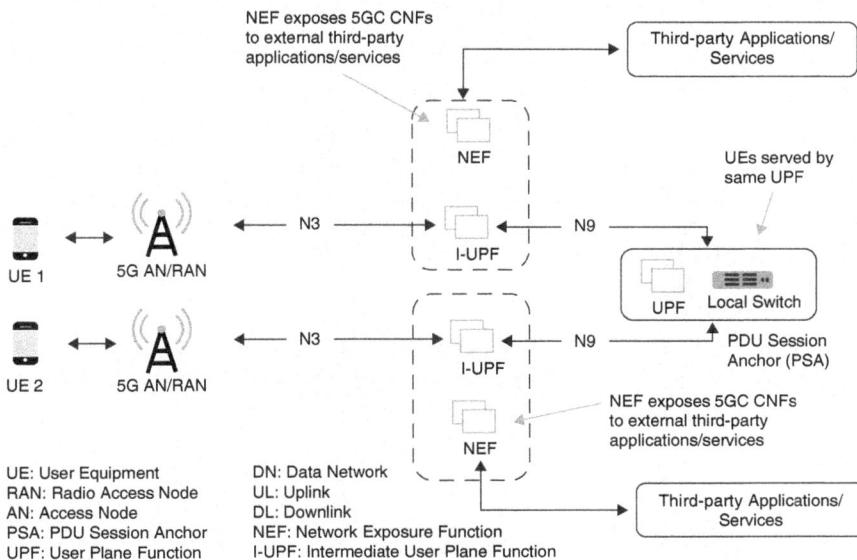

FIGURE 2-19   Local Switch–Based Traffic Forwarding for 5G LAN-Type Service Deployments

■ **N19-based:** In this method, the UL/DL traffic for the 5G VN group communication is forwarded between PSA UPFs of different PDU sessions via N19. N19 is based on a shared user plane tunnel connecting PSA UPFs of a single 5G VN group. Figure 2-20 illustrates the non-roaming user plane architecture to support 5G LAN-type service using N19 tunnel.

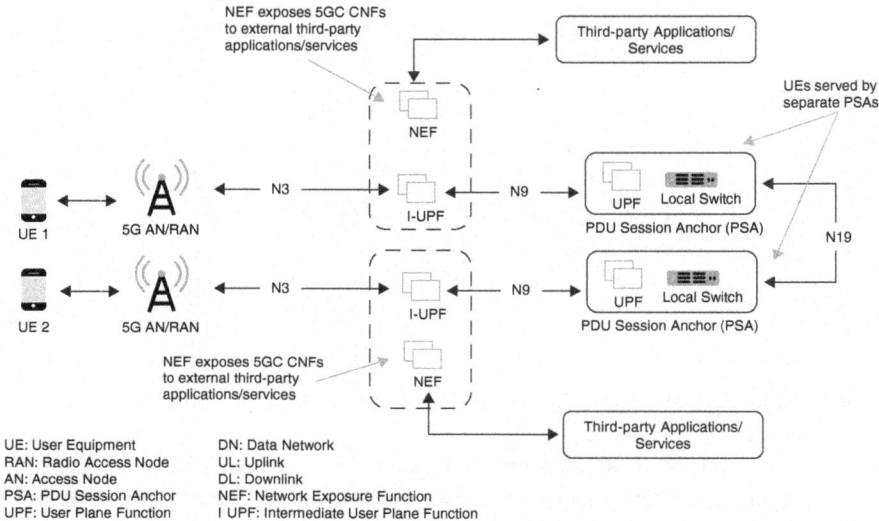

FIGURE 2-20   N19-Based Traffic Forwarding for 5G LAN-Type Service Deployments

Depending on the policy defined by the service provider, the 5G network shall allow trusted third parties to perform dynamic management of 5G VN group identification and membership. The Network Exposure Function (NEF) exposes a set of services to manage (for example, add/delete/modify) the 5G VN group and 5G VN member.

Private 5G/Non-Public Networks

5G SA Option 2 will also serve vertical industry use cases. To address these use cases, the network exposure capability and network slicing feature are defined to ensure openness, SLAs, isolation, and ease of use. 5G deployments that are implemented specifically to fulfill the industry verticals or enterprise use cases are called non-public networks (NPNs). NPNs are also referred to as private 5G networks. NPNs are intended to be used by the private enterprises or industry verticals and are not for general public use.

The two types of NPN networks are as follows:

- **Standalone non-public network (SNPN):** SNPNs are usually operated by an industry vertical such as smart manufacturing factories or critical infrastructure that want to operate their own 5G network and do not rely on traditional mobile network operators/service providers.

- **Public network integrated non-public network (PNI-NPN):** PNI-NPNs rely on traditional mobile network operators/service providers—for example, an enterprise that obtains a slice using an NSaaS offering from the CSP/NSP.

Standalone Non-Public Network (SNPN)

The flexibility and openness offered by 5G allows industry verticals and large enterprises to deploy their own 5G network and become NPN operators. Radio spectrums—sets of frequency ranges for cellular-compatible mobile devices to connect to a cellular network—can be owned by the NPN operator or can be leased from service providers based on the country radio regulations and the chosen operating model.

SNPN Radio Access Network (RAN) broadcasts public land mobile network identity (PLMN ID) and network identity (NID) in the system broadcast, enabling network (re)selection, overload control, access control, and barring. In addition, 3GPP Release-16 also specifies the capability for UE to obtain PLMN services while camping on the standalone non-public RAN. This enables the NPN operators to essentially "whitelist" the subscribers/users/endpoints they want on their network and manage them. The devices will be configured with a subscription permanent ID (SUPI) and will follow the usual registration procedures as being performed in the public networks.

Looking at various deployment models that could be implemented by NPN operators, it would be good to have a simplified comparison with the Wi-Fi deployments in the enterprise. As illustrated in Figure 2-21, a Wi-Fi deployment will consist of many Wi-Fi access points (AP), which are actually radio nodes using the 2.4 and 5 GHz radio spectrums in the Wi-Fi architecture, spread across your campus. These APs are then aggregated by the aggregation switches and forwarded to the wireless LAN controller (WLC). The WLC has integration with AAA servers, which perform authentication, authorization, and accounting for the users and devices. Roaming and handovers are also performed between the Wi-Fi APs based on the mobility of the users and devices.

In summary, the key components of Wi-Fi architecture in the example shown in Figure 2-21 are the Wi-Fi APs and the WLC. Access and aggregation switches are of course important, but the primary components that are "pure" Wi-Fi are still the APs and WLCs.

Now let's look at one of the methods of 5G NPN deployment where the end-to-end 5G component, including the radio spectrum, is catered by an enterprise or industry vertical. Here, along with the Wi-Fi radio node, commonly called a Wi-Fi access point (AP), you can also deploy the 5G radio node, called the gNB. 5G and Wi-Fi technology can co-exist with each other, as they use different radio spectrums, thereby preventing any interference between them.

Enterprise/Industry Vertical Private Network

Primary Wi-Fi Components

IAM — Identity and Access Management and Global Policy Engine

AAA — Authentication, Authorization, and Accounting

IT Applications

WLC — IT Infra

Users and Devices

Wi-Fi Access Points

Access and Aggregation Switches

Wireless LAN Controller

IT DC

Public Cloud Applications and Internet

Enterprise IT and DC

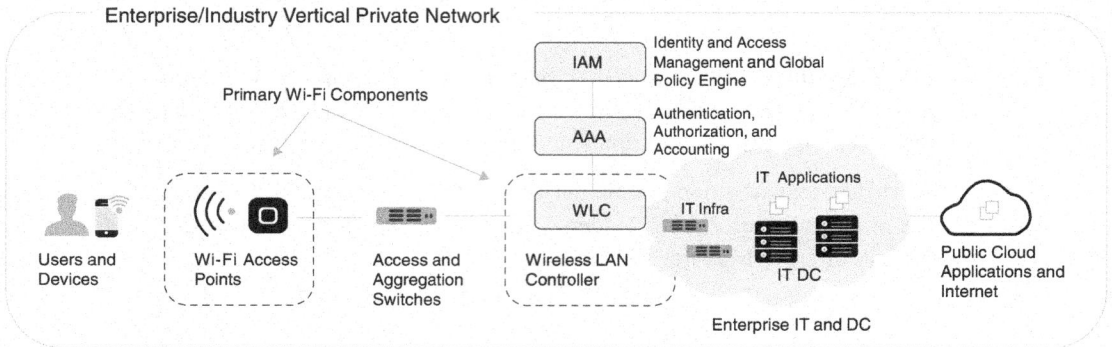

FIGURE 2-21 Wi-Fi Architecture

If you plan to go for the O-RAN/VRAN deployment model, you will use the disaggregated version of the gNB, which is the Distributed Unit (DU) and Centralized Unit (CU), along with the RAN intelligent controller (RIC). The gNBs are then aggregated using the access and aggregation switches and then terminated at the 5G Core (5GC), as illustrated in the Figure 2-22. The authentication, authorization, and accounting of the users and devices are enabled by integrating your existing authentication, authorization, and accounting (AAA) solution. The policy that is to be applied to the users and devices is configured within the Global Policy Engine in your existing Identity and Access Management (IAM) solution. There are also deployment models for SNPN based on shared spectrum, and this would depend on the country regulations and business model of the NPN operator.

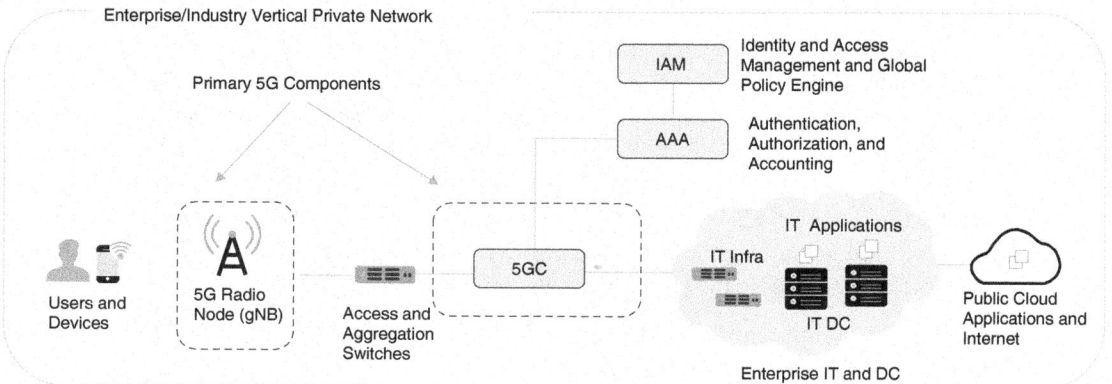

Enterprise/Industry Vertical Private Network

Primary 5G Components

IAM — Identity and Access Management and Global Policy Engine

AAA — Authentication, Authorization, and Accounting

IT Applications

Users and Devices

5G Radio Node (gNB)

Access and Aggregation Switches

5GC

IT Infra

IT DC

Public Cloud Applications and Internet

Enterprise IT and DC

FIGURE 2-22 Non-Public Network Architecture

Figure 2-23 shows a detailed view of 5GC components for standalone 5GC deployments, with the 5GC network functions that are involved in the authentication process for standalone NPN deployments within enterprises or industry verticals.

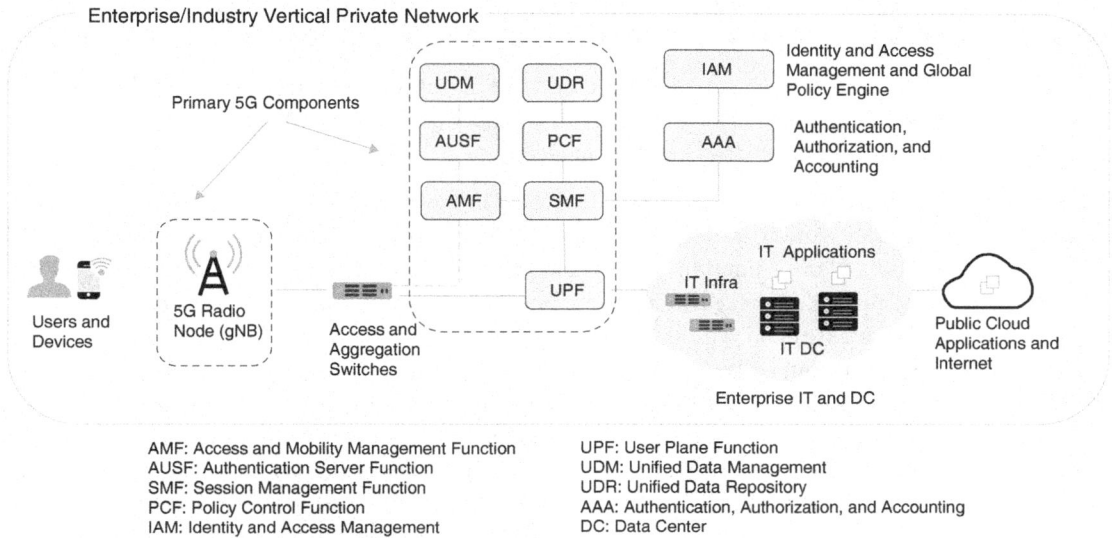

FIGURE 2-23   Non-Public Network Architecture—Detailed View

Public Network Integrated Non-Public Networks (PNI-NPN)

In public network integrated non-public network (PNI-NPN), network selection and reselection are based on PLMN ID. Cell selection and reselection as well as access control could be optionally based on Closed Access Group Identity (CAG ID). The CAG cell shall broadcast information such that only the UE supporting CAG is accessing the cell, which enables the PNI-NPN operator to control the users/devices that can access the network. When an NSaaS offering is obtained by the NPN operator, the user devices can be preconfigured with Single Network Slice Selection Assistance Information (S-NSSAI) to allow access to specific slices serving the NPN.

PNI-NPN could be seen as an option, where some 5GC Network Functions (NFs) such as the User Plane Function (UPF) are deployed by the NPN operator and rely on the public network for other NFs. The primary methods of deployment for PNI-NPN are as follows:

- Hybrid Method A: NPN UPF integrated with control plane from SP

- Hybrid Method B: Control plane shared with SP

- Network Slice method

Hybrid Method A: NPN UPF Integrated with Control Plane from SP

Figure 2-24 illustrates one of the PNI-NPN deployment methods where the UPF and MEC applications are owned and deployed by you in your enterprise IT data center, public cloud, or in a hybrid manner, where the UPF is located in your private cloud/enterprise DC, and MEC applications along with the MEC platform manager and orchestration are deployed in the public cloud.

The radio spectrum used for capacity and coverage can be owned by you as an enterprise/industry vertical or could be shared spectrum with the service provider (SP) in your country, which is dependent on country regulations and the chosen business model.

This deployment method will allow you the flexibility to have complete control over the configuration of the radio access network (RAN) nodes, which are indicated in the Figure 2-24 as gNB, along with the User Plane Function (UPF) and MEC application deployments. The Control Plane Functions of the SP will need to be integrated with your gNB and UPF network functions. Depending on the use case, you will also need integration with the UPF of the SP. In such cases, the UPF deployed in the enterprise infrastructure will act as an Intermediate User Plane Function (I-UPF), which is essentially a relay function that directs the GPRS Tunnelling Protocol (GTP) with header extensions for 5G toward the UPF deployed in the SP MEC. The UPF deployed in the SP MEC will act as a UPF PDU session anchor, which terminates the GTP protocol. Depending on the use case (such as Identifier Locator Separation [ID-LOC] solution), Locator/ID Separation Data Plane (LISP-DP) protocol or identifier location addressing (ILA) could also be used to support migration into ID-LOC native network.

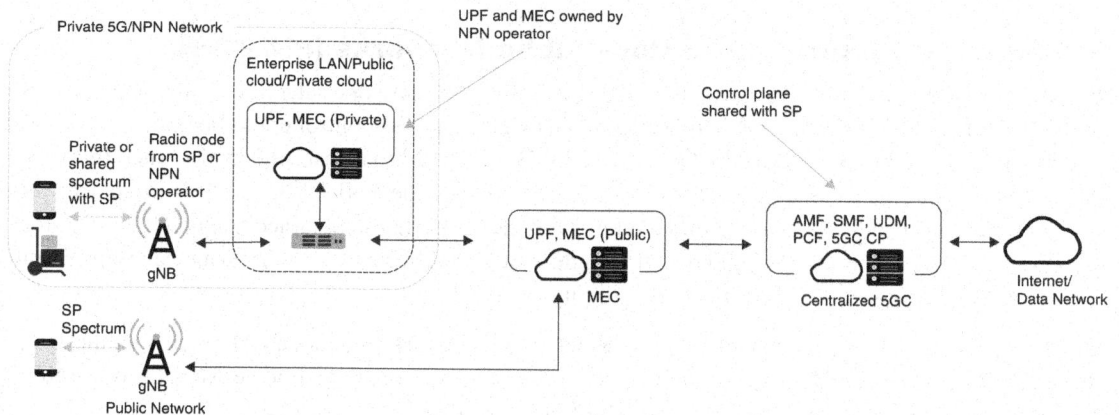

FIGURE 2-24   Hybrid Method A: NPN UPF Integrated with Control Plane from Service Provider

Hybrid Method B: Control Plane Shared with Service Provider

Figure 2-25 illustrates one of the PNI-NPN deployment methods where the UPF, MEC applications, and parts of the 5G control plane are owned and deployed by you in your enterprise IT data center, public cloud, or in hybrid cloud. The control plane in this deployment method is shared with the service provider (SP). Depending on what 5G components are shared with the SP, it can be termed as RAN sharing, core sharing, or RAN and core sharing. In the example shown in Figure 2-25, the control plane part of the 5GC is shared with the SP and can be termed the *core-sharing PNI-NPN deployment method* if the RAN is managed by the NPN operator. If the RAN is also managed by the SP, it can be termed the *RAN- and core-sharing PNI-NPN deployment method.* The option for radio spectrum used for capacity and coverage is the same as the Hybrid Method A, which means that the radio spectrum can be owned by you as an enterprise/industry vertical or could be shared radio spectrum with the service provider (SP) in your country, which is dependent on country regulations and the chosen business model.

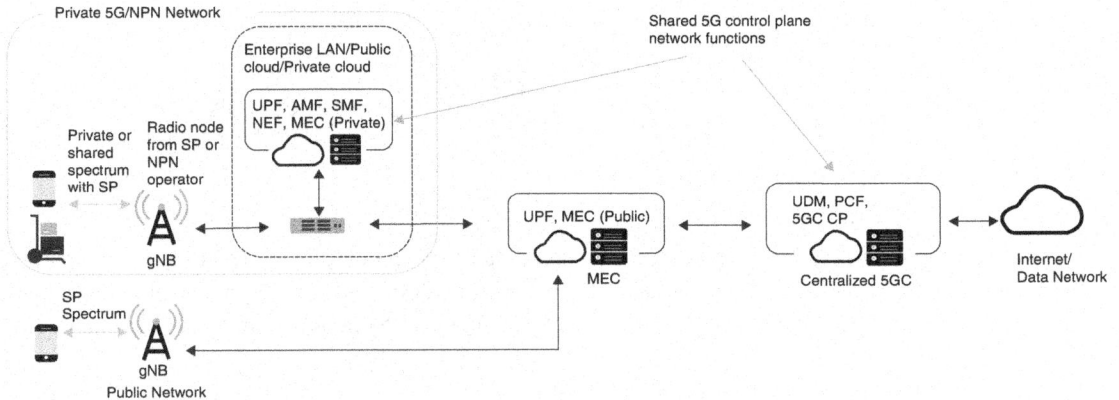

FIGURE 2-25   Hybrid Method B: Control Plane Shared with SP

Network Slice Method

In PNI-NPN deployments using the network slice approach, the NPN network obtains the network slice from the service provider, which manages the control plane and user plane. Some deployments in the network slice method might only utilize the control plane method. This purely depends on the use case. Figure 2-26 illustrates the network slice PNI-NPN deployment method, where an end-to-end network is deployed and managed by the SP. You as an enterprise or industry vertical might choose to deploy the UPF and MEC applications to cater for the URLLC use cases for Internet of Things (IoT) and industrial IoT (IIoT) deployments.

As shown in Figure 2-26, Slice 1 caters for the devices in the public slice and Slice 2 caters for the devices in the NPN network. You can choose to have the Local Break Out (LBO) deployed in your enterprise LAN by implementing MEC applications and user plane function for low-latency requirements of URLLC use cases. This method of deployment can be fulfilled by obtaining the NSaaS offering from the SP. You can agree to the SLAs and the KPIs with the SP to ensure the required QoS.

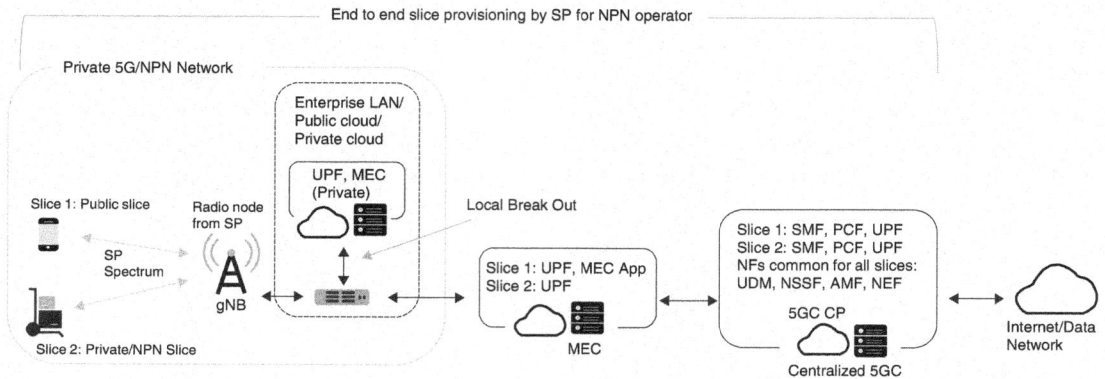

FIGURE 2-26   Network Slice Method

There are many more PNI-NPN deployment methods that vary in complexity and functionality based on the network function being deployed and the openness you plan to have on your deployment. Here are some of the key points you have to consider before choosing the deployment models:

- The key use cases you plan to fulfill using 5G.

- The maturity level of expertise in your organization to manage 5G and evolving architectures.

- Analyze whether your use cases can be fulfilled by using a shared model with service providers for 5G RAN, transport, and core networks.

- Choose whether you plan to own the radio spectrum yourself or share it with the SP.

- Have a plan to train your key personnel regularly on cellular technologies (5G and beyond).

- Decide whether you want a more open network deployment model such as O-RAN or want to use 3GPP-based distributed architecture.

Summary

5G Standalone (SA) is crucial not only for public network deployment but also for enabling industry verticals to deploy private 5G/non-public networks, thereby enhancing the existing industrial networks to smart industry 4.0, smart manufacturing, and addressing other key emerging use cases. 5G also allows you to monetize your network investments by allowing you to offer services to industry verticals using a NSaaS offering.

3GPP provides multiple architectures and deployment models for 5G, allowing service providers and industry verticals to choose the best-suited deployment model. Some choose the option of transitioning from 4G to 5G NSA and then from 5G NSA to SA, thereby taking a step-by-step approach. Some choose to deploy 5G SA directly without transitioning from 4G to 5G NSA.

There are other deployment models where you have the option of implementing 5G NSA or SA network functions on virtualized infrastructure within your premises, such as in your data centers, on a lightweight x86 server in remote sites, on programmable cell site routers, on a hybrid on-premises and public cloud, or entirely on a public cloud.

You can further make full use of the openness and flexibility in 5G architecture by opting to deploy O-RAN models and optimize your investments by choosing the best-of-the-breed vendors for specific parts of your 5G network buildout.

Before you choose any deployment model, it is best to understand the use cases you are trying to achieve, your target customer or industry vertical, and the strengths of your existing team. For example, if you are primarily a cloud service provider, then it might make more sense to build 5G network functions in your cloud infrastructure and then offer your 5G services to your customers.

Acronym Key

Acronym	Expansion
5G LAN-VN	5G local area network–virtual network
5GC	5G Core network
AF-TT	Application function TSN translator
AP	Access point
API	Application programming interface
BH	Back haul
CU	Centralized Unit
DL	Downlink
DS-TT	Device-side TSN translator
DU	Distributed Unit
eMBB	Enhanced mobile broadband
eNB	E-UTRAN NodeB
EPC	Evolved Packet Core
FH	Front haul
gNB	New Radio NodeB
HW	Hardware
M2M	Machine-to-machine

Acronym	Expansion
MBB	Mobile broadband
MEC	Multiple-access edge compute
MH	Mid haul
NPN	Non-public network
NR	New Radio
NSaaS	Network slice as a service
NW-TT	Network-side TSN translator
O-CU	O-RAN-compliant Centralized Unit
O-DU	O-RAN-compliant Distributed Unit
O-RAN	Open RAN
O-RU	O-RAN-compliant Radio Unit
PDU	Protocol Data Unit
PLMN	Public land mobile network
PNI-NPN	Public network integrated NPN
SNPN	Standalone non-public network
TLS	Transport Layer Security
TSN	Time-sensitive network
TT	TSN translator
UE	User equipment
UL	Uplink
UPF	User Plane Function
URLLC	Ultra-reliable and low-latency communications
WLC	Wireless LAN controller
X2-C	X2–control plane
X2-U	X2–user plane
Xn-C	Xn–control plane
Xn-U	Xn–user plane
XnAP	Xn Application Protocol

References

Pramod Nair, "Securing Your 5G Infrastructure to the Edge and Beyond," Cisco Public, April 1, 2021

Pramod Nair, "Securing 5G and Evolving Architectures," Cisco Knowledge Networks, Cisco Public, September 17, 2020

Pramod Nair, "Why 5G Is Changing Our Approach to Security," Cisco Public, July 10, 2020

Cisco Systems, "Reimagining End-to-End Network Mobile Network in the 5G Era," Cisco Public, 2019

3GPP TS 24.301, "Non-Access-Stratum (NAS) protocol for Evolved Packet System (EPS): Stage 3"

3GPP TS 38.300, "NR: NR and NG-RAN Overall Description"

3GPP TS 38.460, "NG-RAN: E1 general aspects and principles"

3GPP TS 38.470, "NG-RAN: F1 general aspects and principles"

3GPP TR. 23.374, "Study on Enhancement of 5G System (5GS) for Vertical and Local Area Network (LAN) Services"

3GPP TS 22.101, "3rd Generation Partnership Project: Technical Specification Group Services and Systems Aspects; Service aspects; Service principles"

3GPP TS 23.222, "Functional architecture and information flows to support Common API Framework for 3GPP Northbound APIs"

3GPP TR 28.822, " Study on charging aspect of 5G LAN-type Services" Rel-17

O-RAN Xhaul Packet Switched Architectures and Solutions 1.0, November, 2020

3GPP TS 32.290, "5G system: Services, operations and procedures of charging using Service-Based Interface (SBI)"

ETSI GS MEC 003, Multi-access Edge Computing (MEC); Framework and Reference Architecture

O-RAN Minimum Viable Plan and Acceleration towards Commercialization, White Paper, June 29, 2021

https://5g-acia.org/whitepapers/integration-of-5g-with-time-sensitive-networking-for-industrial-communications/

GSMA, "5G Implementation Guidelines: SA Option 2"

https://www.3gpp.org/

Chapter | **3**

Securing 5G Infrastructure

After reading this chapter, you should have a better understanding of the following topics:

- Security enhancements in 5G as compared to 4G
- Key 5G security enhancements in 3GPP Rel-16
- Key security challenges in 5G deployment
- Need for an external layer of security for 5G deployments

As seen in previous chapters, not only will 5G technology usher in a new era of improved network performance and speed; it will also offer new connected experiences for users and fulfill various use cases for industry verticals. This chapter discusses the reasons why you should have an external layer of security controls even though 3GPP provides some enhancements in security.

This chapter will be of particular interest to the following individuals and teams from enterprise/5G service providers deploying 5GC and cybersecurity vendors planning product developments for 5G security use cases:

- Security leaders within government organizations who need to prioritize investments in security
- Management consultants advising governments and service providers on 5G security investment
- CSO and CTO teams from service providers
- CSO and CTO teams from enterprises deploying NPN/private 5G
- CTO teams from service providers planning to offer network slice as a service (NSaaS)
- Cybersecurity vendor security architects looking to secure 5GC deployments for their customers
- Cybersecurity vendor product managers looking for use cases or features to enhance security products to cater for secure 5G deployments

■ Teams and personnel from service providers, industry verticals, and military sectors who would like to get a brief understanding of security enhancements and security challenges in 5G

5G use cases will unlock countless applications, enable more robust automation, and increase workforce mobility. Incorporating 5G technology into these environments requires deeper integration between enterprise networks and the 5G network components of the service provider. This exposes enterprise owners (including operators of critical information infrastructure) and 5G service providers to risks that were not present in 4G. An attack that successfully disrupts the network or steals confidential data will have a much more profound impact than in previous generations.

3GPP 5G Security Enhancements

As compared to 4G, 5G brings in a unified, access-agnostic, and flexible authentication framework that caters for 3GPP and non-3GPP users and devices, allowing different types of credentials. This has again been enhanced in 3GPP Release 16 (Rel-16), and further enhancements are being planned in Release 17 (Rel-17) to allow secure integration with non-public networks.

The 5G security architecture is designed to integrate 4G security. In addition, other security threats, such as attacks on radio interfaces, signaling plane, user plane, masquerading, privacy, replay, bidding down, man-in-the-middle, and inter-operator security issues, have also been taken into account for 5G. The enhancements to the security will be seen in upcoming releases, depending on the use cases and features being considered for the release, and this will be an on-going activity.

5G also improves subscriber identity protection by evolving the security for the initial Non-Access Stratum (NAS) messages. 5G also enables 5G Core (5GC) Network Function (NF) to be securely exposed via the Network Exposure Function (NEF). For the roaming interface, 5G introduces Security Edge Proxy Protection (SEPP), which protects the roaming interconnections between the home and visited networks.

The sections that follow discuss some of the key security enhancements in 5G as compared to 4G.

5G Trust Model: Non-Roaming

5G Standalone (SA) architecture evolves the trust model, where the trust within the network is considered as decreasing when moving away from the 5G packet core.

Figure 3-1 illustrates the 5G trust model for non-roaming.

The trust model for the user equipment (UE), Radio Access Network (RAN) nodes, and 5G packet core is detailed in the following list:

■ **UE:** The tamper-proof universal integrated circuit card (UICC) on which the Universal Subscriber Identity Module (USIM) resides as a trust anchor, along with the mobile equipment (ME). The ME and the USIM together form the UE.

■ **RAN:** The RAN is split into Distributed Unit (DU) and Centralized Unit (CU), as shown in Figure 3-1.

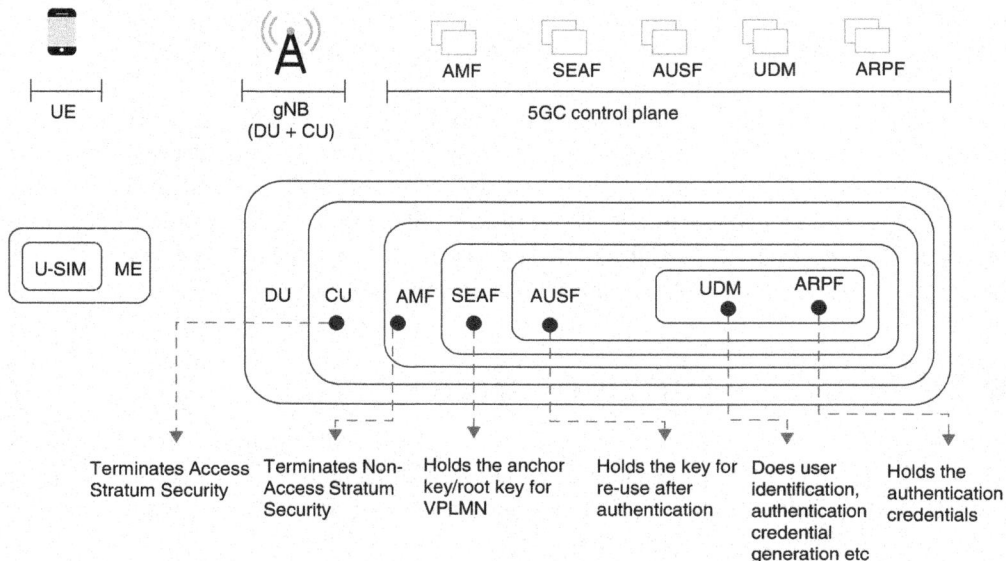

FIGURE 3-1   Trust Model for 5G Non-Roaming Architecture

- **Distributed Unit (DU):** The DU consists of DU-CP and DU-UP. The DU performs rate adaptation, channel coding, modulation, and scheduling as well as provides support for the lower layers of the protocol stack. It can be deployed together with the Radio Unit (RU)/Remote Radio Unit (RRU) or can be deployed in the far edge/edge layer where it serves multiple RUs.

- **Centralized Unit (CU):** The CU consists of CU-CP and CU-UP. The CU caters for the Radio Resource Control (RRC) protocol, Service Delivery Adaptation Protocol (SDAP), and Packet Data Convergence Protocol (PDCP) as well as terminates the Access Stratum (AS) security. the CU can be co-located with the DU, or it can be deployed on far edge or edge layer and aggregates multiple DUs. The CU-UP and CU-CP split can also be performed to enable deployment of CU-UP closer to the DU.

- **5GC control plane:**

  - **AMF:** The Access Management Function (AMF) serves as the termination point for Non-Access Stratum (NAS) security.

  - **SEAF:** The Security Anchor Function (SEAF) supports the primary authentication using Subscription Concealed Identifier (SUCI) and holds the root key (known as anchor key) for the visited network.

  - **AUSF:** The Authentication Function (AUSF) keeps a key for reuse, derived after authentication, in case of simultaneous registration of the UE in different access network technologies (that is, 3GPP access networks and non-3GPP access networks such as IEEE 802.11 wireless local area network).

- **ARPF:** The Authentication Credential Repository and Processing Function (ARPF) keeps the authentication credentials. This is mirrored by the user equipment.

- **UDM:** Unified Data Management (UDM) holds the home network private key for the private/public key pairs used for subscriber privacy.

5G Trust Model: Roaming

In the roaming architecture, the home and the visited network are connected through Security Protection Proxy (SEPP) for the control plane of the internetwork interconnect. This enhancement is done in 5G because of the number of attacks coming to light recently, such as key theft and re-routing attacks in SS7 as well as network node impersonation and source address spoofing in signaling messages in Diameter that exploited the trusted nature of the internetwork interconnect.

Figure 3-2 illustrates the 5G trust model for roaming.

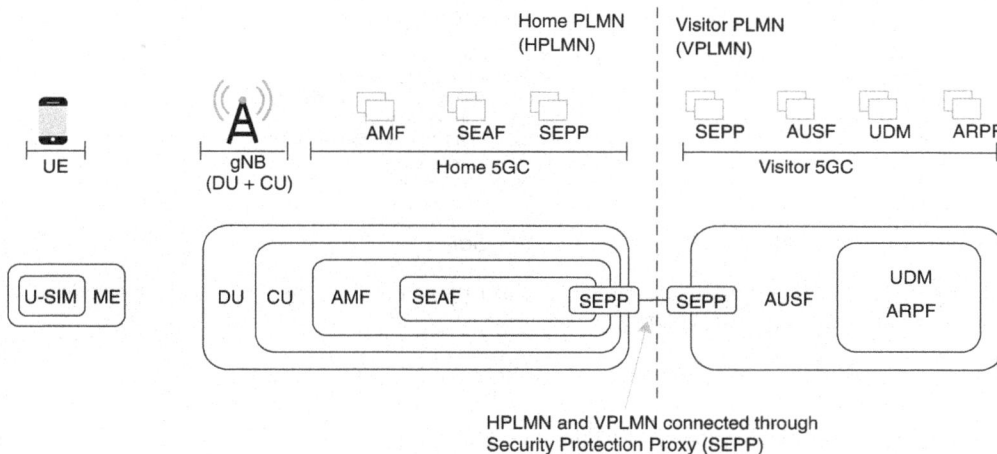

FIGURE 3-2   Trust Model for 5G Roaming Architecture

Integration of Non-3GPP Network to the 5G Core Network

For non-3GPP access networks, 3GPP has introduced the Non-3GPP Inter-Working Function (N3IWF), which functions as a security gateway to allow the UE to access the 5G core over untrusted, non-3GPP networks through IP Security (IPsec) tunnels. Figure 3-3 illustrates the integration of the non-3GPP network to the 5G Core (5GC) network. The AMF terminates the Non-Access Stratum (NAS) security. The SEAF, which is co-located in the Phase 1 specification, holds the root key for the visited network. The Subscriber Identity De-Concealing Function (SIDF) decrypts the Subscription Concealed Identifier (SUCI) to reveal the Subscription Permanent Identifier (SUPI). When the home network public key is used for encryption of the SUPI, the SIDF uses the home network private key, which is securely stored in the home operator's network, to decrypt the SUCI.

The Non-3GPP Inter-Working Function (N3IWF) provides IPsec between the UE and the N3IWF for both user plane and control plane traffic. N3IWF also helps with NAS signaling, initial registration, and authentication.

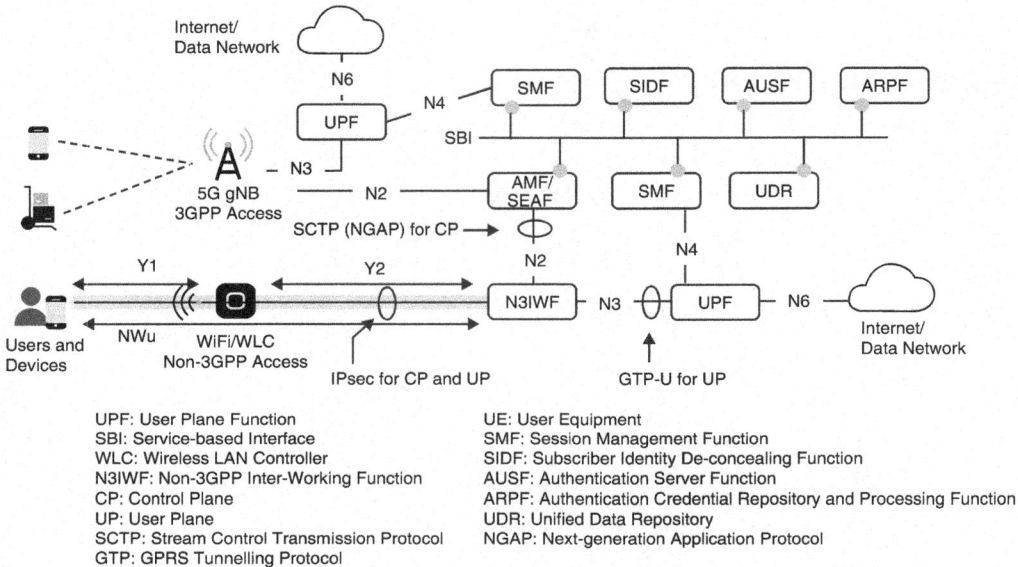

UPF: User Plane Function
SBI: Service-based Interface
WLC: Wireless LAN Controller
N3IWF: Non-3GPP Inter-Working Function
CP: Control Plane
UP: User Plane
SCTP: Stream Control Transmission Protocol
GTP: GPRS Tunnelling Protocol

UE: User Equipment
SMF: Session Management Function
SIDF: Subscriber Identity De-concealing Function
AUSF: Authentication Server Function
ARPF: Authentication Credential Repository and Processing Function
UDR: Unified Data Repository
NGAP: Next-generation Application Protocol

FIGURE 3-3   Integration of Non-3GPP Network to 5G Core

Once authenticated, the N3IWF will also establish the user plane between the UE and the data network using the N3 interface between the N3IWF and the UPF.

The reference points specific to non-3GPP access, as shown in Figure 3-3, are listed as follows:

- **Y1:** Reference point between the UE and the non-3GPP access

- **Y2:** Reference point between the untrusted non-3GPP access and the N3IWF for the transport of NWu traffic

- **NWu:** Reference point between the UE and N3IWF for establishing secure tunnel(s) between the UE and N3IWF so that control plane and user plane exchanged between the UE and the 5G Core Network are transferred securely over untrusted non-3GPP access

Unified, Access-Agnostic, and Unified Authentication Framework

4G use cases were primarily voice and data for mobile subscribers, where the authentication was based on the 4G Evolved Packet System Authentication and Key Agreement (EPS-AKA) and required a physical or software Universal Subscriber Identity Module (USIM). 5G enables industry verticals to deploy new use cases—some of which require authentication with data networks (DNs) outside the service provider's domain. Such use cases might include non-3GPP-based devices.

The 5G authentication framework allows both 3GPP and non-3GPP devices to authenticate and connect to your network. This is made possible due to the multiple authentication methods, such as 5G Authentication and Key Agreement (5G AKA), Extensible Authentication Protocol Authentication and Key Agreement (EAP-AKA'), and Extensible Authentication Protocol Transport Layer Security (EAP-TLS), as shown in Figure 3-4.

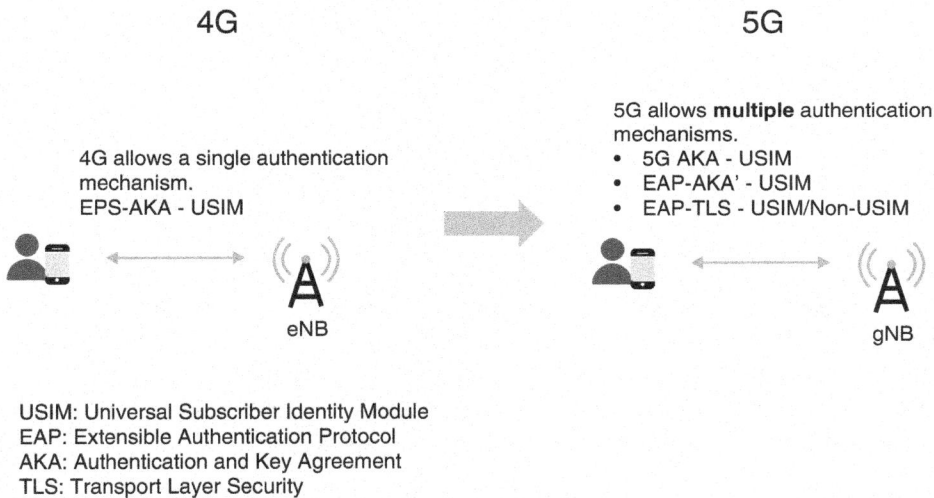

FIGURE 3-4 Comparison Between 4G and 5G Authentication Methods

EAP-AKA' and 5G AKA are mandatory 5G primary authentication methods, whereas EAP-TLS is primarily seen in non-public network (NPN) deployments to enable the industry vertical/enterprise to reuse the existing security infrastructure for network access control for 5G devices. EAP-TLS is also very useful for low-cost massive IoT (mIoT) deployments such as sensors and such, where SIM cards are usually not considered due to their prohibitive cost. These new authentication methods make 5G authentication access-layer agnostic.

To enable the unified authentication framework, 5G has introduced key Network Functions (NFs) such as Subscriber Identity De-concealing Function (SIDF), Authentication Server Function (AUSF), Unified Data Management (UDM), Security Anchor Function (SEAF) in the new 5G service-based architecture (SBA), and a new key hierarchy related to authentication in 5G System (5GS). Figure 3-5 illustrates the 5GS authentication key hierarchy.

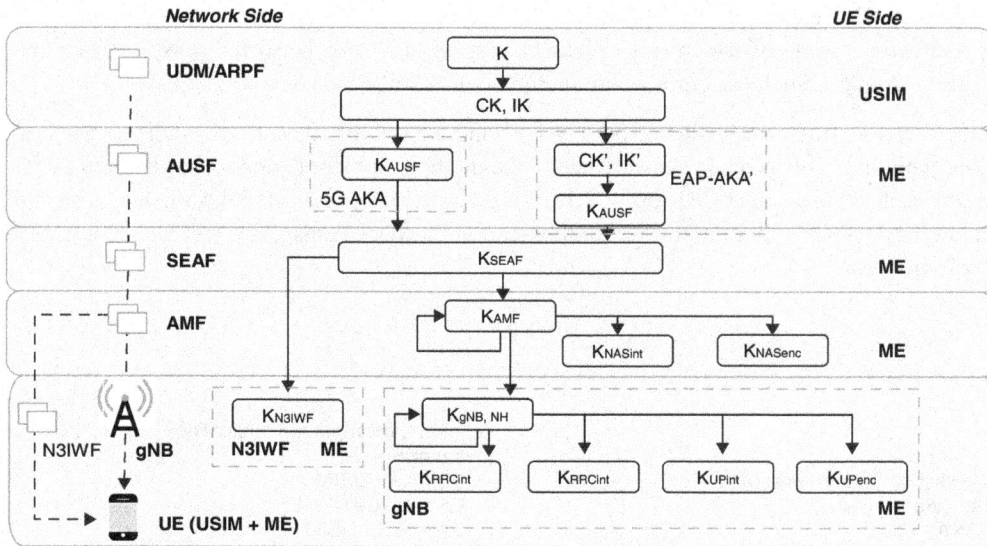

FIGURE 3-5   5GS Authentication Key Hierarchy

As shown in Figure 3-5, the primary source of security context is the long-term secret key (K) provisioned in the USIM and the 5G core network, similar to the 4G key implementation. The differentiator here between 4G and 5G is the primary and secondary authentication. All devices have to perform primary authentication for accessing the mobile network services as well as secondary authentication to an external data network (DN), if so desired, by the external data network.

Another key area of enhancement is the UP integrity protection, which is another key enhancement that is valuable for the IoT-related use cases.

These security enhancements in 5G enable wider acceptability of 5G technology.

Use of Subscription Permanent Identifier (SUPI) and Subscription Concealed Identifier (SUCI)

One of the risks was with the International Mobile Subscriber Identity (IMSI), which consists of 15 to 16 digits. This is the unique identity of the subscribers being occasionally sent in clear text while performing the Non-Access Stratum (NAS) procedures. The key risk here was the IMSI catcher, which is an eavesdropping device used for intercepting mobile phone traffic and tracking the location of the mobile phone users. The eavesdropping is performed by the IMSI catcher by mimicking a service provider cellular base station in order to force all nearby mobile phones and other cellular data devices to connect to it. Once these devices are connected, the IMSI catcher can force the UE to use no call encryption (A5/0 mode) or to use easily breakable encryption (A5/1 or A5/2 mode), making the call data easy to intercept and convert to audio, leading to a man-in-the-middle (MitM) attack scenario.

In 5G, privacy is achieved even before authentication and key agreement, by encrypting the SUPI before transmitting using a home network public key, which is stored in the USIM. 3GPP has specified SUCI, which replaces IMSI and T-IMSI for all air interface interactions, as shown in Figure 3-6.

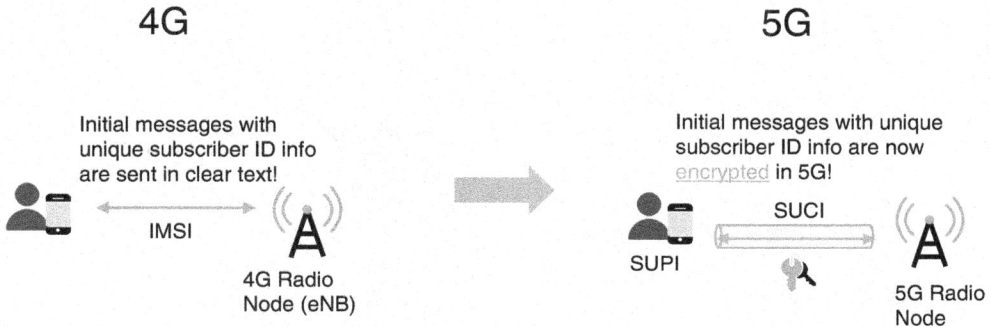

4G   5G

Initial messages with unique subscriber ID info are sent in clear text!

IMSI

4G Radio Node (eNB)

Initial messages with unique subscriber ID info are now encrypted in 5G!

SUCI

SUPI

5G Radio Node

IMSI: International Mobile Subscriber Identity
SUPI: Subscription Permanent Identifier
SUCI: Subscription Concealed Identifier

FIGURE 3-6   Comparison of Subscriber Privacy Between 4G and 5G

Figure 3-7 illustrates the structure of the SUPI.

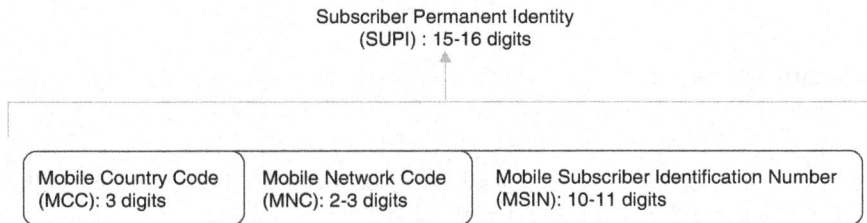

Subscriber Permanent Identity
(SUPI) : 15-16 digits

Mobile Country Code (MCC): 3 digits	Mobile Network Code (MNC): 2-3 digits	Mobile Subscriber Identification Number (MSIN): 10-11 digits

FIGURE 3-7   SUPI Structure

The SUPI contains sensitive subscriber and subscription information that is secured by the UE, which generates the SUCI, using a protection scheme based on Elliptic Curve Integrated Encryption Scheme (ECIES) profiles.

Figure 3-8 illustrates the structure of the SUCI.

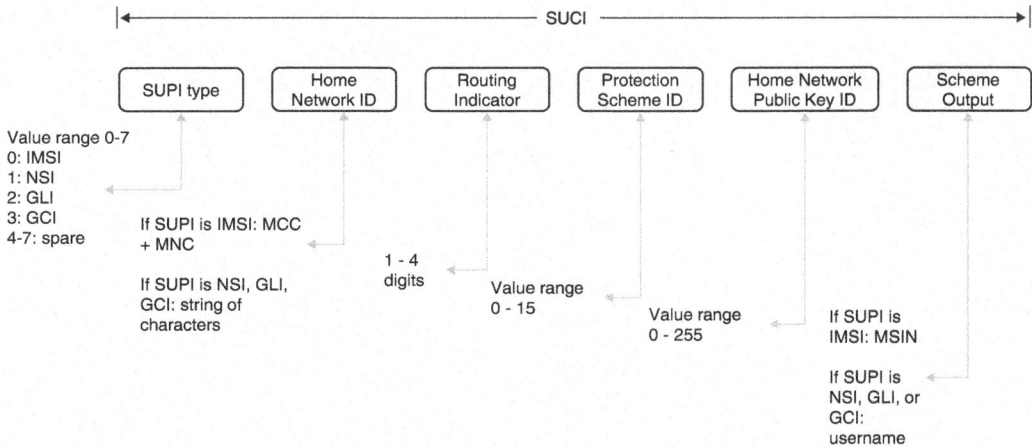

FIGURE 3-8 SUCI Structure

As shown in Figure 3-8, the structure of the SUCI is composed of the following parts:

- **SUPI type:** The value range of the SUPI type is 0 to 7, as detailed next:

  - **0:** International Mobile Subscriber Identity (IMSI)

  - **1:** Network Specific Identifier (NSI)

  - **2:** Global Line Identifier (GLI)

  - **3:** Global Cable Identifier (GCI)

  - **4–7:** For future use

- **Home network identifier:** The value of the home network identifier will depend on the format of the SUPI type. If the SUPI is an IMSI, then the home network identifier will consist of a mobile country code (MCC) and mobile network code (MNC). If the SUPI type is NSI, GLI, or GCI, the home network identifier will consist of a string of characters that is variable in length, representing a domain name, such as user@serviceprovidername.com.

- **Routing indicator:** The routing indicator consists of one to four decimal digits, provisioned in the Universal Subscriber Identity Module (USIM) by the home network service provider. This, along with the home network identifier, is used to route network signaling with the SUCI to the Authentication Server Function (AUSF) and Unified Data Management (UDM) in your 5G packet core to serve the subscriber user or device.

- **Protection scheme identifier:** The protection scheme identifier consists of a value in the range of 0 to 15, which indicates the non-null protection scheme specified by the home network service provider. Null scheme shall be used if the SUPI type is a GLI or GCI.

- **Home network public key identifier:** The home network public key identifier represents the public key provisioned by the home network service provider or standalone non-public network (NPN), which is used to identify the key used for SUPI protection. Its value is in the range of 0 to 255.

- **Scheme output:** The scheme output can be a string of characters with variable length or hexadecimal digits, depending on the type of protection scheme used. If the SUPI type is IMSI, then the scheme output is the mobile subscriber identification number (MSIN), and if the SUPI type is network specific identifier, global line identifier, or global cable identifier, the scheme output is the username part of the network access identifier (NAI).

These enhancements and evolution in authentication in 5G technology, specifically in non-AKA-based authentication mechanisms using EAP-TLS, allow industry verticals to adopt 5G deployments for non-public consumption. For service providers, this also means the enhanced capability to consume non-cellular IoT and IIoT devices as well as devices being served by non-3GPP network access to register and connect to the 5GC using non-AKA protocols such as EAP-TLS and offering services for monetizing 5G deployments.

Enhanced Inter-PLMN Interconnect Using Security Edge Protection Proxy (SEPP)

In 4G, the Inter-PLMN Diameter messages can be exploited, leading to source address spoofing and other threats. 5G introduces SEPP to enhance the security of inter-PLMN interconnect, as discussed earlier in the section "5G Trust Model: Roaming." Figure 3-9 compares the roaming security of 4G and 5G.

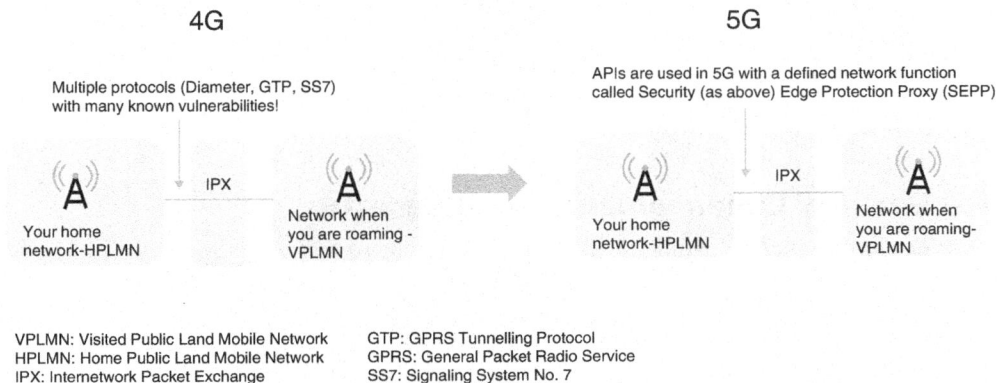

FIGURE 3-9 Comparison Between 4G and 5G Roaming Security

SEPP implements application layer security for all the service layer information exchanged between two NFs across operator domains. This prevents IPX providers from reading sensitive information, such as authentication vectors. SEPP can also detect any unauthorized message modifications, and it performs message source validation, message format verification, and topology hiding.

In addition to Transport Layer Security (TLS), Application Layer Security (ALS) is also needed to protect information elements (IEs) exchanged between Network Functions (NFs) of different operators.

SEPP enforces inter-PLMN security on the N32 interface and is located in the perimeter of the PLMN for protecting inter-PLMN communications. The N32 interface consists of N32-c and N32-f. N32-c is the control plane, which is used by SEPP to negotiate the security capabilities. N32-f is the forwarding interface, also referred to as the end-to-end encrypted application interface. For N32-f, SEPP also provides encapsulation of HTTP/2 core signaling messages using JavaScript Object Signing and Encryption (JOSE) protection.

Figure 3-10 illustrates the SEPP in a 5G interconnect domain.

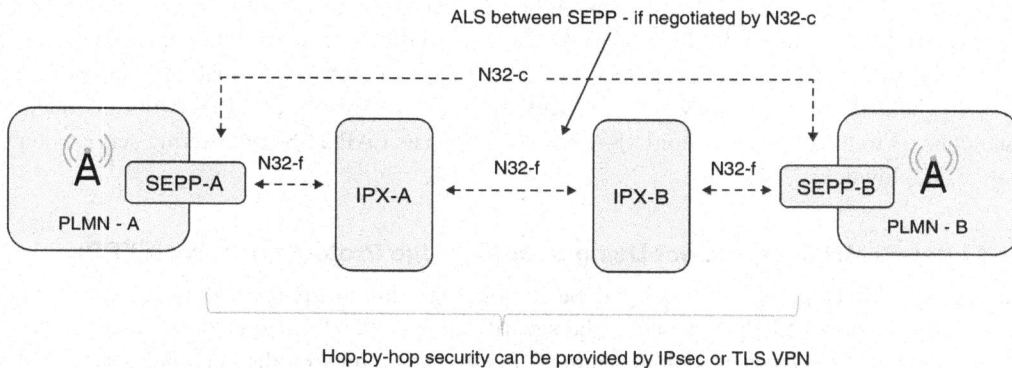

ALS: Application Level Security
TLS: Transport Layer Security
SEPP: Security Edge Protection Proxy
PLMN: Public Land Mobile Network
IPX: Internetwork Packet Exchange
IPsec: Internet Protocol Security

FIGURE 3-10   SEPP in 5G Interconnect Domain

Other Key Security Enhancements in Release 16

Rel-16 brought enhancements to a number of features. Each of these features is in the initial stages and will be subject to later modification or enhancement, over several years, by the means of change requests (CRs). The important Rel-16 features are as follows:

- Enhancement of ultra-reliable low-latency communications (URLLC)
- Support of LAN-type services
- Cellular Internet of Things (IoT)
- Enhanced V2X support
- API-related items

- Coexistence with non-3GPP systems

- Railways and maritime

- Mission critical, public warning

- Conversational services, streaming, and TV

- 5G location and positioning services

- User identities, authentication, multi-device

- Network slicing

- UE radio capability signaling optimization

The list of all the features introduced in Rel-16, including the aforementioned ones, is provided in the 3GPP document TR 21.916, "Summary of Rel-16 Work Items." Some of these new features also bring in enhanced security, which is discussed further in the following section.

Security Enhancements in Northbound API-Related Items

While a basic Common API Framework (CAPIF) is made available in 3GPP Rel-15, there are several enhancements that are considered for developing eCAPIF in Rel-16, which introduces key features such as support for stronger integrations with third-party domains, as shown in the Figure 3-11. Key enhancements are based on the enhanced interconnection between two CAPIF providers, support for distributed deployments, and closer integration between third-party API providers and the CAPIF framework.

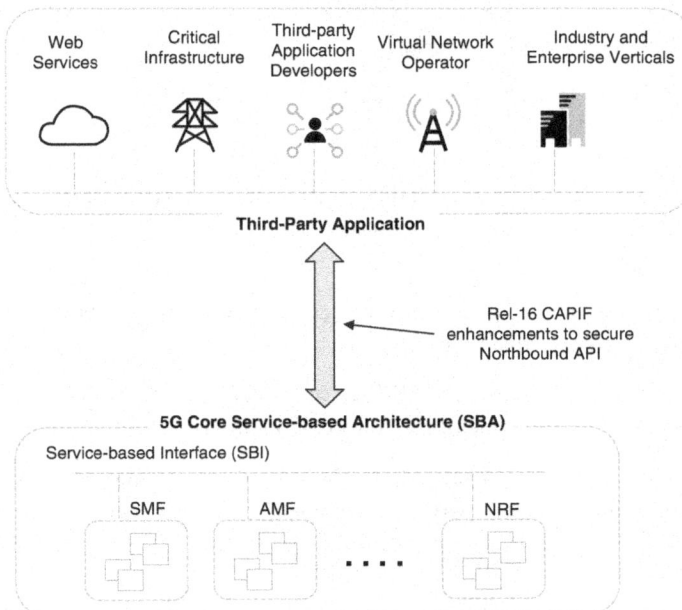

FIGURE 3-11   3GPP Rel-16 Enhancement for Better Third-Party Domain Support for CAPIF

The architecture functional model supports multiple API providers (within and outside the PLMN trust domain), where CAPIF-3, CAPIF-4, and CAPIF-5 were enhanced as CAPIF-3e, CAPIF-4e, and CAPIF-5e, respectively, to enable the API-exposing function, the API publishing function, and the API management function of the API provider domain within the third-party trust domain interaction. The API-exposing function within the PLMN trust domain also interacts with the API-exposing function in the third-party trust domain via CAPIF-7e, as illustrated in Figure 3-12.

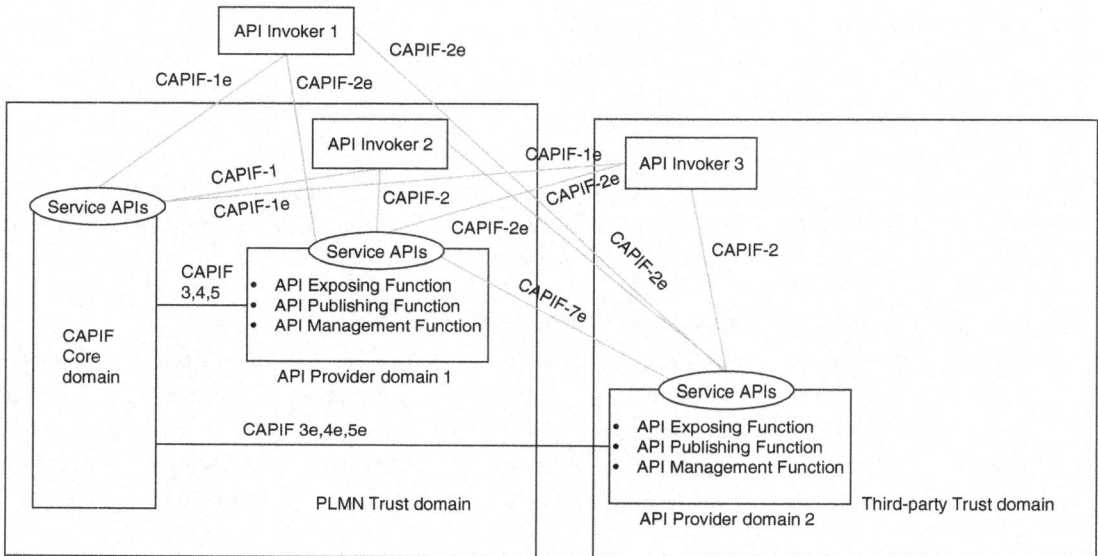

FIGURE 3-12   3GPP Rel-16 CAPIF Security Functional Model

The key security enhancements for CAPIF in Rel-16, as specified in 3GPP TS 33.122, are further detailed next:

■ Rel-16 introduces API topology hiding. When topology hiding is enabled, the CAPIF core function shall respond to service API discovery requests with AEF information, which exposes the service API and acts as a topology-hiding entity.

■ Rel-16 also provides mandatory support of TLS for integrity protection, replay protection and confidentiality protection of CAPIF -1e/2e/7e interface.

■ The API invoker and the API-exposing function using CAPIF-2e will establish a dedicated secure session using a TLS connection based on one of the following methods.

   ■ **TLS-PSK:** TLS-PSK is one of the TLS key exchange methods that uses symmetric keys, shared in advance among the communicating parties, to establish a TLS connection.

This method is generally used in a closed network environment to avoid the need for public key operations.

■ **TLS-PKI:** TLS-PKI is one of the TLS key exchange methods that uses key pairs—public key and private key—to establish a TLS connection. Public keys can only encrypt and cannot be used for decryption. Since the message is encrypted using a given public key, it can only be decrypted by the matching private key. This establishes the ownership of the private and public keys, ensuring the message is only read by the approved parties. This is one of the most scalable and widely used methods to establish TLS by service providers today.

■ **TLS with OAuth token:** In this method, once the TLS is successfully established, the OAuth 2.0 protocol flow is carried out in Figure 3-13. The OAuth protocol flow is illustrated in the Figure 3-13.

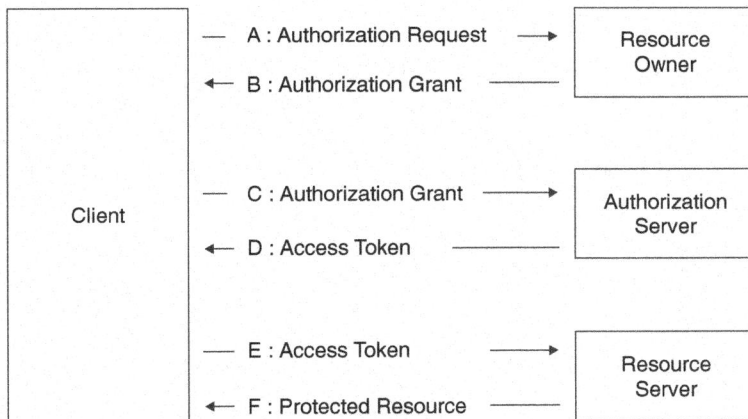

FIGURE 3-13   OAuth Protocol Flow

The OAuth flow is as follows:

A. The client requests authorization from the resource owner (directly or via the authorization server).

B. The client receives an authorization grant (a credential representing the resource owner's authorization).

C. The client requests an access token by authenticating with the authorization server and presenting the authorization grant.

D. The authorization server authenticates the client and validates the authorization grant, and, if valid, issues an access token.

E. The client requests the protected resource from the resource server and authenticates by presenting the access token.

F. The resource server validates the access token and, if valid, serves the request.

After successful establishment of TLS on the CAPIF-2e reference point, the API-exposing function shall authorize the API invoker's service API invocation request based on authorization information obtained from the CAPIF core function. The same method will also be used by the CAPIF-7e interface, which provides communication between API-exposing functions in the PLMN trust domain and third-party trust domain.

CAPIF-3e/4e/5e are secured using TLS and IPsec using SecGW.

Mission-Critical Services

Security enhancements for mission-critical (MC) services provide the confidentiality, integrity, user authentication, service authorization, and overall security architecture for the Release 16 mission-critical features, including MC Push to Talk (MCPTT), MCVideo, MCData, MC Location, MC Inter-working, MC Interconnection, and MC Railway.

Release 16 expands on the mission-critical security architecture already defined in previous releases, along with various mission-critical security clarifications and corrections. In this release, mission-critical security, as defined in 3GPP TS 33.180, adds user service authorization for the mission-critical location service. Similar to user service authorization for the other MC services, an appropriately scoped access token is obtained from the identity management server that permits only authorized users to access the MC location service.

MCData payload protection is enhanced to support separate algorithm types for the MCData payload field, which allows the architecture to meet varying security needs of the mission-critical operator for both on- and off-network MCData operational scenarios.

Within a mission-critical system, the signaling plane should be secured using IPsec for SIP signaling and TLS for HTTP signaling, as shown in Figure 3-14.

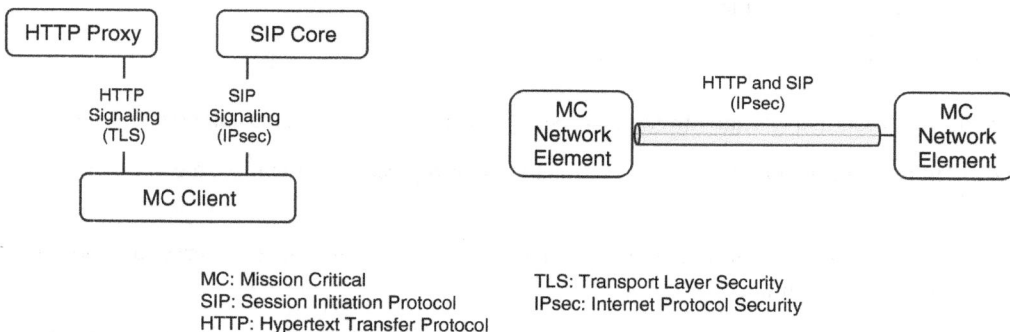

MC: Mission Critical
SIP: Session Initiation Protocol
HTTP: Hypertext Transfer Protocol

TLS: Transport Layer Security
IPsec: Internet Protocol Security

FIGURE 3-14   Signaling Plane Security

Application plane signaling security protects application signaling between the MC client and the MCX server. Initial key distribution for application signaling is performed by sending a client-server key (CSK) from the MC client to the MCX server over the SIP interface. The key is secured using the identity key material provisioned by the key management server. Following initial key distribution, the

MCX server may perform a "key download" procedure to update key material as well as to key the client to allow multicast signaling to be protected, as illustrated in Figure 3-15.

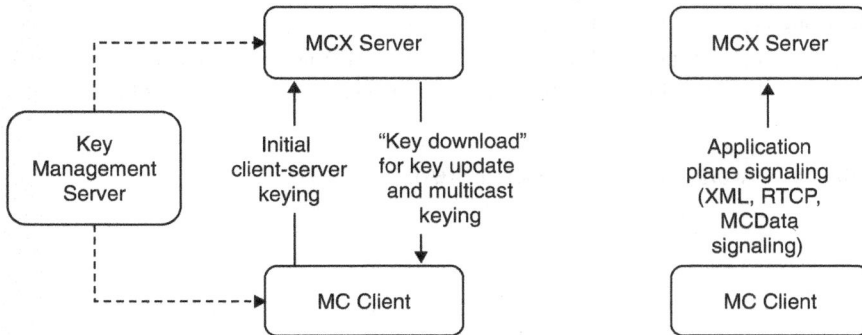

MC: Mission Critical
XML: Session Initiation Protocol
RTCP: Real-time Control Protocol

FIGURE 3-15   Application Plane Security

Rel-16 further introduces enhanced security for interconnection, which now combines the MC gateway, Interconnection Signaling (IS) proxy, and HTTP proxy to provide topology hiding, HTTP protection, and signaling protection for inter-domain MCPTT, MCData, and MCVideo communications between two disparate mission-critical systems for a complete interconnection security solution, as illustrated in Figure 3-16.

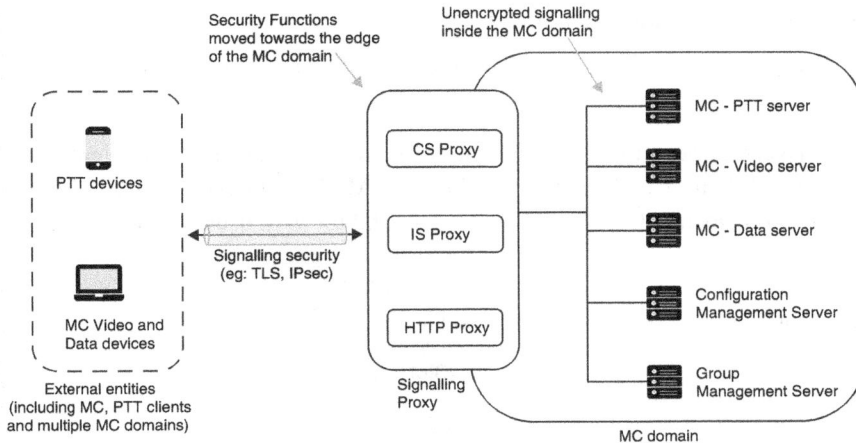

MC: Mission Critical   CS: Client Signalling
PTT: Push to Talk   IS: Interconnection Signalling
Mgmt: Management   HTTP: Hypertext Transfer Protocol
TLS: Transport Layer Security IPsec: Internet Protocol Security

FIGURE 3-16   Interconnection Security

Service Enabler Architecture Layer (SEAL) for Verticals

Similar to the Common Functional Architecture (CFA) for mission-critical services, a set of common capabilities can be utilized by V2X and potentially by multiple vertical industry applications, referred to as the vertical applications layer (VAL) in this section. In Rel-16, this set of common capabilities is developed as a service enabler architecture layer (SEAL) over 3GPP networks and is specified in Stage 2 SEAL in 3GPP TS 23.434. The key security enhancements specified in TS 33.434 are further discussed in this section.

Figure 3-17 illustrates the on-network functional model of SEAL.

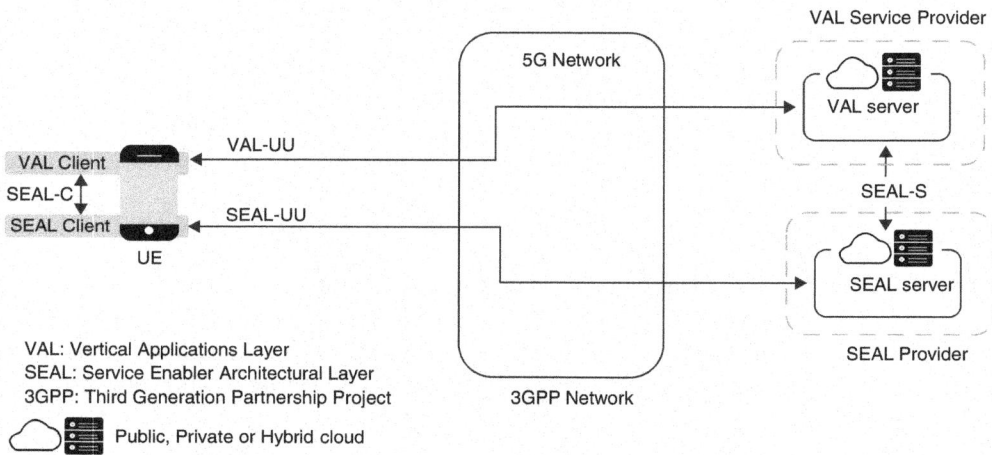

FIGURE 3-17   SEAL On-Network Functional Model

The application plane interfaces, such as VAL-UU and SEAL-UU, should be secured by TLS and IPsec mechanisms.

The SEAL Identity Management Server (SIM-S) and the SEAL Identity Management Client (SIM-C) provide the endpoints for VAL user authentication using the IM-UU reference point, as shown in Figure 3-18. IM-UU supports OpenID Connect 1.0 and OAuth 2.0 for VAL user authentication.

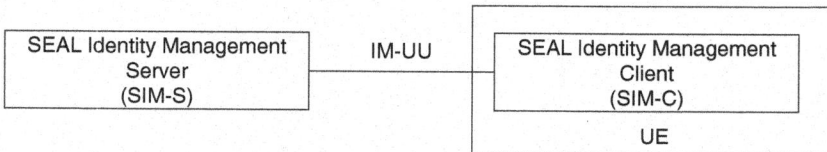

FIGURE 3-18   SEAL Identity Management

Similar to the introduction of signaling proxy functions in the mission-critical services, the SIP core and HTTP proxy layers are introduced for SEAL deployments for securing SEAL signaling control plane messages, as shown in Figure 3-19. The SIP core provides security functions such as authentication and security to the service layer, overload control, and session management.

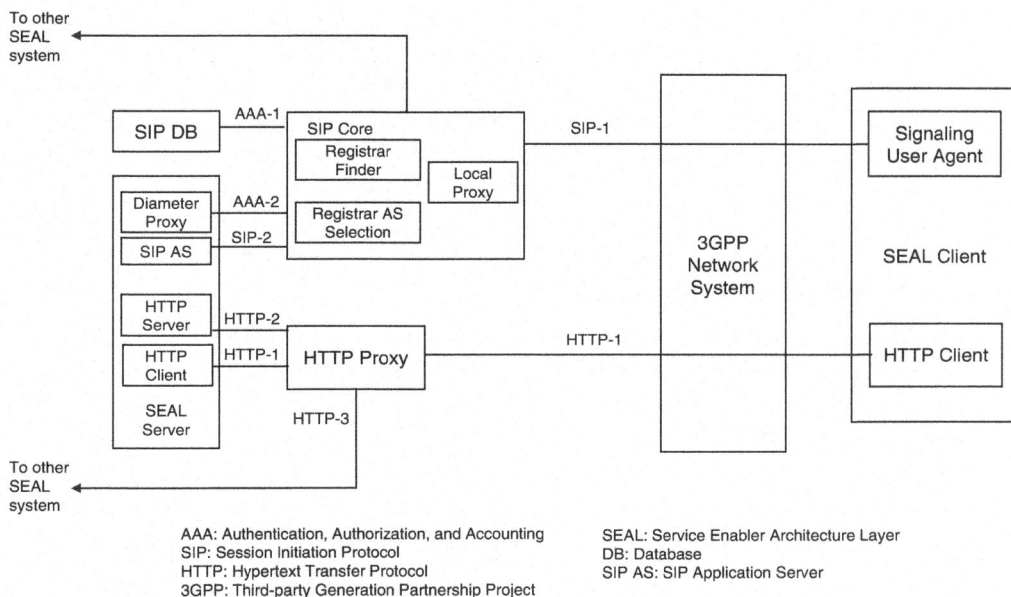

AAA: Authentication, Authorization, and Accounting
SIP: Session Initiation Protocol
HTTP: Hypertext Transfer Protocol
3GPP: Third-party Generation Partnership Project

SEAL: Service Enabler Architecture Layer
DB: Database
SIP AS: SIP Application Server

FIGURE 3-19   SEAL Signaling Control Plane

The HTTP-1 reference point exists between the VAL UE and the HTTP proxy. The HTTP-2 reference point exists between the HTTP proxy and HTTP server. The HTTP-3 reference point exists between the HTTP proxies in different networks. The HTTP interfaces shall be protected using TLS.

The HTTP proxy acts as a proxy for hypertext transactions between the HTTP client and one or more HTTP servers. The HTTP proxy terminates a TLS session on HTTP-1 with the HTTP client of the VAL UE, allowing the HTTP client to establish a single TLS session for hypertext transactions with multiple HTTP servers that are reachable by the HTTP proxy.

Enhancements for User Identities and Authentication in Slicing

This feature allows the use of a user identifier, which is independent of existing identifiers relating to a 3GPP subscription or UE, for network slice-specific authentication and authorization (NSSAA).

In addition to 3GPP-defined network slice selection and authorization, it also defines the interworking of the 3GPP system with an external entity to authenticate and authorize users to access a specific slice. The actual identity provisioning service with creation, management, and authentication of identities is not specified by 3GPP.

Rel-15 network slice selection is based on the NSSAI, which consists of one or more S-NSSAIs. In Rel-16 feature enhancements, subscription information allows the indication of which S-NSSAIs are subject to network slice-specific authentication and authorization (NSSAA). For those S-NSSAIs, the AMF invokes an EAP-based network slice-specific authentication procedure in which the AUSF exchanges AAA protocol messages with a potentially external AAA server (AAA-S) via an optional AAA proxy (AAA-P) to authenticate and authorize the UE for the network slice. Depending on the result of the procedure, the UE is either authorized for a network slice, re-allocated to a different one, or deregistered. A re-authentication and re-authorization procedure may be triggered by the AAA server at any time.

Security Challenges in 5G

3GPP specifies some preventative measures to limit the impact of known threats; however, the adoption of new use cases that require integration with existing technologies from industry verticals, the introduction of new deployment models, and the changes in the 5G architecture, such as the use of APIs in service-based architecture (SBA) and NSaaS, introduces potential new threats for the industry to manage. Adoption of new deployment methods and technologies, such as public cloud deployment models, integration with third-party multi-access edge compute (MEC) applications, and network slicing, will require new security controls to be implemented, other than those introduced by 3GPP.

5G enables openness and flexible deployments due to the use of cloud-native principles, allowing 5GC network functions to be built using polyglot architectures with open source software and implementation of SBA, which uses API-based communication between the 5GC control plane functions. Figure 3-20 illustrates some of the key challenges in 5G technologies.

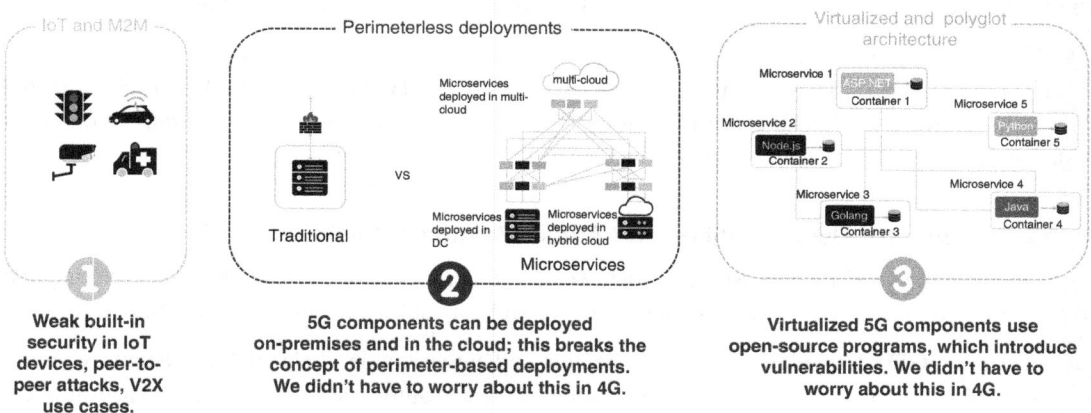

FIGURE 3-20   Key Challenges in 5G

The sections that follow discuss the key challenges shown in Figure 3-20.

IoT and M2M

5G technology will also usher in new connected experiences for users with the help of massive IoT devices and partnerships with third-party companies to allow services and experiences to be delivered seamlessly. For example, in the auto industry, 5G combined with machine learning–driven algorithms will provide information on traffic and accidents as well as process peer-to-peer traffic between pedestrian traffic lights and vehicles in use cases such as vehicle-to-everything (V2X). Distributed denial of service (DDoS) in these use cases is a very critical part of the 5G threat surface.

IoT has been around for quite some time, and some use cases such as wireless point-of-sale (PoS) terminals use 2G in many countries where 2G coverage is prominent. But when we speak about IoT in relation to 5G deployment, it refers to IoT devices of various types. Some IoT devices are not smart devices, and they just send information back to the network in bursts and don't process data or do any analytics within the IoT device itself. Such devices do not have the capability to protect themselves from tampering or any bot taking them over. Some IoT devices are semi-smart and have the intelligence to filter certain data and perform some high-level analysis; however, even these devices have very little or no security, as they were made to fulfill a certain use case with easy and flexible deployment without focusing on the security. True smart IoT devices will help fulfill the critical infrastructure use cases and have the capability to secure themselves and provide security during the information exchange with the management servers. The main threats for IoT and machine-to-machine (M2M) deployments in 5G are related to hardware and software supply chain vulnerabilities and application-level threats, which require more security focus than 3GPP-specified enhancements to 5G.

Chapter 8, "Securing Massive IoT Deployments in 5G," and Chapter 9, "Securing 5G Use Cases," cover the topic of IoT and M2M threat surfaces in more detail. You will also see supply chain threat mitigation covered for domains such as RANs, transport, MEC, and virtualized 5G Core throughout this book, as it is a very extensive and critical part of the end-to-end 5G security architecture.

Perimeter-Less Deployments

5G introduces network decomposition and disaggregation into software and hardware as well as infrastructure convergence, which underpins the emergence of edge computing network infrastructure or MEC. 5G edge computing use cases are driven by the need to optimize infrastructure through offloading, better radio, and more bandwidth to fixed and mobile subscribers. New architectures such as network slicing enable multiplexing of this disaggregated and decomposed infrastructure and help provide isolated end-to-end networks tailored to fulfill diverse requirements being requested by new 5G use cases. The need for low-latency use cases such as URLLC, which is one of several different types of use cases supported by 5G, requires user plane distribution. Certain 5G-specific applications and the user plane need to be deployed in the enterprise network for enterprise-level 5G services for NPN deployments.

The key threats in MEC deployments are mainly due to improper implementation of APIs, the use of third-party applications being developed without security in mind, insufficient testing of software code, malicious/rogue MEC deployments, API-based attacks, insufficient segmentation, and improper access controls on MEC deployed on enterprise premises. MEC deployments with third-party workloads are specifically prone to malicious code injection attacks due to weak inherent security implementation in the software development stage, leading to data exfiltration and other risks during live deployments. Such deployments using third-party workloads should be secured with multi-layered security controls, which is explained in detail in Chapter 5, "Securing MEC Deployments in 5G."

Deploying 5G Network Functions (NFs) in the public cloud brings in the shared responsibility model, where the cloud service provider and the 5G mobile network service providers or industry verticals deploying 5G have to share the responsibility of securing the workloads and the communication of the workloads. In the shared security model, you are responsible for part of the security when moving the workloads, such as 5G Cloud Native Functions and third-party MEC applications, to the cloud. To fulfill the shared security model, you need security controls other than what is specified by 3GPP, such as ensuring that you and the 5G vendor follow strict CI/CD procedures, employing dynamic and automated vulnerability management to secure the runtime environment, ensuring the implementation of the correct security policies when new instances of 5G NF are instantiated by the orchestration layer, and so on.

Virtualized Deployments

Virtualization and cloud-native architecture deployment for 5G are some of the key concerns for service providers. Although virtualization has been around for a while, and you have seen Virtualized Network Functions based on virtual machines in 4G, the container-based deployment model consisting of 5G Cloud-Native Functions (CNFs) is a fresh approach. Apart from the known vulnerabilities in the open source components used to develop the 5G CNFs, most CNF threats are actually unknown, which is riskier. The deployment model of CNFs in the public and private cloud brings in another known yet widespread problem of inconsistent and improper access control permissions putting sensitive information at risk.

The adoption of CNF brings in a whole new ecosystem, such as new methods of development and deployment. One example is the software development method, which follows polyglot architectures. This means that the 5G Network Functions (NFs) are now developed using any open source programming language (Golang, Python, and so on) and will follow the development model used by various enterprise applications. While this opens the door to a range of applications and enables critical use cases, it also brings in the need to identify, track, and apply a robust vulnerability and patch management system to the 5G ecosystem. In previous generations, the software was vendor proprietary, and the vulnerabilities were tracked and immediately acted upon by the vendor, which knew the ecosystem extremely well. In polyglot architectures, this becomes a double-edged sword. While the vulnerabilities and bugs are identified by the open source communities, polyglot architectures are also exploited quite easily by the attackers before they are patched or a workaround is suggested. The deployment model also brings in a new threat surface linked to supply chain vulnerabilities, as the method of downloading the image from the repository and deploying it should follow strict best practices, without which you will be essentially compromising your entire network and subscriber-sensitive information.

The adoptability of 5G CNF and the new SBA, which uses API-based communications, introduces a host of other threat surfaces that are not mitigated by security enhancements brought in by 3GPP for 5G technology. API-based SBA is fragile in nature, as it is prone to weak implementation by developers, and there are many known vulnerabilities in the API web servers today that can be exploited by attackers to gain access to critical 5G infrastructures and exfiltrate sensitive information—or worse, cause a denial of service (DoS) attack that potentially leads to a loss of revenue.

Securing virtualized deployments in 5G is further discussed in detail in Chapter 6, "Securing Virtualized 5G Core Deployments."

Summary

5G technology will introduce advances throughout the network architecture, such as decomposition of RAN, utilizing APIs, container-based 5G Cloud-Native Functions, and network slicing, to name a few. These technological advancements, while allowing new capabilities, also expand the threat surface, opening the door to adversaries trying to infiltrate the network. Figure 3-21 illustrates the key end-to-end threat surface in 5G deployments.

Apart from the expanded threat surface, 5G also presents the security team with an issue of a steep learning curve to identify and mitigate threats faster without impacting the latency or user experience.

Although 3GPP provides security enhancements around the user and privacy, it recommends a shared security model, where apart from the inherent 3GPP-specified mandatory requirements on the 5G network components, entities deploying 5G should also ensure that they have a trusted environment to deploy 5G networks.

Being cloud native and completely software driven, 5G uses open source technologies. Although this is critical for scalability and allowing cloud deployment integrations, vulnerabilities from multiple open source applications could be exploited by attackers. To reduce the attack surface, you will also need to verify the 5G vendor-specific secure development process to ensure hardened software and hardware.

Chapters 4–9 in Part II, "Securing 5G Architectures, Deployment Modes, and Use Cases," will take you through the threat model/threat surfaces on end-to-end components of the 5G architecture, including the threats shown in Figure 3-21, and threats on various deployment models, and then explain methods to mitigate them using vendor-agnostic security controls using principles of zero trust.

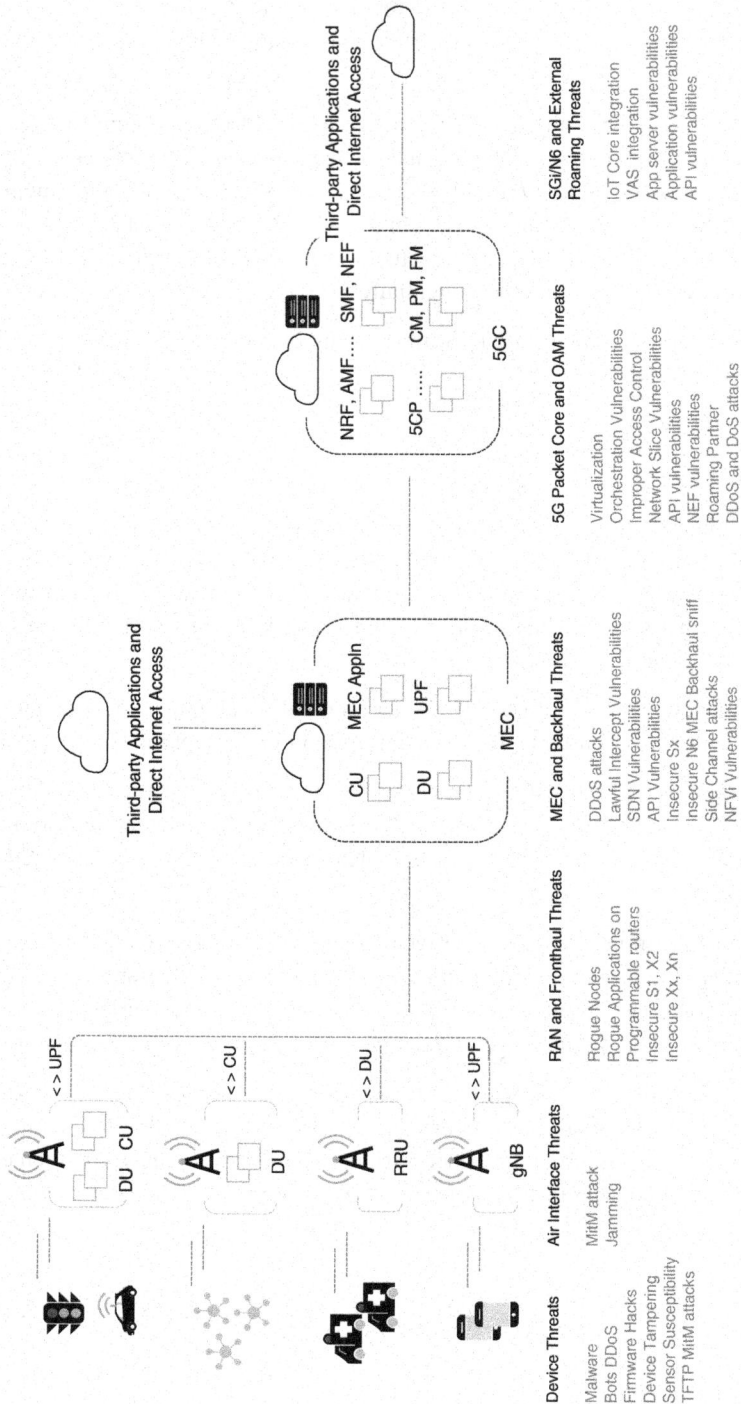

FIGURE 3-21   End-to-End Threat Surface in 5G

Device Threats

Malware
Bots DDoS
Firmware Hacks
Device Tampering
Sensor Susceptibility
TFTP MitM attacks

Air Interface Threats

MitM attack
Jamming

RAN and Fronthaul Threats

Rogue Nodes
Rogue Applications on
Programmable routers
Insecure S1, X2
Insecure Xx, Xn

MEC and Backhaul Threats

DDoS attacks
Lawful Intercept Vulnerabilities
SDN Vulnerabilities
API Vulnerabilities
Insecure Sx
Insecure N6 MEC Backhaul sniff
Side Channel attacks
NFVi Vulnerabilities

5G Packet Core and OAM Threats

Virtualization
Orchestration Vulnerabilities
Improper Access Control
Network Slice Vulnerabilities
API vulnerabilities
NEF vulnerabilities
Roaming Partner
DDoS and DoS attacks

**SGi/N6 and External
Roaming Threats**

IoT Core integration
VAS integration
App server vulnerabilities
Application vulnerabilities
API vulnerabilities

Acronyms Key

The following table expands the key acronyms used in this chapter.

Acronym	Expansion
5GC	5G Core network
AES	Advanced Encryption Standard
AKA	Authentication and Key Agreement
AMF	Access and Mobility Management Function
ARPF	Authentication Credential Repository and Processing Function
AUSF	Authentication Server Function
CIoT	Cellular Internet of Things
CU	Central Unit
DN	Data network
DU	Distributed Unit
DoS	Denial of service
DDoS	Distributed denial of service
FN-RG	Fixed network RG
gNB	NR Node B
GUTI	Globally unique temporary UE identity
IPUPS	Inter-PLMN UP Security
IPX	IP Exchange Service
MEC	Multi-access edge compute
N3IWF	Non-3GPP Access Inter-Working Function
N5CW	Non-5G-capable over WLAN
N5GC	Non-5G-capable
NAI	Network access identifier
NAS	Non–Access Stratum
NF	Network Function
NG	Next-Generation
ng-eNB	Next-Generation Evolved Node-B
NG-RAN	5G Radio Access Network
NR	New Radio
QoS	Quality of service
RES	Response
SCG	Secondary cell Group
SCP	Service communication proxy
SEAF	Security Anchor Function
SEPP	Security Edge Protection Proxy

Acronym	Expansion
SIDF	Subscription Identifier De-concealing Function
SMC	Security Mode Command
SMF	Session Management Function
SUCI	Subscription Concealed Identifier
SUPI	Subscription Permanent Identifier
SBA	Service-based architecture
TLS	Transport Layer Security
TWAP	Trusted WLAN access point
TWIF	Trusted WLAN Interworking Function
UDM	Unified Data Management
UDR	Unified data repository
UE	User equipment
UP	User plane
UPF	User Plane Function
V2X	Vehicle-to-everything
SBA	Service-based architecture

References

Pramod Nair, "Securing Your 5G Infrastructure to the Edge and Beyond," Cisco Public, April 1, 2021

Pramod Nair, "Securing 5G and Evolving Architectures," Cisco Knowledge Networks, Cisco Public, September 17, 2020

Cisco Systems, "Reimagining the End-to-End Mobile Network in 5G Era," Cisco Public, 2019

https://www.3gpp.org/

3GPP TS 23.501, "System Architecture for the 5G System; Stage 2"

3GPP TS 33.501, "Security Architecture and Procedures for 5G System"

3GPP TS 33.402, "3GPP System Architecture Evolution (SAE): Security aspects of non-3GPP accesses"

Chapter | **4**

Securing RAN and Transport Deployments in 5G

After reading this chapter, you should have a better understanding of the following topics:

- 5G RAN and Transport Threats
- Securing 5G RAN and Transport
- Real Scenario Case Study: Examples of Threat Surfaces and Threat Mitigation Techniques

The previous chapter discussed how device-based threats have an impact on a 5G network and different ways of mitigating the threats. This chapter is aimed primarily at the 5G radio access network (RAN) threat surfaces and mitigation techniques for the service provider network. This chapter will be of particular interest to the following teams from service providers and vendors:

- Mobile infrastructure strategy

- Security strategy

- Radio planning and optimization

- Transmission and the packet core team

- Security architects and design teams responsible for securing the mobile infrastructure

- Cybersecurity vendors looking for use cases to secure RAN and the transport layer

- Enterprise teams responsible for securing private 5G deployments

Along with major technological change meeting the scale requirements for the massive Internet of Things (IoT) and the ultra-low-latency requirements for real-time applications, the new network infrastructure for 5G must simultaneously satisfy increasing demands for bandwidth, high scalability, and the ultra-low-latency needs of new applications and services in an efficient, automated, and programmable manner.

In enabling these new use cases, 5G uses technologies such as software-defined networking (SDN), New Radio (NR), virtualization such as Cloud-Native Functions (CNF), Control and User Plane Separation (CUPS), and network slicing. These developments will have significant impacts on the underlying network. For example, depending on the spectrum being used by the service provider, the footprint and the density of the network will need to expand significantly. Hosting new ultra-reliable low-latency communication (URLLC) and massive machine-type communication (mMTC) applications efficiently will require bringing the applications closer to the edge of the network.

Other key attributes of the 5G transport infrastructure will be to easily carve out and reuse capacity, compute, and resiliency as well as to guarantee latency by supporting network slicing. Service providers will have to manage the network using a completely new approach in order to enable the rapid provisioning and service automation required. In addition to the service provider network–enhancing requirements, the network design needs to cater to the different classes of services, such as fixed line and broadband customers, as well as support the different customer types (consumer, small-to-medium business, and enterprise).

The disaggregation and distribution of the 5G RAN using Distributed Unit (DU) and Centralized Unit (CU) functions allow for openness and the consideration of flexible deployment options. This open and flexible architecture also brings in new threats that require evaluating security options.

Although there are several security enhancements in 5G as compared to 4G technology, this chapter considers the pragmatic deployments in 5G where the implementation of the 5G network elements and 5G network functions are not always in trusted environments. Apart from the non-trusted 5G network deployment considerations, there have not been enough validations of live 5G networks to ascertain the robustness of the 5G AKA protocol.

The new use cases in 5G also require features such as network slicing, multi-access edge computing (MEC), integration with SDN, and orchestration and automation layers. This necessitates strict granular access control policies to be applied to these layers as well as the need to secure the communication between the 5G network elements and the components enabling the features.

Service providers deploy different models of transport network. Many service providers follow the model where they own their transport network as well share or lease parts of the transport infrastructure, and security has to be considered to ensure that the data is secured while passing through all the different transport networks.

Government regulators of many countries also request the service providers to encrypt the end-to-end network to better secure the transport and mitigate any occurrences of data exfiltration from the service provider network.

5G RAN and Transport Threats

This section takes you through multiple options of securing the RAN and the transport and evolving the security posture from 4G toward 5G. It also covers the evolution of the security architecture by looking at security controls (for example, the security gateway evolving from 4G to 5G).

Figure 4-1 provides a high-level view of the threat surfaces in the air interface, RAN, and the transport layer for 5G deployments. The vulnerabilities are explained in more detail in this section.

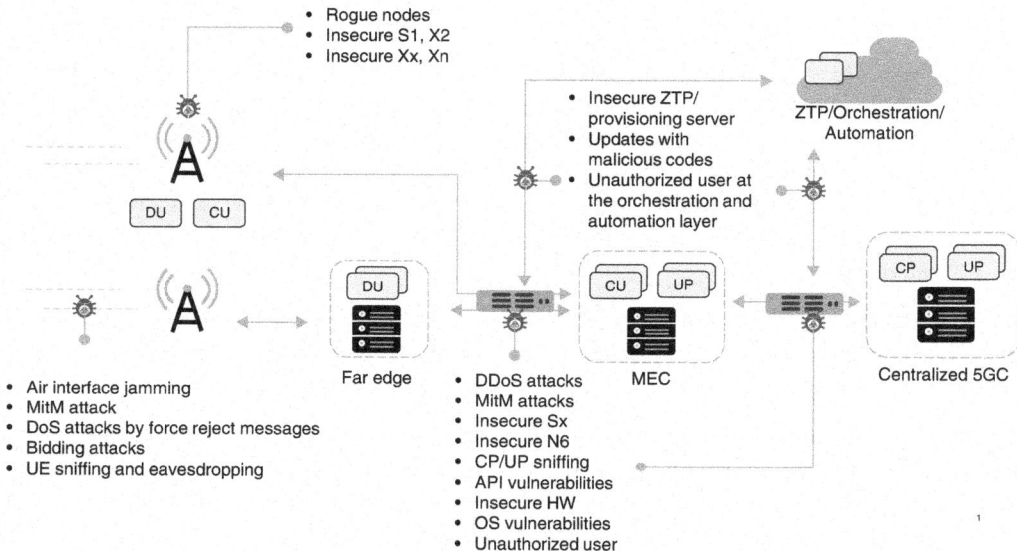

FIGURE 4-1   RAN and Transport Threats in 5G

Vulnerabilities in Air Interface

Although there have been security enhancements in the uplink and downlink air interface communications between the user equipment (UE) and the RAN/packet core in the evolution from 4G to 5G, there are still some unprotected messages in the uplink and downlink. Examples of such messages are Radio Resource Control (RRC) UECapabilityInformation in the uplink and RRC UECapabilityEnquiry and REJECT in RRC/NAS (Non-Access-Stratum) in the downlink. These aforementioned messages being sent unprotected before the Access Stratum (AS) security activation is purely a design choice to enhance the service or connectivity for the user (for example, to provide early optimization for better service connectivity).

An example of such an early optimization would be the gNB sending the UECapabilityEnquiry toward the UE in the RRC message and the UE replying with the UECapabilityInformation, indicating the UE capability. This unprotected message could be exploited by an attacker by using a rogue or fake base station to set up a man-in-the-middle (MitM) attack. Let's discuss how this could be exploited by an attacker.

One of the ways rogue/fake base stations can be set up is by using a Linux box with open-source thin layer of packet core software and radio node software with mod/demod (modulator and demodulator) connected to a power amplifier. Depending on the attack planned by the attacker, an omnidirectional antenna with gain varying from 3 to 6 dBi or a directional antenna with gain around 16 to 18 dBi (depending on the beamwidth of the antenna—the lower the beamwidth, the higher the gain) could be used. For example, if the target is a group of people meeting in a room, a directional antenna would make more impact. However, in a setting such as a crowded area, where the target is not seated and is quite mobile, an omnidirectional antenna provides better results.

The next step is to ensure that the UE is force-camped on to the rogue base station. The procedure of UEs camping on to a base station would depend on whether the phone has just been switched on or whether it's in a cell selection, cell reselection, or handover procedure. The procedures for cell selection and reselection for 5G are specified in 3GPP TS 38.133.

In 5G, Srxlev indicates the signal strength and Squal indicates the quality of the signal. When camped on a cell, the UE continuously performs measurements to search for a better cell, based on the cell reselection criteria. Depending on the cell reselection criteria, the UE would perform the cell reselection to the same radio access technology (RAT) or a different RAT (inter-RAT) as part of the idle mode procedure. Depending on the handover parameters and hysteresis criteria, the UE would perform the handover to the same RAT or a different RAT, specifically 4G (Inter-RAT), as part of the active mode handover procedure. By using the appropriate parameters, the rogue/fake base station can allow MitM attacks to take place by sniffing the air interface for UECapabilityInformation and capturing it. Once captured, the message could be modified to lower the capability level and then forwarded to the real gNB, causing the UE to operate only with restricted or limited radio capability.

A malicious actor might also manipulate the unprotected REJECT messages sent from the network to the UE, which is used to optimize the availability of the system to the connected UE (even in the RRC_INACTIVE state). The REJECT messages could take the UE out of service. This message could be exploited by the attacker to force UE/5G endpoints from the 5G network to the EPC network. Once the UE is forced to the EPC network, the vulnerabilities existing in the EPC network could be used to exploit the UE and 5G endpoints, such as international mobile subscriber identity (IMSI) spoofing and MitM attacks.

A malicious actor might manipulate the RRCResumeRequest message, where the ResumeCause field is unprotected and hence open to MitM attacks, and the field attribute can be changed. An example is where the attribute value "emergency" can be changed to "ran update." This change in attribute can cause a big difference in the user service requested by the UE and the service delivered to it. Whereas the user had requested emergency service, the RAN receives the message with ran update and sends the UE to INACTIVE. These kinds of attacks can be aimed at high-value targets or at use cases for critical services.

A malicious actor might manipulate the weakness in the RRCResumeRequest procedure. When the RRCResumeRequest procedure is initiated by the UE and the gNB is busy, it sends an RRCReject with a wait timer; once the wait timer expires, the UE tries to establish a connection again using the same ResumeMAC-I with the same I-RNTI and Krrcint key, which is the exact same one as the initial request. The attacker can exploit this vulnerability in the air interface and perform MitM attacks such as spoofing the RRCResumeRequest and validating it before the wait timer expiry. This will cause the

RRCResume procedure from the original UE to fail, thus causing a denial-of-service (DoS) attack on the UE.

Apart from the rogue/fake base station attacks, the threat vector could be from a high-powered device trying to jam a part of the spectrum being used for 5G.

Figure 4-2 illustrates an example of air interface jamming.

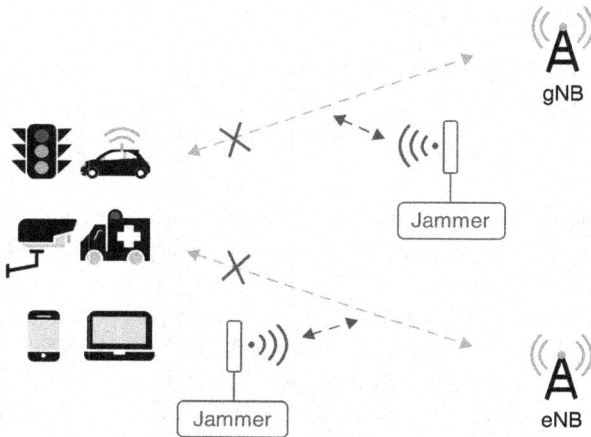

FIGURE 4-2   Air Interface Jamming

What the jammer actually does is to increase the noise to such high levels that the receiver cannot decode the electromagnetic field (EMF) / radio frequency (RF) signal in the air interface. In certain countries, it is illegal to use jammers for the sole purpose of causing interference with wireless communications. However, sometimes it is made available to cater for use cases such as jamming in military and defense systems for the purpose of electronic countermeasures, such as to deceive radar and other detection systems. This could also be deployed in concert halls to disrupt cellular and Wi-Fi services to prevent distractions from cell phones ringing in the audience. Even though jammers are illegal to obtain, attackers will not find it difficult to make one because the required electronic components are easily found on the open market.

Jammers have also become sophisticated and can fine-tune themselves to the frequencies being used by first using a wideband method called *sweeping and combing*. In this process, the key carriers being used in a cluster or in a specific area are identified by first sweeping the entire spectrum with a high sweep speed, such as 2–3 GHz per second, and comb-jamming on fixed frequency signals such as 450 MHz. The jammer can also intelligently prioritize the frequency bands it wants to scan based on the data points of user signals found in previous scans. Once the specific frequency band or group of bands in a specific area is identified, the jammer can then choose to pump high-power RF waves in those selected frequency bands, thereby causing very high noise/interference in the area. This leads to the receivers such as legitimate UE unable to establish reliable communication with the service provider base stations in the area, thereby causing a DoS to the receivers (UE) in that particular area.

Vulnerabilities in the Transport Network

Along with major technological change to the mobile core and radio access network (RAN), operators will also need to evolve their transport networks to deliver a satisfying mobile broadband experience in a cost-effective manner, while simultaneously meeting the scale requirements for the massive IoT and the ultra-low-latency requirements for real-time applications. 5G brings in high densification of the network, which requires higher transport access capacity and rich programmable features.

To understand the vulnerabilities in the 5G transport network in pragmatic deployments, let's look at the deployment type being implemented as well as the lack of adequate planning done by a majority of the service providers, which consists of an overlay design including existing 4G eNBs, 5G NSA, and 5G SA gNBs. The new transport design will include terminations in the edge to enable 5G URLLC use cases. Figure 4-3 illustrates an example of the existing 4G-only transport design, which is less dense compared to the transport deployment, which includes 4G and 5G nodes.

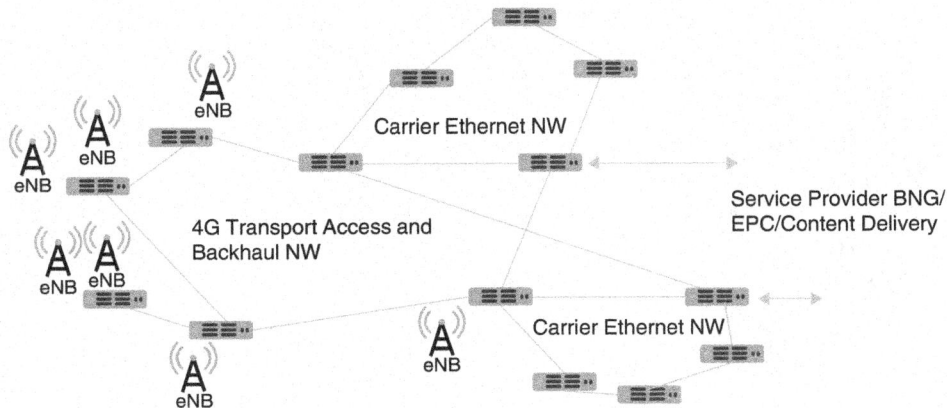

FIGURE 4-3 Example of 4G Transport Network

5G evolution introduces new capabilities, such as New Radio (NR), a strong reliance on virtualized functions, network slicing, and Control and User Plane Separation (CUPS). These developments, in turn, will have significant impacts on the underlying network. For example, as shown in Figure 4-3, to boost capacity, the footprint of the network will need to expand significantly. To host new URLLC and mMTC applications efficiently, the network will need to be able to integrate regional data centers and distributed compute seamlessly—closer to the endpoints in the network.

These edge deployments, consisting of decomposed and distributed 5G CNFs catering for URLLC use cases, coupled with network slicing, require tight time synchronization to support enhanced features such as Coordinated Multipoint (CoMP), which is important in pico cells and small cells aimed at enterprise and industrial IoT (IIoT) deployments. Following the 1588v2-based specifications for synchronization for the transport elements of the network is required. For the frequency sync, 5G Synchronous Ethernet (5G SyncE) and G.8262.1-enhanced Ethernet equipment slave clock (eEEC) requirements have been specified for 5G networks, which have extremely strict demands on frequency drift. This synchronization can be provided by using Global Navigation Satellite System (GNSS) and by delivering sync using transport systems. These GNSS systems are vulnerable because they could be jammed by individuals or malicious entities using channel or spectrum jammers. Any attack on synchronization can impact the network performance and, in many cases, cause a DoS to the impacted node or clusters of nodes.

5G deployments will be quite dense compared to the 4G deployment, as illustrated in Figure 4-4 (refer to Figure 4-3 for a comparison).

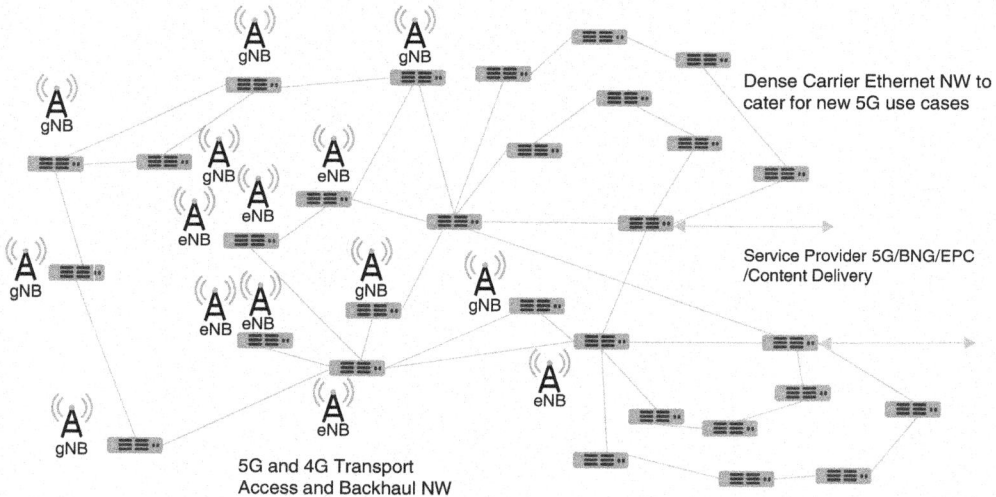

FIGURE 4-4 Service Provider Transport Network Evolution Toward 5G

Figure 4-5 illustrates an example of a rogue transport in the 5G network.

This highly dense deployment in 5G introduces another key threat surface related to the intentional or unintentional deployment of rogue access and aggregation transport devices, as illustrated in Figure 4-5.

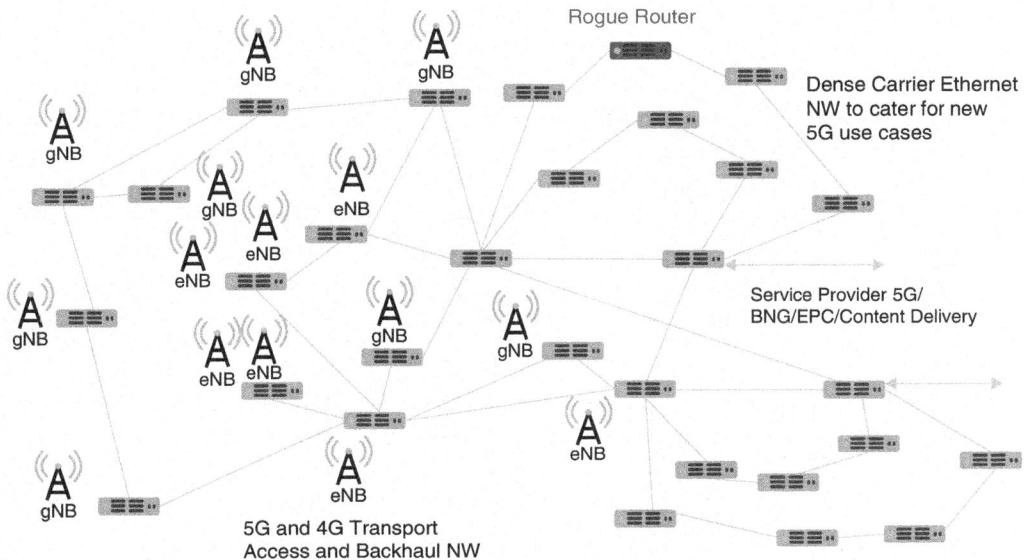

FIGURE 4-5 Rogue Transport Element in Transport Network

These dense deployments of access and aggregation transport devices make it very difficult to identify rogue nodes once the deployment is finished. The rogue nodes can be quite easily identified by following best practices such as running basic OS version and vulnerability patching tests before deployment, but usually such deployments will have quite a tight implementation schedule, and many of the security processes might not be observed by the project managers. The key point during these tight deployment schedules is for the project manager to ensure that the tight timelines are being followed, which pushes security process adherence to be last on the priority list (if existent at all). This is the reality in a majority of network deployments, if not all of them. Once the implementation is completed and acceptance testing (AT) is performed, it will be extremely difficult to identify the rogue nodes if they are intentionally deployed.

Other key attributes of the 5G transport infrastructure will be to easily carve out and reuse capacity, compute, and resiliency and to guarantee latency by supporting network slicing. Operators will have to manage the network using a completely new approach in order to enable the rapid provisioning and service automation required. In addition to the pure mobility requirements, operators must also consider how their network caters to the different classes of services they deliver, such as fixed-line and broadband customers, as well as supporting the different customer types (consumer, small-to-medium business, and enterprise). The new network infrastructure must simultaneously satisfy exploding bandwidth demands, massive logical scale, and the incredibly low-latency needs of new applications and services in an efficient, automated, programmable manner.

Programmability, while being a critical enabler for the 5G network slicing implementation methods, can also be one of the most vulnerable network elements in the 5G architecture, depending on the level of built-in hardware and software security features of the transport pre-aggregation and aggregation routers and the vulnerabilities within the API implementation, such as the Top 10 OWASP (Open Web Application Security Project) threats. If the virtual instances are built with weak security, such as improperly patched software, or upgraded with malicious code within the upgrade image, it will lead to the entire network being taken into control by the attackers. This is one of the most discussed points by any of the service providers planning to enhance their existing transport infrastructure to cater for 5G or deploying a new transport infrastructure to launch a 5G network. Familiar challenges include physical security, management plane security, control plane security, and potentially exploitable bugs on routing devices.

As shown in Figure 4-6, the programmable transport elements have options of deploying virtualized application instances within the transport device itself to enable URLLC use cases.

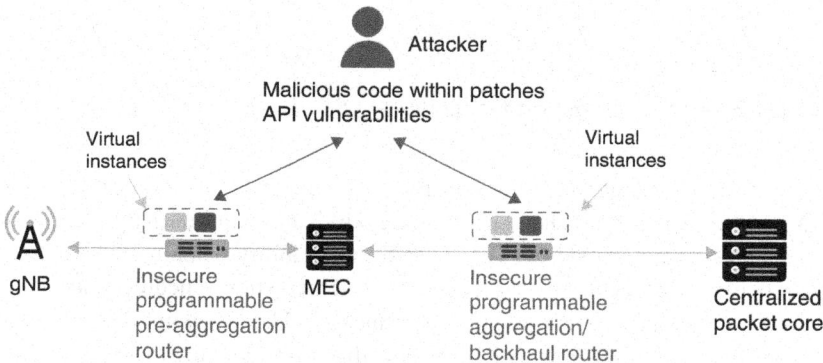

FIGURE 4-6 Attack Scenario in the Transport Network Due to Insecure API

Another threat vector for transport network devices is from attackers using the Plug and Play (PnP) or provisioning server to take control of the transport network elements.

As shown in Figure 4-7, most of the transport routers (front haul, mid haul, and back haul) on the market today use the Zero-Touch/PnP/provisioning server for initial configuration of the transport routers and also to deploy patches and automatically install them.

As Figure 4-8 illustrates, the attackers can take control of the provisioning server and force-download malicious updates to the transport network elements. Once the malicious code is installed on the transport network elements, the entire 5G network is compromised. The impact of such an attack could be very dangerous, including taking several servers down and causing an outage for an extended period of time.

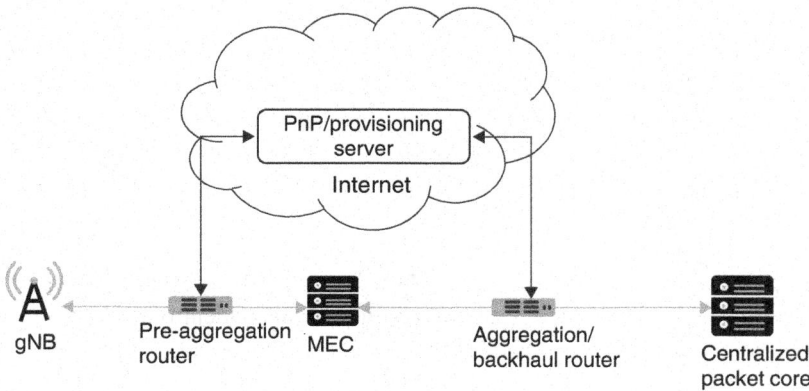

FIGURE 4-7   Configuration of Transport Routers Using Provisioning Servers

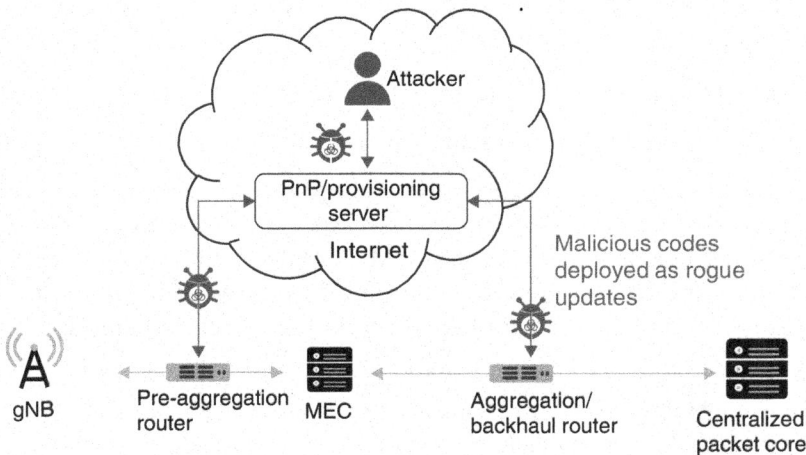

FIGURE 4-8   Attack Scenario in Transport Network Due to Insecure Provisioning Server

Rogue/Fake Base Station Vulnerabilities

One of the key threat vectors in 5G RAN and the transport layer involves rogue/false base stations and IMSI catchers. This section covers the impact on the network and subscribers due to fake/rogue base stations. The 5G system in particular has already made significant security enhancements to combat false base stations—improvements such as guaranteed GUTI (Global Unique Temporary Identifier) refreshment, SUPI (Subscription Permanent Identifier) concealment, protected redirections, and also security features inherited from earlier generations, such as mutual authentication between the UE and network, integrity protected signaling, and secure algorithm negotiations.

Although there are security enhancements in 5G technology as compared to 4G, there are also threat vectors that are not mitigated by those enhancements. The key threat vectors and risks due to fake/ rogue nodes are mainly related to MitM attacks, where the key aim of the attacker is to maliciously perform snooping and gain insight on specific subscribers or groups of subscribers. This method primarily targets network clusters serving 5G devices in critical infrastructure deployments and clusters providing coverage to high-profile industry verticals and embassies. Some of the key threat vectors and risk examples are detailed in the following list:

- A MitM false base station may perform a linkage attack via Subscription Concealed Identifier (SUCI) replay. That is, it replaces a SUCI in a registration request or in an identity response by a previously captured SUCI and observes whether the UE will be authenticated and receive service.

- Using a rogue/false base station, an attacker may launch various types of attacks, such as transporting security-protected messages without any modification while dropping, altering, and/or injecting unprotected messages such as the following:

  - Pre-authentication traffic

  - MAC/RLC layer message headers

  - Lower layer control messages such as buffer status reports

- A MitM attack could also include replaying messages (that is, the MitM sits between the actual base station and the UE and forwards the messages of the base station to the UE and the messages of the UE to the base station). In this position, the MitM might do nothing for a very long time, making it very difficult to detect. However, on certain occasions, the MitM might inject/ alter/drop messages. The basic requirement to defeat MitM attacks is often related to replay protection.

- A MitM false base station might force the UE to camp on to it by passing all the messages between the UE and the real base station. It might then reject or drop service requests, not pass on paging messages from the UE, and so on.

- When UP integrity protection is not used, a MitM false base station might trick the UE into accessing malicious websites or it might even impersonate the UE on the IP layer, which includes the decryption of downlink traffic and performing encryption of faked uplink traffic.

Securing 5G RAN and Transport

Although there are several security enhancements in 5G as compared to 4G technology, this chapter considers the pragmatic deployments in 5G where the implementation of the 5G network elements and 5G network functions are not always in trusted environments. Apart from the non-trusted 5G network deployment considerations, there haven't been enough validations of live 5G networks to ascertain the robustness of the 5G- AKA protocol.

Securing the Air Interface

To bring more openness to 5G, the 5G authentication is made access-agnostic, meaning that it should support both 3GPP access (cellular networks) and non-3GPP access (Wi-Fi and cable) networks.

To make the connections from both 3GPP and non-3GPP more secure, 5G uses a new unified authentication framework. The initial registration procedure used for 4G, called Primary Authentication in 5G, now also supports a widely used protocol in IT called Extensible Authentication Protocol (EAP) Authentication and Key Agreement (EAP-AKA), which allows different types of credentials such as username and passwords, certificates, and pre-shared keys (PSK). This further extends 5G authentication to devices used in IT, factory, and industry verticals.

Apart from the primary authentication, 5G now also supports secondary authentication—a separate authentication and authorization procedure that can be performed by external data networks or the service provider apart from the initial primary authentication performed by your network. EAP is supported for the secondary authentication procedure, which can be executed by the external data network or service provider. The purpose of the primary authentication and key agreement procedures is to enable mutual authentication between the UE and the network and provide keying material that can be used between the UE and the serving network in subsequent security procedures.

As an example, your network will perform the initial primary authentication when the UE tries to connect to your network. When the same UE tries to connect to an external network, your 5G network will use EAP to request the external network to perform a secondary authentication with a separate AAA (authentication, authorization and accounting) server owned by the external network. The UE will be permitted connectivity to the external network only after successful secondary authentication.

The three authentication methods are as follows.

- **5G-AKA**: 5G-AKA enhances EPS-AKA used in 4G by providing the home network with proof of successful authentication of the UE from the visited network. The proof is sent by the visited network in an Authentication Confirmation message. It provides mutual authentication between the UE and the 5G network. Both 5G-AKA and EAP-AKA' use the shared key K_i stored in the USIM for the AKA protocol when the UE connects to the cellular network. The authentication procedure includes the UE requesting access by sending its SUPI to the network. In 4G, this was unencrypted, leading to the risk of interception; in 5G, it is encrypted. The 5G network responds to this request by sending an Authentication Vector to the UE. The UE must encrypt this using the shared-key K_i and send it as the response. Because the home network has a copy of the key, it can validate the decrypted response.

- **EAP-AKA'**: EAP-AKA' is a small revision of the Extensible Authentication Protocol method for the 3G Authentication and Key Agreement (EAP-AKA) method. The change is a new key derivation function that binds the keys derived within the method to the name of the access network. This limits the effects of compromised access network nodes and keys.

- **EAP-TLS**: EAP-TLS is used in 5G when the devices do not support USIM. EAP-TLS uses the TLS public key certificate authentication mechanism within EAP to provide mutual authentication of client to server and server to client. With EAP-TLS, both the client and the server must be assigned a digital certificate signed by a Certificate Authority (CA) that they both trust. When selected as the authentication method, EAP-TLS is performed between the UE and the AUSF. For mutual authentication, both the device and the AUSF can verify each other's certificate.

The 5G technology also allows for attributes such as wait timers to be configured to combat some of the attacks from rogue/fake base stations, such as the REJECT attacks mentioned earlier.

One method of detecting rogue base stations, rogue repeaters, and jammers is by first using key performance indicators (KPIs) of the sectors and cells covering the area to understand the clusters of cells showing high interference or drops due to interference or bad quality signal. Once the 4G or 5G cells or clusters of 4G or 5G cells are identified as being impacted by the interference, a drive test can be carried out in the identified area using drive-test equipment and spectrum analyzers to detect the jamming equipment.

Using Trusted Transport Network Elements

Given that most aggregation devices will be placed in untrusted or partially trusted environments, strong emphasis on tamper-proofing is essential. Beyond this concern, 5G enlarges the problem with network data center compute elements, software running on those elements, and new orchestration capabilities that must also be secured.

Service providers planning to build the transport infrastructure should ensure that the transport vendor has products designed with security as a foundation, starting from the built-in secure silicon to deliver platform trust, network trust, and application trust. There are vendors in the market today using trustworthy systems for securely storing unique cryptographic keys in hardware and capabilities like secure boot.

The foundation of a trusted network is trusted devices, and all trust must begin in hardware by utilizing technologies such as Trust Anchor, which establishes a hardware root of trust for software integrity and strong encryption. This hardware security component provides a unique cryptographic identity of each platform component and should be used as the basis for the advanced secure boot infrastructure by the vendor you select for 5G deployments. Hardware-rooted secure boot infrastructure–based platforms provide significantly stronger protections against compromises of the firmware and operating system than typical firmware-based secure boot infrastructures (such as those used in mainstream x86 platforms). This, coupled with advanced runtime OS protections and control-plane protections, provides the platforms with unique capabilities to establish and maintain trust in exposed environments.

Apart from the built-in security layers mentioned, additional security services such as strong cryptographic protection for secret data and key material within the router should also be catered for. 5G will be increasingly deployed in insecure remote locations, and this built-in hardware-keyed protection for secret data at rest is required to maintain trust and control of critical network services.

Secure Deployments and Updates Using Secure ZTP

As explained earlier in the section, "Vulnerabilities in the Transport Network," 5G use cases and deployment options will bring in higher densification and programmable capabilities to allow seamless integration with SDN.

In most transport network deployments, service providers receive their purchased routers and usually ferry them to pre-staging facilities as part of the first "truck" rollout. At the pre-staging facility, the devices are typically configured manually with the help of technical personnel to place bootstrap configurations on them. These pre-configured boxes are then shipped out to the installation site, often using third-party installers to simply set up and power on the devices that were pre-staged. This constitutes the second "truck" rollout. This might not be the exact workflow for all network operators, but it's fairly representative of the workflows in use across a majority of them.

The use of competent technical personnel at pre-staging facilities, the multiple truck rollouts, the post-installation rollouts, and corrections in case of errors in the bootstrap configurations all contribute to the consistently rising operational expenses (OPEX) that most large-scale service providers incur. As the number of devices in these deployments continues to grow to meet consumer demands and newer 5G architectures, it is imperative to find techniques to reduce the OPEX as much as possible.

Apart from the preceding key points, the mitigation of rogue transport access devices, such as rogue access and aggregation routers, is very critical to the end-to-end security of the 5G network. To allow for optimized OPEX and have secure automation for the deployment and operation of the transport access devices, service providers have to use Zero-Touch Provisioning (ZTP) and Secure ZTP (SZTP) techniques.

ZTP automates the process of installing or upgrading software images and installing configuration files on transport access devices that are deployed for the first time in the network. It reduces manual tasks such as upgrading and configuring the devices. ZTP is one of the most critical features for 5G transport access network elements, as service providers are looking to automate Day 0 provisioning of routers to help reduce the OPEX associated with technical personnel and staging facilities used today. ZTP is also important for the patch and updates process.

SZTP is a bootstrapping strategy that includes updating the boot image, committing an initial configuration, and executing arbitrary scripts to address auxiliary needs to enable devices to securely obtain bootstrapping data with no installer action beyond physical placement and connecting network and power cables. The updated device will subsequently be able to establish secure connections with other systems. For instance, a device might establish NETCONF and/or RESTCONF connections with deployment-specific network management systems. SZTP also enables nontechnical personnel to bring up devices in remote locations without the need for any input from your end.

You should follow SZTP, at a minimum, to ensure the secure deployment and secure lifecycle of the transport access network elements.

As shown in Figure 4-9, provisioning servers following SZTP techniques can be used to provision and manage patches and updates to the entire transport access network. When a device that supports ZTP boots up and does not find the startup configuration (during a fresh install on Day 0), the device then enters ZTP mode.

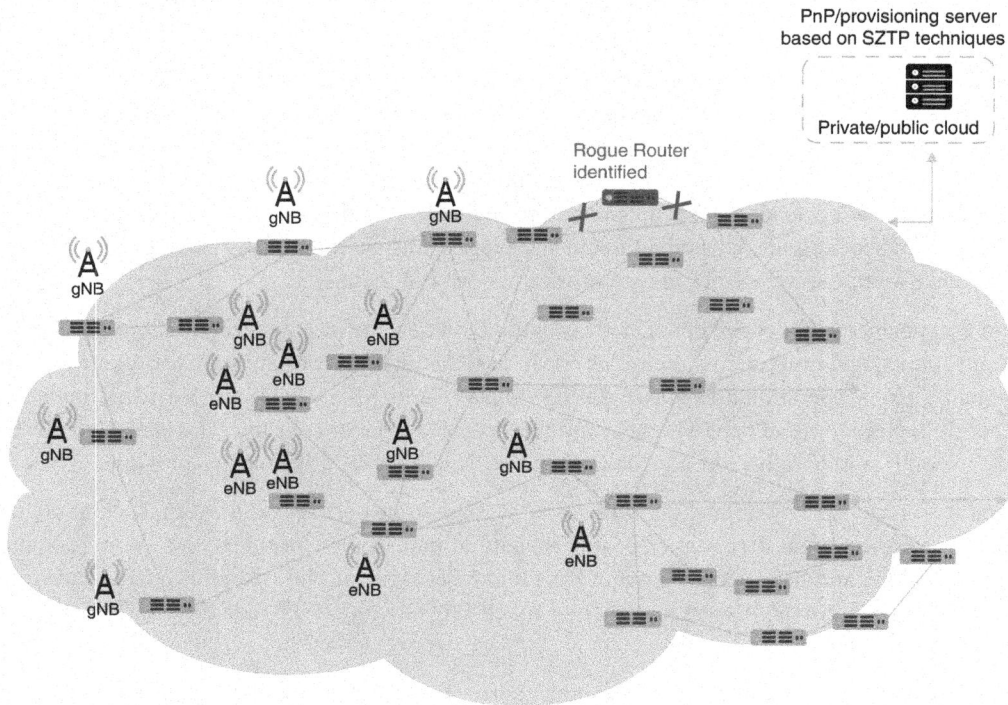

FIGURE 4-9   Mitigation Example of Rogue Transport Device Scenarios

In ZTP mode, the device locates a Dynamic Host Control Protocol (DHCP) server, bootstraps itself with its interface IP address, gateway, and Domain Name System (DNS) server IP address, and enables Guest Shell. The device then obtains the IP address or URL of the provisioning server depending on the deployment method chosen by the service provider.

You should deploy the provisioning server using SZTP techniques to enable configuration of only trusted transport access devices; this will mitigate the threat vectors from untrusted transport network devices. To adhere to SZTP procedures, the service provider's network management system (NMS) would need to authenticate the IDevID certificates (the NMS obtains from the manufacturer the trust anchor certificate for the IDevID certificates during the initial enrolment process), and only the genuine devices would be authenticated. This would allow the detection of the rogue transport access device, which can then be removed from the network, or further investigation can be done to understand how it was present in the network.

By following the described SZTP process and best practices in the list that follows, you can better secure the transport access network and mitigate the threat vectors, such as the deployment of rogue transport access devices, in the service provider network:

- Ensure that the transport infrastructure device has built-in hardening mechanisms such as tamper resistance, tamper detection, tamper response, and tamper evidence to mitigate any hardware tampering of the transport device.

- Ensure that the proper enrollment process is followed by the service provider and the device vendor/manufacturer in the pre-deployment stage.

- Ensure that the device has the pre-configured state and SZTP bootstrapping flag set to "true" to enable SZTP bootstrapping to run when the device powers up for the first time.

- The device should have the private key for the TLS client certificate and the private key for decrypting SZTP artifacts securely stored, ideally in a cryptographic processor such as a trusted platform module (TPM).

- Ensure that the transport device can be validated. This can be achieved by making sure that the SZTP server can validate the certificates provided by the device in the initial request.

- Ensure that the SZTP server can be validated to make sure the device is joining the correct network. This can be done by loading the owner certificate onto the device in the factory provisioning process by the device manufacturer, which is then validated against the domain certificate or the ownership voucher during the SZTP server interactions.

- Ensure that the boot image is securely booted using the principles of Secure Boot provided in the Unified Extensible Firmware Interface (UEFI) specifications.

- Ensure that the user access control to the bootstrapping server and the SZTP server follows strict access control and has very granular access policies applied to it.

- Ensure that the pre-configuration scripts, initial configuration scripts, and post-configuration scripts are thoroughly validated by the NMS teams and by the service provider in a lab or a pilot network before being deployed in a production/live network.

- Ensure that the source of the software update is validated and has a process in place where the software update is first applied to a lab or pilot transport network. Only after thorough investigation of the pre- and post-update behavior should the update be allowed to be deployed in the production or live network.

Using Security Gateway (SecGW/SEG) to Secure the RAN and Transport Layer

One of the main concepts introduced in 3GPP 33.210 is the notion of a security domain, which is a network managed by a single administrative authority. Within a security domain, the same level of security and use of security services is typical. Normally, a network run by a single network operator or a single transit operator will constitute one security domain, although an operator may at will divide the network into separate subnetworks.

The border between the security domains is protected by a security gateway (SecGW or SEG). The SecGWs are responsible for enforcing the security policy of a security domain toward other SecGWs in the destination security domain. The network operator might have more than one SecGW in its network in order to avoid a single point of failure or for performance reasons.

Figure 4-10 depicts a network with a 5G gNB and an LTE eNB. The network has traffic passing through the untrusted and insecure transport layer, which is not owned by the operator, and terminates the traffic in the multi-access edge compute (MEC). The traffic from the MEC is also passing through the untrusted network before it gets terminated in the centralized packet core.

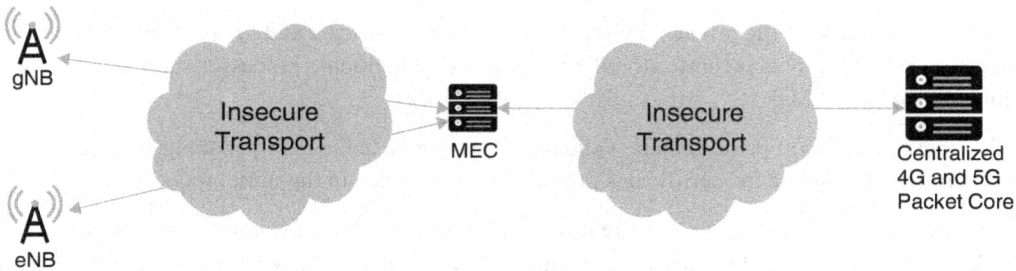

FIGURE 4-10   Transport Network Using an Insecure Public Network

Figure 4-11 shows the deployment of the SecGW to terminate all the IPsec tunnels from the gNB and eNB and also to secure the traffic between the MEC and the centralized packet core. The network collectively uses distributed and centralized security gateways to secure the end-to-end transport of the 4G and 5G infrastructure.

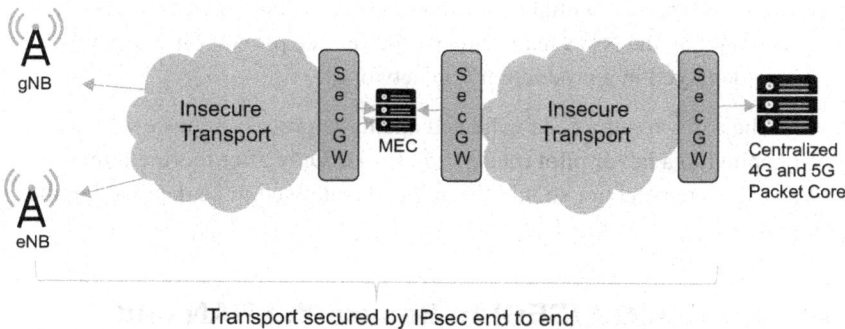

FIGURE 4-11   Securing a Transport Network Using IPsec

The primary concerns for any service provider planning the back haul are the security of the data and preventing eavesdropping when the data traverses the insecure transport infrastructure, be it a public network or a partner network. One way to do this is by encrypting the data by deploying encryption and decryption at each site and introducing authentication mechanisms to ensure devices providing encryption and decryption are legitimate.

IPsec is a suite of protocols developed under IETF to provide authentication, integrity, access control, and confidentiality. The fundamental components of IPsec are defined in the RFC 2401 and described in the following list:

- **Security protocols:** Authentication Header (AH) and Encapsulation Security Payload (ESP)

- **Key management:** ISAKMP, IKE, and SKEME

- **Algorithms:** Enabling encryption and authentication

The security gateway provides the IPsec functionality. Using a security gateway ensures that the data transferred between the 4G and 5G radio nodes (eNB/gNB) and the packet core (EPC/5GC) and the data transferred using an insecure transport network are encrypted and verified.

Centralized SecGW

In more centralized environments, where there are fewer geographic sites or there is a requirement for larger-scale clusters of IPsec termination (that is, massive throughput), the centralized SecGWs should be able to create large pools of IPsec capability and be able to statefully inspect the traffic entering and leaving the tunnel. These SecGWs can also be positioned in a geographically dispersed environment if stateful inspection is required, and they should be able to provide resiliency by configuring it as active/standby, active/active, or clustered, as needed.

Distributed SecGW

The distributed SecGW provides a fully distributed IPsec termination capability to the very edge of the network. This is requisite for the 5G URLLC use cases and 3GPP LTE-A X2 latency requirements when features such as Coordinated Multipoint (CoMP) are enabled. The inclusion of a distributed SecGW does not negate the requirement for a geographically dispersed/centralized SecGW. The main purpose is to ensure that the UPF can be deployed in distributed MEC layers and the X2/Xn/Xx traffic can be forwarded between 4G and 5G nodes with minimal latency. For many service providers, the plan is to deploy both centralized and distributed SecGWs instead of having to choose between the two modes of SecGW deployment.

Interfaces Secured by SecGW

Figure 4-12 shows the key interfaces of 5G, which are secured using the centralized/distributed SecGW deployment. The security services that have been identified as being needed are confidentiality, integrity, authentication, and anti-replay protection. These will be ensured by standard procedures, based on cryptographic techniques provided by the SecGW. Please note that as per 3GPP specification 33.501, IPsec support is mandatory in 5G gNB. So for the N2 and N3 interfaces, the IPsec tunnel can be initiated by the gNB toward the SecGW without any SecGW deployment at the gNB location. 5G gNB can also terminate the IPsec tunnel from the SecGW without any need for SecGW deployment at the 5G gNB location.

For the IPsec tunnel between the MEC and central DC 5GC, specifically for interfaces such as N4, a separate SecGW would be required in the MEC and central DC to establish the IPsec tunnel.

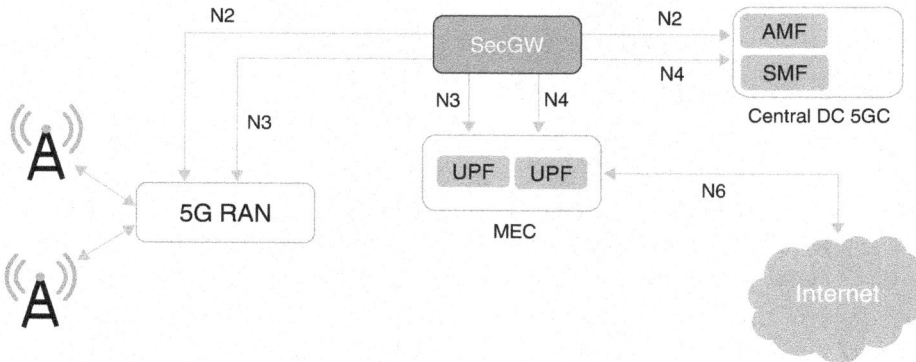

FIGURE 4-12   Securing 5G Interfaces Using a Security Gateway

In Figure 4-12, the following interfaces are secured by using a SecGW:

- The N2 interface, which uses SCTP (NGAP), supports the control plane signaling between 5G RAN and 5G Core covering UE context management and PDU session/resource management procedures. In the figure, N2 is shown as deployed in the central DC 5GC, which most of the service providers are planning to implement, but it can be located at the MEC DC, depending on the deployment option chosen by the service provider. N2 being signaling doesn't impact the low-latency use cases much (depending on the low-latency requirement of the use case) and is hence normally planned to be deployed in the central DC.

- N3, the interface between 5G RAN and the UPF, carries the user plane data from the UE to the UPF. In Figure 4-12, the UPF is located at the MEC, and hence the N3 is terminated at the MEC layer. N3 uses the GTP (GPRS Tunnelling Protocol) with header extensions for 5G, segment routing (SR), and so on. For the service provider, the location of deployment for the UPF is also dependent on the use cases. For example, for some of the IoT use cases, the UPF is still planned to be deployed in the central DC. The location and the level of security in the deployment location would determine the need for securing the N4 interface using IPsec ESP and IKEv2 certificate-based authentication provided by the SecGW.

- N4, the interface between the User Plane Function (UPF) and the Session Management Function (SMF), uses the Packet Forwarding Control Plane (PFCP) protocol, which was also used in the Sx interface in the 4G CUPS architecture. The location of deployment of the UPF (that is, whether the UPF is co-located with the SMF or is distributed) would determine the need for securing the N4 interface using IPsec ESP and IKEv2 certificate-based authentication provided by the SecGW.

Apart from the preceding interfaces shown in Figure 4-12, other interfaces can be secured by the SecGW, as described in the following list:

- The Xn interface connecting the 5G RAN nodes consists of Xn-C, carries signaling and Xn-U, and carries user plane data. In addition to IPsec ESP and IKEv2 certificate-based authentication provided by the SecGW, Datagram Transport Layer Security (DTLS) can also be used to provide integrity protection, replay protection, and confidentiality protection. The decision of whether to terminate the Xn interface in the distributed or centralized security gateway is dependent on the network topology of the RAN nodes' deployment.

- The Xx interface connecting the 5G RAN nodes with the 4G RAN nodes consists of Xx-C, carries signaling and Xx-U, and carries user plane data. In addition to IPsec ESP and IKEv2 certificate-based authentication provided by the SecGW, DTLS can also be used to provide integrity protection, replay protection, and confidentiality protection. The decision of terminating the Xn interface in the distributed or centralized security gateway is dependent on network topology of RAN deployment.

- The F1 interface connecting the gNB-CU to the gNB-DU consists of F1-C, carries control plane data and F1-U, and carries user plane data. The IPsec ESP and IKEv2 certificate-based authentication support on the gNB-DU and gNB-CU is mandatory, but the deployment of cryptographic solutions such as SecGW is purely the decision of the service provider. Due to the dynamic nature of the gNB-DU, it is recommended to have the SecGW at the gNB-CU location, which can terminate the IPsec from multiple gNB-DUs. In addition to IPsec ESP and IKEv2 certificate-based authentication provided by the SecGW, DTLS can also be used to provide integrity protection, replay protection, and confidentiality protection as documented in RFC 6083.

- The E1 interface connecting the gNB-CU-CP to the gNB-CU-UP is used for the transport of the signaling data. The IPsec ESP and IKEv2 certificate-based authentication support on the gNB-DU and gNB-CU is mandatory, but the deployment of cryptographic solutions such as SecGW is purely the decision of the service provider. In addition to IPsec ESP and IKEv2 certificate-based authentication provided by the SecGW, DTLS can also be used to provide integrity protection, replay protection, and confidentiality protection, as documented in RFC 6083. Per the service provider, the gNB-CU-CP and gNB-CU-UP are co-located, sometimes even in the same rack, and hence might not require the SecGW to be deployed. This, of course, would depend on the deployment option chosen by the service provider, the location of the edge/far-edge DC, and who owns the edge/far-edge DC (that is, whether it is a shared infrastructure or deployed in a public or private cloud infrastructure).

The deployment options of the SecGW and the interfaces being secured will be explained in more detail throughout the chapters in this part of the book.

Set of Features Required by the Security Gateway in 5G Networks

For service provider networks, the IPsec security protocol should always be ESP; it is further mandated that integrity protection/message authentication together with anti-replay protection should always be used. All 3GPP and fixed broadband traffic should pass through a SecGW before entering or leaving the security domain. When choosing a security gateway, the service provider can go through the key points covered in the sections that follow to understand the capability of the SecGW and the requirements for the 5G network. Please note that not all the points in the following sections are mandatory, but it would be good to understand both the optional and mandatory requirements for SecGW before choosing any vendor.

Support of ESP

ESP provides a mix of security services in IPv4 and IPv6. It can be applied alone or along with AH. The ESP header is inserted after the IP header and before the next layer protocol header (transport mode) or before an encapsulated IP header (tunnel mode). ESP can be used to provide confidentiality, data origin authentication, connectionless integrity, anti-replay service (a form of partial sequence integrity), and (limited) traffic flow confidentiality. The set of services provided depends on the option selected at the time of security association (SA) establishment and on the location of the implementation in a network topology.

The SecGW the service provider plans to deploy in the network should support ESP security protocol according to RFC 4303. Extended sequence numbers may be supported. For compatibility with earlier 3GPP releases, it is recommended to be backward compatible with nodes supporting only RFC 2406.

Support of ESP Encryption Transforms

The SecGW should follow the implementation conformance requirements for ESP encryption transforms (including authenticated encryption transforms), as stated in RFC 8221.

Support of ESP Authentication Transforms

ESP should always be used to provide integrity, data origin authentication, and anti-replay services, and hence the NULL authentication algorithm is explicitly not allowed for use. Encryption without authentication is not effective and must not be used. IPsec offers three ways to provide both encryption and authentication:

- **ESP with an Authenticated Encryption with Associated Data (AEAD) cipher**: This is the most modern method and handles encryption/decryption and authentication in a single step. In this case, the AEAD cipher is set as the encryption algorithm, and the authentication algorithm is set to none. Examples of this are ENCR_AES_GCM_16 and ENCR_CHACHA20_POLY1305.

- **ESP with a non-AEAD cipher + authentication**: This is a traditional approach, where ESP is used with an encryption and an authentication algorithm. This approach is slower, as the data has to be processed twice: once for encryption/decryption and once for authentication. An example of this is ENCR_AES_CBC combined with AUTH_HMAC_SHA2_512_256.

- **ESP with a non-AEAD cipher + AH (Authentication Header) with authentication**: This is the slowest method and is not recommended. It also takes up more octets due to the double header of ESP+AH, which results in a smaller effective MTU for the encrypted data. With this method, ESP is only used for confidentiality without an authentication algorithm, and a second IPsec protocol of type AH is used for authentication. Examples of this are ESP with ENCR_AES_CBC and AH with AUTH_HMAC_SHA2_512_256.

The SecGW should also have mandatory support of ESP authentication algorithms specifically marked as "MUST" in RFC 8221.

For the IoT use cases, the ESP Encryption Algorithm (ENCR_AES_CCM_8) and ESP and AH Authentication Algorithm (AUTH_AES_XCBC_96) need to be supported by the SecGW if the service provider has plans to deploy IoT as part of the network evolution. This might change as stronger and better algorithms emerge that are more suitable for a wide variety of CPU architectures and device deployments, ranging from high-end bulk encryption devices to small, low-power IoT devices.

Requirements on the Construction of the Initialization Vector (IV)

An initialization vector is a fixed-size input to a cryptographic algorithm. Its primary use is that of a nonce to ensure encrypting the same data with the same key does not result in the same ciphertext. In some cases, it might not be needed to transmit the IV because it can be derived from other information, such as the sequence number in ESP. Some algorithms require the IV to be random/unpredictable (for example, AES-CBC), but in many modern combined mode algorithms (for example, ChaCha20/Poly1305), it can just be a counter.

The requirements for the IV construction for the IPsec deployed using SecGW are as follows:

- **For CBC mode**: The IV field should be the same as the block size of the cipher algorithm being used. The IV should be chosen at random and should be unpredictable to any party other than the originator.

- **For CTR, GCM, CCM, and GMAC mode**: The IV field should be eight octets. The IV should be generated in a manner that ensures uniqueness. The same IV and key combination should not be used more than once. It is explicitly not allowed to construct the IV from the encrypted data of the preceding encryption process.

Profiling of IKEv2 (Internet Key Exchange, Version 2)

The SecGW that's planned to be deployed in the 5G network should support IKEv2 as defined in RFC 7296, "Internet Exchange Version 2 (IKEv2)" parameters. As stated in 3GPP 33.210, the following additional requirements apply for IKEv2:

The following algorithms need mandatory support:

- **Confidentiality:** AES-GCM with a 16-octet ICV with a 256-bit key length
- **Pseudo-random function:** PRF_HMAC_SHA2_384
- **Diffie-Hellman group 20:** 384-bit random ECP group

The following algorithms should be supported:

- **Confidentiality:** AES-GCM with a 16-octet ICV with a 128-bit key length
- **Pseudo-random function:** PRF_HMAC_SHA2_256
- **Integrity:** AUTH_HMAC_SHA256_128
- **Diffie-Hellman group 19:** 256-bit random ECP group

For IKE_AUTH exchange:

- Authentication Method 2 – Shared Key Message Integrity Code should be supported.
- IP addresses and fully qualified domain names (FQDNs) should be supported for identification.
- Re-keying of IPsec SAs and IKE SAs should be supported as specified in RFC 7296.

For the CREATE_CHILD_SA exchange:

- A DH key exchange should be used (giving Perfect Forward Secrecy) and the session keys should be changed frequently.

For reauthentication:

- IKE SAs should be reauthenticated, as specified in RFC 7296.
- An NE should proactively initiate reauthentication of IKE SAs and the creation of child SAs (that is, the new SAs should be established before the old ones expire).
- An NE should destroy an IKE SA and its child SAs when the authentication lifetime of the IKE SA expires.

Support of IPsec Tunnel and Transport Mode

Before we discuss security gateway deployment modes, it is important to understand IPsec modes. IPsec can be configured in tunnel mode or transport mode:

- **IPsec tunnel mode:** In IPsec tunnel mode, the entire original IP datagram is encrypted, and it becomes the payload in a new IP packet. In this mode the SecGW performs encryption on behalf of the hosts. The source SecGW encrypts packets and forwards them along the IPsec tunnel. The destination router decrypts the original IP datagram and forwards it on to the destination system. The major advantage of tunnel mode is that the end systems do not need to be modified to receive the benefits of IPsec. Tunnel mode also protects against traffic analysis; with tunnel mode, an attacker can only determine the tunnel endpoints and not the true source and destination of the tunneled packets, even if they are the same as the tunnel endpoints. For inter-domain use cases, tunnel mode support for IPsec will be required and hence needs to be supported by the SecGW.

- **IPsec transport mode:** In IPsec transport mode, only the IP payload is encrypted, and the original IP headers are left intact. This mode has the advantage of adding only a few bytes to each packet. It also allows devices on the public network to see the final source and destination of the packet. With this capability, you can enable special processing (for example, QoS) on the intermediate network based on the information in the IP header. However, the Layer 4 header will be encrypted, thus limiting the examination of the packet. Unfortunately, transmitting the IP header in cleartext, transport mode allows an attacker to perform some traffic analysis. The option to use tunnel mode or transport mode is a design choice by the customer.

Security Gateway Deployment Modes

There are basically two primary modes of deploying IPsec:

- Single tunnel concept or single child SA concept
- Multiple IPsec or multiple child SA concept

In IPsec transmission between the eNB/gNB and SecGW, Internet Key Exchange (IKE) is used to create a security association. Before the encrypted traffic is sent or received, IKE version 2 (IKEv2) will create a second tunnel, an IKE security association (SA), which will be used for secure transmission of the child SA. This can be a single child SA or multiple child SAs.

The sections that follow discuss the single IPsec and multiple IPsec options.

Single Tunnel Concept or Single Child SA Concept

The use of single IPsec or multiple IPsec will depend on the network architecture. If all or multiple network functions are deployed in the same location, you will use a single IPsec tunnel to encrypt the traffic toward the destination. You can then configure internal VRFs (IVRFs) to route the traffic between the different VRFs, as shown in Figure 4-13, which illustrates the single tunnel concept.

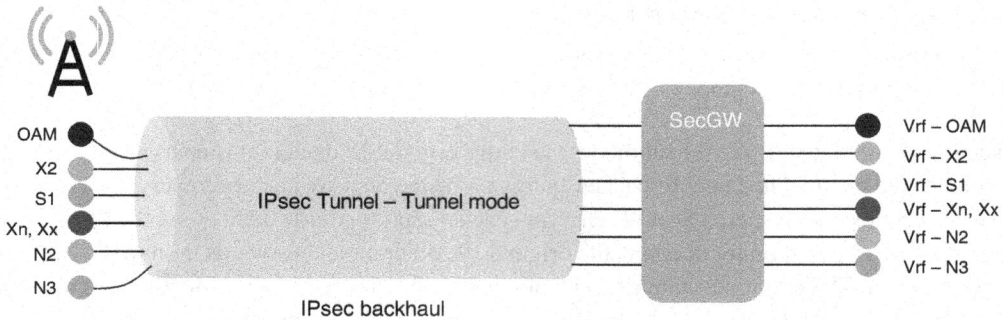

FIGURE 4-13   Single IPsec Tunnel Mode Concept

Here are the key points:

- All traffic, such as OAM, S1-C, S1-U, X2-C, X2-U, Xn-C, Xn-U, Xx-C, and Xx-U, traverses the same tunnel.

- It's difficult to change the architecture in later phases.

> **Note**
>
> Based on my experience of deploying SecGWs with many service providers, the following details are worth knowing beforehand:
>
> - Some RAN vendor radio base stations (eNBs) support single tunnel for everything. In such situations, it would be recommended to have management (OAM) on SSL/SSH and not IPsec for troubleshooting, as maintenance would not be possible if the tunnel is not up.
>
> - Some RAN vendor radio base stations (eNBs) will support two tunnels: the first tunnel for X2 and S1 and the second tunnel for OAM. Therefore, there's no need to configure SSL/SSH for troubleshooting/maintenance purposes.

Multiple IPsec or Multiple Child SA Concept

Multiple IPsec deployments are mostly used in an enterprise architecture where you need to connect multiple enterprise branches to the enterprise's headquarters. In service provider telco deployments, multiple IPsec will be used if you have a very distributed architecture where each of the network functions needs separate SecGWs or multiple instances of SecGWs catering for each of the components. Figure 4-14 illustrates an example of the multiple IPsec tunnel mode where a context within a SecGW is dedicated to a network function.

Having the option of configuring multiple contexts, which allows a single SecGW to behave like multiple standalone SecGWs with options of dedicated interfaces or shared interfaces, will provide you the flexibility to use a single SecGW as multiple standalone SecGWs to enable multiple IPsec deployment, as shown in Figure 4-14.

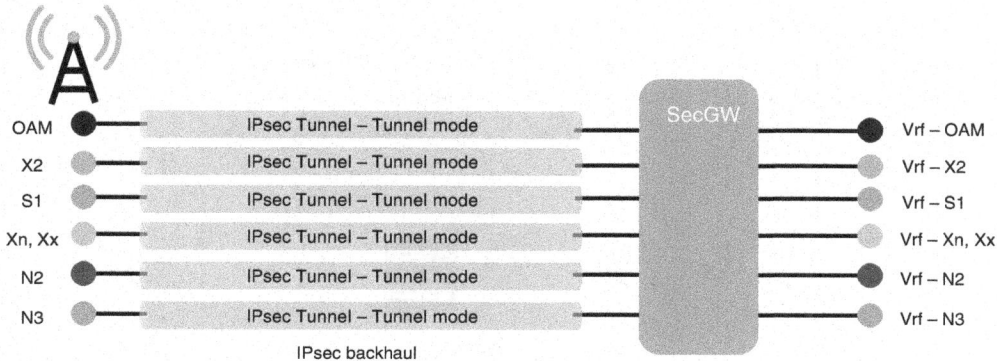

FIGURE 4-14   Multiple IPsec Tunnel Mode Concept

Here are the key points:

- Different traffic types traverse different tunnels/SAs.

- Traffic separation is based on tunnels/SAs.

- Multiple IPsec tunnels increase IKE SAs and the number of outer IP addresses.

Security Gateway in 4G and 5G Networks

This section provides the service provider with information on how the SecGW solution could cater for 4G and 5G deployment use cases. The section starts with the 4G use cases to help you understand the evolution of the security gateway control from 4G to 5G without having to make major changes in the design or replacing the existing centralized SecGW deployment.

4G Network

Many of the service providers who own their own back haul were used to not deploying a SecGW as the eNB network element (NE), and the Evolved Packet Core (EPC) was considered to be in the same trusted secure domain. But in recent times, there has been a surge in threat actors trying to connect rogue NE nodes toward the packet core. Apart from the threat vector of rogue NEs, many service providers have some NEs passing through the non-trusted insecure transport network domain owned by another service provider.

Deploying IPsec using SecGW will provide authentication and encryption between the eNB and the packet core nodes, thereby securing the back haul. The key interfaces that the SecGW would secure are the S1 and X2 interfaces. The SecGW can be deployed as centralized or distributed, depending on whether you have deployed centralized 4G or Control Plane User Plane Separated (CUPS) 4G. In some cases, both centralized and distributed deployments of a SecGW need to be deployed, as you might have S1 and X2 terminating in the edge site as well as the centralized packet core in some instances. The section "4G CUPS Network" discusses distributed SecGW deployment in more detail.

Figure 4-15 shows the option of a centralized security gateway, where the S1 interface and X2 interface are secured using the security gateway. In this case, both the S1 interface and the X2 interface are terminated in the centralized security gateway, which is located in the centralized 4G Evolved Packet Core (EPC) location.

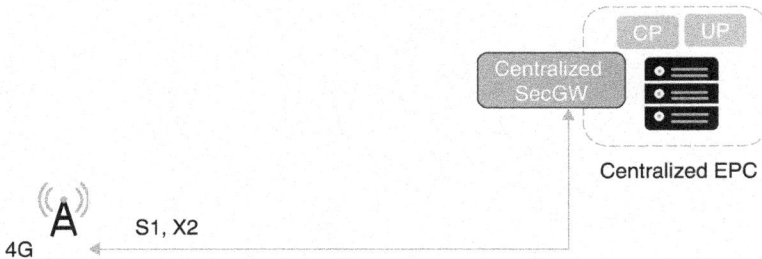

FIGURE 4-15   Securing S1 and X2 Interfaces Using Security Gateway

The SecGW might require large pools of IPsec capability depending on the network size of the service provider and should also be able to statefully inspect the traffic entering and leaving the tunnel. This device might also be positioned in a geographically dispersed environment if stateful inspection is required and configured in different resiliency modes (Active/Standby, Active/Active, and so on). Figure 4-16 illustrates the IPsec tunnel mode deployment for a centralized 4G SecGW deployment, which protects the S1, X2, and OAM interfaces. Once the tunnel is terminated in the SecGW at the central DC location, it can then be routed to MME and SGW. In actual deployments, IPsec tunnels are mapped to Virtual Routing and Forwarding (VRF) instances using a single public IP address. When an IPsec-encapsulated packet arrives at the SecGW from the eNB, IPsec performs security association (SA) lookup for the Security Parameter Index (SPI), destination, and protocol. The packet is decapsulated using the SA, gets associated with the inside VRF (IVRF), and is forwarded using the IVRF routing table.

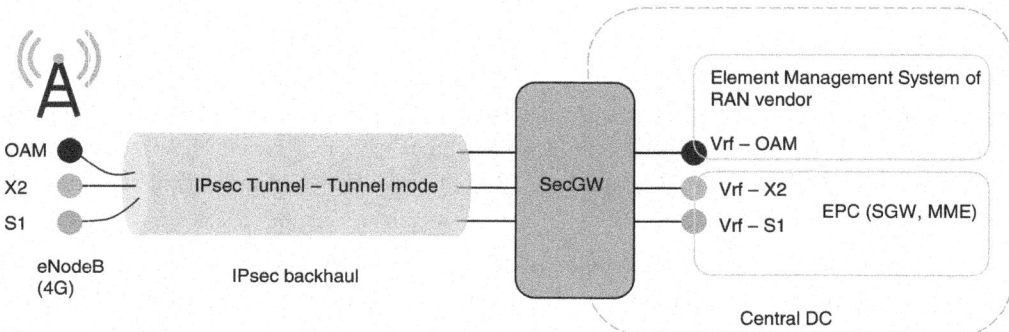

FIGURE 4-16   Detailed View for Securing S1 and X2

The deployment of the SecGW in the centralized deployment mode should provide the security controls mentioned in Table 4-1 and the service provider could choose what traffic (CP only, CP + UP) is required to be secured. Normally in traditional 4G deployment configurations, it's difficult to separate the CP and the UP traffic, unless CUPS is deployed. The section "4G CUPS Network," later in the chapter, discusses this in more detail.

As Figure 4-17 illustrates, the SecGW routes the S1-C traffic to MME and S1-U traffic to SGW. In some cases, such as shared sites between service providers, NAT might be used along with routing to provide topology hiding.

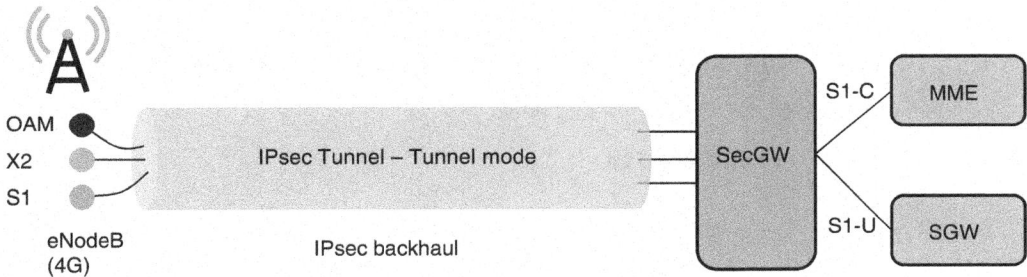

FIGURE 4-17   Routing of S1-C to MME and S1-U to SGW Using SecGW

Table 4-1 outlines the security controls per interface in centralized 4G deployment.

TABLE 4-1   Security Controls per Interface—4G

Interface	Access Node	Core Node	Security Control
OAM	eNB	NMS/EMS	IPsec
S1-C	eNB	MME	IPsec
S1-U	eNB	SGW	IPsec
X2-C	eNB < > eNB	N/A	IPsec
X2-U	eNB < > eNB	N/A	IPsec

If the RAN vendors support single tunnel for everything, it is recommended that you have management (OAM) on SSL/SSH and not IPsec for troubleshooting, as maintenance would not be possible if the tunnel is not up. If OAM is not protected using IPsec, then mechanisms such as NETCONF/TLS with PKIX for certificate-based mutual authentication or PKIX-based SSHv2 should be implemented. Granular access controls should also be implemented to enhance the OAM access control, as discussed later in the chapter in the section "Granular Access Control."

4G CUPS Network

4G CUPS brings in Control Plane User Plane Separation (CUPS) and can be deployed as a co-located CUPS or distributed CUPS. 4G CUPS based on the Packet Forwarding Control Plane (PFCP) protocol was implemented to enable the following functions:

- Reduced latency on application service (for example, by selecting user plane nodes that are closer to the RAN or more appropriate for the intended UE usage type without increasing the number of control plane nodes).

- Support for an increase of data traffic by enabling the addition of user plane nodes without changing the number of SGW-C, PGW-C, and TDF-C in the network.

- Locating and scaling the CP and UP resources of the EPC nodes independently.

- Independent evolution of the CP and UP functions.

- Enabling SDN to deliver user plane data more efficiently.

The SecGW deployment in co-located CUPS mode is similar to the deployment of EPC in non-CUPS mode, as the control plane and user plane, though separated, still exist in the same location or the same DC (data center). In distributed CUPS, the control plane and user plane are not only separated, but also distributed and deployed in edge data centers.

Figure 4-18 shows the distributed deployment of 4G CUPS, where the user plane (UP) is deployed toward the edge to cater for the latency-sensitive use cases, whereas the UP for use cases where latency is not an issue as well as the control plane (CP) is deployed in the centralized 5GC DC. CUPS introduces three new interfaces—Sxa, Sxb and Sxc—between the CP and UP functions of the SGW, PGW, and TDF, respectively, which is collectively stated as Sx in the figure and is protected by the IPsec ESP and IKEv2 certificate-based authentication provided by the SecGW.

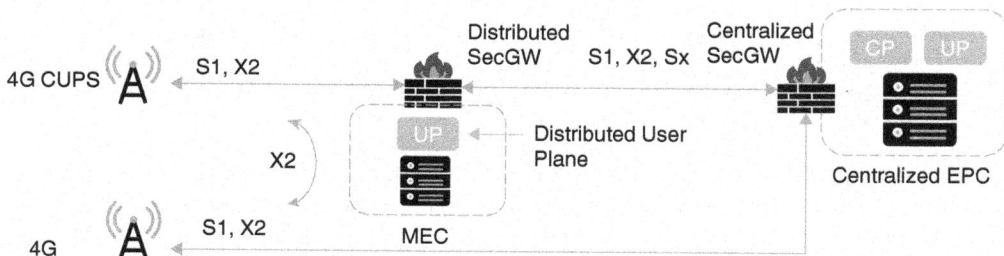

FIGURE 4-18 Evolution of SecGW Architecture to Secure 4G CUPS

As Figure 4-19 illustrates, the S1-U from the eNB can be terminated in the distributed SecGW deployed in the MEC, and S1-C from the eNB is terminated in the centralized SecGW, which will then forward the packets to the SGW-C after decryption. X2, although also shown as being terminated in the centralized SecGW location, can also be deployed on the MEC layer, which is purely the service provider's choice.

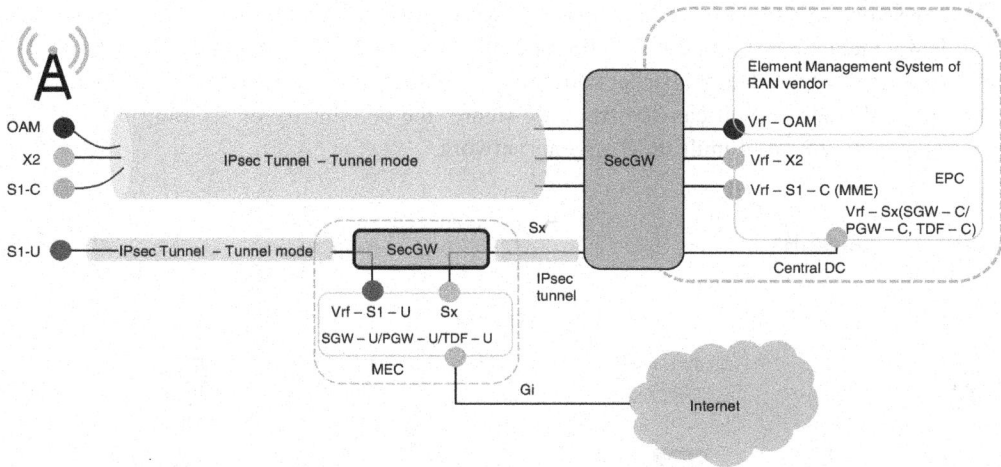

FIGURE 4-19 Detailed View for Securing Distributed 4G CUPS Deployments

In the CUPS, the deployment of the SecGW in the centralized and distributed mode would secure the Access and Packet Core nodes described in Table 4-2 using the IPsec ESP and IKEv2 certificate-based authentication.

TABLE 4-2 Security Controls per Interface—4G CUPS

Interface	Access Node	Core Node	Security Control
OAM	eNB	NMS / EMS	IPsec
X2-C	eNB	NA	IPsec
X2-U	eNB	NA	IPsec
S1-C	eNB	MME	IPsec
S1-U	eNB	SGW-U	IPsec
Sxa	N/A	SGW- U < > SGW-C	IPsec
Sxb	N/A	PGW-U < > PGW-C	IPsec
Sxc	N/A	TDF-U < > TDF-C	IPsec

5G NSA (Non-Standalone) Network

Multiple deployment options are available for the 5G Non-Standalone (NSA) model. One of the NSA options is Option 3, where the RAN is composed of a master node and secondary node. The master node is LTE eNB, the secondary node is New Radio (NR) en-gNB, and the packet core is EPC (4G packet core). NSA Option 3 supports legacy 4G devices, and the 5G devices only need to support NR protocols. The 5G NSA options are purely the choice of the service provider, as you might choose to skip 5G NSA and deploy 5G SA directly. Most of the 5G NSA options are based on the software upgrade on the eNB and the EPC to allow higher throughput capabilities.

Figure 4-20 shows you the different interfaces in the network when you have a multi-RAT (radio access technology) deployment, such as 4G, 4G CUPS, and 5G NSA deployed in the network. The packet core might consist of an upgraded 4G packet core or some elements of the 5G core, depending on the 5G NSA model deployed. As you can see from the figure, the complexity of securing the network increases with the deployment of multi-RAT in your network.

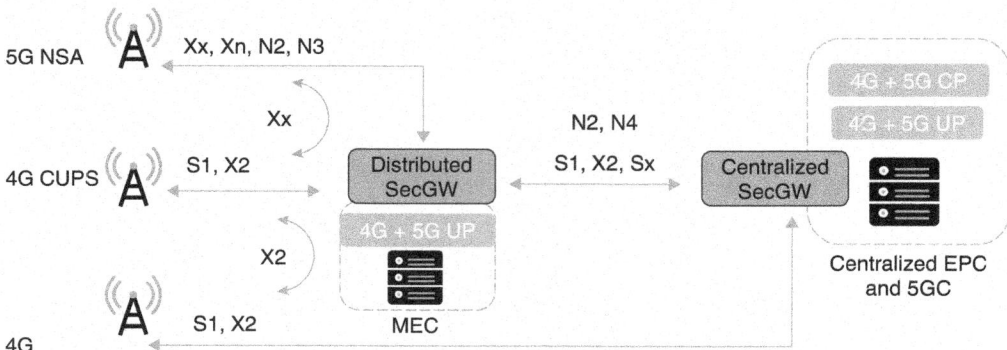

FIGURE 4-20   Evolution of SecGW Architecture to Secure 5G NSA Deployments

As Figure 4-21 illustrates, the deployment of the SecGW in the centralized and distributed mode would secure the Access and Packet Core nodes described in Table 4-3 using the IPsec ESP and IKEv2 certificate-based authentication in 5G NSA and multi-RAT deployments.

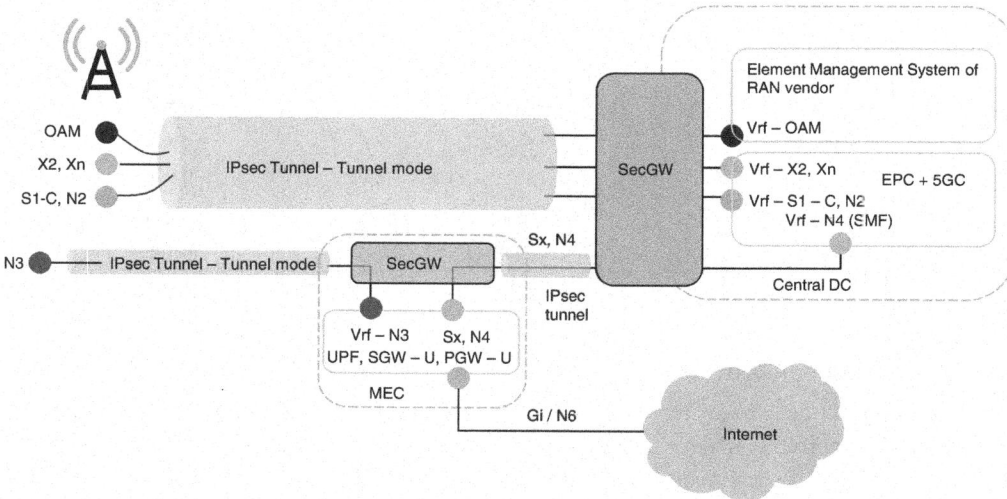

FIGURE 4-21   Detailed View for Securing NSA Deployments

TABLE 4-3   Security Controls per Interface—NSA

Interface	Access Node	Core Node	Security Control
OAM	eNB	NMS/EMS	IPsec
X2-C	eNB < > eNB	NA	IPsec
X2-U	eNB < > eNB eNB < > ngNB	NA	IPsec
S1-C	eNB	MME	IPsec
S1-U	eNB	SGW-U	IPsec
Sxa	NA	SGW-U < > SGW-C	IPsec
Sxb	NA	PGW-U < > PGW-C	IPsec
Sxc	NA	TDF-U < > TDF-C	IPsec
Xn-C	gNB < > gNB	NA	IPsec
Xn-U	gNB < > gNB	NA	IPsec
N2	gNB	AMF	IPsec
N3	gNB	UPF	IPsec
N4	NA	UPF < > SMF	IPsec

5G Standalone (SA) Network

The 5G SA packet core is equipped with several new capabilities inherently built into it so that operators have the flexibility and capability to face new challenges presented by the new set of requirements for various new use cases. The network functions in the new 5G core are broken down into smaller entities, such as the Session Management Function (SMF) and User Plane Function (UPF), which can be used on a per-service basis. Gone are the days of huge network boxes; welcome to services that automatically register and configure themselves over the Service-Based Architecture, which is built with the new functions like the Network Repository Function (NRF) that borrow their capabilities from cloud-native technologies.

In 5G SA, the control plane and the user plane from the 5G radio node (gNB) are terminated in the 5GC. There is no dependency on the 4G infrastructure, such as eNB or EPC. This allows for the complete virtualization of the entire packet core stack for 5G, which can be separate from the 4G packet core (EPC) as illustrated in Figure 4-22.

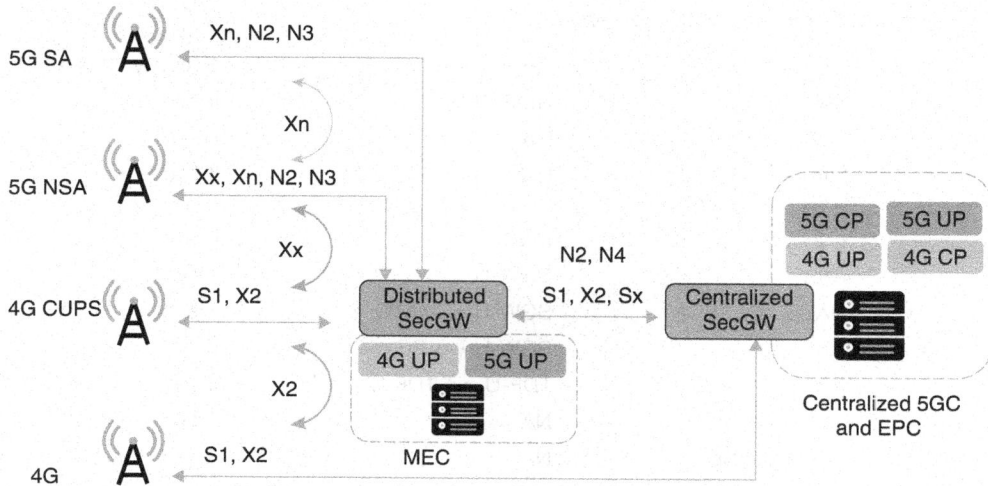

FIGURE 4-22 Evolution of SecGW Architecture to Secure 5G SA

As Figure 4-23 illustrates, the deployment of the SecGW in the centralized and distributed mode would secure the Access and Packet Core nodes described in Table 4-4 using the IPsec ESP and IKEv2 certificate-based authentication in 5G SA and multi-RAT deployments.

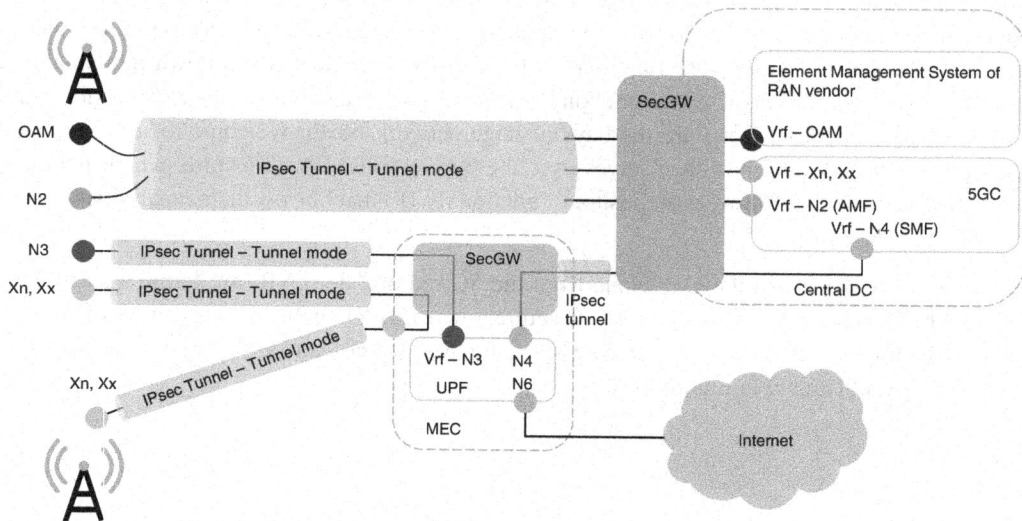

FIGURE 4-23 Detailed View for Securing SA Deployments

TABLE 4-4 Security Controls per Interface—4G Toward NSA

Interface	Access Node	Core Node	Security Control
OAM	eNB	NMS/EMS	IPsec
X2-C	eNB < > eNB	NA	IPsec
X2-U	eNB < > eNB	NA	IPsec
S1-C	eNB	MME	IPsec
S1-U	eNB	SGW-U	IPsec
Sxa	NA	SGW-U < > SGW-C	IPsec
Sxb	NA	PGW-U < > PGW-C	IPsec
Sxc	NA	TDF-U < > TDF-C	IPsec
Xn-C	gNB < > gNB	NA	IPsec
Xn-U	gNB < > gNB	NA	IPsec
N2	gNB	AMF	IPsec
N3	gNB	UPF	IPsec
N4	NA	UPF < > SMF	IPsec

VRAN (Virtualized RAN), O-RAN (Open RAN), and C-RAN (Cloud RAN) Networks

Before we get into the details of O-RAN specifically, let's discuss the differences between VRAN, O-RAN, and C-RAN. These terms are used in conjunction with each other, and though they are closely linked together, there are some differences between them.

Virtualized RAN (VRAN) relates to the decomposition of software from hardware and provides you with virtualized 5G network functions that can be deployed in an x86 server.

Open RAN (O-RAN), which is specified by the O-RAN Alliance, provides you with openness, enabling you to choose different vendors for Radio Resource Unit (RRU), Distributed Unit (DU), and Centralized Unit (CU). The DU and CU can be Virtualized RAN (VRAN) functions or can be deployed as custom-built hardware. So, in theory, you can further split O-RAN into O-RAN based on VRAN and O-RAN based on bare metal (BM).

Cloud RAN (C-RAN), which is also sometimes referred to as Centralized RAN, has the Base Band Unit (BBU) deployed at a centralized location, which enables you with a lightweight deployment of RRU and Antenna at the cell site. C-RAN architectures use a front-haul interface between the BBU in the centralized core and the RRU in the cell site. Most service providers are looking at O-RAN deployments with or without VRAN, as compared to C-RAN deployments specifically, due to the flexibility provided by O-RAN architectures, such as flexibility in deploying CU and DU based on use cases and cost optimization due to different vendor selection for RRU, DU, and CU. This section will discuss O-RAN deployments based on VRAN and BM as well as the open front-haul interface related to both O-RAN and C-RAN.

The O-RAN approach is a modular strategy for designing and building networks that combines the optionality of open interfaces and a robust ecosystem with cloud-scale economics. Following 3GPP architectural principles and leveraging standards being defined by the O-RAN Alliance, the monolithic base station is decomposed into functional elements interconnected by open interfaces. Disaggregation (the decoupling of software and hardware) is maximized to allow the deployment of radio signal processing software in a cloud or mini-cloud environment. The combination of these principles allows operators to utilize a far more diverse supply chain to build an architecture that scales like cloud systems instead of hardware appliances. Modular software solutions enable automation across the full lifecycle, from deployment to management and operations. Figure 4-24 illustrates the O-RAN logical architecture.

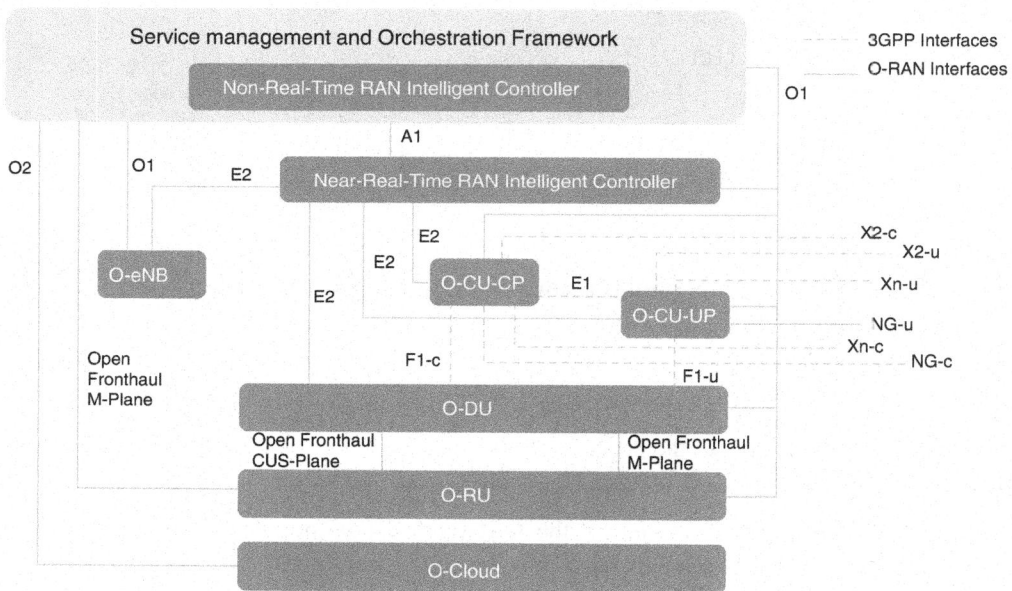

FIGURE 4-24   Logical O-RAN Architecture

As shown in Figure 4-24, the O-RAN architecture includes 3GPP interfaces and the following new interfaces related to O-RAN:

- **3GPP interfaces:** X2 control and user planes, Next Generation (NG) control and user planes, E1 control and user planes, and F1 control and user planes.

- **O-RAN interfaces:** O1, O2, A1, E2, Open Front Haul (FH) control plane, user plane, synchronization plane (CUS-Plane), and Open FH M-plane.

- **Near-real-time RAN Intelligent Controller:** Enables near-real-time control and optimization of O-RAN elements and resources via fine-grained data collection and actions over the E2 interface.

- **Non-real-time RAN Intelligent Controller:** Enables non-real-time control and optimization of RAN elements and resources, AI/ML workflow (including model training and updates), and policy-based guidance of applications/features in near-RT RIC.

- **O-RAN Central Unit – Control Plane (O-CU-CP):** A logical node hosting the RRC and the control plane part of the Packet Data Convergence Protocol (PDCP).

- **O-RAN Central Unit – User Plane (O-CU-UP):** A logical node hosting the user plane part of the Packet Data Convergence Protocol (PDCP) and the Service Data Adaptation Protocol (SDAP).

- **O-RAN Radio Unit (O-RU):** A logical node hosting Low-PHY layer and RF processing based on a lower-layer functional split.

- **O1:** Interface between management entities in Service Management and Orchestration Framework and O-RAN-managed elements, for operation and management.

- **xAPP:** Independent software plug-in to the near-RT RIC platform to provide functional extensibility to the RAN by third parties.

Apart from the aforementioned new interfaces, O-RAN also introduces the following new functions and frameworks:

- **O-Cloud:** O-Cloud is used to host the network functions.

- **Services Management and Orchestration Framework (SMO):** SMO manages the network functions and O-Cloud.

- **Non-real-time and near-real-time RAN Intelligent Controller (RIC):** RIC provides mobility management, admission control, and interference management as x-Apps (near-real-time RIC) or r-Apps (non-real-time RIC) on the northbound interface and enforces network policies via a southbound interface toward the radios.

The network functions using these interfaces, as well as the security controls to secure these interfaces, are discussed later in Table 4-5.

With the introduction of 5G, the industry acknowledged that networks need to be built in a more efficient, scalable, and automated way, and that included the RAN components as well. As a result, the RAN architecture is evolving from a static, monolithic, highly proprietary architecture to a cloud-based architecture, O-RAN/VRAN. O-RAN architecture incorporates key technology trends into the RAN space, such as the following:

- Decomposing the Base Band Unit (BBU) monolithic block into subfunctions, Distributed Unit (DU) and Central Unit (CU), and allowing them to be centralized and flexibly placed

- Disaggregating and virtualizing those subfunctions, enabling RAN components to run on a NFVi infrastructure

- Using fully standardized and open interfaces to enable a truly multivendor RAN system

The benefits that can be derived from this new architecture are multiple. The open multivendor system promotes competition, innovation, diversity of use cases, and time-to-market (TTM) availability of products. Additionally, the technical benefits provided by the O-RAN architecture translate into effective cost savings, namely the following:

- The centralization of DU and CU RAN functions simplify the cell site and, with it, reduce the cost of building the site and the recurrent cost associated with it, such as power and rent. Building so-called skinny sites also increases the number of sites that qualify as a cell site, which is critical for 5G high-density deployments, which translates into operational efficiencies and TTM.

- Centralization of RAN functions also enables better use of RF resources while improving the service to the users.

- Virtualization adds the benefits of hardware (HW) pooling, dynamic resource allocation, and improved resiliency and redundancy in the RAN (which is critical for some SLAs).

- Centralized and virtualized components with open/standard interfaces provide the baseline for full RAN system automation, where multiple vendors and multiple domains can be jointly automated by the same cross-domain automation system for the same service flow (for example, scalable and fast RAN site deployment, scalable upgrades, central monitoring, real-time visibility, and closed-loop operations). Automation is key for accelerated network deployment/network changes and service level assurance.

- O-RAN/VRAN architecture provides a future-proof foundation for critical RAN developments, such as RAN network slicing support with flexible resource allocation and node/function placement per slice type. This can only be achieved by RAN systems that are built to provide cloud-like flexibility.

- Network functions such as the near-real-time RIC, O-CU-CP, O-CU-UP, and O-DU implemented as containerized microservices can leverage cloud-native security advances such as hardware resource isolation, automatic reconfiguration, and automated security testing, which can improve both open source vulnerability management and security configuration management.

As described in the preceding list, the RAN BBU component is disaggregated into the DU and the CU and can be deployed in far-edge and edge DCs, as per the network requirement. The O-RAN architecture brings in new interfaces—notably the F1 interface between gNB-DU and gNB-CU, the E1 interface between gNB-CU-CP and gNB-CU-UP, and the eCPRI open interface between the RU and DU.

Figure 4-25 illustrates the evolution of the SecGW to secure the distributed RAN deployments in the O-RAN architecture.

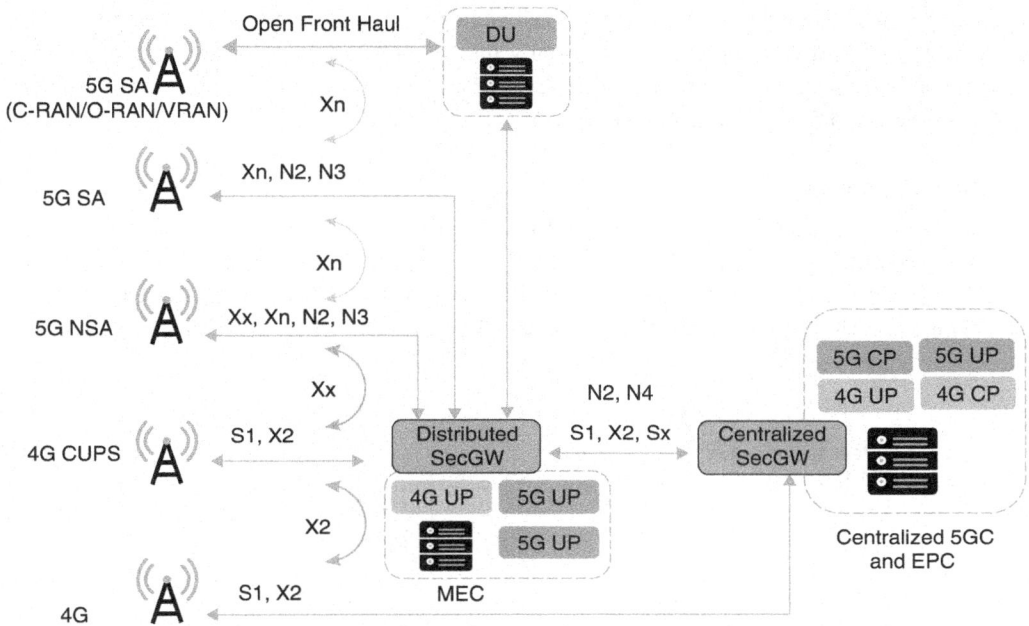

FIGURE 4-25   Evolution of SecGW Architecture to Secure 5G C-RAN/O-RAN/VRAN

Figure 4-26 provides an example of securing the O-RAN deployment using the SecGW.

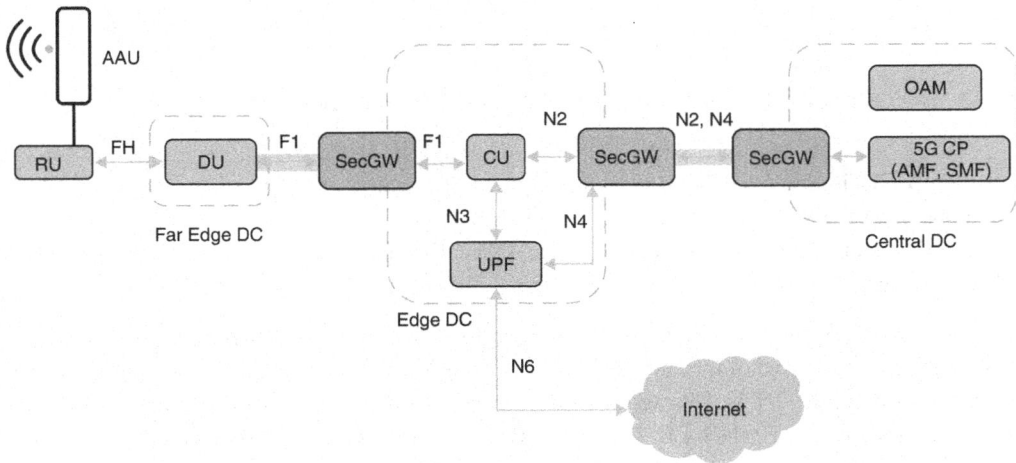

FIGURE 4-26   SecGW Deployment Specific to Secure O-RAN

As you can see from Figure 4-26, both the centralized and distributed SecGW deployment modes are used to secure the 5G O-RAN deployment. Normally the CU-UP and the UPF are deployed in the same rack and are within a trusted zone. But if the CU-UP and the UPF are in separate locations, then the N3 interface needs to be secured using the SecGW.

Securing the F1 Interface

The interface between the gNB-DU and the gNB-CU is called the F1 interface. The F1 interface consists of the F1-C interface and the F1-U interface. F1-C is the interface between the gNB-CU-CP and gNB-DU-CP interface. F1-U is the interface between the gNB-CU-UP and gNB-DU-UP interface. Figure 4-27 illustrates the deployment of SecGW in the location of the CU to aggregate multiple DUs.

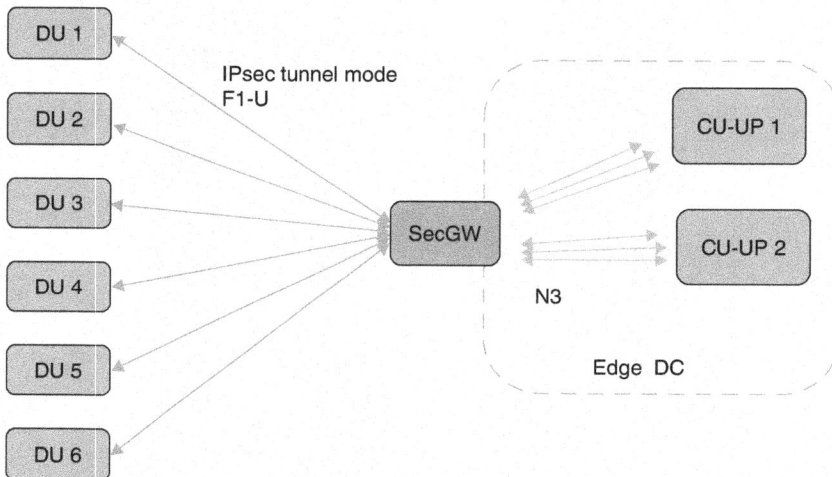

FIGURE 4-27   SecGW at CU Location Aggregating IPsec from Multiple DUs

Usually, the gNB-CU is deployed in the topology as an aggregator of gNB-DUs. The number of gNB-DUs supported by gNB-CU is dependent on a number of factors, but primarily its attributed to the max CPU allocated to the virtual instance of CU and the CPU consumed per virtual instance of DU. The F1-C interface gNB-DU-CP and gNB-CU-CP usually has IPsec tunnel support built in.

F1-U has a mandatory requirement for IPsec support, and the deployment of cryptographic solutions like SecGW is purely the choice of the service provider. The gNB-CU-UP scales in and out depending on traffic. Correspondingly, F1-U EPs (endpoints) are added and removed dynamically, and any gNB-DU can interface with any gNB-CU-UP F1-U EP for a given UE session. This makes the implementation of IPsec with mapping per gNB-DU-UP and gNB-CU-UP impractical because it needs the service provider OAM to configure the IPsec tunnel between the gNB-CU-UP and gNB-DU-UP pair.

The recommended way to deploy cryptographic solutions like SecGW to provide ESP and IKEv2 certificate-based security in the F1-U interface is to have a SecGW deployed at the DC, as shown in the preceding figure.

Table 4-5 outlines the security control for C-RAN/O-RAN/VRAN deployment. Please note that the RAN components, such as CU and DU, are all O-RAN compliant and that the naming convention is kept as gNB-CU instead of O-CU to be consistent across O-RAN, C-RAN, and VRAN architectures.

TABLE 4-5 Security Controls per Interface O-RAN/C-RAN/VRAN

Interface	Access Node / O-RAN Node	Core Node	Recommended Security Control
OAM	gNB-DU, gNB-CU	NMS/EMS	IPsec
N2	gNB-CU-CP	AMF	IPsec
N3	gNB-CU-UP	UPF	IPsec
Xn-C	gNB-CU-CP	N/A	IPsec
Xn-U	gNB-CU-UP	N/A	IPsec
E1	gNB-CU-CP < > gNB-CU-UP	N/A	IPsec
F1-C	gNB-DU < > gNB-CU-CP	N/A	IPsec
F1-U	gNB-DU < > gNB-CU-UP	N/A	IPsec
O1	Service Management and Orchestration < > gNB-CU-CP, gNB-CU-UP, gNB-DU-CP, gNB-DU-UP	N/A	TLS
A1	RIC < >RIC	N/A	TLS
Open FH M-plane	gNB-DU < > RU	N/A	NETCONF/TLS with PKIX for certificate-based mutual authentication and SSHv2 with RFC 6187 support for PKIX-based SSHv2.
Open FH CUS-plane	gNB-DU < > RU	N/A	802.1X and MACsec. (O-RAN is discussing the use of 1588-2019 with the newly defined security TLV to protect announce messages.)
E2	RIC < > CU, DU	NA	TLS
x-r Apps	RIC < > third-party applications		RBAC, Segmentation, API protected by TLS.

Table 4-6 outlines the security controls for multi-RAT deployments.

TABLE 4-6   Security Controls from 4G Toward 5G O-RAN/C-RAN/VRAN

Interface	Access Node	Core Node	Recommended Security Control
OAM	eNB, gNB, DU, CU	NMS/EMS	IPsec
X2-C	eNB < > eNB	NA	IPsec
X2-U	eNB < > eNB	NA	IPsec
S1-C	eNB	MME	IPsec
S1-U	eNB	SGW-U	IPsec
Sxa	N/A	SGW-U < > SGW-C	IPsec
Sxb	N/A	PGW-U < > PGW-C	IPsec
Sxc	N/A	TDF-U < > TDF-C	IPsec
Xn-C	gNB < > gNB	N/A	IPsec
Xn-U	gNB < > gNB	N/A	IPsec
N2	gNB	AMF	IPsec
N3	gNB	UPF	IPsec
N4	NA	UPF < > SMF	IPsec
N2	gNB-CU-CP	AMF	IPsec
N3	gNB-CU-UP	UPF	IPsec
Xn-C	gNB-CU-CP	N/A	IPsec
Xn-U	gNB-CU-UP	N/A	IPsec
E1	gNB-CU-CP < > gNB-CU-UP	N/A	IPsec
F1-C	gNB-DU-CP < > gNB-CU-CP	N/A	IPsec
F1-U	gNB-DU-UP < > gNB-CU-UP	N/A	IPsec

DDoS Protection

Evolving the back-haul and RAN architectures necessitates enhanced transport security, which includes DDoS security to cater for interface protection between multi-RAT communications.

There are three primary types of DDoS attacks:

- **Volume-based attacks:** Some examples of volume-based DDoS attacks follow:
  - **UDP flood example:** A User Datagram Protocol (UDP) flood targets random ports on a remote server with requests called UDP packets. The host checks the ports for the appropriate applications. When no application can be found, the system responds to every request with a "destination unreachable" packet. The resulting traffic can overwhelm the service.

- **ICMP (ping) flood example:** An Internet Control Message Protocol (ICMP) flood sends ICMP echo request packets (pings) to a host. Pings are common requests used to measure the connectivity of two servers. When a ping is sent, the server quickly responds. In a ping flood, however, an attacker uses an extensive series of pings to exhaust the incoming and outgoing bandwidth of the targeted server.

- **Application attacks:** Some examples of application-based DDoS attacks follow:

  - **HTTP flood example:** An HTTP flood is a Layer 7 application attack that uses botnets, often referred to as a "zombie army." In this type of attack, standard GET and POST requests flood a web server or application. The server is inundated with requests and may shut down. These attacks can be particularly difficult to detect because they appear as perfectly valid traffic.

  - **Slowloris example:** Named after the Asian primate, the Slowloris attack moves slowly. It sends small portions of an HTTP request to a server. These portions are sent at timed intervals, so the request does not time out, and the server waits for it to be completed. These unfinished requests exhaust server resources such as memory and affect the server's ability to handle legitimate requests.

- **Protocol attacks:** Some examples of protocol-based DDoS attacks follow:

  - **SYN flood example:** In a SYN flood attack, the attacker sends seemingly normal SYN requests to a server, which responds with a SYN-ACK (synchronized-acknowledgment) request. Typically, a client then sends back an ACK request, and a connection is made. In a SYN flood attack, the attacker does not respond with a final ACK. The server is left with a large number of unfinished SYN-ACK requests that burden the system.

  - **Ping of Death example:** In a Ping of Death attack, the attacker tries to crash or freeze a server by sending a normal ping request that is either fragmented or oversized. The standard size of an IPv4 header is 65,535 bytes. When a larger ping is sent, the targeted server will fragment the file. Later, when the server formulates a response, the reassembly of this larger file can cause a buffer overload, resulting in a crash.

The transport layer of the 5G network is more susceptible to volume-based attacks and protocol attacks. With 5G having customer premises equipment (CPE) with capacity of 1Gbps, a huge DDoS attack could occur if the CPE becomes the victim of a malicious actor who then attacks the network infrastructure, forcing the content servers, which are susceptible to high-volume and protocol attacks from CPE, to crash. The high traffic coming from the CPE also consumes the bandwidth of the back haul, thereby reducing the bandwidth available to the legitimate traffic.

DDoS protection/anti-DDoS mechanisms are usually deployed at the Internet peering or Gi/N6 interface, as illustrated in Figure 4-28. This deployment would help you secure your network from inbound DDoS attacks from the Internet.

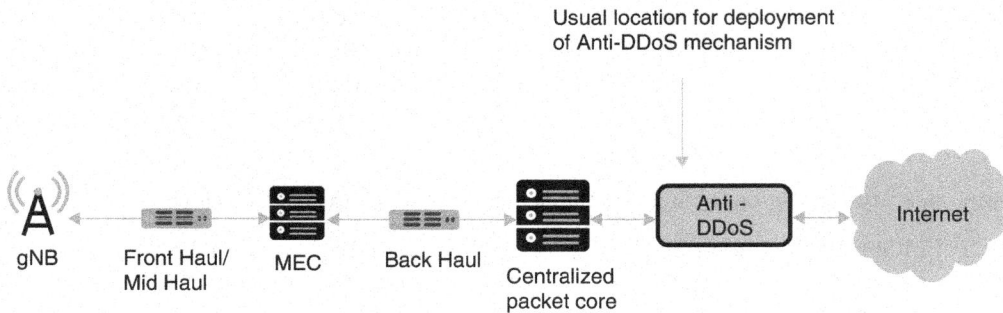

FIGURE 4-28   Typical Location of DDoS Protection Solution

One way to mitigate the DDoS threats from the CPE toward the content servers and inbound from the Internet, both at the edge and central DC layers, is to have anti-DDoS solutions deployed both at the MEC and central DC layer to protect the back haul, as illustrated in Figure 4-29.

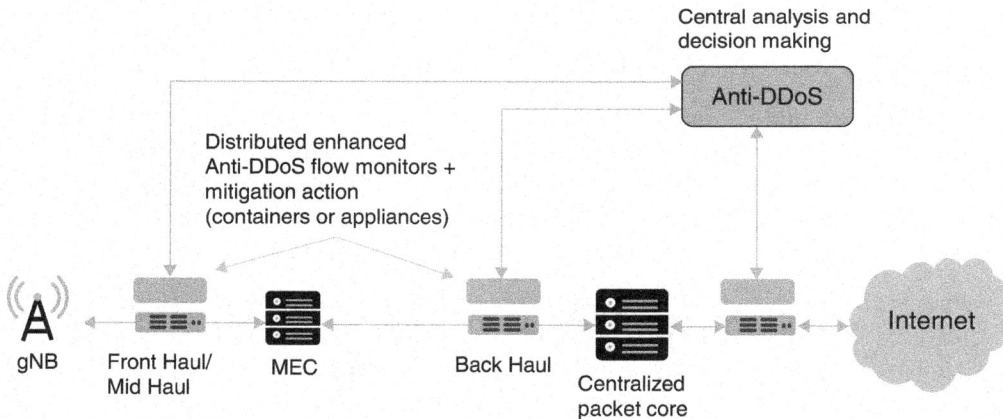

FIGURE 4-29   Distributed DDoS Protection Solution

For 5G-related DDoS attacks from various types of IoT devices (consumer, industrial, and so on), it is important that the service providers choose a DDoS solution that not only can mitigate the threats from the edge of the network but also have high accuracy of detection and mitigation as well as strong automation to ensure multivector mitigation actions.

Attacks are becoming completely automated and more sophisticated, making it difficult to defend against them manually. New techniques like burst attacks and advanced persistent denial of service (APDoS) demand advanced detection and mitigation and underscore the need for automation. Automation is at the core of a successful DDoS attack mitigation solution. To withstand the evolving threat landscape, service providers need to have anti-DDoS solutions with enhanced algorithms in place to shorten the response time and automatically respond to attacks.

With continuous monitoring and intelligence, service providers can detect a wide range of attacks before they wreak havoc on the network. These attacks/symptoms include the following:

- Behaviors related to zero-day malware

- Insider threats

- Advanced persistent threats (APTs)

- Distributed denial-of-service (DDoS) attempts

To enable detection of such malicious traffic, monitoring is required, not only on the traffic going in and out of the network but also on the lateral (east-west) traffic inside the network to identify network abuse and insider threats. Traditionally, known insider attacks are perpetrated on the network by individuals working for the organization. In 5G deployments, insider threats extend to IoT devices and industrial IoT devices (IIoT) that are deployed as a part of the Non-Public Network (NPN) or Public Network Integrated Non-Public Network (PNI-NPN), such as critical infrastructure verticals. These IoT/IIoT devices are authenticated to the network using 5G authentication mechanisms and are seen as part of the trusted network within a network slice, such as an IoT slice. Even if one of these devices or the network gateway controlling the devices is upgraded with an exploited patch that includes modified malicious code, the entire IoT/IIoT slice can cause forced paging, thereby consuming the resources and preventing legitimate devices from accessing the network.

Using the continuous monitoring method, an anomaly in the network traffic in the front haul, mid haul, back haul, or Internet peering can be detected. Once the anomaly is detected, the information can be sent to the central analysis and decision-making center, which can take a deeper look at the anomaly and, depending on the analysis, can command the distributed anti-DDoS devices to control the flow of the anomalous traffic. These distributed anti-DDoS devices, which are deployed at the edge of the network (such as the transport and MEC layer) are also referred to as *distributed flow controllers*, as their main function is to control the flow of the data traffic across the switches and routers in the transport and MEC layer.

For mitigating DDoS attacks based in UDP flood, you should use the anomaly detection solution, which detects any behavioral anomaly within your network, to identify the large number of inbound UDP packets on irregular ports and then decide the mitigation methods such as dropping/logging the packets automatically before the attack impacts the servers. This is not required if DDoS protection solutions are deployed in your network. The DDoS solution would automatically identify and mitigate DDoS attacks such as UDP flood attacks.

Real Scenario Case Study: Examples of Threat Surfaces and Their Mitigation

This section of the chapter takes you through some real scenarios seen in labs, proof-of-value deployments, and attacks seen in networks very similar to 5G deployments. At press time, 5G is still being deployed, and most of the networks are in nascent stages. Having this real-world view might provide you with that extra edge in understanding the exact threat vector so you can make the necessary reviews of the design you plan to deploy and then make the necessary enhancements to the security architecture.

Figure 4-30 shows three primary threat vectors:

- **A:** The attacker takes control of IoT devices with weak security and launches a DDoS attack.

- **B:** The attacker uses the vulnerability in insecure transport to use rogue eNBs/fake base stations and uses attacks such as MitM in the 5G deployment.

- **C:** The attacker uses the insecure transport and carries out MitM attacks in back haul.

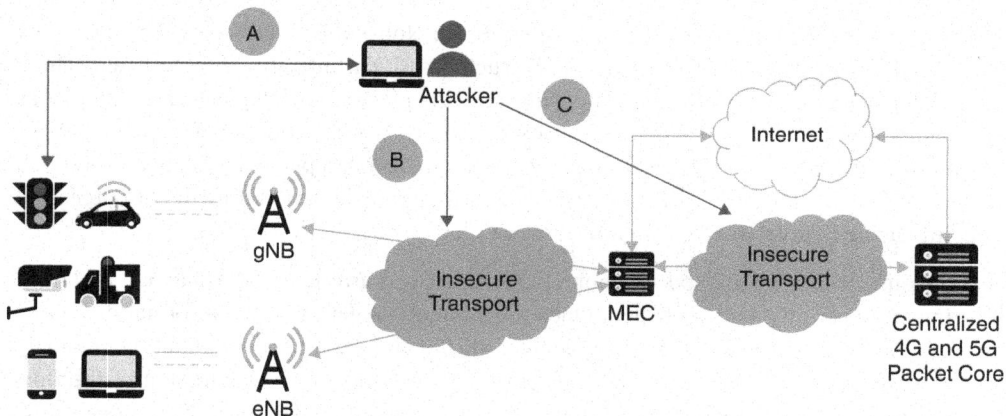

FIGURE 4-30   Example of Threat Vectors

The sections that follow analyze the threat vectors (A, B, and C) in more detail.

A: The Attacker Takes Control of IoT Devices with Weak Security and Launches DDoS Attack

DDoS attacks can be from subscriber 5G IoT terminals that are being served by the service provider, or they can be from external bots from the Internet trying to take down the service provider network infrastructure such as the DNS servers (also called inbound attacks), as illustrated in Figure 4-31.

Figure 4-31 shows the two key DDoS threat vectors: inbound DDoS attacks and DDoS attacks from internal DDoS devices served by the 3GPP and non-3GPP technologies served by the service provider.

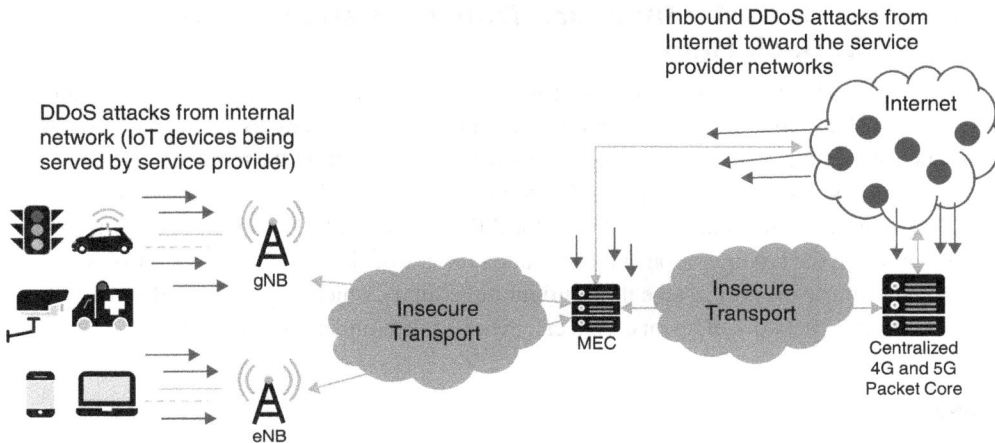

FIGURE 4-31   Internal Devices and Inbound DDoS Attacks

There are many variants of the inbound DDoS attacks. The UDP flood attack is one of them, where the attacker uses a botnet to send large amounts of traffic to the service provider's server, which could be a server deployed in the public or telco cloud infrastructure serving the subscribers with a specific application. UDP attacks are generally very rapid in nature, and the aim of such attacks is primarily to consume the available bandwidth allocated to the server rather than to exhaust server resources, which allows the attacker to prevent the bandwidth being allocated to the legitimate user. Successful UDP attacks usually follow a trend where a large number of packets arrive, each with different destination ports, and the server is forced to process each of the packets and also respond to each of the requests. The servers in such attacks could be taken down very fast, sometimes in minutes, leaving very little time for incident response teams to react to such attacks.

B: The Attacker Uses the Vulnerability in S1 and Insecure Transport to Use Rogue eNBs and Uses MitM Attacks in the 5G NSA Deployment

The capabilities of rogue/false base stations vary depending on whether the mobile network is GPRS, UMTS, LTE, or 5G. The 5G system in particular has already made significant improvements to combat false base stations, but as mentioned earlier in this section, there are still facets of this threat vector that need further validation to understand the new protection mechanisms and detect vulnerabilities within the new authentication mechanisms in 5G. Early 5G commercial networks are being deployed using 3GPP's NSA 5G specifications, meaning these networks are required to use the LTE control plane protocols and the LTE Evolved Packet Core network. 5G NSA networks based on LTE core networks have been deployed in many service providers and will continue to operate in the networks for more years, so any LTE threats and vulnerabilities will also exist in the 5G NSA network.

In Figure 4-30, the attacker exploits the vulnerabilities in the LTE S1 interface and the vulnerabilities in the insecure and untrusted transport network to attack the service provider's infrastructure.

C: The Attacker Uses the Insecure Transport and Carries Out MitM Attacks in Back Haul

In Figures 4-30 and 4-31, the attacker uses the insecure hardware and software within the transport network elements, such as a pre-aggregation or an aggregation router, leading to threats based on MitM threat vectors such as sniffing, data exfiltration, and modifying the packets in the transport. The programmable transport network devices also support API for integration with the SDN controllers and fabric layers. If the API is insecure, the attacker could use the API capability of the transport network element to make modifications to the configuration and cause it to drop traffic, leading to DoS or network outage. The attacker could also use the API to take the configuration dump, which might have information such as the IP address of other network elements. This can then be taken under control by the attacker.

Mitigation

This section takes you through the security controls to mitigate the threats shown in Figures 4-30 and 4-31.

Detection of Rogue Base Stations and False Base Stations

Figure 4-32 illustrates one of the mechanisms to detect the rogue/fake 4G or 5G base stations (eNB/gNB). This mechanism uses the key performance indicators (KPIs) calculated using the UE measurements. The UE would normally provide the received signal strength and quality, along with the location information, in measurement reports. These measurement reports are normally used by various tools, such as the RAN vendor's performance reporting solution, radio optimization tools, or self-organizing network (SON) solutions, to optimize the network, such as improving intra-RAT/inter-RAT handovers, reducing dropped call rates, improving signal quality by antenna down tilt and azimuth correction, and so on, thereby improving the subscriber experience.

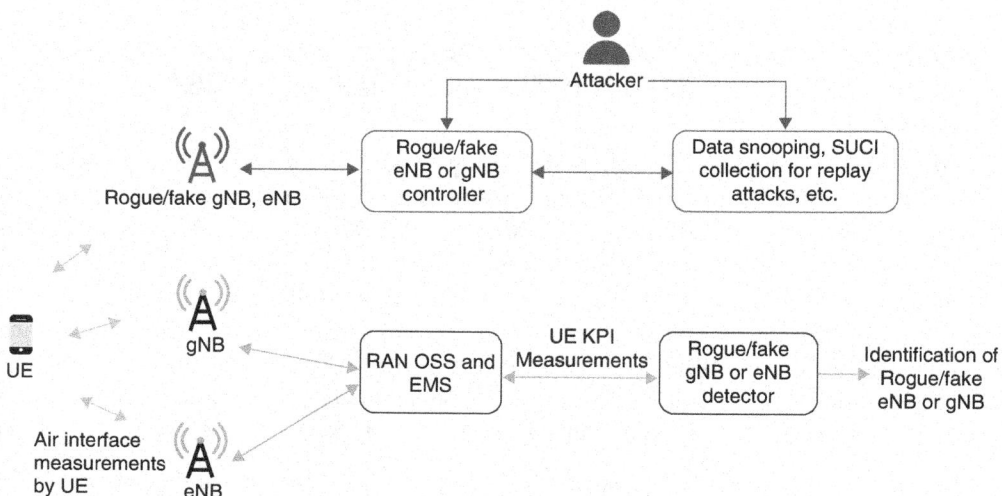

FIGURE 4-32   Detection of Rogue/Fake Base Stations

These UE measurements can also be put to use to detect any rogue or fake base stations. Some methods are explained in the list that follows:

- One example of detection could be because of random spikes in interference, specifically uplink (UL) interference. UL interference could be due to various reasons, including faulty repeaters or a high Voltage Standing Wave Ratio (VSWR) in legacy deployments. Rogue/fake base stations also cause high UL interference due to operating on the same frequency channel as the legitimate base station. Depending on the RAN vendor being used by the service provider, a customer KPI showing UL interference in a cluster or a KPI on the cell level would help to narrow down the possibility of a rogue/fake base station.

- Another method is to use the UE measurements to identify base stations (eNB/gNB) that are transmitting at an incorrect frequency or a higher power using the service provider's PLMN (public land mobile network). This could be due to unintentional misconfiguration or could be a rogue/fake base station trying to pump more power and forcing the UE in a specific location to select or hand over to the rogue/fake base station.

Once the cluster or approximate location is known, a spectrum analyzer can pinpoint the location of the suspected base station.

Deploy Trusted Hardware and Software

Service providers who plan to deploy programmable transport network elements for the 5G network rollout should ensure that the transport vendors being chosen to follow an internal secure development lifecycle, including software (including API) and hardware hardening to mitigate any possibility of attackers trying to take control of the programmable network elements. Figure 4-33 illustrates how trusted hardware and software along with SZTP implementation protects your transport deployments.

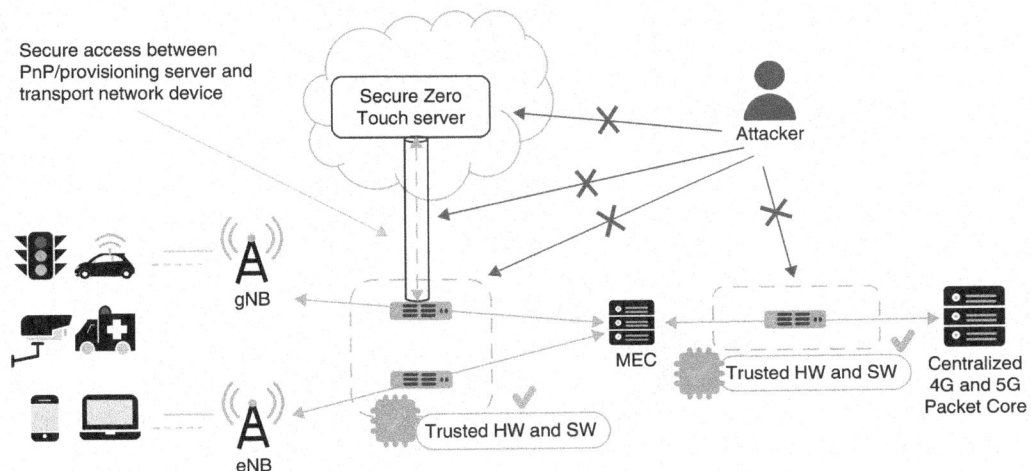

FIGURE 4-33   Secure ZTP and Trusted HW and SW

A majority of transport solution vendors today support different variants of ZTP and API interfaces to allow seamless integration with the SDN network and to allow easier deployment and provide regular updates of transport layer network elements.

Following the SZTP process during pre-deployment, deployment, and operation explained in detail earlier in this section, service providers would mitigate the threat vector of attacks from attackers trying to exploit vulnerabilities in the transport access devices. If the service provider overlooks security controls like using trusted hardware and software, secure SZTP servers, and secure communication between the transport access device and SZTP servers, it would be very difficult to contain the rising threat vectors.

In other words, it is imperative that the security controls discussed are deployed by the service providers planning to deploy new transport access devices for 5G. If there are existing transport access devices that do not have trusted components or use a secured bootstrap or provisioning servers, it's recommended not to use them for 5G deployments or integration with the SDN controller used for 5G network slicing.

Before selecting the 5G equipment vendor, it is very important that you check whether the vendor provides you with the mechanism to ensure that you can verify the integrity of the hardware and software being deployed. One of the mechanisms would be a management layer of the solution that shows you if the secure booting process is successful and maintains a list of the upgrade and patch history for the equipment.

Granular Access Control

Apart from the access control for the RAN and transport elements, service providers must also ensure that the RAN and transport devices support role-based access control (RBAC) and that granular policies can be applied to the RAN and transport network elements. The RBAC controls should be further integrated with multifactor authentication (MFA) solutions to provide granular user and device policy enforcement. Having this layer of protection will provide granular mapping between the vendors and the configuration engineers, thereby only allowing configuration engineers of specific vendors access to their equipment. One way the access controls can be implemented is by using a global Identity and Access Management (IAM) solution integrated with a global policy engine where you can configure the user policies. The IAM solution should then be integrated with an MFA solution to ensure that each and every user goes through the secondary authentication mechanism to provide an added layer of access security. The RAN disaggregation into Distributed Unit (DU) and Centralized Unit (CU) requires a deeper look into the dynamic capabilities of deployment, which will require integration with the orchestration and automation layer. Orchestration and automation, though relevant in 4G, are more relevant in 5G due to the disaggregation and virtualization of the 5G network functions. The security controls around virtualization are better explained in the chapters "Securing MEC Deployments in 5G" and "Securing Virtualized 5G Core Deployments." Orchestration is very important for the 5G-related transport access network due to the closer integration required with the SDN fabric for network slicing. Here is where proper strategy needs to be made around orchestration capabilities and dependencies around the transport network access devices. Due to the relevance of orchestration in the RAN and transport layer, the access control on the orchestration should be applied at a granular level. Figure 4-34 illustrates the granular control required, which should cater to different parts of the network.

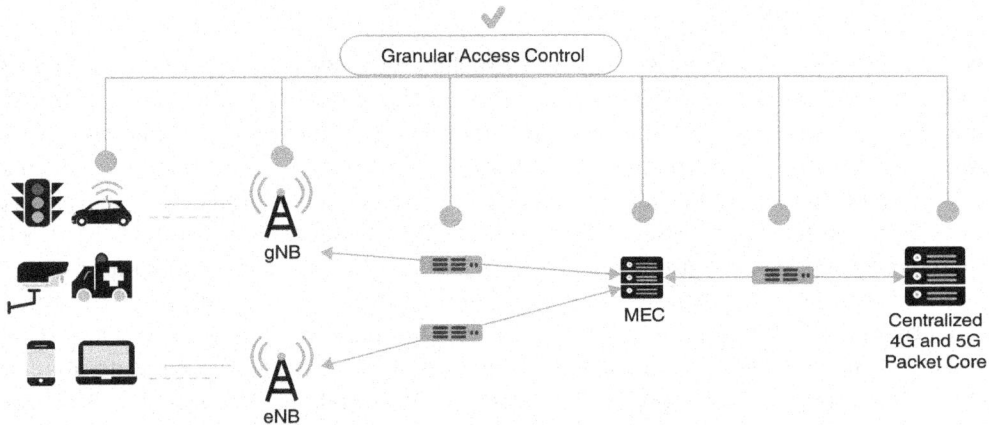

FIGURE 4-34   Granular Access Control

Security Gateway

The topic of rogue/fake 4G and 5G RAN nodes has been discussed in detail in earlier sections. A security gateway uses digital certificates provided by the service provider's PKI infrastructure for authentication. Figure 4-35 illustrates the mitigation of fake/rogue gNB/eNB attacks by using a SecGW.

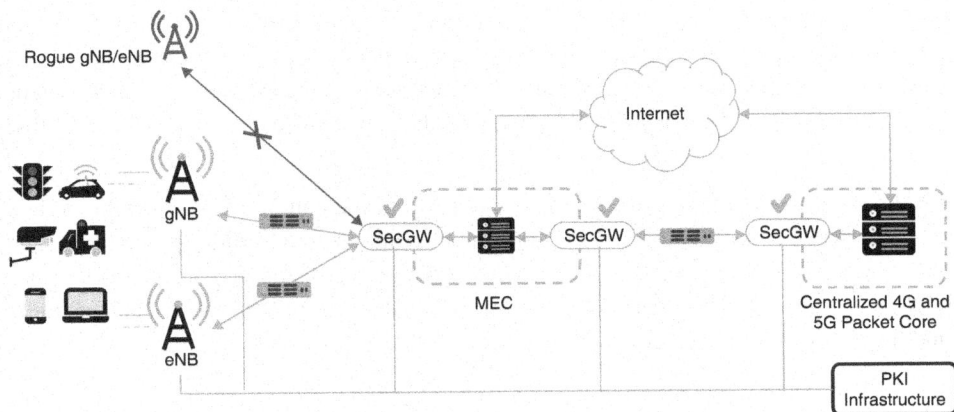

FIGURE 4-35   Security Gateway Mitigating Rogue/Fake RAN Nodes

SecGW deployment between the radio access nodes (eNB, gNB, pico cells, small cells, and so on) and the mobile packet core provides both authentication and encryption. 3GPP has specified mandatory support of the IPsec stack for 4G and 5G radio access nodes, which means you would be required to deploy SecGW only at the mobile packet core location. SecGW is not required at the radio access node location, as the IPsec stack within the radio access node will perform the IPsec tunnel initiation and termination to and from the SecGW.

The radio nodes (eNB/gNB) are pre-provisioned with a public-private key pair by the RAN vendor and have the vendor-signed certificate of their public key pre-installed. Public key infrastructure (PKI) is used to deploy a service provider–signed certificate within the radio node and the service provider–root certificate within the SecGW. The service provider normally uses an RA/CA, which can be a standalone CA with Certificate Management Protocol version 2 (CMPv2). A key point here for you to check is the support of CMPv2 protocols by the SecGW vendor. The CMPv2 protocol, based on RFC 4210, is a PKI protocol for managing X.509 certificates. CMPv2 is used between Certification Authorities (CAs), Registration Authorities (RAs), and end entities (EEs). In CMPv2, most communication is initiated by the EEs, where every CMP request triggers a CMP response message from the CA or RA. This provides a means for initial registration of end entities, key pair update and certificate updates for end entities and CAs, cross-certification between CAs, certificate revocation management, and discovery of certificates and certificate revocation lists (CRLs). CMPv2 provides better lifecycle management of the SecGW by providing functionalities such as certificate revocation and re-enrollment of certificates. For example, if any of the RAN nodes is found to be exploited, the certificate of all RAN nodes can be revoked, and new certificates can be enrolled by using CMPv2 to prevent the attackers getting access to the network using the previously exploited certificate.

The mutual authentication process is carried out between the radio nodes and the SecGW using the X.509 certificates during the IPsec tunnel setup process. In the mutual authentication process, the radio nodes will validate the certificate of the SecGW to ensure that it is connecting to the right network. The SecGW also validates the certificate of the radio node to verify its identity. Once the SecGW and the access nodes are authenticated, they will treat each other as trusted nodes and will establish an IPsec tunnel between them and provide encryption to the traffic between the radio access nodes and the mobile packet core network. This process of mutual authentication cannot be passed by the rogue access nodes, as it does not have the certificates deployed in the legitimate access nodes by your PKI infrastructure.

IKE, ESP, and AH security associations use secret keys to encrypt the data traffic for a finite period, which limits the entire security association (SA). To ensure interruption-free traffic, a procedure called *rekeying* is used, where a new SA takes the place of the expiring SA. This rekeying procedure also makes it much harder for the attacker to derive the secret keys used to encrypt the traffic between radio access nodes and the SecGW.

DDoS Protection

As discussed earlier, 5G brings in a growing number of massive IoT (mIoT) and industrial IoT (IIoT) use cases, which means that the attackers can use them to attack the service provider's network. DDoS attacks can be inbound and outbound, as explained in the section "The Attacker Takes Control of IoT Devices with Weak Security and Launches DDoS Attack."

To secure the network from the inbound attacks from the Internet and the attacks from the consumer IoT devices, which are getting more sophisticated by the day, it is recommended to have a central and distributed protection against DDoS attacks. As illustrated in Figure 4-36, the DDoS protection provided at the edge (such as the MEC layer or the access and aggregation layer) is called distributed Anti-DDoS protection, while the DDoS protection delivered at the centralized DC is called the centralized Anti-DDoS protection.

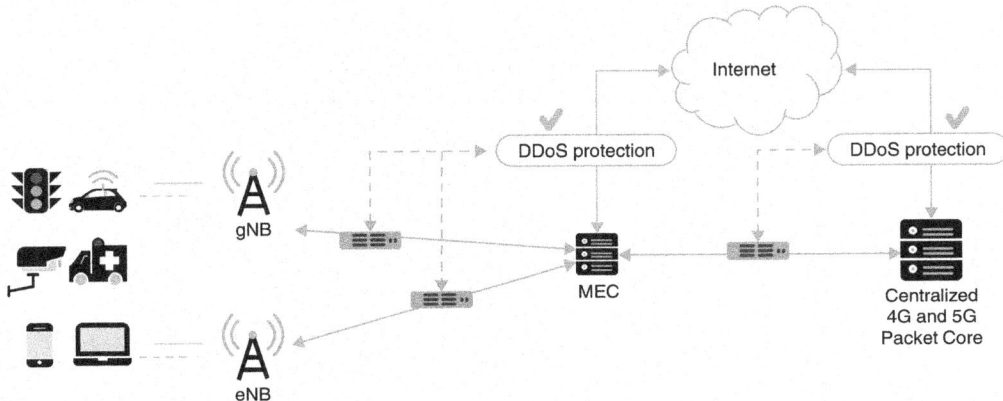

FIGURE 4-36   Centralized and Distributed Anti-DDoS Protection Solution

Figure 4-37 illustrates the distributed Anti-DDoS mechanism to mitigate the DDoS attack. It allows only legitimate traffic toward the network, thereby securing the 5G infrastructure from inbound attacks from the Internet.

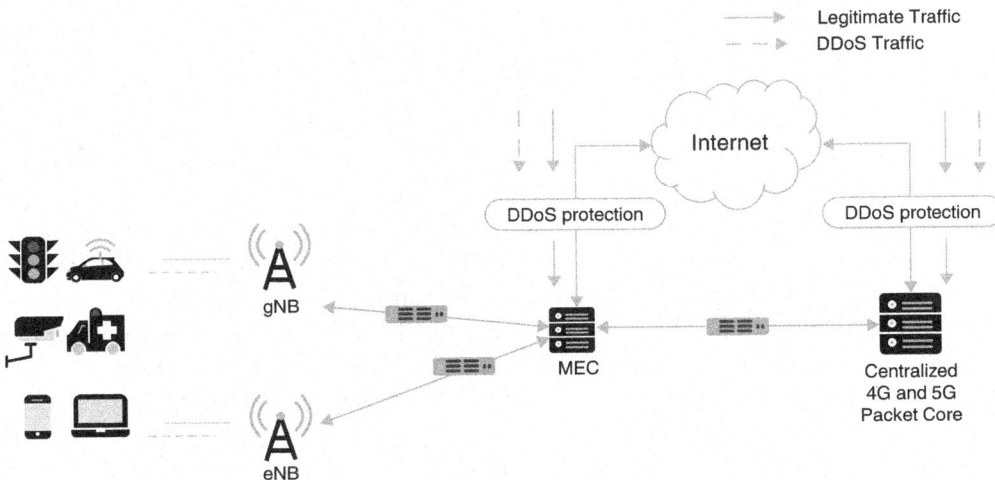

FIGURE 4-37   Inbound DDoS Attack Mitigation

As seen from Figure 4-37, the DDoS protection solution should allow mitigation against volumetric and application-based DDoS attacks in different portions of the service provider's network and prevent impacting legitimate traffic.

As seen in Figure 4-38, the distributed DDoS protection solutions using behavioral-based detection and real-time signature creation algorithms that mitigate the DDoS threat vectors at the edge are recommended to be deployed at the edge of the network, such as in the aggregation and pre-aggregation layers of transport or the MEC layer. These algorithms should create baselines of normal network, application, and user behavior and use these baselines to notice abnormal traffic and accurately detect attacks. When a new, previously unknown zero-day attack is detected, the solution should create a signature in real time that uses the attack characteristics and detects known and unknown attacks in a short timeframe, with minimal false positives.

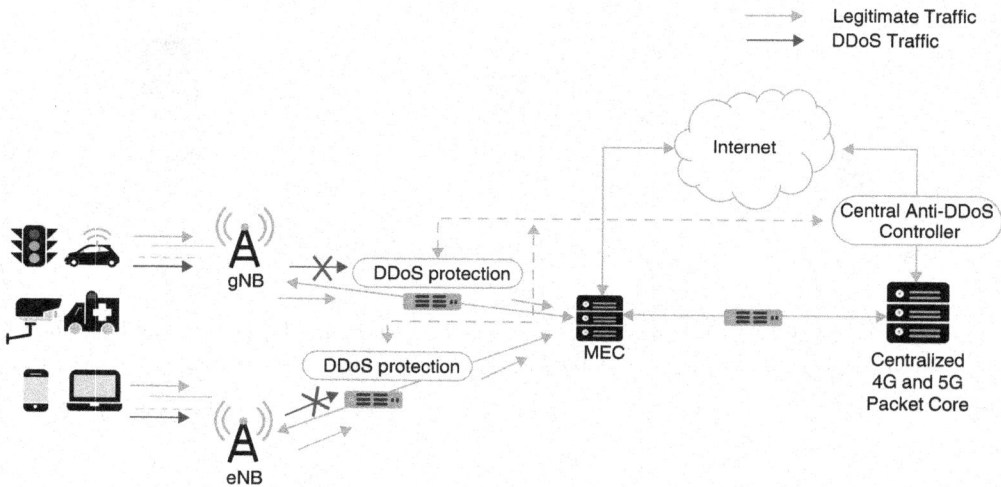

FIGURE 4-38   DDoS Attack Mitigation Using Distributed Anti-DDoS Protection

Security Analytics and Monitoring

5G deployments will see stronger adoption of APIs, not only for inter-5G core network function communication, but also for monitoring use cases. It can be seen as an actual replacement of Simple Network Management Protocol (SNMP). The traditional methods of monitoring service provider networks using polling methods such as SNMP, Syslog, and CLI use a pull model to request information at regular intervals. The data collected might help to efficiently monitor the network initially; however, as the network grows in complexity and scale, this method of data collection would be insufficient for efficient and effective monitoring. Additionally, the polling methods are resource intensive, and network operators face information gaps in the data they collect.

To mitigate the example threat vectors mentioned in this section, the recommended practice is to have the multivendor telemetry collection from the transport and the RAN infrastructure. Generally, the anomaly detection systems would consume the data collected for around 15 days and generate a baseline for the network. Once the baseline is established, the security analytics and monitoring solution can detect the anomalies. For example, as shown Figure 4-39, by detecting the IP address being allocated toward the RAN infrastructure, any new host requesting a new IP address can be detected because it shows up as an anomaly. The same applies for the transport layer anomalies.

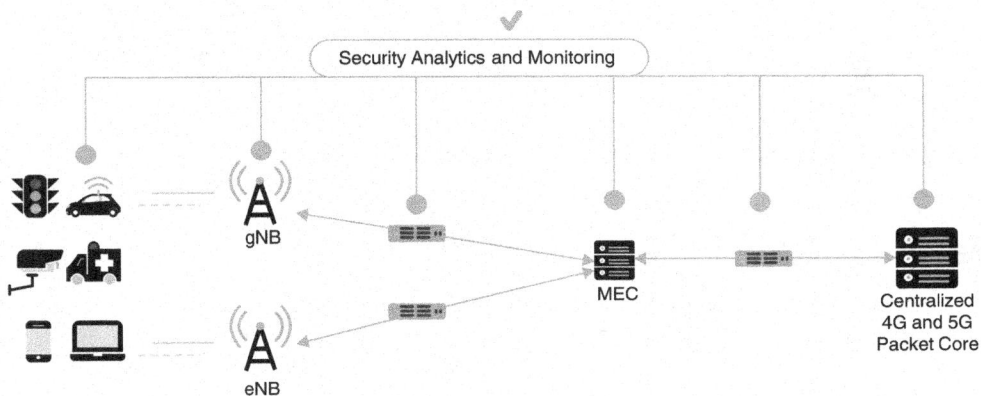

FIGURE 4-39   Security Analytics and Monitoring

To avoid each and every new IP address as an anomaly, the anomaly detection system has a list of known IP address allocated in the network and also updates itself with any known or newly configured IP address using its integration with the existing Network Planning and Provisioning solutions already implemented in your network. These solutions would typically have all the IP addresses allocated to network devices after being authenticated by your Network Access Control (NAC) solution, which enforces policies on devices that access your network or are deployed in your network. The anomaly solution then compares any new IP address to the list of planned or known IP addresses and identifies only the unknown or unplanned IP addresses as anomalies. To reduce the collection of network telemetry from all network switches and routers, it is recommended to collect telemetry from switches and routers that are deployed at aggregation points.

Figure 4-40 shows the network telemetry collection from the network devices in the access and aggregation layer, MEC layer, and the back haul devices. The security analytics and monitoring solution should also create alerts based on any anomaly in network behavior.

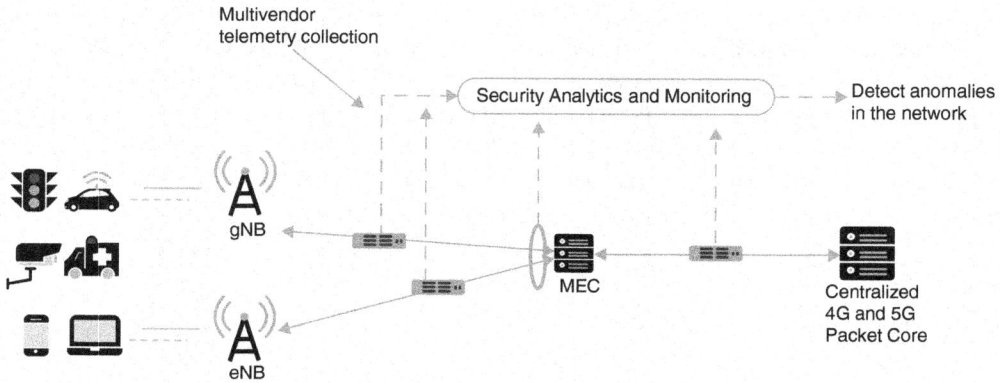

FIGURE 4-40   Security Analytics and Monitoring in RAN and Transport

Summary

As a summary for securing the RAN and Transport infrastructure, the highlighted security controls in Figure 4-41 are recommended.

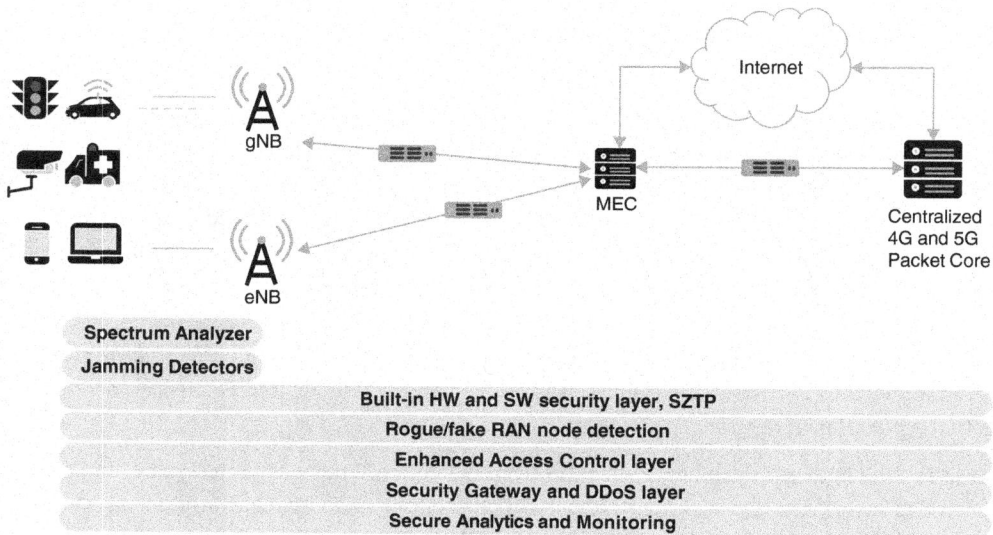

FIGURE 4-41   Multilayered Security Controls for RAN and Transport

As Figure 4-41 illustrates, a range of security controls is required to secure the air interface, RAN, and Transport layer. Just having a single security control is not enough to cater for the RAN and transport security. The security controls highlighted in Figure 4-41 are not unique only to the RAN and Transport layer but can be re-applied to other domains such as DC and 5GC.

Here's a summary of the controls for securing the air interface, RAN, and transport layer:

- **Built-in hardening:** The built-in security of the device based on hardware (HW) and software (SW) hardening is quite critical to prevent DDoS attacks happening within the network. This could happen due to updates that could be infiltrated with malicious code, which might then cause vulnerabilities in all the transport devices being updated. This vulnerability could then be exploited by the attackers. This makes inherent hardware and software hardening by vendors very critical and is strongly recommended to be used by the vendors being chosen for the transport access devices.

- **Spectrum analyzer and the jamming detectors:** These security controls are unique to the air interface layer and would help in the detection of the air interface jamming systems.

- **Rogue/fake node detection:** The detection of the rogue/fake RAN node is made possible by monitoring the KPI and the UE measurements and then detecting the anomalies, as mentioned earlier in the chapter. The detection of fake nodes applies to the transport nodes, and security policies such as communication only between trusted devices should be applied in the network. This is provided by some vendors offering transport access and aggregation routers that check whether or not the next-hop router is trusted. The data packet would be passed to the next hop only if it's trusted.

- **SZTP:** It is recommended that Secure Zero-Touch Provisioning (SZTP) be used for all devices being deployed in the 5G infrastructure. For this section, it includes the RAN components depending on deployment, such as the gNB, DU, and CU, and the transport elements, such as front-haul, mid-haul, and back-haul routers.

- **Granular access control layer:** Granular access controls following Zero Trust principles are recommended to be included as part of the initial deployment plan for the RAN and transport layer. Role-based access control (RBAC) should be provided by the vendor, which can be integrated with the multifactor authentication (MFA) solution. The MFA solution should check the health of the device before the user and the device are allowed access into the network. Specifically, for virtualized components, such as any applications deployed on the programmable access and aggregation routers, it is important to have the user-to-application mapping. It should have a default rule of rejecting any request by default and allowing only specific user requests, which includes rules for API calls.

■ **Secure analytics and monitoring:** The secure analytics and monitoring layer should include the anomaly detection algorithms and should be able to ingest global threat intelligence. This solution should be able to consume multivendor flows such as NetFlow, jFlow, sFlow, IPFIX, Syslog, and so on, and baseline the network after observing the network for 2 to 3 weeks. This includes the user accessing specific devices or applications. Once the baseline is finalized, it will used as a threshold for normal behavior of the network. Any variation from this baseline will be considered an anomaly and should be brought to the attention of the relevant teams. The monitoring should also be catered for encrypted traffic without having to decrypt it. This is provided by vendors in the market today and is basically done by understanding more details in the header of the packet.

Acronym Key

Acronym	Expansion
5GC	5G core
AH	Authentication Header
AKA	Authentication and Key Agreement
AS	Access Stratum
CNF	Cloud-Native Function
CP	Control plane
CU	Centralized unit
CUPS	Control Plane User Plane Separation
C-RAN	Cloud RAN or Centralized RAN or Central RAN
DDoS	Distributed denial of service
DL	Downlink
DoS	Denial of service
DU	Distributed Unit
EAP	Extensible Authentication Protocol
ESP	Encapsulation Security Payload
FH	Front haul
GUTI	Global Unique Temporary Identifier
IKE	Internet Key Exchange
IMSI	International Mobile Subscriber Identity
IPsec	IP Security
ISAKMP	Internet Security Association and Key Management Protocol
MAC	Medium Access Control
MEC	Multi-access edge compute

Acronym	Expansion
MitM	Man in the middle
mMTC	Massive machine type communication
NAS	Non–Access Stratum
NAT	Network Address Translation
NGAP	Next-Generation Application Protocol
NR	New Radio
NSA	Non-standalone
OAM	Operations Administration Management
O-RAN	Open RAN
O-RU	O-RAN Compliant Radio Unit
O-DU	O-RAN Compliant Distributed Unit
O-CU	O-RAN Compliant Centralized Unit
PFCP	Packet Forwarding Control Protocol
RAN	Radio access network
RAT	Radio access technology
RLC	Radio Link Control
RRC	Radio Resource Control
SA	Standalone
SCTP	Stream Control Transmission Protocol
SDN	Software-defined network
SecGW	Security gateway
SEG	Security gateway
SUCI	Subscription Concealed Identifier
SUPI	Subscription Permanent Identifier
SZTP	Secure Zero-Touch Provisioning
UE	User equipment
UL	Uplink
UP	User plane
URLLC	Ultra-reliable low-latency communication
VRAN	Virtual RAN
ZTP	Zero-Touch Provisioning

References

IETF RFC-4303, "IP Encapsulating Security Payload (ESP)"

IETF RFC 8221, "Cryptographic Algorithm Implementation Requirements and Usage Guidance for Encapsulating Security Payload (ESP) and Authentication Header (AH)"

3GPP TS 33.501, "Security architecture and procedures for 5G system," Release 16, v 16.5.0, December 2020

3GPP TS 33.201 V16.4.0, "Network Domain Security (NDS); IP network layer security," Release 16

3GPP TS 33.809 V0.11.0, "Study on 5G Security Enhancement against False Base Stations (FBS)," Release 17

O-RAN Alliance white paper, "O-RAN Minimum Viable Plan and Acceleration towards Commercialization," June 2021

Pramod Nair, "Securing 5G and Evolving Architectures," Cisco Knowledge Networks, Cisco Public, September 17, 2020

Pramod Nair, "Why 5G is changing our approach to security," Cisco Public, June 10, 2020

https://www.cisco.com/c/dam/en/us/solutions/collateral/service-provider/mobile-internet/solution-overview-c22-741906.pdf

O-RAN Security Threat Modeling and Remediation Analysis 1.0, March 2021

Chapter 5

Securing MEC Deployments in 5G

After reading this chapter, you should have a better understanding of the following topics:

- Key threat surfaces in different models of 5G MEC deployments
- Key security controls for securing 5G MEC
- Real scenario case study examples on MEC threat surfaces and threat mitigation techniques

In the previous chapters, we discussed the security controls to mitigate threat surfaces in 5G radio access network (RAN) deployments. This chapter will take you through the threat surfaces in 5G multi-access edge computing (MEC) deployments and mechanisms to mitigate the threats.

This chapter will be of particular interest to the following teams from enterprise 5G MEC consumers, 5G service providers, and cybersecurity vendors providing security layers for MEC deployments:

- Mobile infrastructure strategy teams of service provider and enterprise verticals deploying 5G
- Security strategy teams within service provider and enterprise verticals deploying 5G
- Radio planning and optimization teams within service providers and private 5G enterprises
- Transmission and the packet core team within service providers and private 5G enterprises
- Cloud computing and data center teams involved with 5G strategy and deployment
- Security architects and design teams looking at securing the mobile infrastructure
- Enterprise solution and security architects deploying 5G MEC on enterprise premises
- Enterprise solution and security architects using 5G MEC from a service provider
- Defense personnel deploying 5G MEC for defense use cases
- Government departments deploying 5G MEC

- Cybersecurity vendor teams looking to secure MEC deployments for their customers

- Product managers of cybersecurity vendors

- Consultants working on MEC strategy

Multi-access edge computing (MEC) is an architecture concept defined by the European Telecommunications Standards Institute (ETSI) that enables deployment of cloud computing and IT services at the edge of the network. The fundamental idea here is that moving the application and the processing of the traffic closer to the user reduces the latency for the end user of the application.

MEC is technology-agnostic and opens up the ecosystem to many different areas, such as cloud computing, software-defined networking (SDN), application developers working on artificial intelligence (AI), Internet of Things (IoT), and the delivery of content with local cache and real-time analysis of data, by opening up the applications deployed on the network edge to the third-party application ecosystem in the public cloud. This is done as MEC architecture specified by ETSI uses open interfaces such as Representational State Transfer (REST) application programming interfaces (APIs) for the exchange of data.

MEC discussions are becoming more prevalent within service providers due to 5G use cases such as massive IoT (mIoT) and machine-to-machine (M2M). MEC enables disaggregation and distribution of 5G RAN and 5GC (5G packet core). You can now deploy the 5G RAN functions such as Distributed Unit (DU) and Centralized Unit (CU) and the user plane 5GC functions such as User Plane Function (UPF) in the edge of the network to provide faster and high-bandwidth applications to its subscribers. These 5G functions require to be deployed dynamically for optimal use of resources, which is enabled with features such as virtualization, SDN, and orchestration.

MEC is also becoming more prevalent in enterprise networks, as the enterprise infrastructures are growing in complexity, accommodating more and more diverse, end-user devices and IoT connections. Today's applications are more interactive and bandwidth hungry, generating massive amounts of data that supports real-time analytics and problem solving. Many enterprises have adopted multicloud strategies with unified management solutions to support microservices and containerized applications at the network edge with more distributed and intelligent edge network capabilities with constantly evolving security. Edge networks and computing allow enterprise architectures to optimize processing for business-critical analysis of data sets from IoT applications and communications.

The MEC ecosystem also includes container-based platforms implemented on OpenStack and other cloud-native container platforms, enabling open source edge computing and IoT cloud platform optimized for low-latency and high-performance applications. There are lots of innovations happening in MEC related technologies such as MEC service-chaining. As an example of MEC service-chaining, instead of having separate hardware equipment specific for security controls such as Firewalls, DDoS protection, load balancers and so on, all these security controls and network functions such as UPF can be deployed as virtual components on a scalable commercial off the shelf (COTS) server with integrated orchestration and automation components. This brings in advantages related to lower energy consumption when compared to traditional network deployments.

Here are the key advantages of following the MEC architecture deployment:

- Enabling optimized DC deployment using distributed cloud computing and IT service in edge functionalities of MEC

- Enabling ultra-reliable low-latency communication (URLLC) use case deployment, which targets latency of 1ms

- Reducing the traffic in the back haul by optimized traffic offload in the MEC design, allowing Direct Internet Access (DIA) and Local Break Out (LBO)

- Better quality of experience (QoE) for end customers by utilizing the edge cache functionality using MEC

Two primary MEC architectures are being planned by service providers and enterprises:

- Service provider network-based MEC

- Enterprise network-based MEC

The sections that follow describe these in more detail.

Service Provider Network-Based MEC

This model is used by service providers to enable 4G and 5G networks. Here, user plane functions and certain MEC applications are deployed in the edge to allow better QoE for subscribers as well as optimized traffic volume in the back haul by using the DIA and LBO functionalities of the MEC. In telco environments, OpenStack is seen as the main platform to host the container running in the environment.

Figure 5-1 shows the key components of service provider network-based MEC.

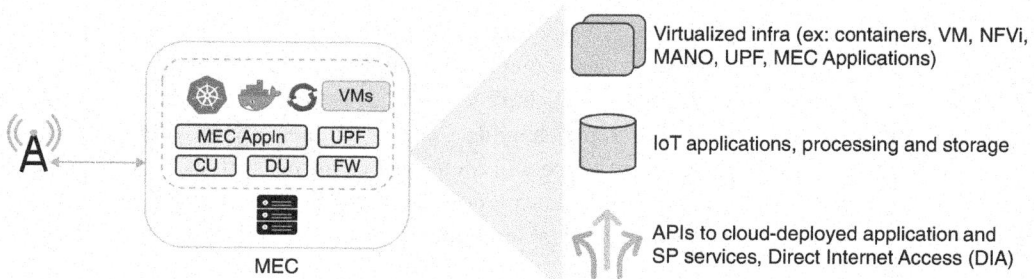

FIGURE 5-1 Service Provider Network-Based MEC in 5G

Examples of use cases include the following:

■ Enabling distributed CUPS deployment of 4G

■ Enabling distributed, disaggregated RAN components—DU and CU

■ Enabling distributed deployment of 5GC by enabling virtualized UPF deployment closer to the edge

■ Efficient deployment of IoT by enabling data processing and "noisy signaling" IoT use cases at the edge instead of the cloud

■ Deployment of URLLC use cases at the edge

■ Deployment of vehicle-to-everything (V2X) applications in the edge

■ Deployment of applications and services at the edge for monetization

■ Enabling high-bandwidth use cases such as video streaming in eMBB

Enterprise Network-Based MEC

The enterprise network-based MEC model is used by enterprises to fulfill use cases such as deploying private 5G (non-public 5G networks) to reuse the service provider's infrastructure. It can also be used by service providers to offer certain services from the edge to the enterprises.

Figure 5-2 shows the key components of enterprise network-based MEC.

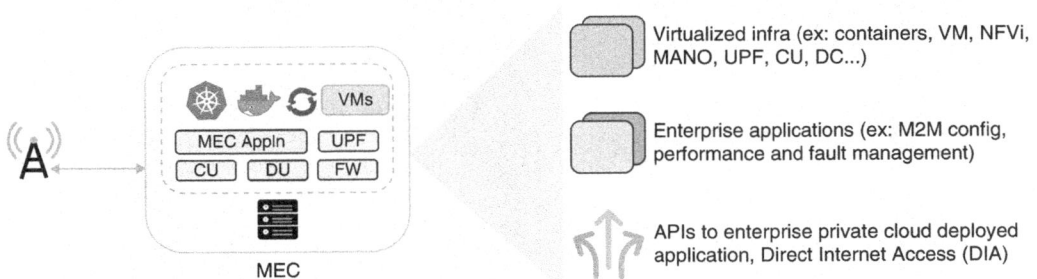

FIGURE 5-2   Enterprise Network-Based MEC in 5G

Example of use cases include the following:

■ Edge deployment of IoT devices with high-traffic volume in enterprise networks.

■ M2M/robotics use cases in manufacturing sector.

■ Industrial IoT use cases such as real-time information.

■ Streaming content use cases in the enterprise, including augmented reality (AR) and virtual reality (VR).

■ Location-limited use cases such as assets restricted to certain locations.

■ Healthcare use cases where data privacy is important and deployed on the edge network of the healthcare vertical. Only non-private data is sent between the edge and the central DC.

■ Smart city use cases.

■ AI and edge computing use cases focusing on leveraging AI to enhance automated processes and analytics data interpretation necessary to support ultra-low latency application delivery and to alleviate network congestion in core data centers.

MEC Deployment Models

MEC is not a new concept for 5G. It was used in distributed mode deployment of 4G Control Plane User Plane Separation (CUPS). 4G CUPS gave us the option of separating the user plane (UP) and control plane (CP). If the UP and CP are deployed in the same location, it is called a *co-located CUPS* architecture. It could also be deployed in distributed mode; if the user plane is deployed closer to the edge of the architecture and the control plane is deployed at the centralized data center (DC), it is then called a *distributed CUPS* deployment. The control plane, which is less sensitive to latency, was deployed in the centralized data center, and the latency-sensitive user plane was deployed in the edge. Some use cases of IoT, which are not latency sensitive, can have the user plane deployed at the central DC as well.

Figure 5-3 shows the key control plane and user plane components of 4G architecture from a RAN integration point of view.

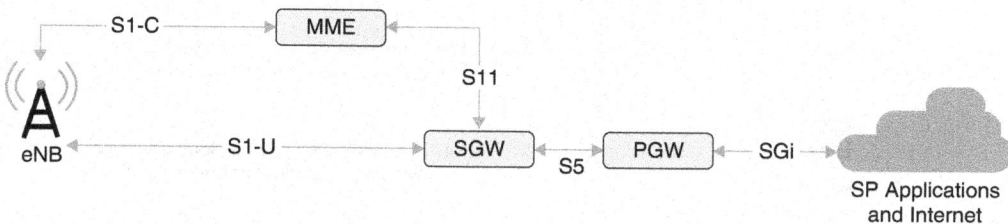

FIGURE 5-3   Key Control Plane and User Plane Connections for 4G RAN

Figure 5-3 depicts the control plane and user plane in 4G architecture. The S1 interface in LTE is used between LTE RAN nodes eNodeB (eNBs) and the Evolved Packet Core (EPC) nodes MME and S-GW. The S1 interface includes the S1-CP (control plane) and S1-UP part (user plane). The signaling transport on S1-CP will be based on SCTP. The signaling protocol for S1 is called S1-AP. The user plane interface is based on GTP User Data Tunneling (GTP-U).

Figure 5-4 shows the key Control Plane and User Plane Separation (CUPS) of the 4G architecture.

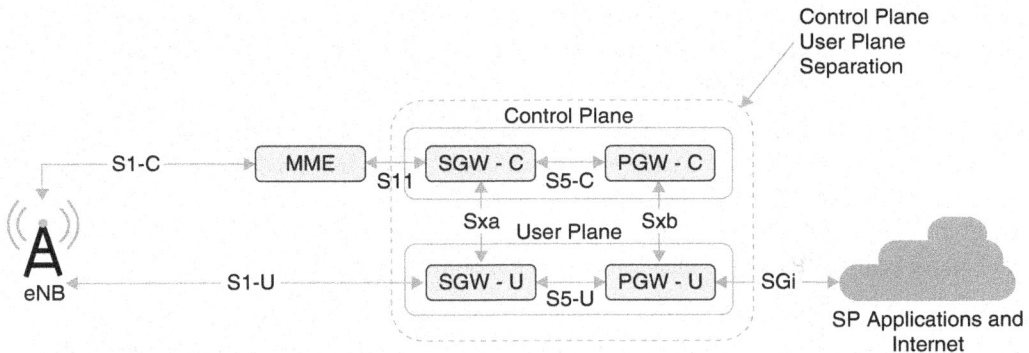

FIGURE 5-4   Control Plane and User Plane Separation in 4G

The mobile network today started seeing increased variation in resource demands across the control and user planes. Some devices are high control plane resources (signaling), while other devices might require very little signaling resources but substantial user plane and data resources. As Figure 5-4 illustrates, the introduction of CUPS addressed this to a certain degree. CUPS offers the capability to truly separate and independently scale the control plane and user plane of the mobile network on an as-needed basis and in real time. You now have the flexibility to specialize the user plane for different applications, without incurring an associated cost of a dedicated control plane for each application. New use cases such as remotely operated machinery, edge computing, and ultra-high per-user throughput will now be possible due to this distribution of control plane and user plane as well as the flexibility to deploy closer to the edge. Finally, CUPS enables you to optimize data center costs by hosting the control plane and the user plane in different geographic locations as well as save on back-haul costs by terminating data at the edge of the network.

In a CUPS deployment, the SGW and PGW are decomposed into separate control plane and user plane components. The SGW is decomposed into SGW-C and SGW-U, and the PGW is decomposed into PGW-C and PGW-U. Figure 5-4 also illustrates the involvement of new network elements for the control plane, such as SGW-C and PGW-C, and the user plane, such as SGW-U and PGW-U. The functions of these new network elements in 4G CUPS are listed here:

- **SGW Control Plane (SGW-C):** SGW-C performs the control plane functionality for the assigned SGW-U in a 4G CUPS deployment. The user equipment (UE) is served by a single SGW-C, but multiple SGW-Us can be selected for different Packet Data Network (PDN) connections. SGW-C also supports triggering downlink data buffering while the subscriber is being paged, PGW pause of charging management based on implemented policies (failed paging, abnormal radio link release, and dropped packets/bytes at SGW-U), and lawful interception of subscriber traffic. It supports the Sx interface, which enables CUPS between SGW-C and SGW-U.

- **SGW User Plane (SGW-U):** SGW-U performs the user plane functionality of the SGW in a CUPS deployment. It consists of an S1 data plane stack to support the S1 user plane with eNodeB. SGW-U also supports triggering downlink data buffering while the subscriber is being paged and lawful interception of subscriber traffic. It supports the Sx interface, which enables CUPS between SGW-C and SGW-U.

- **PGW Control Plane (PGW-C):** PGW-C performs the control plane functionality of the PGW in a CUPS deployment. When a subscriber establishes an EPS (Evolved Packet System) bearer to a given PDN, the PGW-C selects and controls the point of attachment to that PDN for the life of the EPS bearer. PGW-C caters for resource management for bearer resources, bearer binding, subscriber IP address management, and mobility support.

- **PGW User Plane (PGW-U):** PGW-U serves as user plane functionality of the PGW in a CUPS deployment. When a subscriber establishes an EPS bearer to a given PDN, the PGW-U under the control of the PGW-C serves as the point of attachment to that PDN for the life of the EPS bearer.

For use cases that are low-latency dependent, you can deploy the user plane functions such as SGW-U and PGW-U closer to the edge in a remote data center or MEC node so that the control plane functions can be deployed in the centralized DC. This is because the latency is user plane latency dependent and less sensitive to control plane latency. This CUPS deployment will allow you to implement the use case by keeping the cost in check, as the resources in the remote DC or MEC node are not consumed for control plane functions. Expensive upgrades such as back-haul upgrade can be avoided while using CUPS architecture, as the remote DC or MEC can implement Direct Internet Access (DIA) or Local Break Out (LBO), where the user plane traffic exits out to the Internet from the edge DC using SGW-U and PGW-U and does not require the traffic to be back-hauled to the centralized DC.

Figure 5-5 shows a deployment example of 4G CUPS architecture.

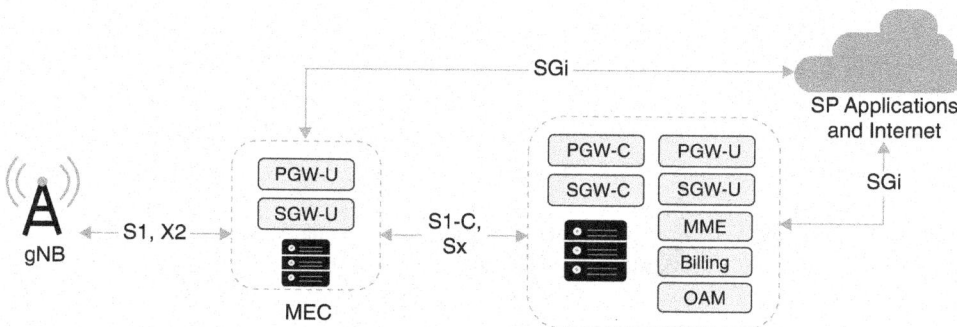

FIGURE 5-5   CUPS Deployment in 4G

Figure 5-5 shows the distributed deployment of user plane and MEC applications. It provides you with the capability to truly separate and independently scale the control plane and user plane of the mobile network on an as-needed basis and in real time, providing the flexibility to specialize the user plane for different applications, without incurring an associated cost of a dedicated control plane for each application. New use cases, such as remotely operated machinery and ultra-high per-user throughput, are now possible by deploying the user plane on the MEC. Finally, CUPS enables you to optimize data center costs by hosting the control plane and the user plane in different geographic locations as well as save on back-haul costs by terminating data at the edge of the network.

The introduction of 5G brought huge improvements in bandwidth and latency and the capability to tailor service experiences across different vertical markets and individual users. Applications that harness these capabilities can transform industries such as manufacturing, automotive, healthcare, and transportation. To deliver the high-quality experiences that consumers of 5G services expect and demand, there is a need to fundamentally reimagine the service edge. 5G also introduces disaggregation and distribution of RAN components. 5G RAN can be decomposed into CU and DU, which can be deployed as virtual elements in the service edge nodes/MEC. As illustrated in Figure 5-6, the control plane functions are Access and Mobility Management Function (AMF) and Session Management Function (SMF), and the user plane function of 5GC is the User Plane Function (UPF).

Figure 5-6 shows the key control plane and user plane components between 5G RAN and 5GC.

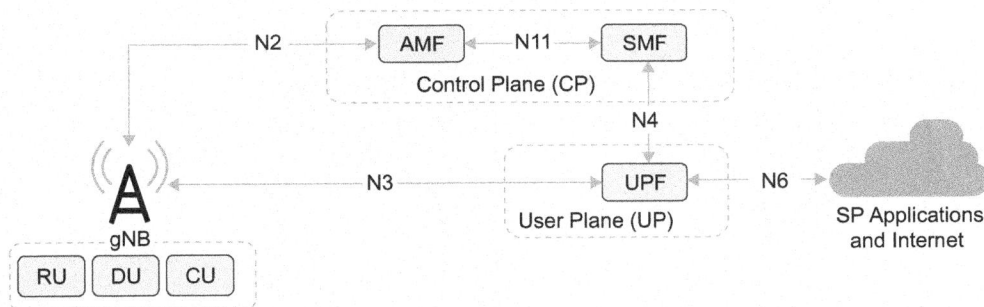

FIGURE 5-6   Key Control Plane and User Plane Connections from Next-Generation Node B (gNB)

The list that follows describes the interfaces in Figure 5-6:

- **Control plane:**

    - **N2:** This interface uses SCTP (NGAP) and supports control plane signaling between 5G RAN and 5G Core, covering UE context management and PDU session/resource management procedures. N2 being signaling doesn't impact the low-latency use cases much (depending on the low-latency requirement of the use case) and is therefore normally planned to be deployed in the central DC.

■ **N4:** The interface between the UPF and the SMF. N4 uses the Packet Forwarding Control Plane (PFCP) protocol, which is also used in Sx interface in the 4G CUPS architecture. The location of deployment of the UPF (that is, whether the UPF is co-located with the SMF or is distributed) determines the need for securing the N4 interface using IPsec ESP and IKEv2 certificate-based authentication provided by the SecGW.

■ **N11:** The interface between the Access and Mobility Management Function (AMF) and the Session Management Function (SMF). N11 uses the service-based interfaces (SBIs) for the interactions employing RESTful API over HTTP/2.

■ **User plane:**

■ **N3:** The interface between 5G RAN and the UPF. N3 carries the user plane data from the UE to the UPF. N3 uses the GPRS Tunneling Protocol (GTP) with header extensions for 5G, Segment Routing (SR), and so on.

■ **N6:** The interface between the UPF and the data network.

The actual deployment location (far edge/edge/regional/central, and so on) of the control plane and user plane is dependent on the use cases. The key deployment models being considered by service providers are explained in more detail in the section that follows.

Distributed UPF and MEC Application Deployment

Figure 5-7 shows an example of 5G deployment with separated control plane and user plane.

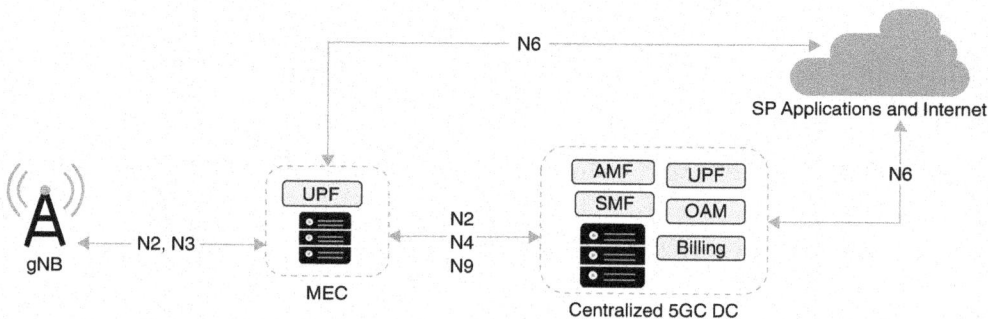

FIGURE 5-7   Example of Deployment Locations for User Plane and Control Plane Functions in 5G

Figure 5-7 shows the deployment of the UPF (User Plane Function) in the MEC. The 5G control plane functions, such as AMF and SMF, are deployed in the centralized 5GC DC. For services/use cases that are not low latency sensitive, the UPF can be deployed in the centralized DC. This is an example of how service providers are deploying 5G networks with the distribution of the UPF catering for the low-latency use cases closer to the edge, while the control plane functions, which are less susceptible to latencies, are deployed in the centralized DC.

C-RAN/O-RAN/Open VRAN Deployment Enabled by MEC

Figure 5-8 shows an example of distributed 5G RAN and 5GC deployment using components of Open RAN (O-RAN), Open VRAN (Virtual RAN), and Cloud RAN (C-RAN).

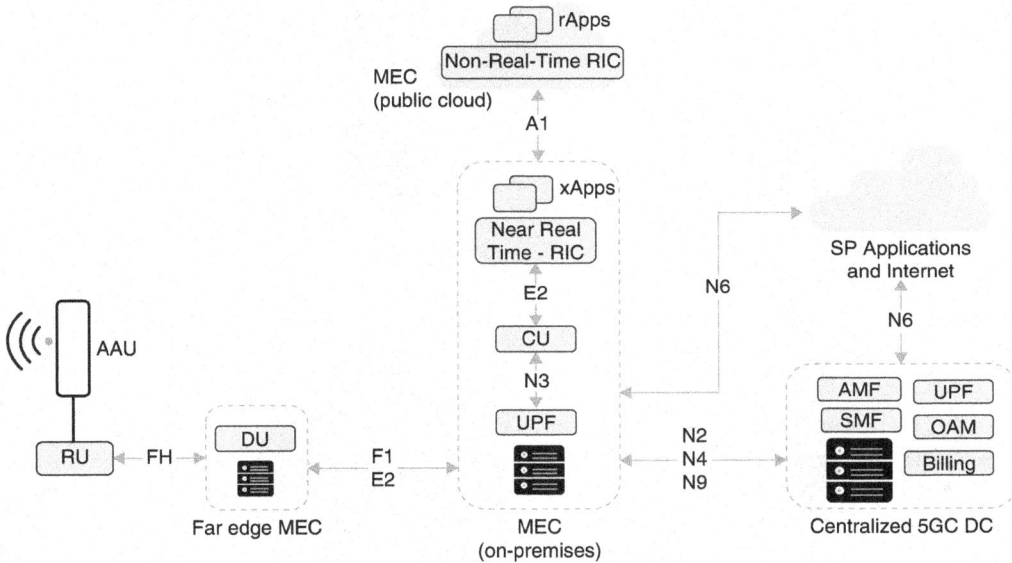

FIGURE 5-8   Example of C-RAN/O-RAN/VRAN Deployment

Note that the DU can be co-located with the CU but is shown as separate in this figure, as it could be one of the deployment options. This architecture using MEC is becoming more popular as service providers explore options of flexible deployment of RAN and reduce cost of densification in terms of transport deployments. The new cloud RAN architecture addresses, among other things, the challenges of building multivendor networks and harmonizing to a common feature set. Applications represent an extension of a RAN Intelligent Controller, which can be deployed in the on-premises or public cloud MEC based on the use case. The applications can be either an xApp for the Near-Real Time RAN Intelligent Controller (Near-RT RIC) or an rApp for the Non-Real-Time RAN Intelligent Controller (Non-RT RIC). xApps are the applications for near real time, such as radio resource management (RRM) and radio optimization algorithms to improve the near-real-time functionalities, which have a latency dependency between 10ms and 1 second. rApps are the applications for non–real time that can support a latency requirement of more than 1 second, such as enriching the information collected by xApps to provide enhanced optimization to the RAN. Applications in these two different environments have different platform mechanisms; however, they share a common behavior of allowing an abstraction layer to provide the capability for the application to be developed such that it can be deployed to either environment from the same package.

Enterprise MEC Deployment

Figure 5-9 shows an example of MEC deployment in the enterprise.

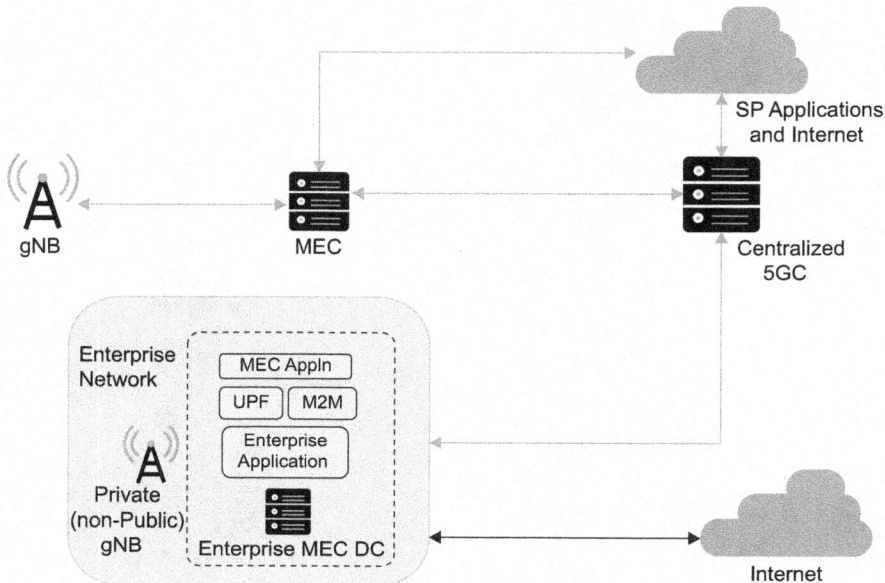

FIGURE 5-9   Example of Enterprise MEC Deployment

Apart from the service provider–based MEC, the key use cases of MEC are to enable key 5G use cases such as M2M. Depending on the M2M use case, the enterprise might require a non-public deployment of 5G RAN nodes, user plane functions, and the MEC application such as application interacting with the M2M use case to be deployed within the enterprise perimeter, most probably within the existing enterprise data center. The control plane functions could be deployed in the centralized DC of the service provider or within the same enterprise DC where the UPF are deployed. A majority of service providers are leaning toward the model of having the CP functions of the 5G deployed in the centralized DC and the UP functions within the enterprise DCs.

A majority of enterprises also seem to be leaning toward this model as they have to cater only for their primary services, which require the UP and MEC applications to be deployed in the enterprise DC. The operations, administration, and management (OAM) of the M2M devices is provided by the M2M vendor, and hence the need for the integration arises only for the CP function messages between the enterprise DC and the service provider DC. Few enterprises are also planning to have the entire 5GC ecosystem, including the gNB, UP, and CP, all deployed within the enterprise DC or using a public cloud provider to host all the 5G UP and CP network functions.

While technically all the mentioned options are viable, the key criteria are costs along with scalability and flexibility. Having the entire 5G network functions owned by the enterprise brings in additional cost for deployment and day-to-day operations, while having a split model allows enterprises to realize the use cases by deploying only the necessary 5G network functions while offloading the non-necessary network functions to be catered by the service provider.

Hybrid MEC Deployment

Figure 5-10 shows an example of hybrid MEC deployment that utilizes both public cloud and private cloud infrastructures to make the best use of both infrastructures.

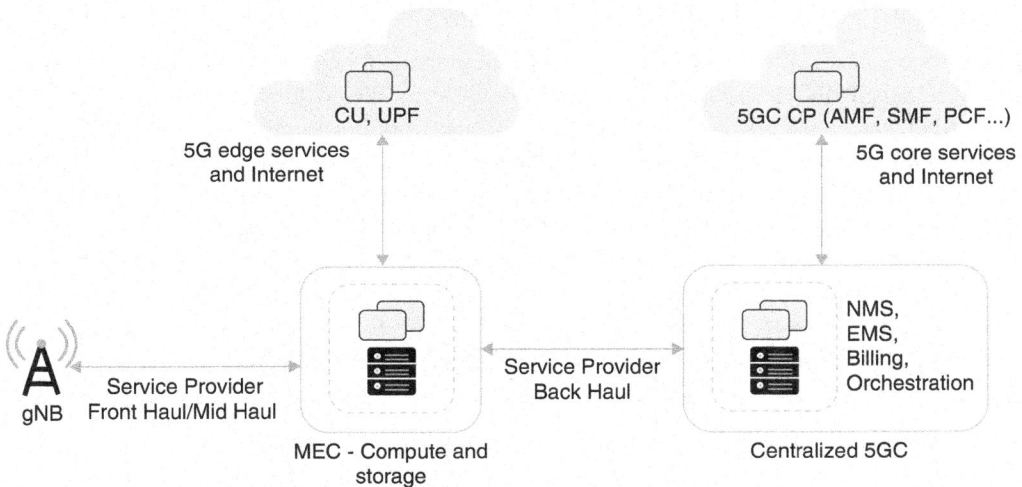

FIGURE 5-10   Example of Hybrid MEC Deployment

5G also provides the option of closer interaction between public cloud providers and the service providers due to the cloud-native function (CNF) capability of 5G network functions. All the 5G network functions should be capable of cloud native, which means it could be deployed in any infrastructure, including cloud. A majority, if not all, of the 5G CNFs are based on open source components. This allows any cloud provider to create their own version of 5GC network functions or partner with any of the 5GC vendors.

Depending on the public cloud provider, there are different options of deploying the 5G network functions using public cloud. One of the options of deploying the 5GC in the public cloud is to deploy the UP and CP 5GC network functions in the public cloud and reuse the service provider's front haul/mid haul and the back-haul transport. This deployment option also provides the opportunity to reuse the DC infrastructure of the service provider for compute and store being implemented by the public cloud provider. Critical 5GC management functions and billing/charging functions could be deployed in the telco DC of the service provider network.

Threat Surfaces in 5G MEC Deployments

When the MEC architectural concepts outlined in the previous section are combined into the 5G architecture, the threat surface for 5G expands. The threat surface expands for a number of reasons; among the main ones are the following:

- The physical structure of the network is changing, and applications and use cases will require compute and storage locations closer to the edge for reasons of localization and latency. This requires separating certain network functions such as the UPF from the protected perimeter of the service provider's centralized data center and deploying it closer to the edge of the network.

- 5G allows use cases with URLLC to be realized by enabling the third-party applications to be deployed in the MEC. These third-party applications, also called external Application Functions (AFs), can be integrated to the 5G Core (5GC) network functions using API, with the help of Network Exposure Function (NEF). NEF exposes the 5GC network functions to the external AFs using APIs. These external AFs can also be integrated with other third-party applications using APIs to fulfill requirements of certain use cases, specifically in 5G non-public network (NPN) industry verticals and enterprise-related deployments. These new API-based deployment options in MEC make API security very important. If API implementation best practices and the right security controls like granular user access controls for APIs are not followed, it will make your MEC deployment risky due to threat vectors related to insecure APIs such as user- and function-level authorization, excessive data exposure, broken object level, and so on.

- The structure of the networking functions has changed from physical to virtual implementations, and the functions' virtualized components can be placed across distributed edge and centralized core clouds. Virtualization, specifically cloud-native functions (CNF), although providing flexibility and scalability for 5G deployments, brings in threat vectors based on the vulnerabilities related to untrusted images, open source software vulnerabilities, improper segmentation, and so on. This makes the MEC more prone to such attacks.

- There is an emphasis on flexible software-based architecture enablers such as SDN, software-defined access (SDA), software-defined radio (SDR), and orchestration to enable use cases related to MEC. Integration of MEC applications with these enablers brings in threat vectors related to improper access control and spinning up rogue virtual instances, among other threats.

- Aside from traditional attacks against servers and caches (for example, via HTTP response splitting), new threat vectors arise. For instance, denial-of-service (DoS) attacks can cause major disruption to the service level agreements (SLAs) committed to by operators. For example, you might have agreed to an SLA with a certain maximum downtime to a content provider who depends on your 5G network to provide content to their customers. In this scenario, because a very large number of caches at the edge of the network would be deployed to cater to a large number of subscribers using low-latency applications (for example, video caching), attackers will be able to easily overwhelm these caches with requests for content not likely to be used by non-malicious users. This situation would result in filling local caches with "useless" content unusable by the subscriber and thereby increasing the downtime period and leading to disrupting the SLA you had agreed upon earlier with the content provider. The vulnerabilities that might cause this to happen would be through traditional attacks on hardware components of the infrastructure, application vulnerabilities, APIs that are not properly secured, and rogue nodes within the architecture.

Figure 5-11 shows an example of threat vectors in 5G MEC deployments.

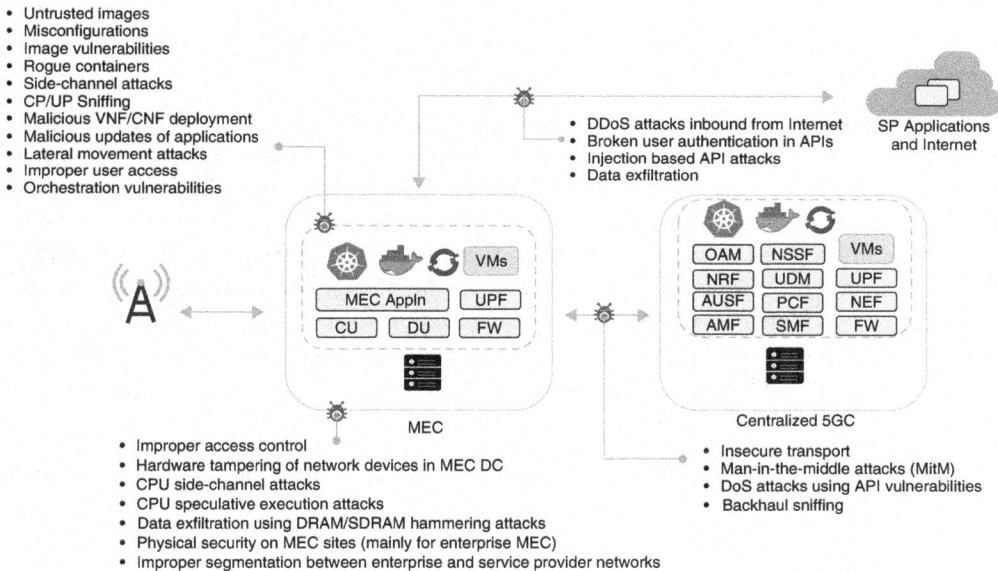

- Untrusted images
- Misconfigurations
- Image vulnerabilities
- Rogue containers
- Side-channel attacks
- CP/UP Sniffing
- Malicious VNF/CNF deployment
- Malicious updates of applications
- Lateral movement attacks
- Improper user access
- Orchestration vulnerabilities

- DDoS attacks inbound from Internet
- Broken user authentication in APIs
- Injection based API attacks
- Data exfiltration

SP Applications and Internet

VMs

MEC Appln   UPF

CU   DU   FW

MEC

OAM   NSSF   VMs
NRF   UDM   UPF
AUSF PCF   NEF
AMF   SMF   FW

Centralized 5GC

- Improper access control
- Hardware tampering of network devices in MEC DC
- CPU side-channel attacks
- CPU speculative execution attacks
- Data exfiltration using DRAM/SDRAM hammering attacks
- Physical security on MEC sites (mainly for enterprise MEC)
- Improper segmentation between enterprise and service provider networks

- Insecure transport
- Man-in-the-middle attacks (MitM)
- DoS attacks using API vulnerabilities
- Backhaul sniffing

FIGURE 5-11   Threats Surface 5G MEC

As Figure 5-11 illustrates, a wide range of threat vectors exists, depending on the MEC deployment model and the 5G network functions deployed on the MEC.

Physical Security

The centralized DC of the service provider or enterprise is usually well-secured with many layers of protection, including physical security layers such as 24×7 manned security. In MEC, the deployment of an edge DC could be in the public cloud or a mini/micro data center with limited space and power resources. A majority of the MEC deployments (even if the hybrid cloud model is used for MEC) will contain network devices like switches, routers, gateways, firewalls, application delivery systems, virtual workloads such as virtual machines, cloud-native functions, and storage servers. Service providers and enterprises deploying MEC should take stringent measures for securing the site and servers physically. If the MEC site is not physically secure, malicious adversaries might be able to sabotage the site and cause a service outage.

Hardware and Software Vulnerabilities

While many of the service providers and enterprises invest in perimeter security such as firewalls and intrusion detection and prevention systems, very few are actually looking at built-in hardware and software hardening. The reality is that if the firmware of the server being used to host all the data center applications is attacked and exploited, it is a very difficult and resource-consuming process to contain the attack and purge all the impacted malicious hosts. Built-in hardening verification is one of the key security checks that should be carried out on the vendor before choosing them as the preferred vendor for MEC (or any data center for that matter).

Key areas that impact hardware and software in data centers are malicious software and firmware updates, firmware vulnerabilities, and supply chain vulnerabilities.

MEC sites might also host applications that store user data. The cost of a breach of security can have severe consequences on both the company managing the data center and on the customers whose data is exfiltrated. This makes it very important for you to ensure data loss prevention (DLP) measures, such as monitoring, detecting, and blocking sensitive data while in use, in motion, and at rest, are implemented to adhere to privacy and security laws such as General Data Protection Regulation (GDPR). The GDPR is the toughest privacy and security law in the world. Though it was drafted and passed by the European Union (EU), it imposes obligations onto organizations everywhere, so long as they target or collect data related to people in the EU. The regulation was put into effect on May 25, 2018. The GDPR will levy harsh fines against those who violate its privacy and security standards, with penalties reaching into the tens of millions of euros.

Figure 5-12 shows an example of data exfiltration by an attacker deploying a malicious code within a software update.

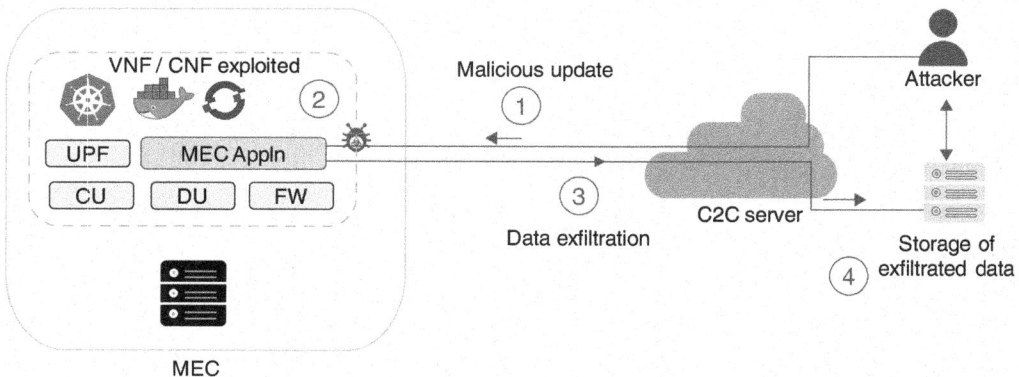

FIGURE 5-12 Example of Data Exfiltration in 5G MEC Deployment

As shown in Figure 5-12, the following steps could be used by the attacker to exfiltrate data:

Step 1. The attacker chooses to inject malicious code into one of the updates for the CNF component hosted in the MEC site as the first step.

Step 2. The malicious code that is implanted into the CNF could have a sophisticated method, such as establishing an encrypted layer of communication with the targeted application or host system, as the second step.

Step 3. The data can be exfiltrated silently using existing pin holes or open ports in the firewall or by reusing the ports being used for encrypted messages, such as port 443 for TLS.

Step 4. This exfiltrated data can then be stored in an external server by the attacker and then sold to other malicious actors looking for stolen data.

Figure 5-13 shows an example of data exfiltration by an attacker deploying bugs directly into the chipsets during the manufacturing process, thereby creating a backdoor entry into the hardware server and the infrastructure where the hardware is deployed. If such hardware is deployed in the MEC infrastructure or used by cloud providers in their data center farm where MEC applications are hosted, it can provide a backdoor to the MEC applications and cause mass data exfiltration or, worse, modification of the data without being noticed.

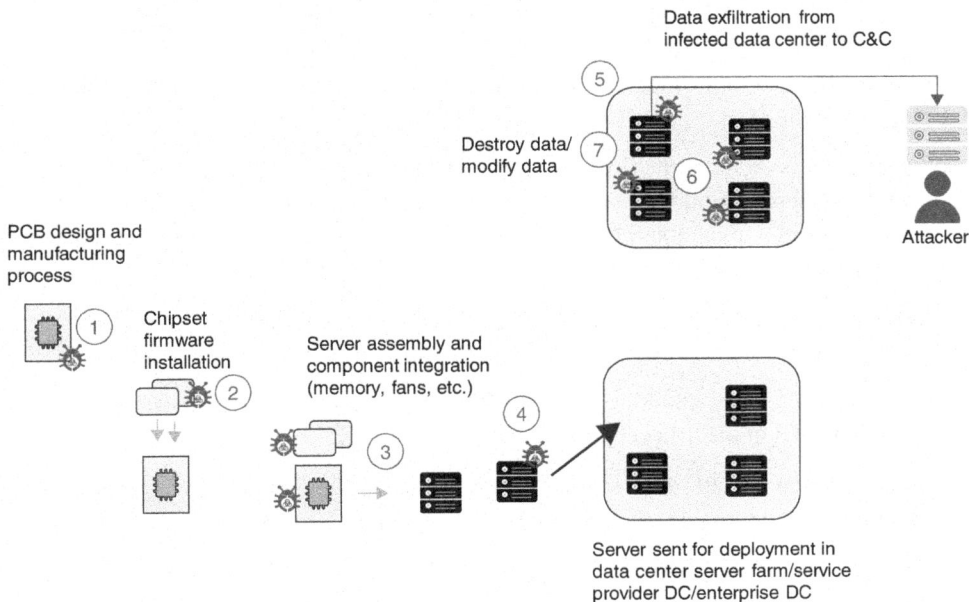

FIGURE 5-13 Supply Chain Threat Vectors

As shown in Figure 5-13, the following steps could be used by the attacker/attacking entity in the supply chain.

> **Note**
>
> One of the key differentiators between the supply chain threat vector and other threat vectors is the amount of resource and planning required to pull off an attack. It would also require the support of hardware and software engineering teams of the company manufacturing the chipset and assembling the server.

Step 1. The attacking entity could infiltrate the printed circuit board (PCB) manufacturing process and could perform the following actions:

    a. The PCB manufacturing process includes the design phase, which includes the blueprint diagram of the electronic circuit. This blueprint diagram is then used to create tracks on the PCB by computerized systems for mass-manufacturing the PCB boards. The attacking entity can change the schematic to include micro-chipsets for data eavesdropping and sniffing. Another example could be an attack on the BIOS flash memory to modify the hardware configurations.

    b. Once the manufacturing process is completed, the quality assurance (QA) team usually checks the PCB for any anomalies, such as over/under value of voltage, current, impedance of the track within the circuits, and so on. The QA team has to follow certain method of procedure (MoP) documents for the checks. The attacking entity has to ensure that the MoP documents are modified to prevent checks on the malicious chipsets.

Step 2. The attacking entity could infiltrate the software teams catering for the firmware installation for the chipsets in the PCB.

The software process includes the firmware coding and installation of firmware on the chipset. The attacking entity could modify the code to add malicious lines of code. The attacking entity could also deploy malicious code during the firmware installation on the chipsets, making the chipsets capable of malicious functions such as overclocking.

Step 3. The server assembly process is when the PCBs are assembled into the server, along with memory and such.

The attacking entity could exploit the memory chips or power controller chips, causing issues when deployed in the data center server farm.

Step 4. The server could then be part of the servers being ordered by the customer, which could be a service provider, cloud provider, or an enterprise.

Step 5. When deployed in the data center, the server with malicious chipsets and firmware can now create an encrypted connection to the attacker's command and control (C&C) center and cause data exfiltration.

Step 6.   The malware installed within the server might also be an advanced persistent threat (APT), which might be passive and cause lateral movement within the data center using privilege escalation to look for specific target hosts. Privilege escalation is a method of exploiting a bug or gaining elevated access to the network elements.

Step 7.   The malware installed within the servers by the attacking entity in the supply chain could also cause wipe-out attacks, destroying sensitive data within the data center.

5G MEC Infrastructure and Transport Vulnerabilities

Along with major technological change to the mobile packet core and RAN, you will also need to evolve your transport networks to cost-effectively deliver a satisfying mobile broadband experience, while simultaneously meeting the scale requirements for massive IoT and the ultra-low latency requirements for real-time applications. This becomes more complicated when MEC use cases are realized. Depending on the MEC use case and mode of network deployment, the third party MEC applications and the 5GC network functions could be deployed on the public cloud or on the on-premises telco data centers, and the back haul used could be the service provider's or the public network's. The MEC infrastructure, be it an on-premises telco cloud deployment, public cloud deployment, or hybrid mode deployment (using both on-premises and cloud infrastructures), should be well secured.

Figure 5-14 shows an example of the 5G transport traffic traversing through an insecure transport network such as the public Internet.

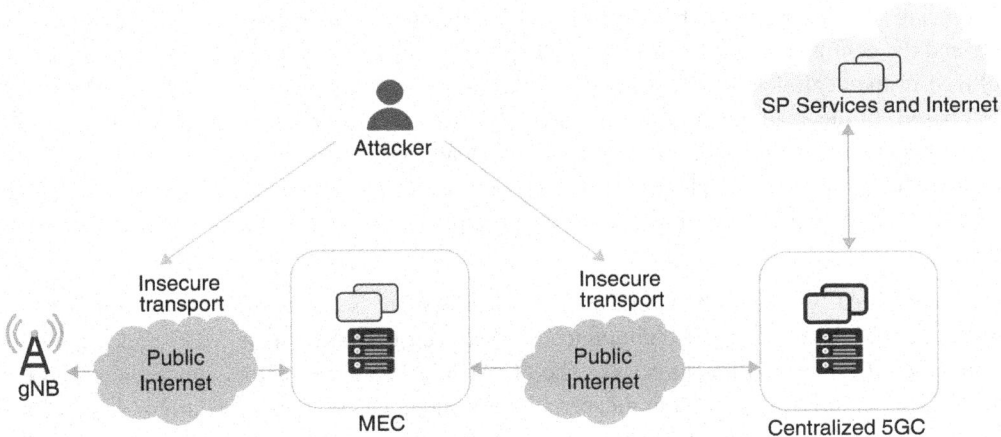

FIGURE 5-14   Attacks Due to Insecure Transport in On-premises 5G MEC Deployments

As shown in Figure 5-14, the attacker can sniff the traffic coming into the MEC layer from the insecure public network, and the traffic flowing out to the insecure network is both susceptible to eavesdropping and man-in-the-middle (MitM) attacks. Depending on the applications deployed on the MEC infrastructure, the severity of the eavesdropping could be critical.

Figure 5-15 shows an example of a transport layer threat vector in the 5G MEC hybrid deployment model.

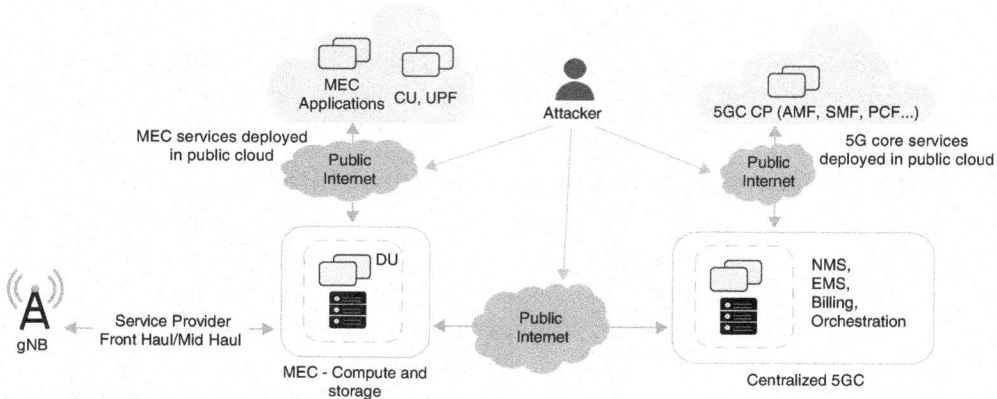

FIGURE 5-15 Attacks Due to Insecure Transport in Hybrid 5G MEC Deployments

As shown in Figure 5-15, the CU, UPF, and the MEC applications are deployed in the public cloud environment, and the DU, along with compute and storage, is deployed on premises in an edge data center. The interfaces from the on-premises MEC components to the public cloud–deployed network functions, including the MEC applications, are using an API or non-API interface (depending on the application) using the unprotected public network. The communication between the on-premises edge data center and the centralized 5GC is also using the public Internet. Depending on the network functions deployed in the centralized 5GC data center, the mode of communication could be API based or non-API based. In the example shown in Figure 5-15, the network functions deployed in the 5GC are mostly management, billing, and orchestration, which are primarily API based. A majority of the hybrid deployments in discussion plan to have the control plane in the public cloud with the 5GC components handling user-sensitive information being deployed at the on-premises centralized 5GC data center.

The key transport threat vector in the hybrid mode of deployment is due to unprotected traffic while traversing the public cloud. The API communication between the on-premise MEC and the public cloud is vulnerable to multiple threat vectors. The 5G MEC API attack vectors are explained in the "5G MEC API Vulnerabilities" section in this chapter. For the non-API communications, some of the key threat vectors are eavesdropping and replay attacks, which lead to DoS attacks or theft of sensitive user information.

Figure 5-16 shows an example of a transport layer threat vector in the 5G MEC public cloud deployment model.

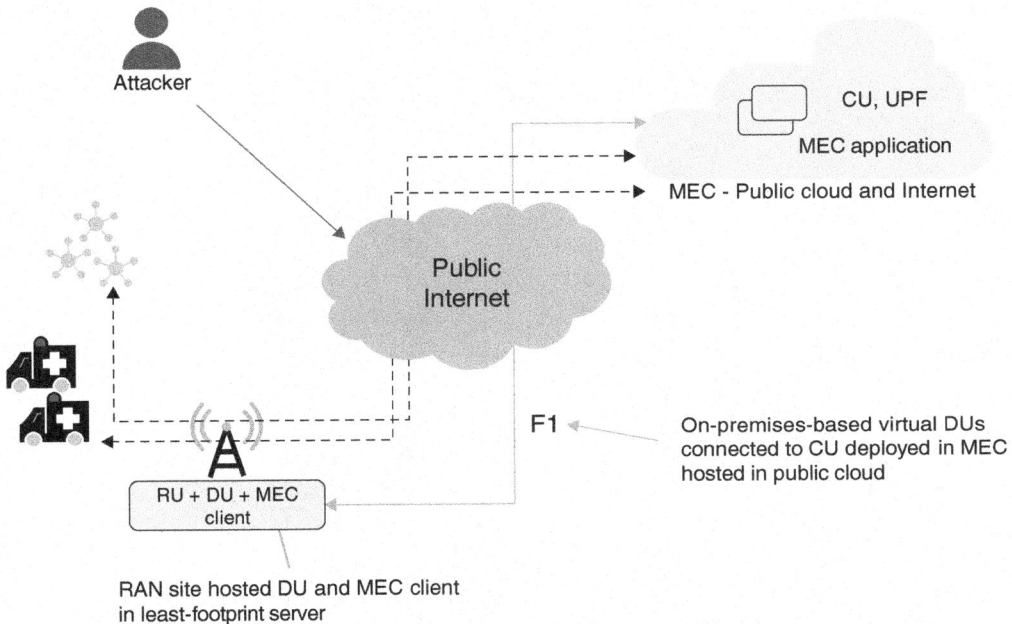

FIGURE 5-16 Attacks Due to Insecure Transport in Hybrid 5G MEC Deployments

As Figure 5-16 illustrates, in the public cloud deployment model, the CU, UPF, and all control plane network functions are deployed in the cloud. The 5G radio node gNB is co-located with a least-footprint MEC server, which has a virtual instance of DU and MEC clients running on it. This model can typically be used when there is a need to deploy ultra-reliable low-latency communications (URLLC) use cases. The deployment of virtual instances in the low-footprint server in the same location as the gNB allows you to customize the chipset requirements, such as accelerated processing of packets based on hardware acceleration as compared to traditional field-programmable gate array (FPGA) chips.

In this model of deployment, the traffic between the least-footprint servers and the public cloud–hosted 5G network functions are both API and non-API based traffic. The gNB Distributed Unit – user plane (gNB-DU-UP) and the gNB Distributed Unit – control plane (gNB-DU-CP) from the least server hosted MEC toward the CU hosted in the public cloud are non-API-based traffic, while API is used between MEC client hosted on premises and MEC applications/MEC server hosted in the public cloud. Most of the VRAN/O-RAN/C-RAN vendors provide built-in DU-CP IPsec protection, whereas the DU-UP protection is purely optional and depends on the deployment model.

The key threat surfaces in this model are API vulnerabilities for the on-premises least-footprint server-hosted 5G MEC client interactions with the cloud-deployed MEC applications. Some of the key API threat vectors for this deployment model are improper rate limiting, brute-force attacks, broken authentication, improper configuration of the Docker daemon, and so on. The key threat surfaces in non-API communication, such as between on-premises-based virtual DUs and the cloud-deployed CU is the unprotected DU-UP interface, which is prone to replay attacks, eavesdropping, and MitM-based attacks.

Figure 5-17 shows an example of a transport layer threat vector in an enterprise-based 5G MEC deployment model.

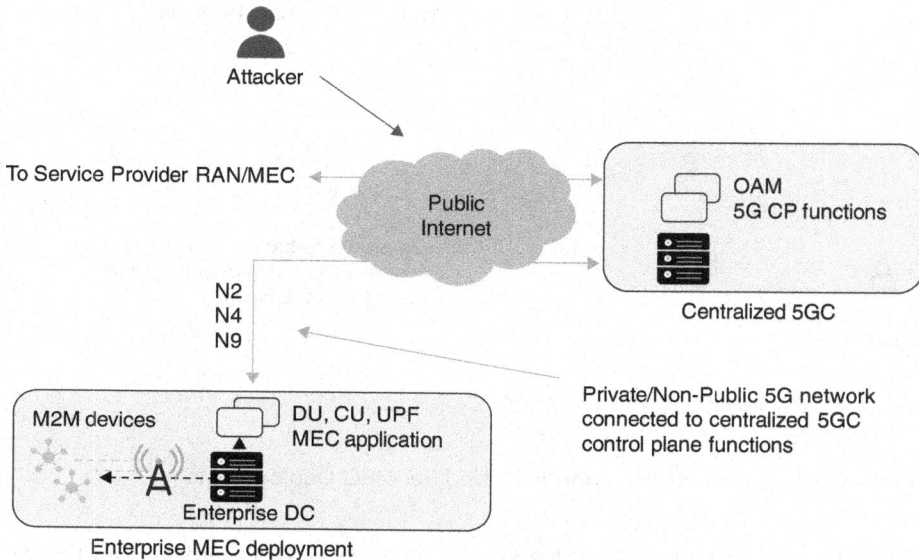

FIGURE 5-17   Attacks Due to Insecure Transport in Enterprise 5G MEC Deployments

In the enterprise 5G MEC deployment model shown in Figure 5-17, the 5G radio network functions, user plane functions, and MEC applications are deployed in the enterprise data center. The centralized 5GC control plane functions are hosted by the service provider. The N2, N4, and N9 interfaces have to traverse through the unprotected public Internet in this example. The details of the interfaces are as follows:

- **N2:** The interface between gNB and AMF in a centralized 5GC or between gNB-CU-CP and AMF

- **N4:** The interface between the edge UPF and SMF in a centralized 5GC

- **N9:** The interface between the edge UPF and UPF in a centralized 5GC

The key threat vector is the N2, N4, and N9 traffic traversing the unprotected public network toward the 5GC control plane functions and the unprotected API between the MEC applications deployed in the enterprise's premises and the public network. Another key threat is the attacker accessing the service provider's network functions due to weak segmentation between the enterprise MEC network and the service provider's centralized network.

Due to the unprotected public network, the N2, N4, and N9 interfaces are vulnerable to eavesdropping, MitM attacks, and reply attacks. The unprotected API interface in the MEC applications is also open to API-related attacks due to security misconfigurations and injection attacks (such as SQL injection attacks) leading to the exposure of sensitive data.

Figure 5-18 shows an example of lateral movement attacks in MEC infrastructures.

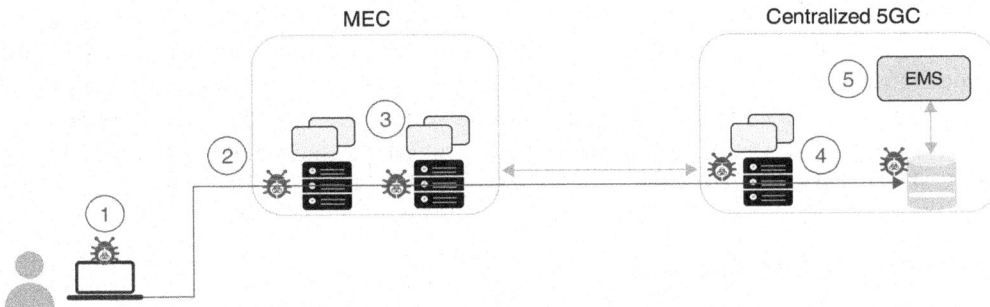

FIGURE 5-18   Lateral Movement Attack in MEC Infrastructure

As shown in Figure 5-18, the attacker infiltrates the network with malicious code or malware, which then moves laterally from the MEC to the centralized 5GC data center. This technique moves deeper into the network, searching for specific sensitive data or a key database of a critical asset. Once the critical asset is found, the malware can then launch a ransomware attack or a wipe attack. Ransomware attack, as the name suggests, is where the malware encrypts the data and then a ransom is negotiated before the files are decrypted. The other malware attack of wiping data is used when the malware finds the sensitive data it is looking for and then wipes it out. There are sophisticated advanced persistent threat (APT) attacks today where the malware even infects the stored data when it is copied from cold storage (archive), thereby infecting each and every copy of the sensitive file.

The attacker could use the following steps in a lateral movement attack, where the user is an internal employee in your organization:

Step 1.   The attacker infects the laptop of the user with malicious tools for password extraction and malware.

Step 2.   The legitimate user uses an infected endpoint unintentionally, such as using the infected laptop to configure the MEC application.

Step 3.   The malware within the laptop is programmed to steal specific targeted data. A credentials-sniffing tool available in the market today can extract plaintext passwords, hashes, and so on from memory. Once the credentials are sniffed, the malware uses the known credentials to move laterally within the network.

Step 4.   The malware can also move into the 5GC centralized DC using the east-west configuration between data centers.

Step 5.   Once the targeted host/data is found, the malware will infect the target. Depending on the type of malware, the attack varies. If the attack is motivated by ransom, then the data within the infected host is encrypted. If the attack is to cripple the network or data, the malware can destroy the data and wipe it out completely.

One of the reasons the sophisticated lateral attack is difficult to detect is that it can mimic the admin and network traffic behavior. For example, if the admin always logs in around 8 a.m. every day, the malware with the user credentials will also start its activity around 8 a.m. If the network has traffic peaks and reads/writes into the database around 9:30 a.m., the malware will also start modifying the database around that time. This is one of the key aspects of an APT—it is can be passive, understand the network behavior, and then launch the attack, such as a complete data wipe, and remain undetected. A majority of the attacks that include lateral movement and passing malware payloads between multiple hosts also include data exfiltration.

Virtualization Threat Vectors

5G permits complete programmability and flexibility, allowing the 5G network functions to be deployed as cloud-native functions (CNFs), which are basically container and virtual machine based environments that can be deployed in any infrastructure, such as on a container within the router connected to the radio node, on an on-premises data center, on an enterprise data center, on a public cloud, or based on a hybrid model of on-premises and public cloud deployments. It brings in true openness in network deployment and implementation of network functions. You can do away with proprietary hardware and software radio and packet core network elements and can use open source software and any commercial off-the-shelf server, or you can use the public cloud to deploy MEC. Although virtualization has been around for a while (since the 1960s), containers based on Docker and orchestrated by Kubernetes are relatively new. In short, while we understand the threat vectors of virtual machines after using them for so many years, we are yet to understand and uncover all the threat vectors of containers.

While virtual environments consisting primarily of virtual machines were prevalent in 4G, they were mainly deployed in centralized data centers secured by a perimeter defense and sometimes deployed in an air-gapped infrastructure. In 5G, the CNF, based mostly on open source programs, is deployed as close to the edge as possible, sometimes even in a multi-cloud stack. This new model of creating network functions with open source programs, having a dynamic runtime environment, and having a new method of deployment will need a serious reexamination into the way security is planned, implemented, operated, and managed.

Typical 5G network deployment will consist of thousands of virtual instances or workloads. If the RAN infrastructure is virtualized, you will have tens of thousands of workloads. While it is good to have virtualization to optimize OPEX/CAPEX and have flexibility of deployment, it does introduce a few new complexities.

One of the complexities is proper segmentation between all these different types of workloads. Previously, before virtualization became more prevalent, the procedure was to interview or talk to people of different departments, such as radio planning, optimization, transport planning, and implementation, and understand who needed what access, what network components needed what kinds of services (DNS and so on), and what traffic needed to exit the infrastructure. The access to specific applications was then planned and configured in the Identity and Access Management (IAM) layer. This method of having manual segmentation/micro-segmentation is inefficient and not scalable for managing tens of thousands of workloads. The solution is to provide micro-segmentation based on rules that can be maintained using behavior analysis of the application. The method for providing automated/semi-automated micro-segmentation is located in the "Securing Virtualized Deployments in 5G MEC" section in this chapter.

Another area of importance in 5G MEC virtual deployments is vulnerability assessments. 5G allows open source components to be used to create an MEC application leading to a polyglot microservice architecture. *Polyglot* generally means that an application developer can use the program language of their choice to create an application.

Figure 5-19 shows the polyglot nature of a microservice-based 5G network function or 5G MEC application.

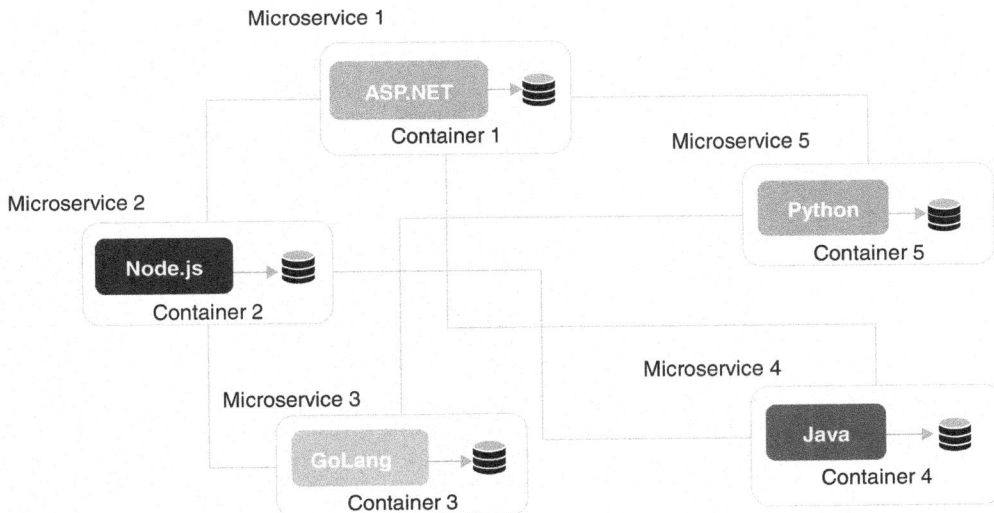

FIGURE 5-19   Polyglot Nature of Microservices

As shown in Figure 5-19, the microservice is polyglot in nature, which allows MEC application developers to pick a programming language of their choice in order to build products in more efficient ways. In 5G, virtualization of the RAN and 5GC network functions is cloud native based and can be deployed on the cloud. This allows you to use different container deployment models, such as containers on virtual machine, containers on bare metal, or deployed on multiple public clouds. The 4G subsystems might be still deployed on virtual machines (VMs) due to the industry shortcomings on 4G components, but for 5G you might actually skip the container on virtual machine deployment and have it deployed on bare metal or on the public cloud due to the early development of 5G network functions based on containers. For the MEC applications, there have been quite good developments in the containerized space, and there are already applications available today using open source programs that can be deployed in any infrastructure, including cloud. You should be very careful before choosing to deploy any open-source components without validating them for vulnerabilities.

One of the key issues is the lack of visibility in checking if any of the programs has a vulnerability that needs to be patched, as shown in Figure 5-20. If it's only one microservice that needs attention, it can be checked and updated manually. But in 5G, when we have tens of thousands of MEC applications, it becomes very difficult to conduct a vulnerability assessment and compliancy check.

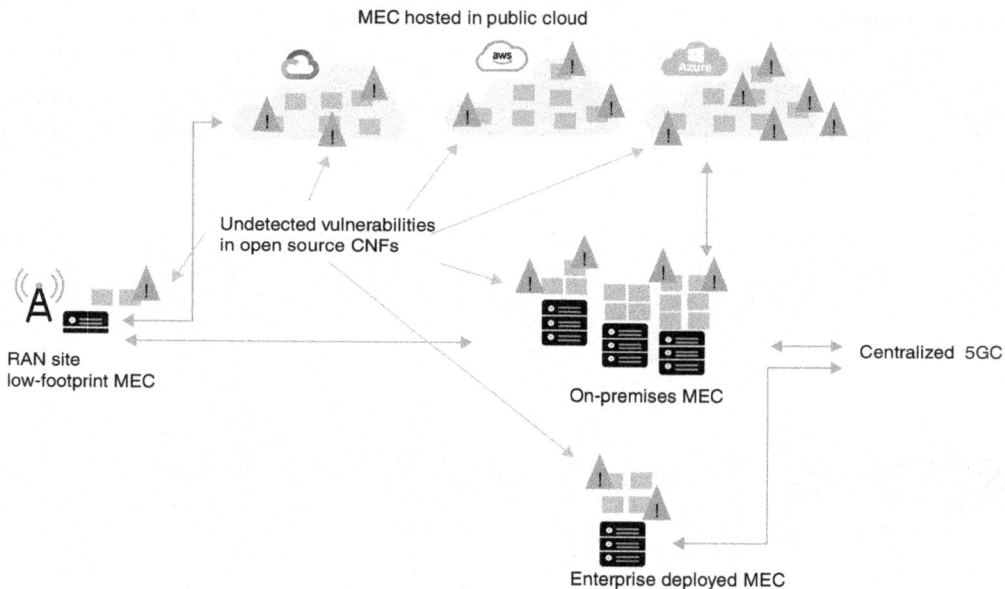

FIGURE 5-20 Vulnerabilities in Open Source Cloud-Native Deployments

As shown in Figure 5-20, it becomes really tricky if you need to have a visibility into the vulnerabilities of all open source programs across the entire stack of MEC deployments, whether deployed in a multi-stack cloud, RAN site low-footprint MEC, on-premises MEC, or enterprise-deployed MEC.

Attackers are constantly innovating methods to exploit the container ecosystem in order to exploit the vulnerabilities/improper configurations to launch attacks. Docker is one of the most widely used platforms for managing containers. Over time, it became the de facto standard for development and deployment of web and cloud applications; however, anybody who has worked with Docker quickly realizes that the learning curve is quite steep. Therefore, Docker installations can be easily misconfigured and the Docker daemon exposed to external networks with a minimal level of security.

Figure 5-21 shows crypto-mining malware spreading through the system via an attacker exploiting an incorrectly configured Docker API for third-party MEC applications.

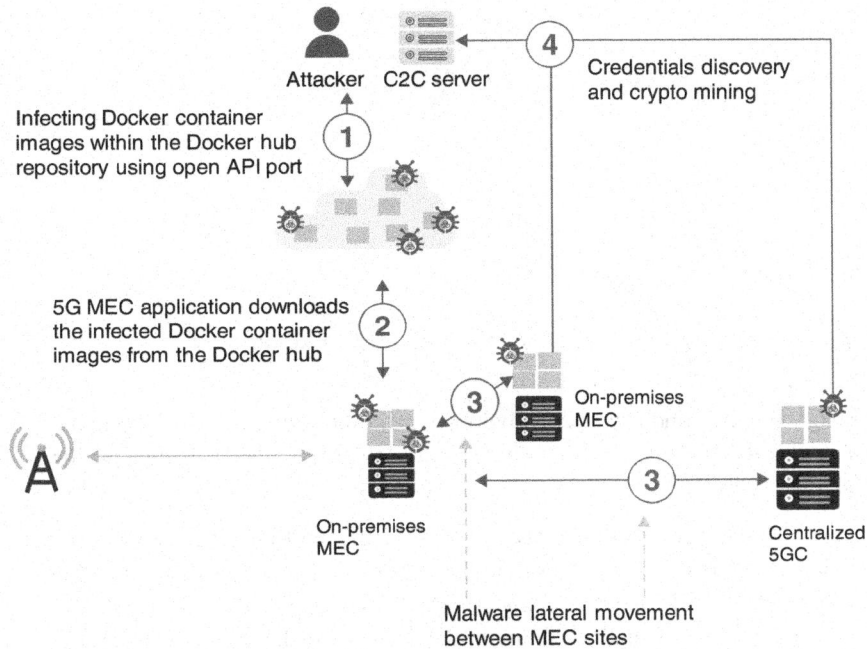

FIGURE 5-21   Attacker Exploiting an Incorrectly Configured Docker API

As shown in Figure 5-21, the main threat vector here is an incorrect configuration of the Docker API, which is exploited by the attacker to gain access to the system that hosts the third-party 5G MEC application. The attacker then laterally travels into another MEC site, a centralized 5GC site, infecting other systems and using it for crypto-jacking (stealing credentials and other sensitive data). For the sample attack shown in the Figure 5-21, the steps are as follows:

Step 1.   The attackers scan the Internet for Docker containers implemented with an unprotected or incorrectly configured API. In case of Docker, the Docker daemon listens for API requests and also manages Docker objects, such as images, containers, networks and volumes.

Figure 5-22 shows the Docker architecture and the API the attackers are trying to exploit.

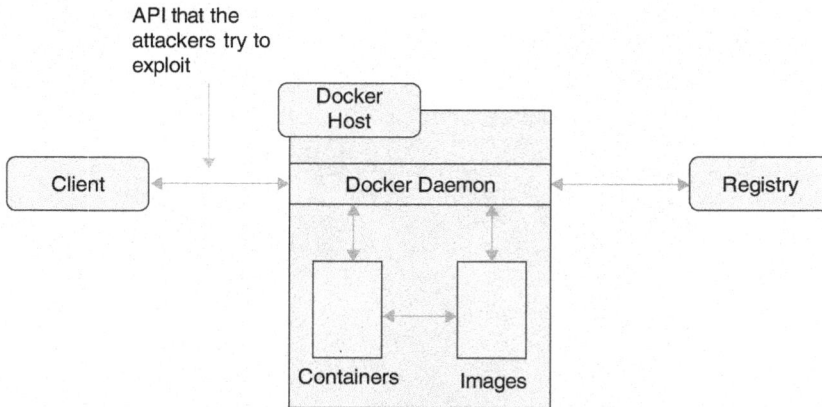

FIGURE 5-22   Docker Architecture and the API That Attackers Try to Exploit

Step 2.   This infected image is pulled and run as containers on the compromised host. This can be easily accomplished if there is malware already present in the MEC host causing a privilege escalation attack.

Step 3.   The infected code within the container can also cause the malware to spread deeper within the local and remote networks. The main module could also attempt to spread to other hosts by stealing the client-side certificates and connecting to them without requiring a password. The other modules within the infected container image can also include scripts for actions such as terminating security services such as anti-malware solutions and removing a competitor's botnets. It can also be programmed to protect itself and hide the mining process from process enumeration tools.

Step 4.   Once ready, the scripts within the infected container image can now extract sensitive information, such as number of CPUs and user credentials, to the C2 server, which can now be used for calculating the mining capabilities. The crypto-jacking worm can now be used for mining. The infected container image can query the C2 server for other hosts it can infect.

Among other threats linked to virtualized deployments, the side-channel threat vector is quite important because it can impact any virtual function sharing CPU memory. There are many variants of side-channel attacks, such as hyper-jacking, Spectre, Meltdown, Cross-VM cache side-channel attacks, and so on. These are generally aimed at leaking the crypto keys, which are then used for eavesdropping or exfiltrating the data out of the infrastructure. Cross-VM cache side-channel attacks are well-planned sophisticated attacks focused on a particular organization. The side-channel attack method chosen by the attacker depends on many factors of the target network implementation, such as installed MEC infrastructure components, the vendor and model of chipsets being used in the MEC data center, the type of network devices present in the targeted MEC network, and so on.

Figure 5-23 shows an example of data exfiltration due to CPU processes used in virtualization.

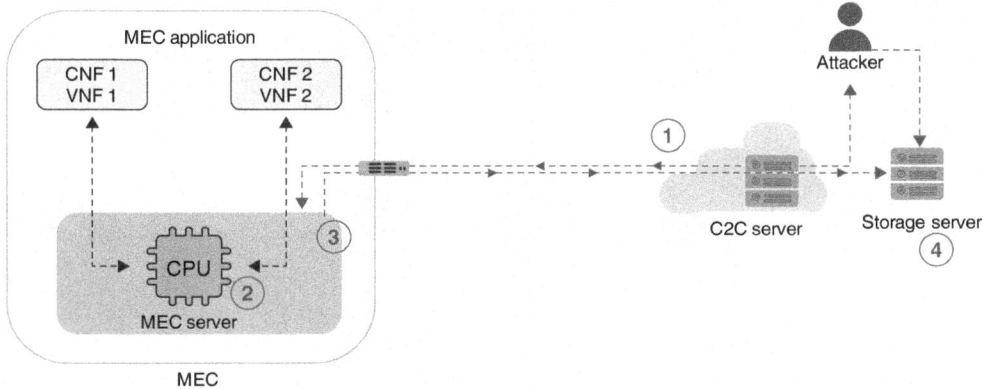

FIGURE 5-23   Data Exfiltration Due to Side-Channel Attack

As shown in Figure 5-23, which shows the attack vector aimed at the virtualization infrastructure's firmware, the attacker can use the following steps:

Step 1.   The attacker uses side-channel attack vectors such as a cross-VM cache side-channel attack and utilizes the underlying hardware CPU vulnerabilities.

Step 2.   The attacker's side-channel attack vector allows one process to extract sensitive information by exploiting shared cache memory between VMs and analyzing the data, such as the relation between the software process and the power consumption of the hardware for that process.

Step 3.   The data can be exfiltrated silently using existing pin holes or open ports in the firewall or by reusing the ports being used for encrypted messages, such as port 443 for TLS.

Step 4.   The exfiltrated data is analyzed to understand the software process and the type of software being used, among other things. Once the software is understood, it can be used to launch a crippling attack at a later stage.

5G MEC API Vulnerabilities

5G architecture has been designed to be cloud native, as it brings elasticity, scalability, and creation of rich services in the edge. 5G also allows you to expose rich services through the cloud and RESTful APIs. APIs make it very easy to integrate between different solutions and information sharing to enable a seamless user experience. API introduction for MEC applications also brings in the expertise of web application developers in developing 5G MEC applications, but it also brings in new threat vectors.

Figure 5-24 shows an example of API threat vectors in 5G MEC deployments, where the attacker uses an injection attack causing DoS to the devices being served by the MEC applications hosted on the MEC server.

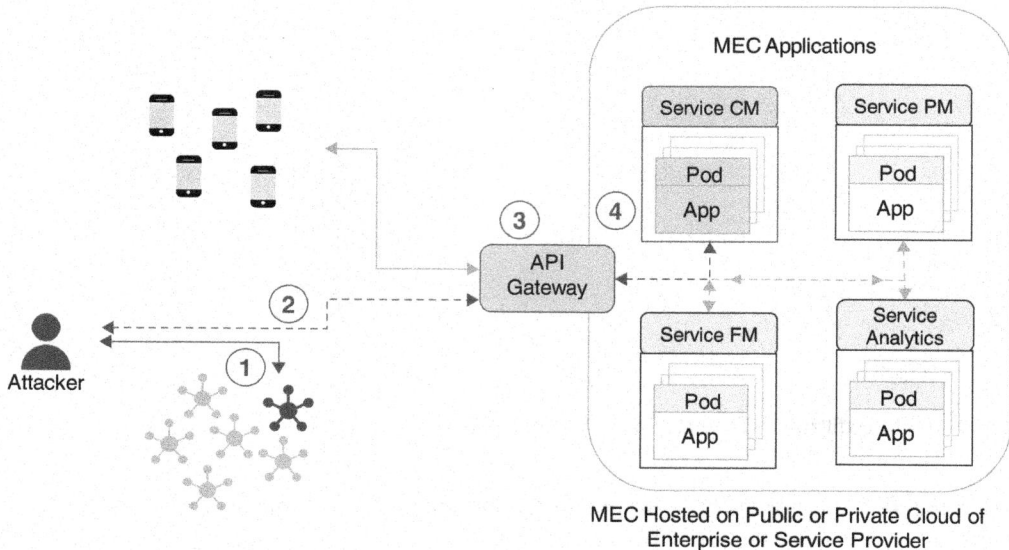

FIGURE 5-24   API Injection Attack in Applications Deployed in MEC

As shown in Figure 5-24, the attacker uses an injection attack, which occurs as follows:

Step 1.   The attacker gains control of a weak IoT device and gets the credentials of the target server to be exploited.

Step 2.   The attacker constructs API calls that include command injection, NoSQL, and so on, toward the CM server.

Step 3.   The API gateway does not perform any checks and passes the calls to the CM server.

Step 4.   The CM server blindly executes the calls, which will crash the CM server, causing DoS to all the devices being served by the CM server.

In this threat vector, the attacker is hidden completely and uses the existing device to launch an attack toward the MEC. This attack could be launched directly by the attacker using the API.

Figure 5-25 shows an example of API excessive data exposure attack.

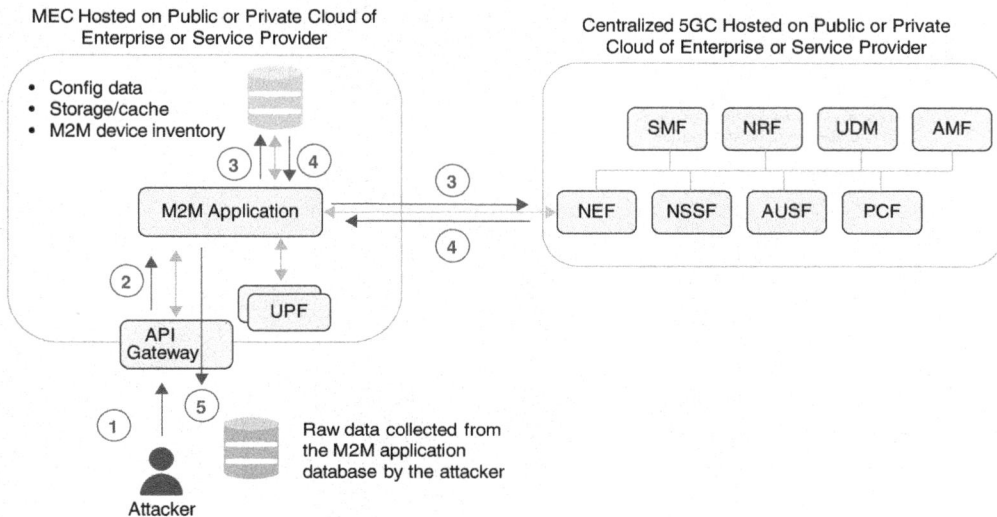

FIGURE 5-25   API Excessive Data Exposure Attack

As shown in Figure 5-25, the attacker uses an unprotected API to gain excessive data exposure from the configuration management. In one of the attack vectors, the following steps could be used by the attacker:

Step 1.   The attacker requests data using "API Get" toward the API gateway.

Step 2.   The API gateway relies on the client to do the filtering.

Step 3.   The API gateway passes the query to the database. Depending on the data requested by the API Get request, the request could also be forwarded to the 5GC CP objects using NEF (Network Exposure Function). The NEF can see the request coming from the MEC applications as a legitimate request, which in turn can respond with the attribute values/bulk configs requested by these API calls.

Step 4.   The API returns full data objects as stored in the M2M application database. Depending on the data requested by the API Get request, the data could also be purged from the 5G CP objects using the Network Exposure Function (NEF) API calls.

Step 5.   The attacker collects all the information, which might include sensitive data such as device IDs, Day 0 configuration templates, username and passwords, and so on.

Figure 5-26 shows an example of excessive data that can be exposed if filtering polices are not used on the MEC application server side.

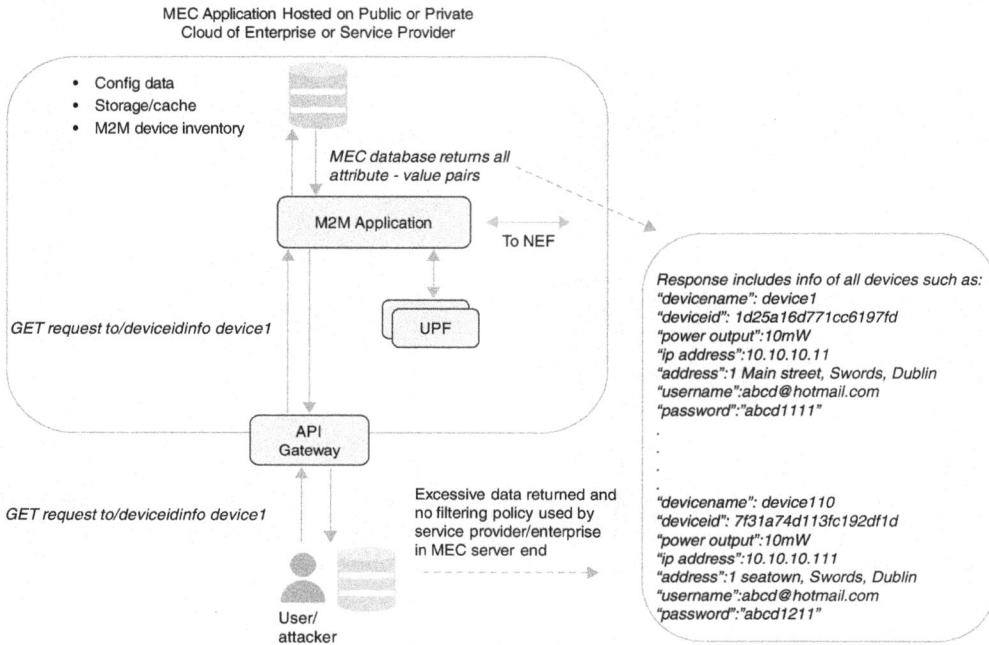

MEC Application Hosted on Public or Private
Cloud of Enterprise or Service Provider

- Config data
- Storage/cache
- M2M device inventory

MEC database returns all attribute - value pairs

M2M Application

To NEF

UPF

GET request to/deviceidinfo device1

API Gateway

GET request to/deviceidinfo device1

User/ attacker

Excessive data returned and no filtering policy used by service provider/enterprise in MEC server end

Response includes info of all devices such as:
"devicename": device1
"deviceid": 1d25a16d771cc6197fd
"power output":10mW
"ip address":10.10.10.11
"address":1 Main street, Swords, Dublin
"username":abcd@hotmail.com
"password":"abcd1111"
.
.
.
"devicename": device110
"deviceid": 7f31a74d113fc192df1d
"power output":10mW
"ip address":10.10.10.111
"address":1 seatown, Swords, Dublin
"username":abcd@hotmail.com
"password":"abcd1211"

FIGURE 5-26   API Excessive Data Exposure Attack

As shown in Figure 5-26, the excessive data can be obtained by the user or an attacker. This attack, aimed at getting sensitive data out of an MEC application database, can cripple industry verticals such as smart cities and smart manufacturing and is very critical for defense and military use cases. The threat vector here is due to the weak security controls, where the API exposes a lot more data than the client requires and relies on the client to do the filtering.

Figure 5-27 shows an example of Broken Object Level authorization and Broken Authentication API attacks.

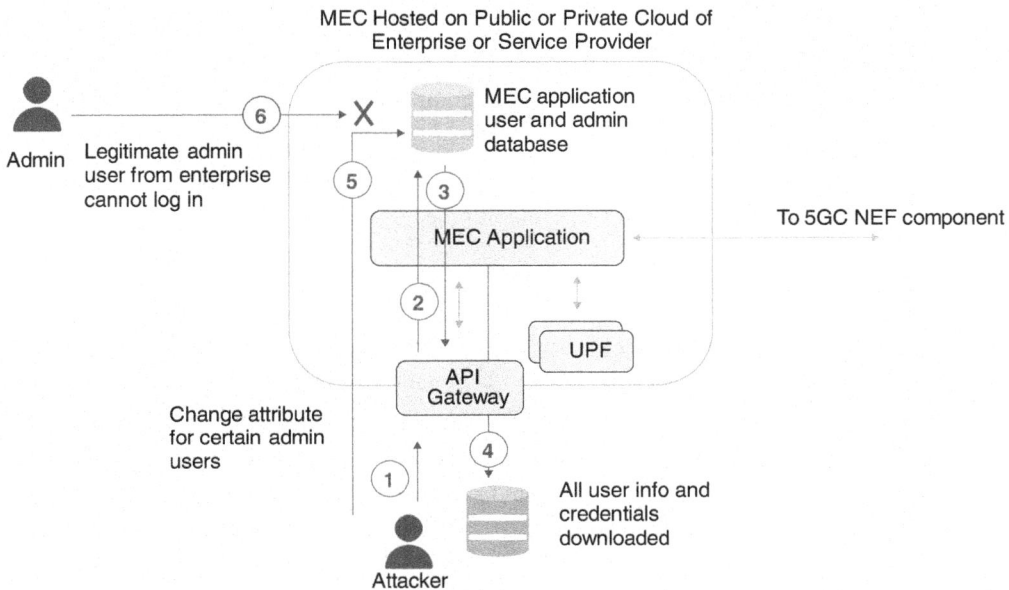

FIGURE 5-27 Broken Object Level Authorization and Broken Authentication API Attacks

As shown in Figure 5-27, the attacker uses Broken Object Level authorization and Broken Authentication to attack the application deployed in the MEC.

The following steps could be used by the attacker:

Step 1. The attacker sends an API call with the admin-level APIs requesting all user information from the database. For example, instead of using

/api/users/MEC_application/user/ my_personal_info

the attacker could use

/api/admins/MEC_application/users/all_personal_info

Step 2. The API gateway relies on the client/user to use a user-level or admin-level API, and hence the request is seen as a valid request from a client/user.

Step 3. The database responds to the API call with all user information, which might include PII (personally identifiable information), which is the personal data of all the users using the application.

Step 4. This information is collected and analyzed by the attacker.

Step 5.   The attacker can now use the ID of another user and launch an attack toward the MEC application user and admin database and remove the admin users.

Step 6.   The legitimate admin users cannot access the MEC application and are denied any service.

This attack can have multiple critical impacts on the service provided by the MEC application and its users. The attack also prevents the administrators from accessing the MEC network, and hence the network or service outage can extend to a longer period of time.

DDoS Attacks

5G use cases, as mentioned in the previous section, will change the way computing elements and IT services are deployed, and this includes application deployment, which is virtualized and highly distributed to enable ultra-reliable low-latency communication (URLLC) use cases.

One of the key DDoS threat vectors that's very specific to 5G is the *zero-day DDoS attack*. The main reason behind this is that it's the first time in the standards that cloud-native architecture has been mandated for a 3GPP technology, and that technology is 5G. The network functions in 5G are based on cloud-native technology, which relies on open source software components and open interfaces. When the service providers and enterprises adopting 5G use these open source components and open interfaces, such as REST API methods, the developers will follow suit, and the MEC applications developed by various third-party and in-house software developers will all use the same open source components and open interfaces so that all the 5G network functions and the third-party MEC applications can co-exist in the same environment. This is the first time in any 3GPP standard so far that such openness and flexibility are allowed.

This new application deployment method using open source components and cloud-native-based architecture also brings in threat vectors like zero-day DDoS attacks, which will consist of new attacks based on the vulnerabilities of these open virtualized environments where the patches have not been created yet or are not yet available. Once the vulnerabilities are known by an attacker or a group of attackers, they can be traded among malicious individuals who can be sponsored to take down the critical infrastructure of service providers, enterprises, or even critical security services deployed by the country to protect *important government infrastructures*. In this section, we will look at the 5G-specific DDoS attacks to better understand the actual threats.

Figure 5-28 shows an example of inbound DDoS attacks from the Internet using container/image vulnerabilities and causing a DoS in a 5G MEC deployment.

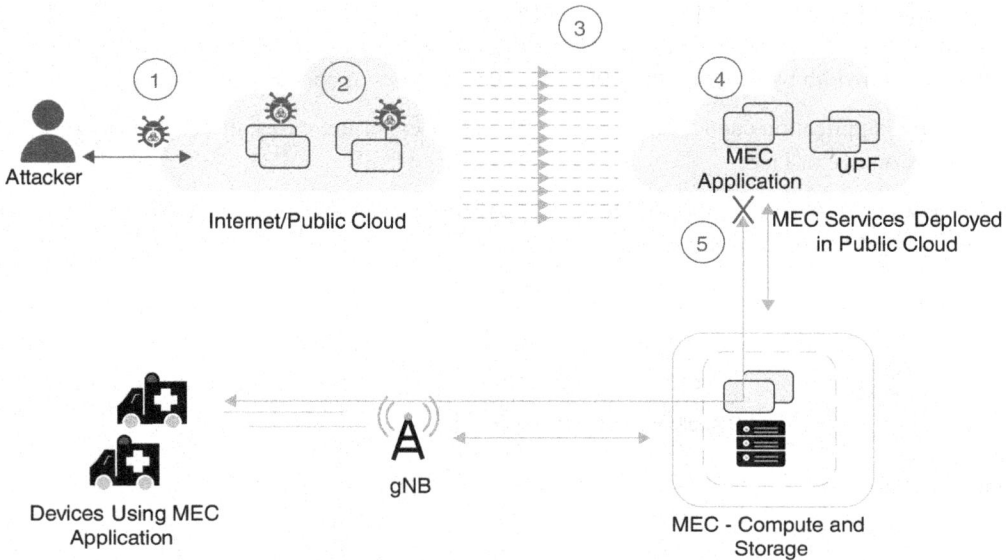

FIGURE 5-28 Inbound DDoS Attack in 5G MEC Components Deployed in Public Cloud

As shown in Figure 5-28, the attacker causes inbound attacks on the applications deployed on the public cloud MEC layer, leading to the DoS to legitimate users of the MEC application. These attacks could also be a type of zero-day attack.

The following steps could be used by the attacker:

Step 1. The attacker scans cloud repositories for unprotected images or a runtime environment. Cloud repositories are where people share container images. Some of the commonly used cloud repositories also allow people to create, manage, and deliver container applications and so on.

Step 2. The unprotected runtime environment could then be assigned malicious code by the attacker. The attacker could then build an image using the runtime environment and upload it into the cloud repository and then start running it. As per the modified command string or the assigned malicious code, the images then start infecting other open and unprotected daemons to create a botnet.

Step 3. The botnets are all assigned a specific target host (MEC application deployed in the public cloud) to attack.

Step 4.   The targeted host (MEC application) is under DDoS attack. Depending on various DDoS methods, the MEC application would be busy processing the request or artificial consumption of bandwidth by fully utilizing the maximum number of concurrent sessions possible.

Step 5.   The MEC application cannot respond to legitimate users or use cases, thereby causing network outage or DoS.

Figure 5-29 shows an example of inbound DDoS attacks from the Internet using and causing DoS in a 5G MEC deployment for legitimate requests.

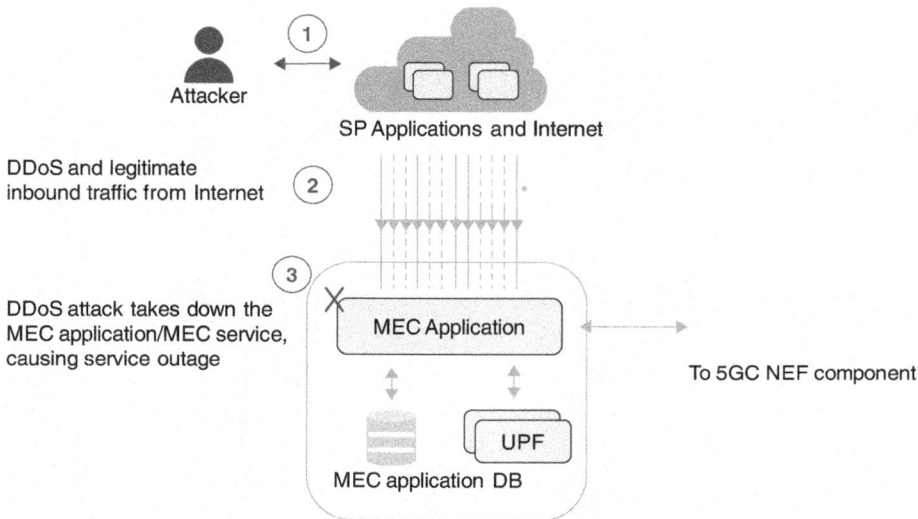

FIGURE 5-29   Inbound DDoS Attack on MEC Applications Deployed in an On-premises MEC

As shown in Figure 5-29, the attacker creates an artificial exhaustion of resources and prevents legitimate requests to be processed.

The following steps could be used by the attacker:

Step 1.   The attacker exploits the cloud-based open source programs using existing vulnerabilities and turns them into bots. These bots, in turn, scan for any other similar vulnerable programs deployed in the cloud, such as in cloud-based repositories, and turn them into bots.

Step 2.   Once the botnet has been established, the attacker is able to orchestrate an attack by issuing remote instructions.

Step 3.   The MEC application server becomes overwhelmed by the number of requests, preventing it from processing legitimate requests and thereby causing DoS for legitimate requests.

Figure 5-30 shows an example of DDoS attacks caused by lack of resources and rate-limiting thresholds on the API calls.

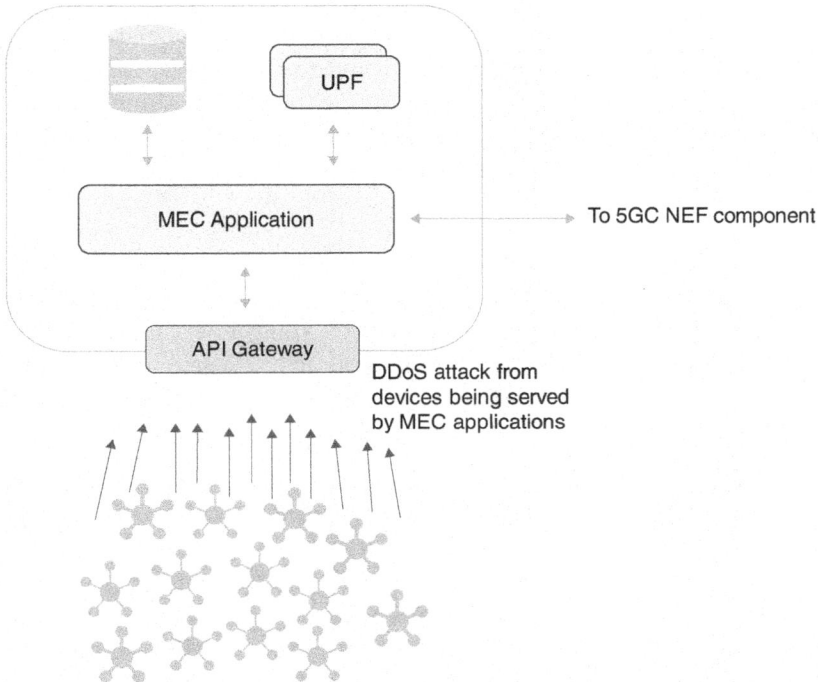

FIGURE 5-30   DDoS Attacks Due to Improper Rate Limiting on APIs

As shown in Figure 5-30, an attack could be initiated by directing a large volume of devices toward the MEC infrastructure hosting the MEC application serving the devices/users. This attack vector can be seen as an application layer attack, as it includes GET/POST floods and targets the vulnerabilities of the API and/or the operating system. These threat vectors are aimed at taking down the servers and causing DoS, and when aimed at critical services and government infrastructure, they can lead to very challenging situations.

Usually, these types of API communications are based on the devices reaching out to the MEC applications or certain network functions such as functions catering for real-time analysis behind the API gateway for use cases such as uploading certain values and getting certain attributes modified by the configuration management of the MEC applications. If these APIs are not protected by rate limitation and payload size, the threat vectors, such as an excessive amount of API calls causing DDoS, can take place, causing DoS to the legitimate API calls from devices. These unprotected APIs can also allow brute-force attacks, which can take down the MEC applications.

Securing 5G MEC

5G edge deployments will supply virtualized, on-demand resources with an infrastructure that connects servers to mobile devices, to the Internet, to the other edge resources, and an operational control system for management and orchestration. These deployments should have the right security mechanisms in the back haul to prevent rogue deployments and the proper security controls to prevent malicious code deployments and unauthorized access. As these MEC deployments will include the dynamic virtualized environments, securing these workloads will be critical. Securing transport, SDN, and orchestration will ensure that service providers are able to provide their services securely from the edge of the network.

Physical Security

Security teams must prepare for new threats that emerge with little to no warning. Part of the solution to this problem lies in technology. It is increasingly common to include the ability to detect and mitigate attacks on various devices, including routers, switches, security appliances, and end hosts. Proper device hardening and the use of security features on these devices can go a long way toward stopping a major outbreak before it occurs.

MEC enables the deployment of containerized 5G network functions in locations closer to the user. Locations closer to the user include on-premises data centers (DCs) of varying footprint and public multi-cloud deployments. Apart from public cloud deployments of MEC, the on-premises deployment of a data center would require layered physical security to ensure that the MEC is secured. Physical security is more critical in an enterprise DC, where the 5G network functions and MEC applications might be deployed in the same DC as the enterprise applications. Some of the key physical security measures are as follows:

- **Monitoring:** One of the key elements is to avert the attacker and prevent any unauthorized entry into the premises. Having a visible CCTV camera, signboards, drone surveillance, automatic light based on motion detection, and so on might ward off an attacker, causing them to look for less-protected areas. The monitoring system should have a cold storage method of housing the recorded surveillance.

- **Access controls:** MEC DCs should include access control systems using card swipes or biometrics. High-resolution video surveillance and analytics can identify the person entering and also prevent tailgating. Video surveillance should also be able to read license plates and conduct facial recognition. You should use multiple different techniques to provide multiple security layers and restrict unauthorized entry.

Here are some of the key best practices for physically securing the on-premises MEC data center:

- Access controls for the data center should have multiverification options, such as access card/biometric/fingerprint to provide an extra layer of security.

- Audit mechanisms and tools should be in place to verify policy enforcement and to determine how data is stored, protected, and used, to validate service.

- Internal audits should be carried out regularly to check for any vulnerabilities in the process for access as well as to ensure proper working of the CCTV cameras and related storage of the captured camera footage.

- Risk management programs should be in place to deal with the continuously evolving deployment modes of MEC.

- Access controls should be regularly checked and updated to reflect the changes in the job roles of the personnel accessing the data center.

Hardening Hardware and Software

Being cloud native and completely software driven, 5G uses open source technologies. Although this is critical for scalability and allowing cloud deployment integrations, vulnerabilities from multiple open source applications could be exploited by attackers. To reduce the attack surface, you need to verify the 5G vendor-specific secure development process to ensure hardened software and hardware. Vendors should follow the secure development lifecycle (SDL) process, adhering to standards such as ISO27034 and ISO9000 for compliance requirements. ISO27034 provides an internationally recognized standard for application security, while ISO9000 helps ensure you meet the statutory and regulatory requirements. One way of doing this is to ensure that you choose vendors that have undergone testing for Federal Information Processing Standard (FIPS) compliance. This standard specifies the security requirements that will be satisfied by a cryptographic module. The standard provides four increasingly qualitative levels of security intended to cover a wide range of potential applications and environments. Level 1 provides the lowest level of security, while Level 4 provides the highest level of security. The security requirements cover areas related to the secure design and implementation of a cryptographic module. The public cloud providers for 5G MEC should be authorized by FedRAMP.

FedRAMP is a U.S. government program to standardize how the Federal Information Security Management Act (FISMA) applies to cloud computing services. FedRAMP provides a standardized approach to security assessment, authorization, and continuous monitoring of cloud-based services. Using a "do once, use many times" framework, FedRAMP reduces the cost of FISMA compliance and enables government entities to secure government data and detect cybersecurity vulnerabilities at unprecedented speeds.

FedRAMP was developed in collaboration with the National Institute of Standards and Technology (NIST), General Services Administration (GSA), Department of Defense (DOD), and Department of Homeland Security (DHS). Other government agencies, working groups, and industry experts participated in providing input into the development of FedRAMP. It was launched in 2011, with the main aim to promote the adoption of secure cloud services across federal agencies.

FedRAMP assessment is quite stringent and takes around 6 to 9 months to complete, and the certification is valid for 1 year. While not all solutions that have achieved FedRAMP authorization might be available for public consumption, by selecting vendors that have achieved FedRAMP authorization on cloud solutions for 5G deployments, you will be taking a step in the right direction.

Just choosing vendors having a robust SDL process and following standards for hardened platforms is not enough to ensure you have a hardened system in place. SDL, in fact, is a continuous journey, as you ensure the secure process is catered for during design, initial deployment, software updates, hardware upgrades, maintenance, and migration.

The first step in having a hardened process is to have a secure boot process where the OS boot images are authenticated against a hardware root of trust. Having the secure boot process authenticating against the hardware root of trust makes it very difficult for an attacker to cause tampering attacks and expose the encryption keys. Without the hardware root of trust, third parties can tamper with BIOS, boot loader, or ROM monitor (ROMMON) boot code to load modified software images; bypass hardware, authenticity, and licensing checks; or perform additional functions with malicious intent. Tampered code can also result in data manipulation or data theft, and it can provide a platform to launch attacks, including DoS.

Hardware root of trust is provided by many vendors by using the Trusted Platform Module (TPM), also known as ISO/IEC 11889. TPM was conceived by the Trusted Computing Group (TCG). The primary scope of TPM is to ensure integrity of the platform. This section primarily details the hardening of the MEC infrastructure. Using embedded security within the MEC servers will help protect the keys, passwords, and digital certificates. One of the key advantages of TPM is the ability to indicate any actions of compromise and the relevant action of denying system access. In some cases, this action of denying system access can be automatic or can be manual based on the alert created for indications of compromise. One of the methods by which TPM takes this action is due to the unique key that is created by taking a baseline fingerprint of the server and its network functions as it boots. This baseline is then compared with measurements that are periodically taken from the system processes. If any change in the boot characteristic changes, then system access can be denied.

Although TPM solves part of the problem in MEC deployments, cloud-native deployment brings in a whole different angle to the topic of "trust" in 5G.

This is one of the most important topics with my discussions with many customers.

The answer to this is to have multilayered security controls for the virtualized functions irrespective of where they are deployed. The applications (network functions and MEC applications) should have the flexibility to be deployed on premises, in the public cloud, or in hybrid mode (on premises and public cloud) and still be able to verify for trust. When data is transferred among networked systems, trust is a central concern. In particular, when communicating over an untrusted medium such as the Internet, it is critical to ensure the integrity and the publisher of all the data a system operates on. Some of the key security controls are discussed in the sections that follow.

Image Signing

Image signing helps you validate the source of the software image and the authenticity of the image, thereby ensuring that the image has not been tampered with. You should ensure that the vendor you are choosing follows the image-signing process for both on-premises deployments and public cloud deployments, including VMs and containers. For on-premises deployments, some vendors require their dedicated hardware for their own specific software image verification, which might limit your deployment models, but it still ensures that you have a trusted image in your network. For container-based deployments (on premises and public cloud), the vendor should provide some sort of container registry mechanism where you can sign images to ensure their integrity in your registry namespace. This container registry should also provide an option for you to push and pull signed images. This will help you to validate whether the images are being pushed to the container registry by the Continuous Integration (CI) tools you own. One of the most used platforms in the container world today is Docker. In Docker, the Docker Engine is used to push and pull images to and from the public and private registry. Docker Content Trust (DCT) provides the ability to use digital signatures for data sent to and received from remote Docker registries. These signatures allow client-side or runtime verification of the integrity and publisher of specific image tags. Trust for an image tag is managed through the use of signing keys.

When DCT is enabled, only signed versions of the image can be pulled by you. For example, if you need an image of a 5G network function by Vendor A, Vendor A uploads a couple of image versions in the DCT. Once Vendor A signs the images, they can then be pulled. This ensures that you can pull only tested and signed images.

Figure 5-32 shows a method of pulling signed images from the Docker Trusted Registry (DTR).

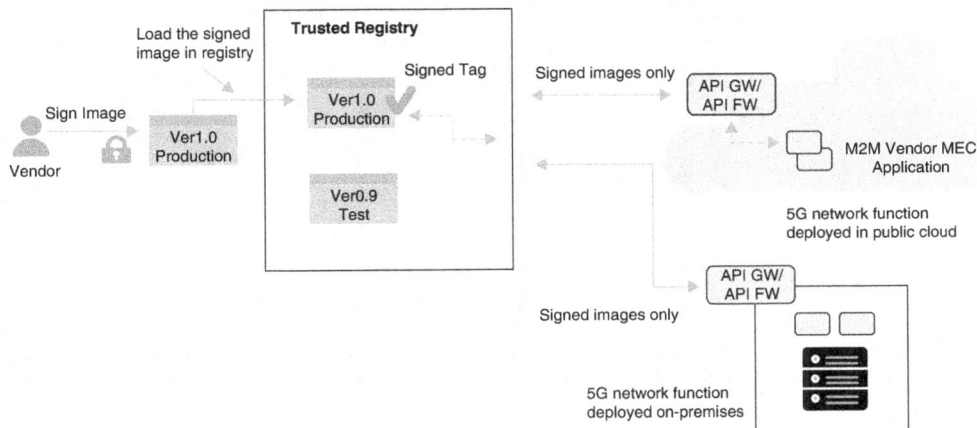

FIGURE 5-31 Trusted Images Being Pulled from the Trusted Registry

As shown in Figure 5-31, the vendor/developer signs the images and loads them into the registry. The registry contains both the signed and unsigned images. You can pull any 5G application-related images from the registry. When you request a pull for an image, only the latest signed image can be pulled from the registry. Once the signed image is pulled from the registry, it needs to be verified. This option is applicable to both the cloud and on-premises deployment models.

Secure Storage

While using the on-premises servers for MEC, you should ensure that the server provides highly secure storage for user credentials, private encryption keys, and other critical security information for the device. There are also very specific vendors providing a hardware security module (HSM), which is a physical device to protect and manage data at rest. The HSM you choose should be able to perform encryption and decryption functions for digital signatures, strong authentication, and other cryptographic functions. There are many varieties of HSMs; some are external devices that physically connect to devices such as PKI infrastructure, and some are plug-in cards typically used to back up the keys handled outside of the HSM.

HSMs are usually designed to execute specially developed modules within the HSM's secure enclosure, which is required in use cases where special algorithms or business logic has to be executed in a secured and controlled environment, such as for critical IoT 5G use cases in on-premises deployments of MEC. Due to the application of HSM in critical use cases, it is very important to ensure that you choose HSM vendors who have Level 4 testing and certification achievement in FIPS 140 (minimum FIPS 140-2). The use of a security gateway to protect MEC transport is explained in detail in the section "MEC Infrastructure and Transport Security." Security gateways use PKI implementation for taking part in IPsec tunnel establishment and must have an RSA key pair and a certificate issued by a trusted Certificate Authority (CA). In PKI implementations, HSMs can be used by the CA to generate, store, and handle asymmetric key pairs. For 5G network functions, PKI-based certificate authentication can be used for TLS. If HSMs are used for 5G PKI-based use cases, they should support the following features to provide enhanced security:

- Provide granular role-based access control (RBAC)
- Secure key backup
- Support for multipart authorization schemas
- Blockchain readiness
- Support for multi-cloud integrations (at least the top three cloud providers in your region)
- Implement quantum-safe algorithms
- FIPS 140 Level 4 certified (minimum FIPS140-2, but FIPS 140-3 is even better)

Details on securing 5G CNFs in production, securing your network from the 5G NF and vulnerabilities, and compliancy of 5G CNFs are covered in the section "Securing Virtualized Deployments in 5G MEC" later in this chapter.

MEC Infrastructure and Transport Security

One of the main reasons to use MEC deployment models is to enable URLLC use cases and allow the flexibility in deployment of 5G-related MEC applications as close to the edge of the network as possible. There are different models of MEC deployment, such as deploying the 5G network functions and MEC applications in the multistack public cloud, in the same location as the gNB using servers with very low footprint, in far-edge data centers, in regional data centers, in hybrid cloud and on-premises models, and also enterprise data centers. As you have seen in the "MEC Infrastructure and Transport Vulnerabilities" section earlier in this chapter, the key vulnerabilities are due to MEC using the public network for the transport layer. This section will take you through the key security controls required to secure the 5G MEC deployments and methods to mitigate the MEC infrastructure-related threats.

Figure 5-32 shows one of the methods to secure transport for on-premises deployment of 5G MEC.

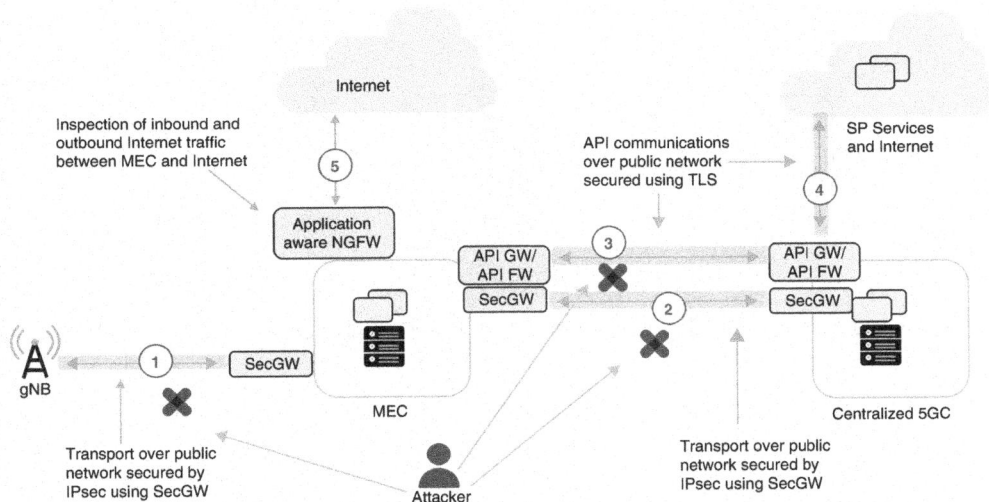

FIGURE 5-32　Securing Transport for On-premises Deployment of 5G MEC

The key security controls recommended for on-premises deployment of 5G MEC, as shown in Figure 5-32, are as follows:

1. **IPsec between gNB and MEC:** Security gateway (SecGW) can be used to provide an IPsec tunnel between the gNB and MEC if an untrusted public network is used as the transport between gNB and MEC. gNB requires mandatory support of the IPsec stack, which can establish IPsec tunnels natively; therefore, you don't need a SecGW at the gNB location. You will require a SecGW at the MEC and centralized 5GC location.

2. **IPsec between MEC and Centralized 5GC:** Depending on the location of the DU, CU, serving User Plane Function, and the control plane functions such as Access and Mobility Management Function (AMF) and Session Management Function (SMF) and the use of an untrusted public network, an IPsec tunnel will be required to secure the traffic traversing between the MEC and centralized 5GC. An example would be the traffic between the DU and CU, as shown in Figure 5-33.

Figure 5-33 shows the SecGW deployment between the DU and CU and between the CU and the centrally located UPF.

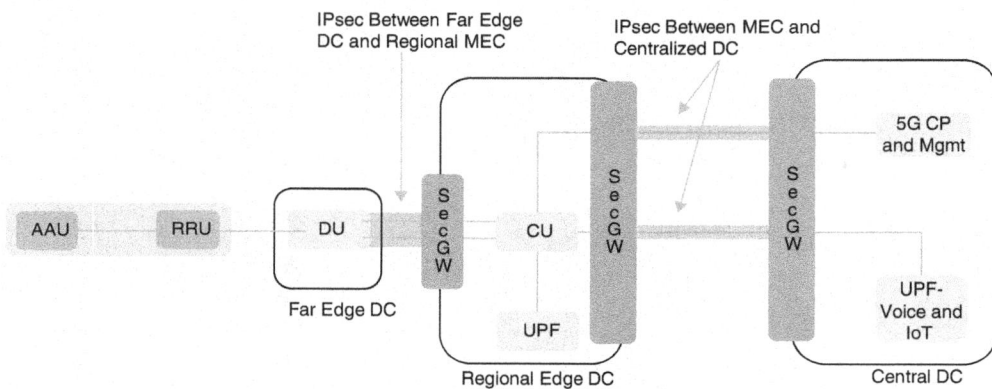

FIGURE 5-33   Example of SecGW Deployment between DU, CU, and UPF

As shown in Figure 5-33, the traffic passing through the untrusted public network is protected by the SecGW:

1. **API GW/API FW between MEC and Centralized 5GC:** To secure API communications such as between the 5G CNF components and orchestration, an API gateway (API GW) or API firewall (API FW) is recommended between the MEC and centralized 5GC. API security has been explained in detail in the "Securing API" section of this chapter.

2. **API GW/API FW between centralized 5GC and services deployed in the Internet:** To secure the API communications between the centralized 5G network functions and services deployed in the Internet, an API gateway (API GW) or API firewall (API FW) is recommended between the MEC and centralized 5GC. An example could be data analytics layer API security, which has been explained in detail in the "Securing API" section of this chapter.

3. **NGFW at the MEC location:** To inspect the traffic from the MEC applications and 5G CNF, such as UPF, a next-generation application-aware firewall (NGFW) is recommended. This firewall is aimed at ensuring that 5G CNFs are using only authorized ports and protocols and also inspects the incoming packets for any abnormalities. This NGFW should also be able to

provide URL filtering and have integrations with threat intel for creating automatic rules based on indicators of compromise (IoCs).

Figure 5-34 shows one of the methods to secure transport for hybrid deployment of 5G MEC.

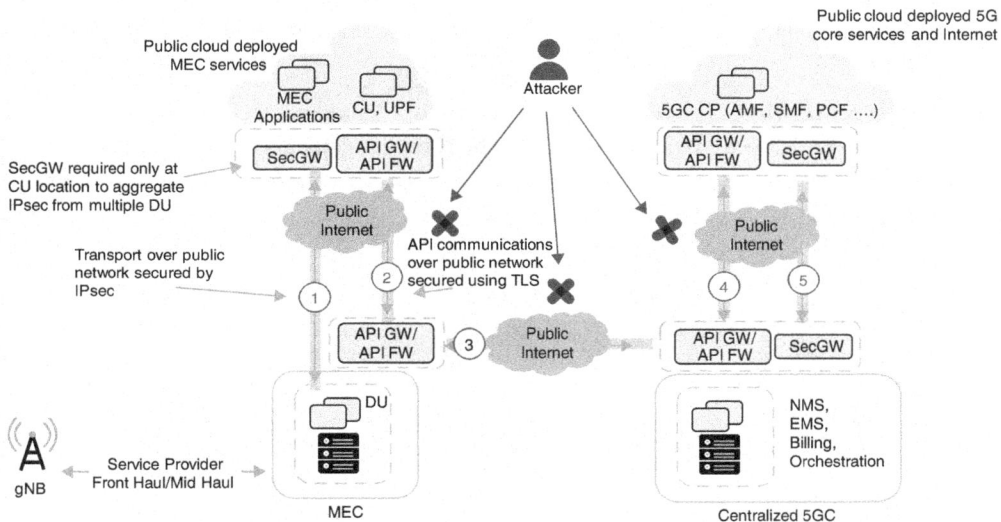

FIGURE 5-34   Securing Transport for Hybrid Deployment of 5G MEC

The key security controls recommended for hybrid 5G MEC deployment, as shown in Figure 5-34, are listed here:

1. **IPsec between MEC and public cloud:** In this deployment, the DU is deployed in the far edge DC, while CU is deployed in the public cloud. The public cloud virtual instances also include MEC applications and the user plane functions. IPsec implemented by SecGW is recommended to secure the traffic in the F1 interface, which is between the CU and the DU. F1-C is the interface between DU-CP and CU-CP, and FI-U is the interface between DU-UP and CU-UP. Only the CU locations need the SecGW implementation, as the DU supports an embedded IPsec stack to support initiation and termination of IPsec tunnels between the DU and SecGW at the CU location.

2. **API GW/API FW between MEC and public cloud:** To secure the API communications using an untrusted public network between any MEC clients deployed on the MEC and the MEC application server deployed in the public cloud, it is recommended to implement an API GW/ API FW. The API FW is also sometimes called Web Application Firewalls (WAF) by firewall vendors.

3. **API GW/API FW between MEC and centralized 5GC:** To secure the API communications using an untrusted public network between any MEC and centralized 5GC, such as API calls from the orchestration layer, it is recommended to implement an API GW/API FW.

4. **API GW/API FW between centralized 5GC and public cloud:** To secure the API communications using an untrusted public network between centralized 5GC and the public cloud, such as API calls between the billing and 5GC charging functions, it is recommended to implement an API GW/API FW.

5. **IPsec between centralized 5GC and public cloud:** In this deployment, the 5GC control plane functions are deployed in the public cloud. The IPsec implementation is recommended to secure non-API communication between 5GC centralized network functions and 5G control plane functions deployed in the public cloud.

Figure 5-35 shows one of the methods to secure transport for RAN location–deployed 5G MEC.

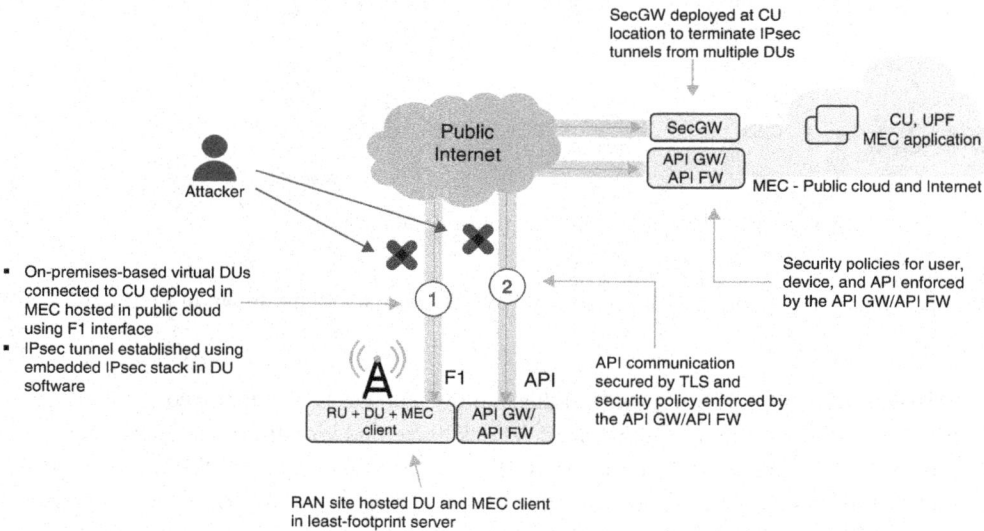

FIGURE 5-35   Securing Transport for 5G MEC Deployed in Low-Footprint RAN Node Server

The key security controls recommended for 5G MEC deployed in a RAN location, as shown in Figure 5-35, are as follows:

1. **IPsec between DU in MEC and CU in the public cloud:** In the example shown in Figure 5-35, the DU control plane and user plane are deployed in the RAN location within a low-footprint server, and the CU control plane and user plane are deployed in the public cloud along with the UPF. Another variant could be DU–UP, DU-CP, and CU-UP deployed in the RAN node location and the CU-CP deployed in the public cloud. But as the CU allows aggregation of multiple DUs, and if you want to have the server as low footprint as possible, it's best to have the CU-CP and CU-UP deployed in the public cloud. In this case, the F1 is protected by the IPsec tunnel, which is supported by the DU software stack, which is terminated by the SecGW deployed in the public cloud, which is aggregated toward the CU-CP and CU-UP.

2. **API GW/API FW between MEC and the public cloud:** To secure the API communications using an untrusted public network between any MEC clients deployed on the 5G MEC deployed on the RAN location and the MEC application server deployed in the public cloud, it is recommended to implement an API GW/API FW. The API can be protected by using TLS.

Figure 5-36 shows one of the methods to secure 5G MEC deployed in enterprise data centers.

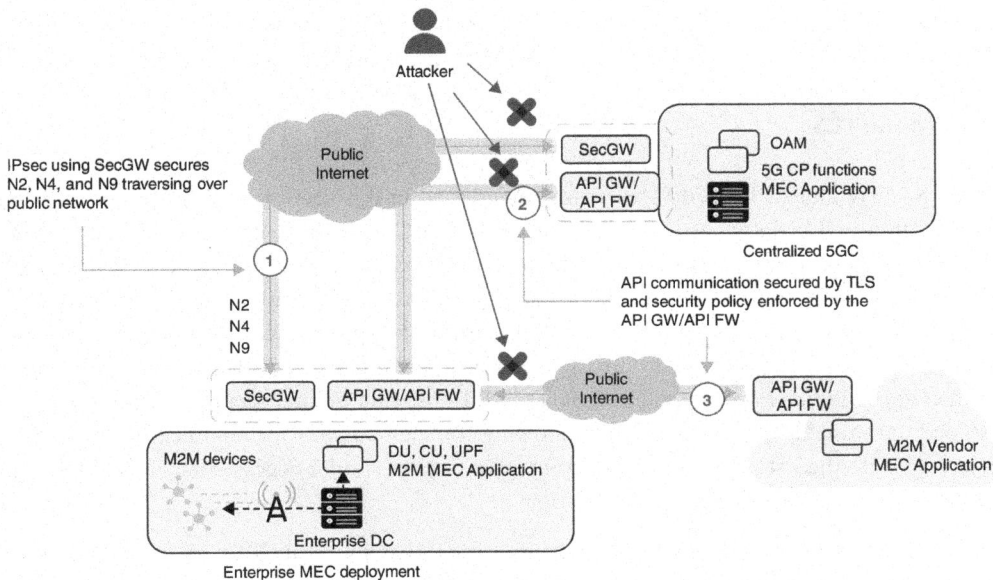

FIGURE 5-36   Secure Transport for 5G MEC Deployed in Enterprise Data Centers

The key security controls recommended for enterprise network–based 5G MEC deployment as shown in Figure 5-36, are as follows:

1. **IPsec between enterprise MEC and centralized 5GC:** The DU control plane and user plane, along with the CU control plane and user plane, are deployed in the enterprise location. The following interfaces would traverse the untrusted public network:

   - **N2:** Interface between gNB-CU-CP and AMF or gNB and AMF (depending on deployment)

   - **N4:** Interface between UPF and AMF

   - **N9:** Interface between UPFs

   To protect these interfaces from any eavesdropping or MitM attacks, it is recommended to use IPsec catered by SecGW between the enterprise MEC and 5GC.

2. **API GW/API FW between enterprise MEC and centralized 5GC:** To secure the API communications using an untrusted public network between any MEC clients deployed on the 5G MEC deployed on the enterprise location and the MEC application server deployed in the centralized 5GC, it is recommended to implement an API GW/API FW. The API can be encrypted using TLS.

3. **API GW/API FW between enterprise MEC and the public cloud:** To secure the API communications using an untrusted public network between any MEC clients deployed on the 5G MEC deployed on the enterprise location and the MEC application registry hosted by the MEC application vendor, it is recommended to implement an API GW/API FW. The API can be encrypted using TLS.

Apart from the SecGW and API GW/FW use cases mentioned in this section, it is also important to provide segmentation and isolation between the enterprise MEC and the centralized 5GC. This can be provided by configuring zones within the data center network and then implementing granular access control policies. You must write specific security policy rules to allow traffic to pass between zones, so only traffic that you explicitly allow can move from one zone to another. The isolation can be further enhanced by using Security Group Tags (SGTs) functionality provided by vendors in the market and then apply policies based on those SGTs. SGTs provide unique security group numbers that are assigned to the security groups, thereby allowing for segmentation without needing VLANs. The security groups are the list of users and endpoints that require similar access control. Having this flexibility will enable you to deploy granular isolation.

Malware detection and response is another key security control required in MEC deployments. For 5G deployments, you will require next-generation malware implementation. The chosen anti-malware solution should build a full real-time context around every process executed and use machine learning models, which identify patterns known as malware characteristics and various other forms of artificial intelligence.

Securing Virtualized Deployments in 5G MEC

5G standalone (SA) deployments are primarily cloud-native network functions (CNFs). This means that the containerized functions should be able to be deployed in any infrastructure and public cloud provider without impacting the interoperability and specific functions. While it sounds really flexible and scalable, care has to be taken while deploying such functions. Apart from the embedded security controls discussed in the section "Hardening Hardware and Software" in this chapter, there needs to be multiple layers of security to ensure that the cloud-native deployments of 5G network functions and MEC applications are protected. We should also ensure that these virtual functions are also secured, and any data exfiltration is detected and stopped.

Securing the cloud native functions and containers is discussed in detail in Chapter 6, "Securing Virtualized 5G Core Deployments." This section is aimed at securing virtualized deployments specific to 5G MEC.

The location of these functions within the network is also important before deciding the security controls. Some of the locations, such as low-footprint servers, are very lightweight deployments that sometimes have no separate rack or shelter to accommodate the security appliances. The different locations for MEC deployments are as follows:

- RAN node location
- On premises (far-edge, edge, regional data center)
- Cloud-based deployment
- Hybrid deployment
- Enterprise location

The virtualized 5G-related network functions expected to be deployed in MEC are as follows:

- Virtualized DU
- Virtualized CU
- Virtualized user plane functions
- Virtualized control plane functions (some functions only, in specific circumstances)
- Subscriber authentication and policy functions (IoT related, apart from the 5G control functions)
- IoT applications
- M2M applications

- Enterprise applications

- MEC applications

All the virtualized network functions described in the list will be mainly cloud-native containerized functions deployed on virtual machines, bare metal, or in the public cloud. In this section, we will discuss the different security controls to secure the virtualized network functions and security controls to secure the MEC infrastructure from vulnerability in the virtual functions.

Figure 5-37 shows the recommended security controls to secure the virtualized network functions and applications in MEC deployments.

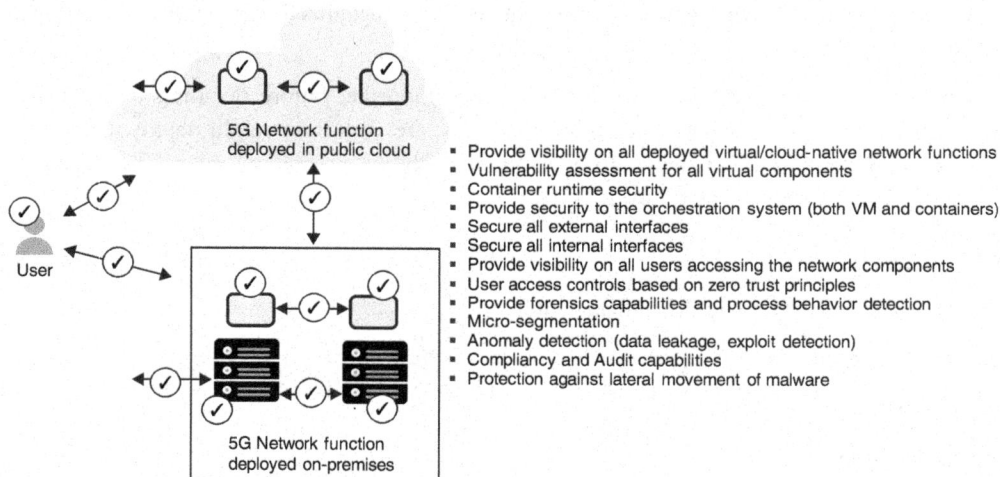

FIGURE 5-37   Recommended Security Controls to Secure the Virtualized Network Functions in MEC

As shown in Figure 5-37, true security for the virtualized deployments in MEC also needs to consider non-virtual components as well. This includes internal and external interfaces of the on-premises server where the virtualized components are deployed on MEC.

Figure 5-38 shows an example and method for image scanning, analytics, and the application policy enforcement layer for securing the virtualized deployments in 5G MEC. The recommended practice is to use the private repository for your 5G RAN and core CNFs and use the public repository only for the 5G MEC applications.

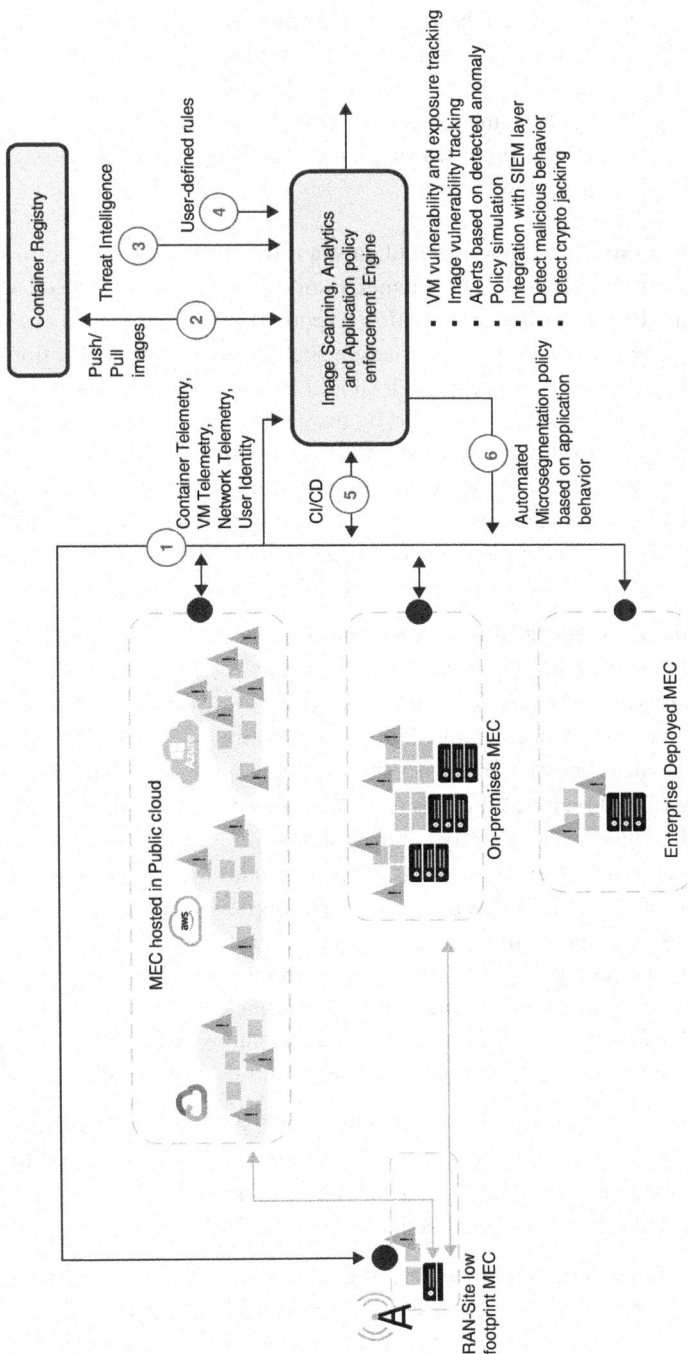

FIGURE 5-38 Multilayered Security Control for Securing Virtualized Deployments in 5G MEC

As shown in Figure 5-38, securing virtualized deployments would require multilayered security controls. The security controls should be able to be enforced irrespective of the location (public cloud, on-premises MEC, enterprise data center) the virtualized network function/MEC application is deployed. The security policies should be consistent wherever the 5G network functions and MEC applications are deployed, be it bare-metal servers, virtual machines, or dynamic environments like containers and microservices in on-premises or public cloud. The security controls and capabilities shown in the Figure 5-38 are as follows:

1. **Telemetry collection:** The solution should have a robust telemetry collection to create an accurate baseline of the network. The telemetry collection should include telemetry such as NetFlow, Internet Protocol Flow Information Export (IPFIX), sFlow, and other types of flow data from your multivendor MEC infrastructure such as routers, switches, firewalls, endpoints, and other network infrastructure devices. It should also receive and collect telemetry from proxy data sources, which can be analyzed by the cloud-based, multilayered machine learning engine for deep visibility into both web and network traffic; this includes container sidecar components, inter- and intra-microservices communication, and it should be able to consume logs from the public cloud. The security controls should provide visibility on all deployed virtual components. The virtual components include virtual components such as VMs, CNF as containers deployed on a VM, bare metal, public cloud, and hybrid cloud environments.

2. **Image scanning and vulnerability assessment:** The security controls should provide the vulnerability scanning for test and production MEC environments. Any images of 5G CNF and MEC applications should be verified and checked for any vulnerability before being deployed on your production pods. Once deployed, the vulnerability assessment should include vulnerability tracking against known Common Vulnerabilities and Exposures (CVEs). The vulnerability scanning should include all forms of virtual components (VMs, CNFs, containers on VMs/BMs/public cloud). The security controls should provide a secure build, deploy, and run process; this includes any DevOps process, which includes the process of signing and tagging any images built by you and loaded in the container registry, GitHub, or serverless deployments for MEC applications. The platform providing the security control should include compliancy controls and runtime policy. User-defined policies should be able to be enforced on the trusted and verified images. Validation of compliancy should include benchmarks from PCI-DSS, NIST800-190, NIST800-53, CIS, and so on, which should be automated. The audits should also be carried out for the orchestration component used for your deployment.

3. **Threat intelligence:** The platform providing the security control should have a global threat intelligence feed to stop any threats that are seen in any other part of the world. The platform should also maintain an up-to-date CVE data feed from multiple sources, including NIST and OS vendor data packs, which contain the latest vulnerability and exposure information.

4. **User-defined policies:** You should be able to configure user-defined policies for specific 5G CNFs being deployed on the MEC. Once the image is verified and trusted, specific policies should be able to be enforced on the image. The security controls should provide granular

access control and include or be able to integrate with controls such as multifactor authentication (MFA) and role-based access control (RBAC). The user-to-application mapping should be part of the baseline, and any anomaly should be identified and alerts/events created for it.

5. **CI/CD integration:** The Continuous Integration / Continuous Delivery (CI/CD) workflow integration is crucial for robust security control, specifically if you have a DevOps team uploading images for MEC applications that might be developed in house by your team. Any images being uploaded by your development team in a registry, GitHub, and so on should go through the vulnerability scanning process to detect any vulnerabilities.

6. **Micro-segmentation and anomaly detection:** The security controls should enable automatic micro-segmentation based on the initial baseline considered for deployment. The micro-segmentation process should also consider 5G CNF-to-services mapping. An example of services is DNS requirements. The micro-segmentation layer should be able to enforce the micro-segmentation policy across all virtual components and deployment locations (on-premises/public cloud/hybrid). The security controls should provide anomaly detection based on behavioral analysis on the initial baseline. The baseline should also include a process baseline and be able to identify deviations for faster detection of indicators of compromise (IoCs). The anomaly detection engine should be integrated or have built-in machine learning, behavior analysis, and algorithmic approaches to ensure that lateral movement can be identified and contained.

Build, Deploy, Run

It can be overwhelming to track all vulnerabilities. The solution is to have a good process to determine which vulnerabilities are relevant to your organization. This prevents overreaction and the mistakes that can be inherent to rapid, poorly planned upgrades or configuration changes.

Figure 5-39 shows the secure runtime environment with secure build, deploy, and run processes, further described in the list that follows.

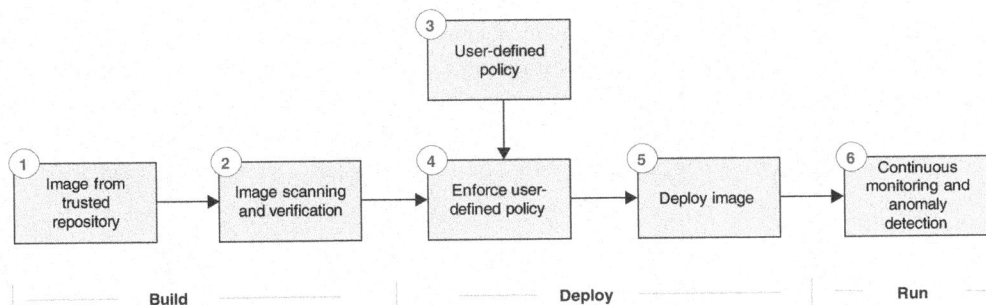

FIGURE 5-39 Secure Runtime Environment with Secure Build, Deploy, and Run Processes

Step 1.   As described in the section "Hardening Hardware and Software" within this chapter, only trusted registries should be used for loading the signed images by the vendor. You should push/pull only signed images from the trusted registries.

Step 2.   The push and pull images should undergo scanning as part of the CI/CD pipelines and run-time deployments to detect vulnerable images. If any vulnerable images are found, then the security control should prevent the creation of a pod.

Step 3.   For 5G MEC deployments, it is important that you can define and enforce policies for the 5G network functions and MEC applications, such as for user plane functions (UPFs), Centralized Unit (CU) and Distributed Unit (DU), and so on.

Step 4.   The verified and trusted images can now have user-defined policies being enforced so that you can deploy images with the right polices.

Step 5.   Once the images are verified and trusted and specific policies are enforced for the images, the next step should have the automated configuration check before it is deployed.

Step 6.   The deployed images should be under continuous monitoring and be tracked for any vulnerability against the databases of known CVEs. Having anomaly detection engines would help you detect any lateral movement between the MEC virtual components.

Side-channel attacks occur due to vulnerabilities or due to methods of exploiting the measurements of the chipsets (mostly crypto chipsets), which leads to leaking of information. There are many variations of side-channel attacks, such as based on time attacks, which are based on the fact that different operations or input values have a significant time variance. Using this information, the attacker can deduce the secret key from the cryptosystem. Another attack vector is based on the attacker performing waveform and amplitude of voltage peaks used by the system, including the idle values and active (during operation) values, which are then compared with the known values for certain operating system software.

These attacks can be passive or active attacks. A passive attack is when the attack simply involves listening to the environment emissions caused by the circuit. In a passive attack scenario, the attacker uses passive monitoring devices such as a receiver antenna in close vicinity of the equipment to monitor the clock frequency of the CPU used in the circuit and capture signal waveforms to retrieve secret information. An active attack is when the attack involves tampering with the operations of the circuit. In an active attack scenario, methods such as power tracing based on the measurement of the current drawn by the chipset from the supply voltage is traced and then compared to the database consisting of known current consumption values of targeted software and hardware circuits and chipsets. In such attacks, the attacker would require a power-tracing chip or circuit between the targeted chipset and the power supply.

There are multiple ways of mitigating these attacks. Mitigation of some of these attack vectors will require software and firmware updates and patches, a secure supply chain, and a very secure physical facility to cater for many of the known vulnerabilities. But the key threats are from the unknown vulnerabilities or vulnerabilities that are not yet exposed. Quantum computing brings in a whole different perspective and requirement of threat and risk analysis required to be done to the entire MEC infrastructure, but that is better covered in Chapter 12 "5G and Beyond."

Figure 5-40 shows an example of one of the methods for mitigating side-channel attacks.

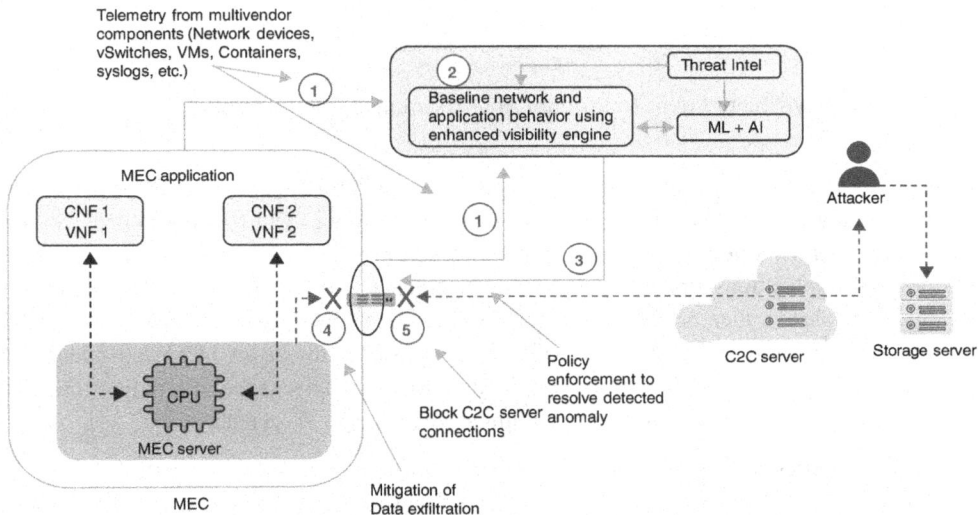

FIGURE 5-40   Example of Side-Channel Attack Mitigation

Figure 5-40 shows one of the countermeasures for side-channel attacks. The key preventive method for side-channel attacks is to use chipsets that are resistant to or have built-in capabilities against known side-channel and fault injection attacks. But the fact is that the chipsets are never fully resistant against side-channel attacks. It is important that the service providers and enterprises build a security mechanism to detect data exfiltration and contain the threat by not allowing it to get out to the network.

Let's look at the steps in one of the methods for mitigating side-channel attacks by using the detection and mitigation system illustrated in Figure 5-40.

Step 1.   Network and user metadata is collected from various sources (network devices, virtual machines, container hosts, containers) and fed into the enhanced visibility solution.

Step 2.   As shown in Figure 5-41, for the enhanced visibility layer to identify any anomaly in the network, there needs to be relevant data to be fed into the system.

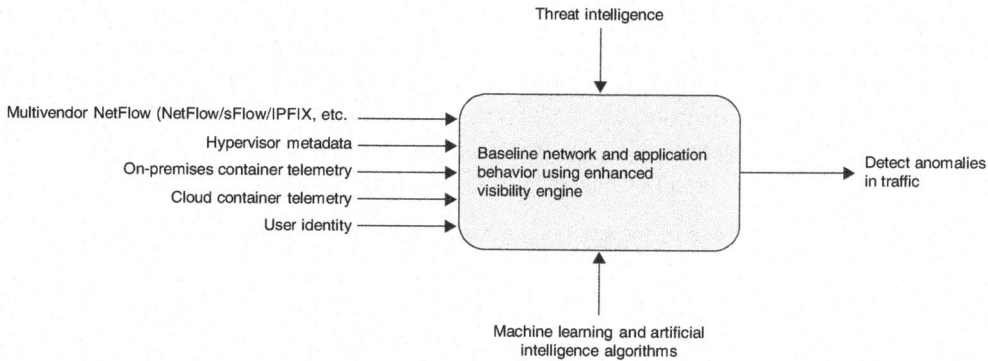

FIGURE 5-41   Enhanced Visibility Layer to Identify Any Anomaly in the Network

An enhanced visibility solution needs to be able to use a combination of different methods. It begins by collecting data from multiple layers and combining them for comprehensive visibility. Once data is collected from the right sources, a baseline of normal behavior for network devices, applications, and users could then be created. The next step is to apply machine learning. The solution then needs a robust threat intelligence input that is aware of malicious campaigns and maps the suspicious behavior to an identified threat for increased fidelity of detection. This should be the bare minimum required to thwart attacks that might have crossed the perimeter, or even threats originating within or are hiding in encrypted traffic.

■ **Network telemetry:** This is composed of data sources that can provide useful insights about who is connecting to the network and what they are up to. It includes a multivendor collection of packet flow, syslog, SNMP information from network device exporters, switches, virtual switches, and includes formats such as NetFlow, jFlow, sFlow, IPFIX, Netstream, SNMP, NSEL (NetFlow Secure Event Logging), and syslog.

■ **Hypervisor metadata:** Hypervisor metadata is the metadata collected from various hypervisors deployed in the infrastructure. The most common sources of hypervisor metadata is from hypervisors such as ESXi and KVM.

■ **On-premises container telemetry:** On-premises container telemetry includes the container's telemetry deployed on the premises of the service provider or enterprise MEC. One of the sources for container telemetry is Kube-state-metrics, which is a simple service that listens to the Kubernetes API server and generates metrics about the state of the objects, such as deployments, nodes, and pods. The other common sources for Kube-state-metrics for Kubernetes hosts running on Kubelet are from the node exporters like Prometheus, which would expose telemetry data for system resources on an HTTP endpoint.

- **Cloud container telemetry:** Cloud container telemetry is the container telemetry for MEC applications deployed on the public cloud. Public cloud providers normally provide out-of-the-box monitoring for the containers deployed in their cloud infrastructure. In many instances the service providers and enterprises going for the cloud deployment of the MEC functions require the logs to be collected and visualized on an on-prem solution to have end-to-end visibility. In such circumstances, the chosen vendor should be able to collect flow logs from the cloud provider and perform various types of flow analysis.

- **User identity:** User identity information is the information gathered by an existing IAM (Identity and Access Management) solution. This information will further enhance the intelligence gathered from the network device, virtual machines, and containers. The mapping of the users/user devices and the network and application flow information would help the host quarantine mechanisms in case any hosts get infected.

- **Threat intelligence:** Threat intelligence is the global information about threats and threat actors, which can be fed into the enhanced visibility engine to act against known threats and threat surfaces. The threat intelligence is usually collected from multiple sources, including the deep and dark webs. The main intention is to detect threats that are previously known and make informed decisions about them.

- **Machine learning and AI algorithms:** This is one of the most important parts of the enhanced visibility engine. The attack vector, which would be quite unique to 5G due to its deployment methods (open-source cloud-native functions), is zero-day attacks. These are attacks that have never been seen before, hence the name "zero day". In 5G networks zero day attacks are typically due to vulnerabilities in the virtual and open source components used for 5G network functions. Machine learning and AI algorithms could apply predictive analysis and various mathematical models to the collected data and threat intel for any prediction of attacks. A fine-tuned model, along with the collected network and user information, will help determine any attacks beforehand.

Step 3. Once the analysis of the data is completed and any anomaly is detected, the solution can take an appropriate action. In this example, the enhanced visibility solution finds a thin stream of data being continuously exported out of the network. The enhanced visibility solution also understands from the analysis and threat intel that the server is a part of a malicious network. The enhanced visibility engine now comes to the conclusion that the data being sent out is part of the data exfiltration procedure being carried out by malicious actor.

Step 4. Enhanced visibility engine then sends out the command to the network device (switch/router/vSwitch) to terminate the connection to the external server and suspend any communication with it.

Step 5. Enhanced visibility engine also detects that the inbound traffic from the malicious server (if any) can be blocked or directed to a sinkhole/quarantine-segmented network to prevent alerting the malicious actor.

As explained in this example, the data exfiltration can be detected using an enhanced visibility engine with the help of data collection from the right sources, robust threat intelligence, and a well-trained machine learning and artificial intelligence algorithm.

Securing API

API security is a very critical security control for securing the MEC and the 5GC cloud-native functions. Although TLS is used for protecting the API and is recommended as part of securing service-based architecture (SBA), there are threat surfaces where malware or DDoS attacks could be orchestrated using the encrypted layer as well. Let's look at the enhancements that 3GPP is working on as well as other mechanisms to secure the API.

3GPP is working on a framework for the unified northbound API framework across several 3GPP functions, called Common API Framework (CAPIF), to help secure the APIs and have a common service platform for 5G implementations and use cases, many of which would be deployed on the MEC layer. Another activity specific for 3GPP edge deployments is the application architecture for edge apps (EDGEAPP).

CAPIF was delivered in Release 15, with enhancements coming in Release 16, and is integrated with the northbound APIs developed by 3GPP SA2 (SCEF/NEF) and 3GPP SA4. The CAPIF architecture is specified in 3GPP specification TS23.222, "Common API Framework for 3GPP Northbound APIs," and Figure 5-42 illustrates the functional model (version 16.8.0, Release 16).

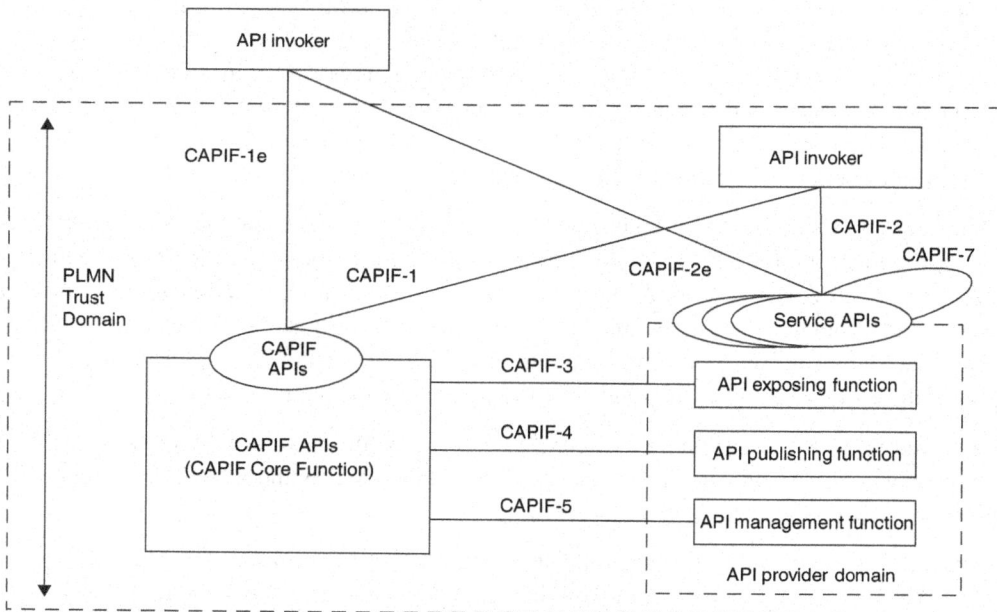

FIGURE 5-42   Functional Model for CAPIF (Source: 3GPP TS23.222)

Figure 5-42 shows the functional model for the common API framework (CAPIF), which is organized into functional entities to describe a functional architecture that enables an API invoker to access and invoke service APIs. The key functional entities are CAPIF Core Function (CCF), API Exposing Function (AEF), and the API Invoker. The CCF is the central repository for all of the APIs, taking care of authentication, on-boarding, and aspects related to logging and charging. Key details specified for these functions in TS23.222 are listed next:

- CAPIF Core Function (CCF):
    - Allows discovery of stored APIs by the API Invoker using CAPIF-1 (PLMN trust model), CAPIF, 1e (outside PLMN trust model), and with API-exposing functions using CAPIF-3
    - Provides authentication and authorization for API invokers
    - Provides logging and charging the API invocations
- API Exposing Function (AEF):
    - Validates the authorization and provides the service to the API invokers
    - Logs the invocation on the CCF and requests charging for the service
- API Invoker:
    - Discovers the service APIs from the CCF
    - Seeks authorization for API invocations
    - Avails the services provided by AEFs

The CAPIF functional model can be adopted by any 3GPP functionality-providing service API. In the functional model, API invoker within the PLMN trust domain interacts with the CAPIF via CAPIF-1 and CAPIF-2. The API invoker from outside the PLMN trust domain interacts with the CAPIF via CAPIF-1e and CAPIF-2e. The API-exposing function, the API-publishing function, and the API-management function of the API provider domain (together known as API provider domain functions) within the PLMN trust domain interacts with the CAPIF core function via CAPIF-3, CAPIF-4, and CAPIF-5, respectively. The CAPIF framework by itself is not enough to mitigate the API attack vectors; the following security controls are recommended to secure the API communications.

EDGEAPP is a new activity that 3GPP has just initiated. The details of EDGEAPP are mentioned in the 3GPP Technical Report TR23.758, which identifies the key issues and corresponding application architecture and related solutions with recommendations for the normative work.

Figure 5-43 shows the application architecture for enabling edge applications.

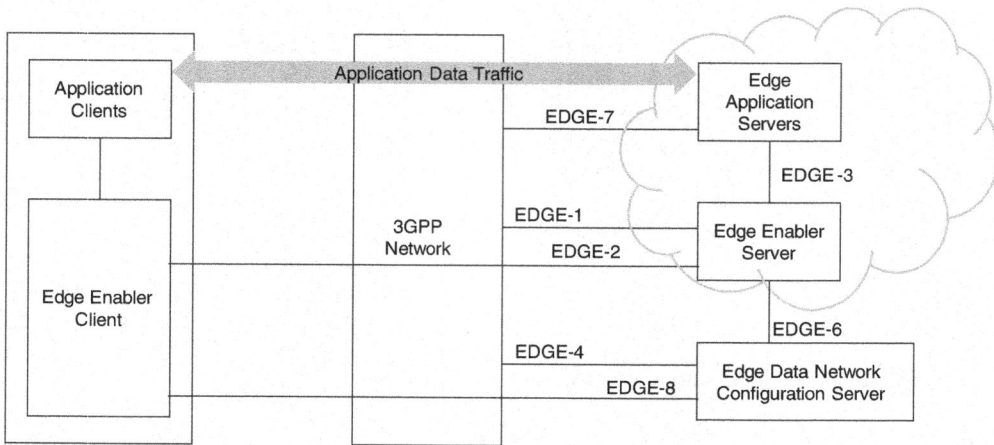

FIGURE 5-43 Application Architecture for Enabling Edge Applications
(Source: 3GPP TR 23.758 V17.0.0)

Although the EDGEAPP architecture is at a very early stage, the key security requirements are as follows:

- The edge-enabling application architecture shall provide mechanisms for the mobile network operator to authorize the usage of edge computing services by the UE.

- The edge-enabling application architecture shall provide mechanisms for the mobile network operator to authorize the usage of edge computing services by the edge application servers.

- Communication within the edge-enabling application architecture shall be protected.

- The authentication and authorization for the use of edge computing services shall support the deployment where the edge enabler functional entities providing the edge computing services are in the same or different trust domains with the 3GPP system.

- The edge-enabling application architecture shall support the use of either 3GPP credentials or application-specific credentials for different deployment needs, for the communication between the UE and the edge enabler functional entities providing the edge computing services.

- The edge-enabling application architecture shall support mutual authentication between servers (the Edge Data Network Configuration Server and the Edge Enabler Server, the Edge Enabler Server, and the Edge Application Server).

- The edge-enabling application architecture shall support authentication between clients and servers to support edge-enabling service authorization.

Apart from the studies being conducted by 3GPP on optional features, it is recommended that the service providers and enterprises using MEC applications use API gateways/API firewalls and follow best practices to secure APIs. The API gateway can be used as one method to mitigate the API attacks. The API gateway should have the following capabilities to protect the 5G applications and network functions from API-based attacks:

- Review all API responses and adapt them to match what the API consumers really need

- Should allow definition of schemas for all the API responses and error responses

- Enforce response checks and server-side data filters to prevent accidental leaks of data or exceptions

- Authorization checks with user policies and hierarchy

- Check authorization for each client request to access database

- Check multiple ways to authenticate to APIs

- Be able to use standard authentication, token generation, password storage, and multifactor authentication (MFA)

- Be able to filter server-side data

- Enforce usage of short-lived access tokens

- Allow authentication of MEC applications

- Allow use of stricter rate limiting for authentication and implement lockout policies and weak password checks

- Allow restriction on administrative access

- Define and enforce all outputs, including errors

- Validate, filter, and sanitize all incoming data

- Define, limit, and enforce API outputs to prevent data leaks

- Allow rate-limiting enforcement to match what API methods, clients, or addresses need or should be allowed to get

- Enforce rate limiting
- Enforce limit on payload sizes
- Add checks on compression ratios
- Deny all access by default
- Only allow operations to users belonging to the appropriate group or role
- Do not automatically bind incoming data and internal objects
- Allow configuration of parameters and payloads
- Allow enforcing specific schemas for runtime
- Strictly define all input data, such as schemas, types, and string patterns, and enforce them at runtime
- Integrate with SIEMs and other dashboards, monitoring, and alerting tools

Apart from the security controls provided by API gateway/API firewall, the following best practices will enhance the security of the APIs:

- Properly documenting all functions and software development kits (SDKs)
- Enforce authentication and authorization access by integration with IAM (Identity and Access Management) solutions
- Use RBAC
- Use access policy that denies all access by default
- Audit the API implementation at least every 3–4 months, along with an audit whenever a new MEC application is deployed
- Configure maximum resource usage by a container
- Prevent the use of personally identifiable information (PII) and any non-public information
- Use segmentation to segment production and non-production data
- Ensure MFA and strict access polices are implemented
- Properly design and test authorization
- Ensure secure hardening and patching processes for all MEC applications

- Automate locating configuration flaws

- Disable unnecessary features

- Keep an up-to-date inventory all API hosts

- Restrict use of old versions of APIs or backport security fixes to them

- Implement additional external controls, such as API gateways/API firewalls

- Log failed attempts, denied access, input validation failures, or any failures in security policy checks

- Ensure secure storage of logs as logs might have sensitive information

Figure 5-44 shows the mitigation of a threat attack surface from API injection attacks using an API gateway/API firewall.

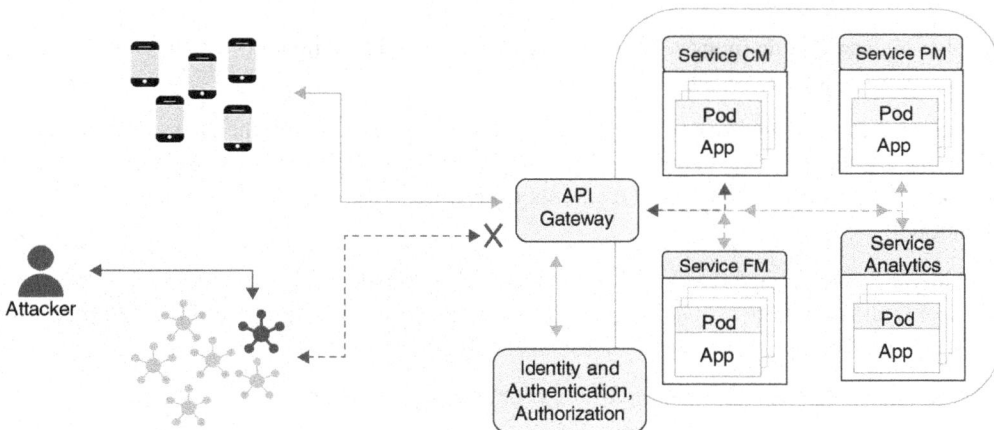

FIGURE 5-44   Mitigation of API Injection Attacks

Figure 5-44 shows an example of mitigating the API injection attacks in the 5G MEC deployments. The attack vector in this example is due to the credentials of the server being exposed to a weak device interacting with the MEC application. This credential is then used by the attacker to exploit the weak API security controls on the 5G MEC and obtain sensitive data from the MEC application database. If we look at the threat surface for this attack, it is multi-dimensional and would require security best practices to be followed along with deploying API security controls.

To mitigate an attacker obtaining credentials of the server from the weak device and mitigating the API injection attacks, the following best practices need to be followed:

1. Ensure that the credentials of the server are not exposed. This requires the devices interacting with the MEC application to have a strong access control mechanism and not allow any unauthorized user to access the devices.

2. If the credentials of the server need to be stored in the device, then it should be accessed only by authorized users having the right level of privilege and depending on the type of the device, it is best stored in a secure encrypted non-removable storage within the device.

3. The user and the device should be authenticated (who the person is) and an authorization (what the person can do) mechanisms should be applied to ensure that the user is restricted to a limited set of API calls.

4. Authentication and Authorization mechanisms should be applied at each request as often as possible.

5. Ensure having server-side input validation as well as client-side validation. Client-side validation is easily bypassed.

6. Ensure using parameterized queries instead of string concatenation with the query. Having a parameterized query restricts the input of the API user to a specific type of individual parameters and prevents the API user from inputting entire SQL statements intentionally or unintentionally.

7. The input validation should also be carried out at different layers, as an example in the GUI (Graphical User Interface) – if GUI is used, MEC application data validation if the API query is directed toward the application and database data validation.

8. Conduct regular audit and penetration test of your API deployments.

Figure 5-45 shows the mitigation mechanism for the attack based on excessive data exposure using an API Gateway.

MEC application hosted on public or private
cloud of enterprise or service provider

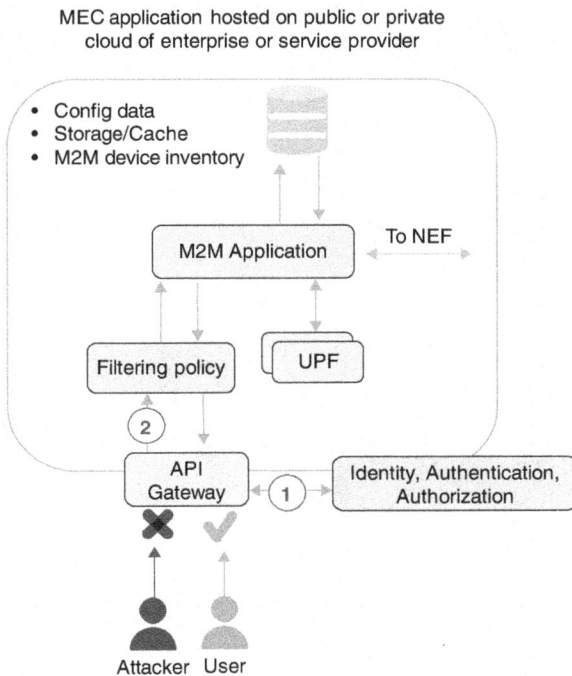

FIGURE 5-45   Mitigation of API Excessive Data Exposure Attacks

In 5G, the MEC applications can be used for a variety of use cases. Many of the use cases can be for the public sector, defense, and healthcare sector. The database of MEC applications aimed at these critical use cases might hold very important data that should not be exposed. Many developers program APIs with simple codes that rely on clients to perform filtering without much consideration given to security for extra checks on the API request or response messages. In the majority of the instances, the programmers do not realize the sensitivity of the data being transferred. As shown in Figure 5-45, the following mitigation mechanisms can be used to identify any excessive data exposure and mitigate it:

Step 1. This step should always be the authentication and authorization of the user and device. Device authentication and authorization policies are important in the M2M use cases.

Step 2. Ensure the right filtering of the data within the API response. This is to ensure that the response of the API call is only for the requested data and that excessive and sensitive data is not exposed. For the right filtering to take place, it is necessary that the API gateway include the policies for API requests, API response, and the error codes. As 5G network functions and MEC applications would communicate mainly using the REST API, filtering policies should be applied as a mandatory feature in any 5G MEC application being deployed by the service provider/enterprise to limit any exposure of sensitive data and mitigate any data being exfiltrated and hoarded by the attacker.

Figure 5-46 illustrates a method of applying filtering policies.

MEC application hosted on public or private
cloud of enterprise or service provider

- Config data
- Storage/Cache
- M2M device inventory

Response includes info of all devices such as:
"devicename": device1
"deviceid": 1d25a16d771cc6197fd
"power output":10mW
"ip address":10.10.10.11
"address":1 Main street, Swords, Dublin
"username":abcd@hotmail.com
"password":"abcd1111"

.
.
.

"devicename": device110
"deviceid": 7f31a74d113fc192df1d
"power output":10mW
"ip address":10.10.10.111
"address":1 seatown, Swords, Dublin
"username":abcd@hotmail.com
"password":"abcd1211"

Excessive data exposure
prevented—only basic attribute
values are returned for the API
query

To NEF

M2M Application

UPF

Filtering policy

API Gateway

Identity,
Authentication,
Authorization

User

Policy filters API response output to the below:
devicename": device1
"deviceid": 1d25a16d771cc6197fd
power output":10mW

deviceidinfo: device1
"deviceid": 1d25a16d771cc6197fd
"power output":10mW

GET request to/deviceidinfodevice1

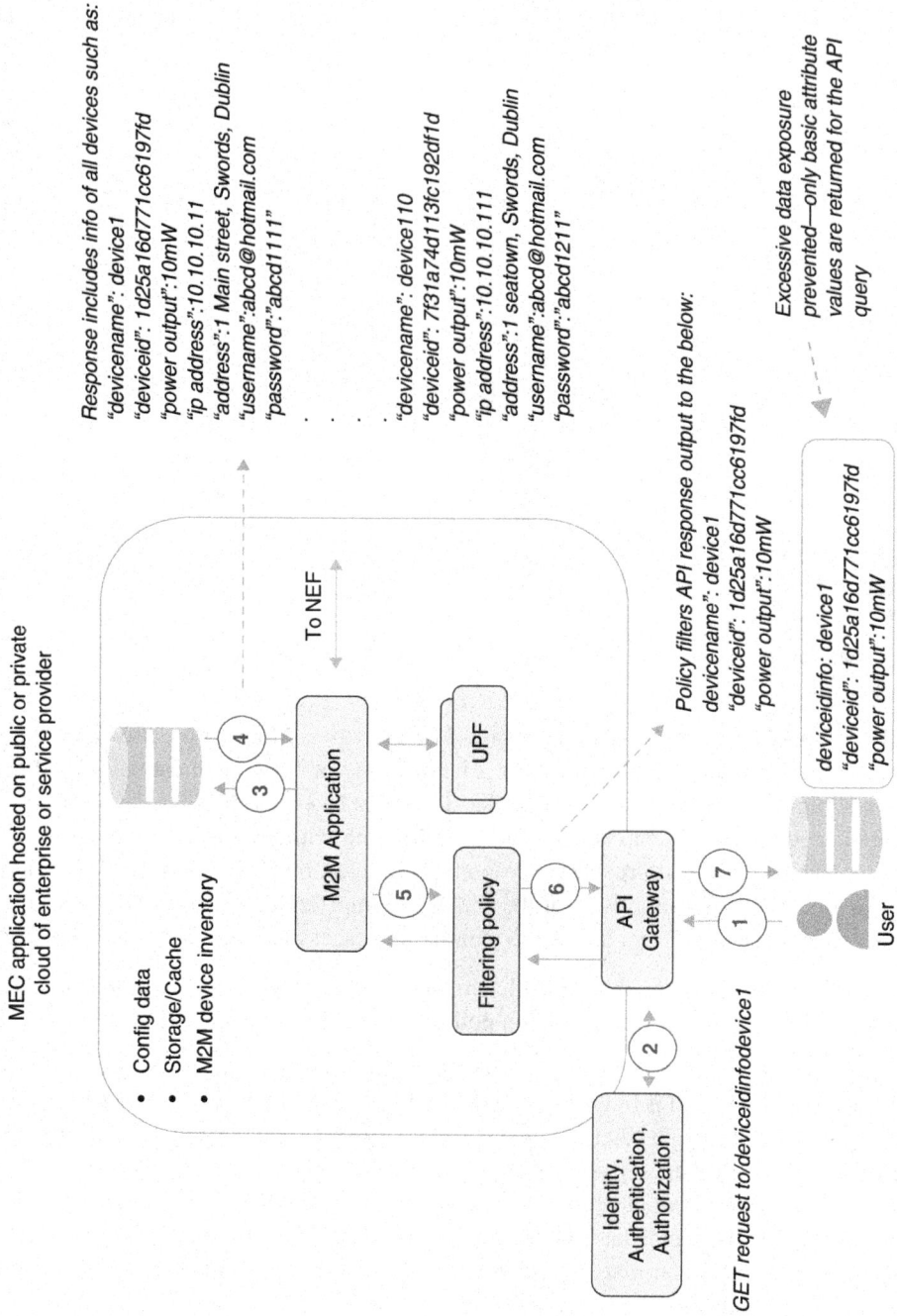

FIGURE 5-46   Applying Filtering Policies to Mitigate Excessive Data Exposure

As shown in Figure 5-46, a method of mitigating excessive data exposure is to use the right filtering policy and is indicated in the following steps:

1. The user or an attacker issues a query using an API GET request for getting the information for device1.

2. The checks are carried out for authentication and authorization.

3. Once the authentication and authorization checks are successful, the API query is sent toward the MEC application database.

4. The MEC application database, which holds the configuration data and the storage and device inventory, then responds with all the device info, which includes sensitive data such as device IP address, username, passwords, and so on.

5. The service provider/enterprise vertical deploying the MEC application also includes a filtering policy to ensure that only nonsensitive data is allowed to be exported. This can be achieved by carrying out checks in the "attribute: value" pairs and restricting certain attributes to be exported. The filtering policy is an API gateway functionality provided by a few API gateway vendors today. If you have an existing API gateway that does not support this function, you can use an intrusion prevention system (IPS) or a firewall to carry out the aforementioned attribute value checks.

6. Depending on the filtering policy applied, only nonrestricted attributes are allowed to be exported. In the example in Figure 5-46, only the attributes "devicename", "deviceid", and "power output" are allowed to be exported, thereby preventing sensitive attributes such as "ipaddress", "address", "username" and "password" from being exported.

7. The user or the attacker now receives only the nonsensitive attributes, and excessive data exposure is avoided.

It has to be noted that varying numbers of MEC applications can have different levels of sensitive information in the attribute values, and where the attribute is present depends on the Managed Object Model of the device. For some MEC applications, even the "deviceid" might be sensitive information. So the filtering policy needs careful consideration and should not be a generic deployment of rules; it should be considered on a case-by-case basis and audited regularly, at least every 3 to 4 months and each time a new MEC application is deployed.

In another variant of the attack, the attackers can exploit MEC applications that are vulnerable to Broken Object Level authorization by manipulating the ID of an object that is sent within the request. An example of this follows:

- **Intended use:** GET/MEC_application/*user*/personal_info

- **Manipulated use:** GET/MEC_application/*adminuser*/personal_info

Here, attackers have substituted the ID (user) in the API call with an ID belonging to another user (adminuser) to exploit the MEC application. The attack is carried out by an attacker observing the pattern of the API call and then inserting certain ID values by guessing and modifying the IDs. This issue is extremely common in API-based applications because the server component usually does not fully track the client's state and instead relies more on parameters like object IDs that are sent from the client to decide which objects to access. Generally, the application developers do not include code to have proper authorization checks, allowing attackers to access sensitive information related to another user or information related to an admin user and then launch a privilege escalation-based attack on the network. The attacks can also be launched to delete or modify certain information that might lead to service outages and loss of revenue for the entity providing the services related to the MEC application.

To mitigate the attacks based on Broken Object Level Authorization and user authentication, it is important to have granular access and authorization controls implemented in the MEC application deployments to ensure that sufficient checks have been carried out on the access rights of the user. The ID within the API should also be masked and shouldn't be easily guessable.

Figure 5-47 shows an example of one of the methods of mitigation of threats based on Broken Object Authorization. In this example the M2M application developer has ensured that the M2M application database has included the user-to-object access checks to ensure that the user can access or query the objects within the database. Using the object access checks will also mitigate the attack vector whereby an attacker tries to use the name of a legitimate user with incorrect credentials. However, if the attacker has the correct username and the related correct credentials, they can have an unauthorized request using the API calls. This can be mitigated by using MFA security controls, which allows access to the users after multiple layers of authentication to verify the identity of the user accessing the API.

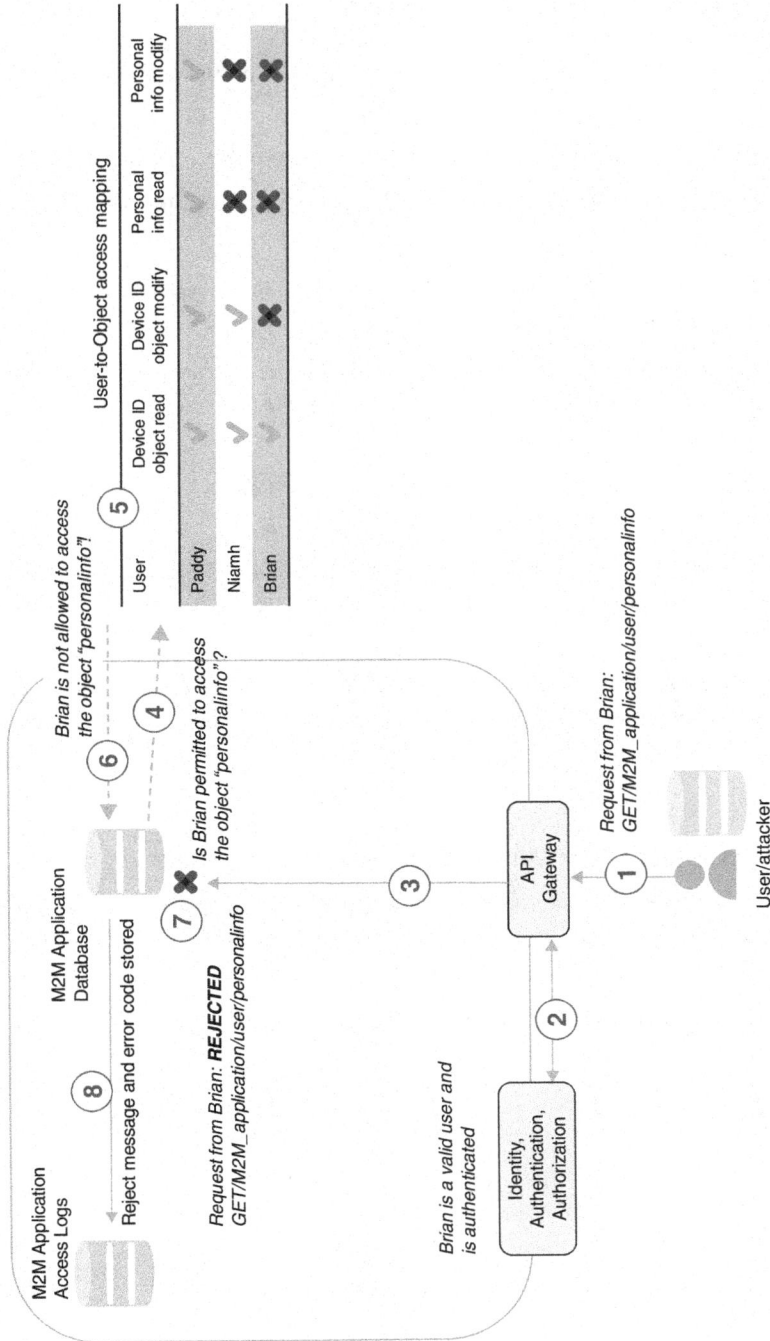

FIGURE 5-47 User-to-Object Access Checks

Figure 5-47 illustrates one of the methods for mitigating a Broken Object Authorization–based attack, which is explained further in the following steps:

1. The user sends an API GET request for user personal info. In this example the name of the user is Brian.

2. An authentication and authorization check is carried out to check whether Brian is a valid user.

3. The user Brian is authenticated and the GET request is sent toward the M2M application database.

4. The database queries the user-to-object access mapping table to verify if the user Brian is allowed to access the object personal info.

5. The user-to-object access rules are checked to verify the access rules. This is the critical step which should be included for verifying the API request and verifying the authorization rules as to what object the user is allowed to access.

6. It is confirmed that the user Brian cannot access the user-to-object access mapping, and Brian is not authorized to access the requested object.

7. The request from Brian to read the attribute value is rejected and the requested information is not passed on to Brian. An added step that could be optional in this process could be a reply with error code.

8. The rejected request from Brian is logged into the M2M application logs, which can then be used by the monitoring entity or security operations center (SOC) to further investigate the activity.

Validating Both Read and Write Requests

Another technique to mitigate API vulnerabilities based on Broken Object Authorization is to validate the read and write request. Here, the user/attacker tries to use the DELETE API request to create or update the resource or attribute.

Figure 5-48 shows another example of mitigation of Broken Object Authorization–based threats.

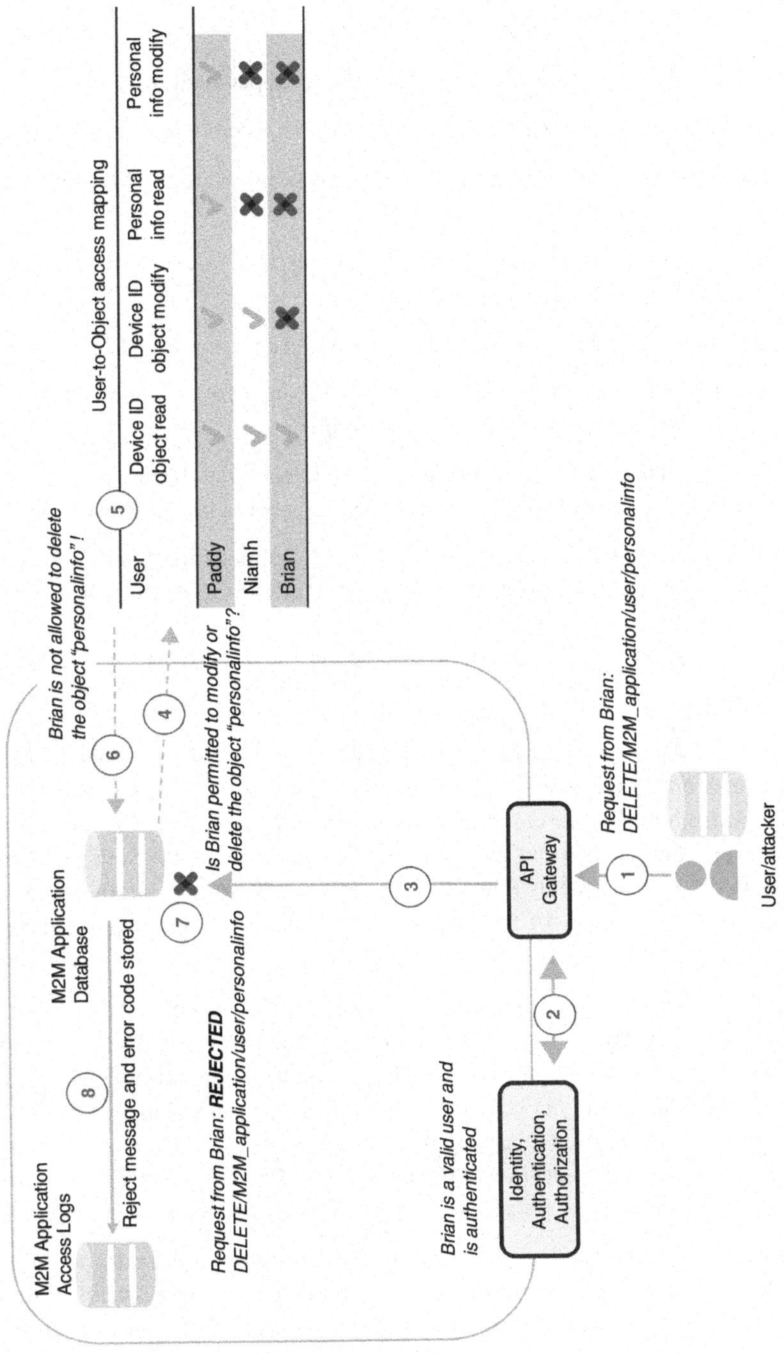

FIGURE 5-48 Validate Read and Write Request

Figure 5-48 illustrates another method for mitigating Broken Object Authorization–based attacks, which is explained further in the following steps:

1. The user sends an API DELETE request for deleting user personal info.

2. The authentication and authorization check is carried out to see whether or not the user is valid.

3. The user is authenticated, and the DELETE request is sent toward the M2M application database.

4. The database queries the user-to-object access mapping table to verify if the user Brian is allowed to modify or delete the object personalinfo.

5. The user-to-object access rules are checked to verify the access rules.

6. It is confirmed that the user Brian cannot modify or delete the user-to-object access mapping and Brian is not authorized to access the requested object.

7. The request from Brian to delete the attribute is rejected and the deletion action is not varied out. An added step that could be optional in this process could be a reply with an error code.

8. The rejected request from Brian is logged into the M2M application logs, which can then be used by the monitoring entity or security operations center (SOC) to further investigate the activity.

DDoS Protection

In order to properly secure the "full stack" in the MEC layer that delivers a connected application, two fundamental elements can be applied: visibility and control. *Visibility* refers to the ability to see and correlate information from the carrier cloud to baseline proper behavior and then to measure deviation from that norm. Simply said, "If you can't measure it, you can't manage it." Sources of visibility come from traditional network measurements (NetFlow, open flow, and so on), but the need to measure all aspects of a flow, from all elements of the carrier cloud to the application to the end customer, has changed what data is collected and where we get it.

Cybersecurity vendors are now providing enhanced visibility, which includes the use of application-level probes that are synthetically generated and travel through the network to get a clear picture of how an application is behaving. Another example is where the Path Computation Element, which has a near real-time database representing the network topology, is queried programmatically to determine the impact of a potential mitigation action on critical service classes for DDoS. Once all of the telemetry is gathered, a security controller and workflow will analyze it and determine, based on policy, suggested mitigation and controls to be applied. Of course, we have an iterative loop of constant learning.

For determining the anomalous traffic, the DDoS protection solution would normally baseline the network by monitoring for a couple of weeks. This baseline would help the DDoS protection solution understand what traffic pattern can be considered normal and what traffic patterns should trigger the activation of DDoS mitigation steps.

Depending on the MEC deployment model, an option of on-premises, cloud, or hybrid DDoS protection could be used. The key is to have an effective security architecture that includes effective DDoS protection. To have effective DDoS protection, it is important to have an IP protection database and a strong threat intelligence input to the DDoS solution, primarily to minimize any chances of increased frequency of zero-day attacks.

There are different approaches to DDoS attack mitigation; the key approaches are as follows:

- **On-premises:** When on-premises mode of DDoS protection is chosen, service providers and enterprises benefit from an almost real-time and automated DDoS attack detection and mitigation. Within seconds of the start of an attack, the MEC deployments are protected. On-premises mode of the DDoS solution, though effective, becomes less effective when a large-volume DDoS attack is launched, as an on-premises DDoS solution is normally deployed based on the DDoS traffic volume, which needs to be mitigated. The larger the volume decided during deployment, the higher the cost. As a rule of thumb, many DDoS vendors recommend the DDoS traffic mitigation volume be around 30–40% of the inbound traffic. This makes on-premises anti-DDoS solutions more favorable to be used against application-level DDoS attacks. Due to the high volume of the DDoS traffic in volumetric DDoS attack scenarios, cloud and hybrid anti-DDoS solutions would be more effective in mitigation, as compared to on-premises DDoS solutions, which are limited in the volume of DDoS traffic that can be mitigated.

- **Cloud:** When cloud mode of DDoS protection is chosen, it is mainly aimed at mitigation of the DDoS attack and not so much as the detection part of the DDoS solution.

- **Hybrid:** Hybrid DDoS solutions offer the best mitigations by combining advantages of the on-premises and cloud deployments of the DDoS solution. Attack detection and mitigation start immediately using the on-premises mitigation device, and the cloud deployment caters for the mitigation by diverting the traffic to the cloud, where it is scrubbed before being sent back to the service provider or enterprise. To have an effective mechanism of DDoS protection, it is recommended to have a tight integration for information exchange between the on-prem and cloud-deployed DDoS solutions to enhance the DDoS attack mitigation.

Some DDoS attacks also require a web application firewall (WAF). WAF would protect the network from application layer attacks such as SQL injection. WAF would normally be deployed between the Internet layer and the application layer and analyzes the HTTP communications. For mitigation of DDoS attacks, WAF could be deployed for managing an IP reputation database that tracks and blocks malicious traffic.

Figure 5-49 shows an example of DDoS protection against HTTP attacks.

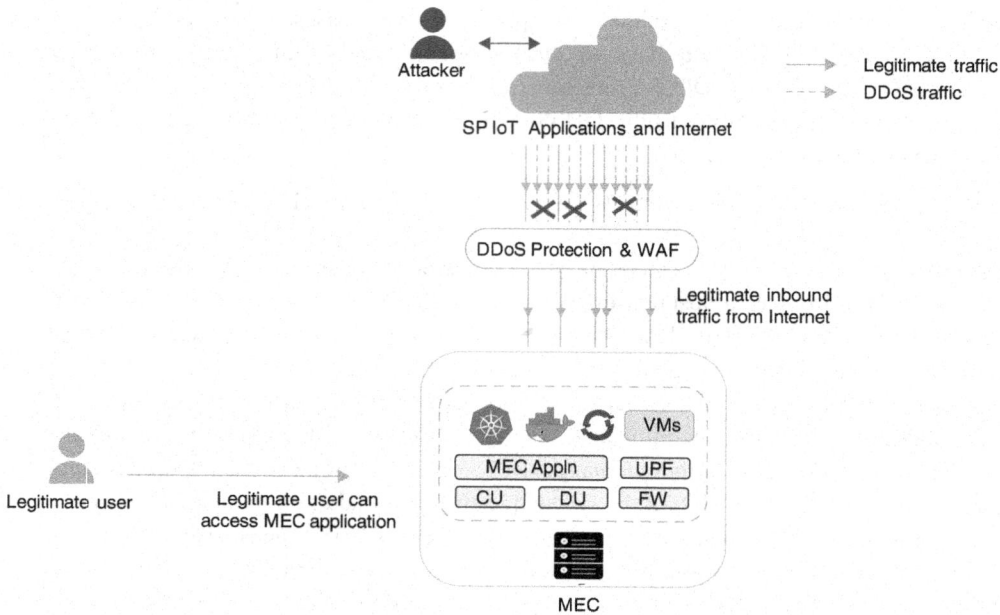

FIGURE 5-49   DDoS Protection Against HTTP Attacks

As shown in Figure 5-49, inbound DDoS attacks could be mitigated by using DDoS protection using anti-DDoS and WAF between the Internet/web layer and the MEC layer.

In this attack vector, the attacker aims to cause DoS for the MEC application users by using a form of HTTP flood attack. Here, the aim is to overwhelm the MEC application with HTTP requests mostly with the help of bots, which makes it unable to respond to legitimate user requests.

The mitigation of the HTTP attacks requires sophistication, as the aim is not to drop all the HTTP requests because there would be legitimate requests to and from the MEC application that should not be blocked. The mitigation of such attacks would require a multilayered security control. One of the security controls used is a DDoS protection solution that has WAF capability. The service provider/ enterprise can now selectively allow only legitimate traffic to communicate with the MEC applications, while the rest of the DDoS traffic can be blocked or sent to a DDoS scrubbing center for more analysis.

Figure 5-50 shows another example of mitigating the DDoS attacks on the MEC application deployed on the public cloud.

As shown in Figure 5-50, in this MEC deployment model, the MEC applications are deployed on the public cloud and the storage and compute are deployed on the on-premises infrastructure of the service provider/enterprise. The security controls considered for mitigating the DDoS attack surface are vulnerability assessment, patch management, enhanced visibility, an anti-DDoS solution, and a threat intel mechanism with artificial intelligence (AI) and machine learning (ML) to proactively detect threats.

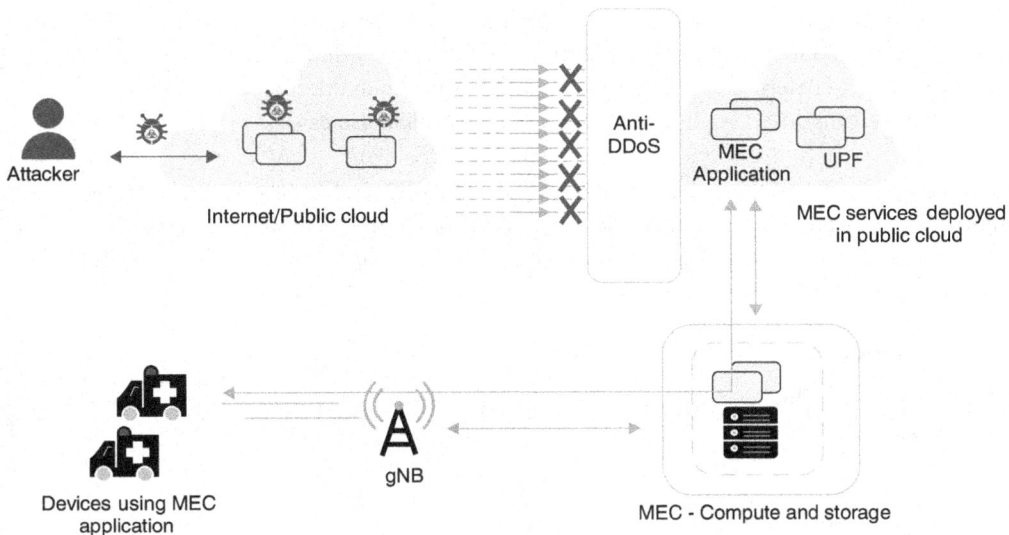

FIGURE 5-50   Multilayered Security Controls to Mitigate DDoS Attacks

There are many variations in the way that the MEC applications deployed in the public cloud could be attacked. The attack vector example in the diagram is due to vulnerabilities in the application deployed in the cloud, which is exploited by the attacker. The application vulnerability might be due to many open source components used to create the microservice. This is of particular interest in 5G, as many of the applications being deployed in 5G will be "CNF created" by using open source components. These open source components might have vulnerabilities not yet patched or widely understood. These vulnerabilities are then exploited by the attacker or attacking entity to cause DDoS or DoS for legit-imate users. In Figure 5-50, the devices using MEC applications cannot be served by the application deployed in the cloud due to the attacker consuming the resources of the application by overwhelming it with "fake" requests.

To mitigate these kinds of DDoS attacks, it is recommended to follow a baseline security control and process, as described in the list that follows:

■ Deployment of an MEC application specific to 5G use cases should have a vulnerability assess-ment system in place that checks the applications for any vulnerabilities and patches that can be applied to mitigate the vulnerability. The vulnerability assessment should include the MEC applications deployed in cloud and on premises.

■ There should be an enhanced visibility layer providing visibility at the application level, such as application-to-service mapping. The service could include DNS for NTP, which is being used by the application. A baseline should be created to understand the normal behavior of the appli-cation, including the memory being consumed.

■ The critical applications serving important use cases such as government, defense, utility sector, and so on should have enhanced visibility and alerting procedures. An example of enhanced visibility layer is to determine if there is any malicious traffic in the encrypted communication between the web and MEC application catering to the critical use cases.

As shown in Figure 5-51, the communication between the MEC application deployed in the cloud and on premises could have different modes of communication. Some could be based using API calls, and others could be non-API calls. The API messages could be secured by using HTTPS. To secure non-API communication, IPsec is recommended.

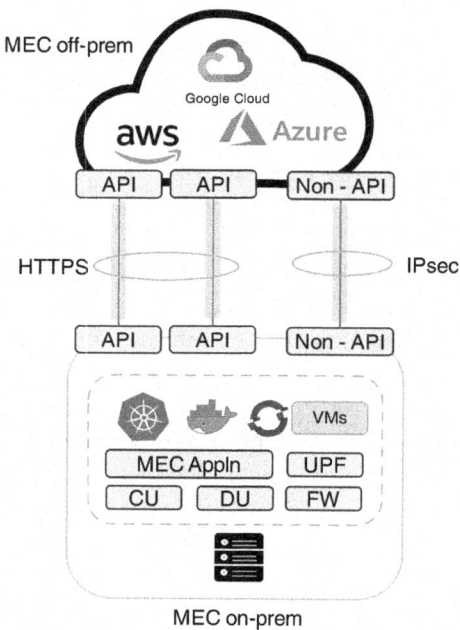

FIGURE 5-51   Securing API and Non-API Interfaces

Using DDoS protection algorithms that employ AI and ML is important for detecting and proactively mitigating zero-day attacks. Instead of having the AI and ML restricted to DDoS solutions only, they should instead be applied to the threat intelligence layer, which can then be used as a threat intel for other security controls as well. This will help the end-to-end security architecture have better efficacy.

All of the preceding methods described are effective if the attack has been seen before in other networks and there is a known mechanism to mitigate it. For mitigation of zero-day attacks, however, there needs to be a bit more stringent security controls in place to ensure proactive mechanisms to deter the attacks.

One of the mechanisms to mitigate the zero-day attacks would be to follow a method to create a baseline of the network and then look out for any anomalies. Any traffic showing an anomaly should then be further analyzed to understand the nature of the traffic and then generate automated rules to block it if it's seen again anywhere else in the network. The best practice of considering all traffic untrusted and then allowing traffic further in the network after going through mechanisms to trust it is very effective. DDoS mitigation mechanisms that can be deployed in the MEC or the edge of the network are very effective at mitigating DDoS attacks.

Real Scenario Case Study: MEC Threats and Their Mitigation

The preceding sections describe the threats and mitigation techniques for specific areas of concern while deploying 5G MEC. This section focuses on a multidomain attack and discusses methods to deploy security controls at different parts of the infrastructure to mitigate the multidomain attack in a real-life environment. Although 5G was not widely deployed at the time of writing this book, this section is based on real-life attacks in similar environments or proof of concepts and lab tests done by various customers.

To explain the attack scenarios, an example of the MEC deployment, as shown in Figure 5-52, is considered in this section.

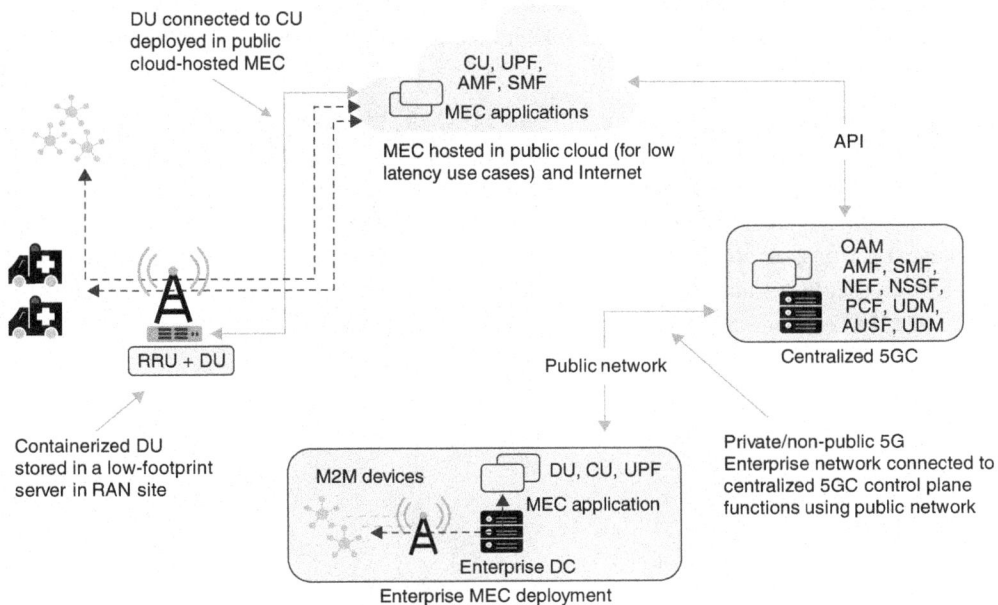

FIGURE 5-52   Example of 5G MEC Deployment

As shown in Figure 5-52, the MEC deployment in this example consists of two layers of MEC.

One of the MEC options is realized by implementing it in the public cloud and hosts the CU and UPF, enabling URLLC use cases aimed at IoT devices and healthcare.

In this deployment option considered by the service provider, the Radio Resource Unit (RRU) and the DU are deployed using a low-footprint server. The low-footprint server hosts the containerized DU, which could be open source or from a RAN vendor. One of the reasons a service provider might use the option of having a containerized version of DU deployed on the RAN site could be due to specific use cases that require extremely low latency and hardware acceleration, where certain computing tasks are offloaded to specialized hardware components. This mode of deployment is particularly helpful when Multiple Input Multiple Output (MIMO) and massive MIMO are deployed. In some cases, even the CU is deployed on the RAN site, but it would limit the aggregation capability of the CU, put pressure on the scalability of the DU, depending on the server footprint, and needs careful consideration. As shown in Figure 5-52, the DU is configured to connect to the CU, which is implemented in the MEC hosted in the public cloud. Other components hosted in the public cloud MEC are the 5G user plane functions (such as UPF), 5G control plane functions, such as Access and Mobility Function (AMF) and Session Management Function (SMF), and MEC applications that would usually include the application configuration, storage, performance, and fault management.

The second MEC in Figure 5-52 is implemented in the enterprise environment. It is based on private/ non-public deployment of 5G, where the enterprise has deployed the RAN infrastructure, user plane functions of 5G, and the MEC application within the enterprise perimeter. Usually, the virtualized components, such as DU, CU, UPF, and MEC applications, would be a containerized form factor and deployed in the enterprise data center.

The control plane functions of 5G for the enterprise MEC deployment are located in the service provider's data center and will host the AMF, SMF, Network Exposure Function (NEF), Network Resource Function (NRF), Network Slice Selection Function (NSSF), Policy Control Function (PCF), Unified Data Management (UDM), and Authentication Server Function (AUSF), among other 5G control plane functions, as required by the service provider. The operations, administration, and management are also deployed in the centralized 5GC data center. Apart from the SMF and AMF for the public-hosted MEC option, the rest of the control plane functions are deployed in the centralized 5GC data center.

Now that we have an overview of the deployed network illustrated in Figure 5-52, let's look at some examples of multidomain threats and methods to mitigate the threats.

Threats: Case Study

In this section, we will see some examples of threats that will impact multiple domains, such as edge compute deployed in the cloud, enterprise, and centralized data center. Out of all the multiple threat vectors illustrated in Figure 5-53, threat vector A and threat vector C are real-life use cases. Threat vectors A and C were not seen in a 5G environment, but the side-channel attack was seen in a service provider data center environment very similar to the one explained in this case study. Threat vectors B and D are added here to show the real scenarios you might face once 5G is deployed.

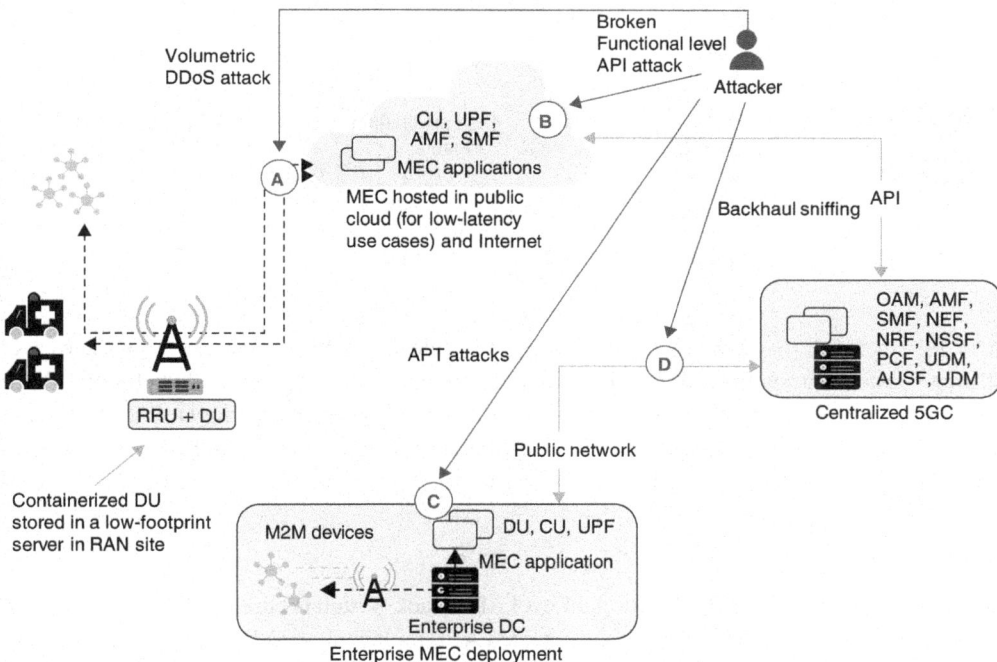

FIGURE 5-53 Example of Multiple Threat Vectors in the Deployed Network

As shown in Figure 5-53, there are four key threat vectors we will look at in this section.

Threat Vector A

This threat vector is a DDoS attack caused by the attacker taking control of the unprotected IoT devices.

Figure 5-54 depicts an attack based on volumetric DDoS.

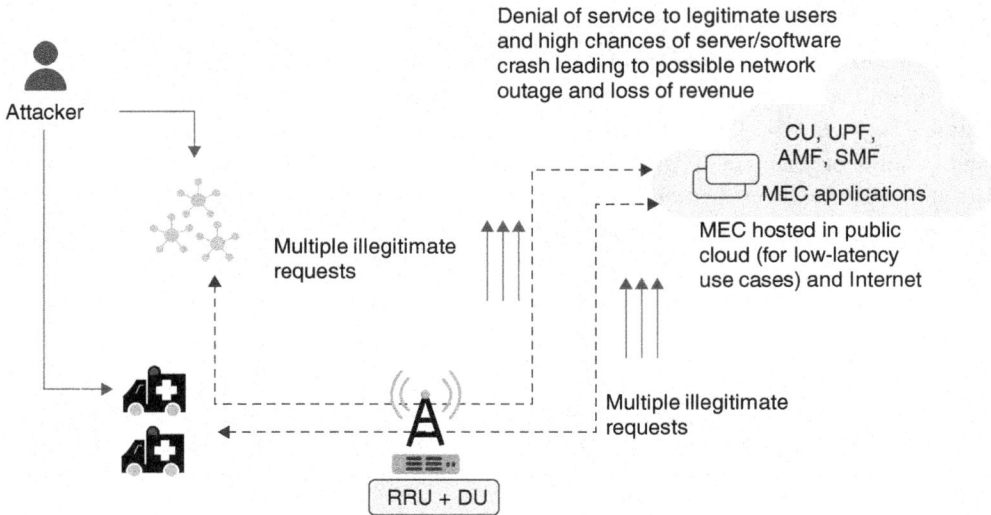

FIGURE 5-54   Example of Attack Based on Volumetric DDoS

As shown in Figure 5-54, the attacker exploits the unprotected IoT devices to cause an excessive number of illegitimate requests toward the MEC application database. This can overwhelm the MEC application database such that it cannot respond to legitimate requests, thus causing DoS to legitimate users. This can also lead to server or software crashes, leading to service outage and possible revenue impact.

Threat Vector B

This threat vector is called the Broken Functional Level API attack, which is based on a vulnerability in the API implemented by the developers or an unprotected API implementation by the service provider.

Figure 5-55 shows an example of Broken Functional Level attack in an unprotected API.

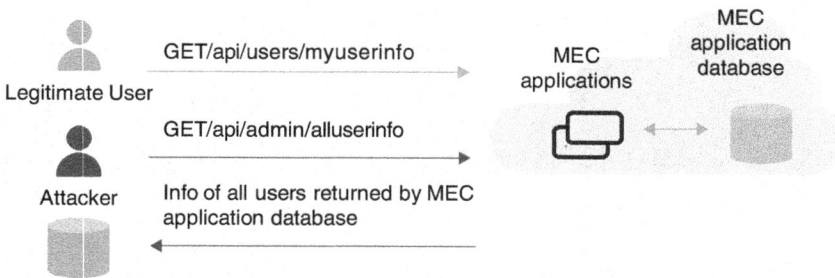

FIGURE 5-55   Example of Broken Functional Level Attack

In the Broken Functional Level attack shown in Figure 5-55, the developer has used a generic implementation of the API, with the API relying on the client to use user-level or admin-level APIs. The threat vector is possible as the attacker figures out the "hidden" admin API methods and then uses them. As shown in Figure 5-55, the user is allowed to query their own user information from the MEC database. But due to the unprotected implementation, the attacker is now able to invoke some commands that are available to only privileged users. Now the attacker uses the admin privileges and exploits them to export the sensitive data, as seen in this example.

Threat Vector C

This threat vector is based on advanced penetration threat (APT) attacks. The main aim of APT attacks is to gain information on the victim or targeted host illegally. The gathered information can be used to launch a large-scale attack on the target organization or to insert malware to encrypt all the sensitive files and destroy the goodwill of the organization by exposing the sensitive files on the Internet.

Figure 5-56 shows one of the methods of an APT-based attack.

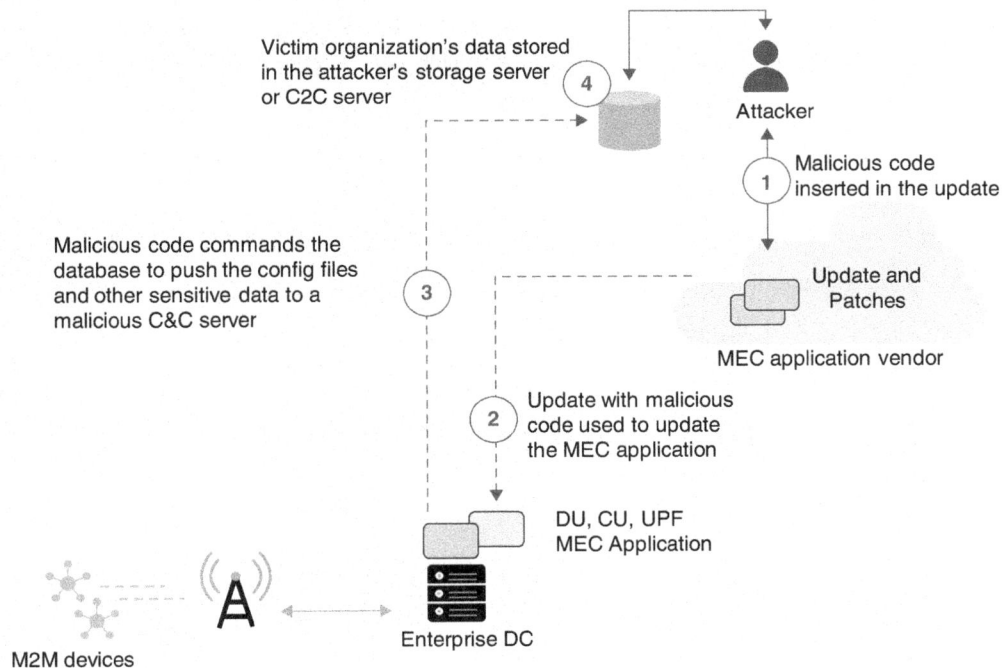

FIGURE 5-56　Example of APT-Based Attack

The steps of this attack shown in Figure 5-56 are as follows:

1. The attacker infiltrates the MEC application vendor and inserts malicious code within the update files or patch files intended to be implemented by the MEC application. This method ensures that the file available in the legitimate vendor's website for download is infected by the malicious code.

2. The MEC applications are notified about the available update and then download the install file from the MEC application vendor's site. The MEC applications will now undergo the update process and get infected by the update file with the malicious code.

3. Based on the intent of the malicious code within the update file, the MEC applications will behave accordingly. In our example, the intent of the malicious code within the update file is to open an encrypted tunnel to an external C&C server and start exporting the sensitive files.

4. All the sensitive files exfiltrated from the victim's organization are now stored in the storage or C&C server, which can be accessed by the attacker. The attacker can now use the sensitive information in a variety of ways. One of the methods is to extort money by providing proof to the victim organization about the stolen data. The other method is to gather information about the organization and then carry out a surgical but devastating attack that causes maximum impact.

Threat Vector D

This threat vector is based on attacks due to transport vulnerabilities when public networks are used.

Figure 5-57 shows an example of back-haul sniffing.

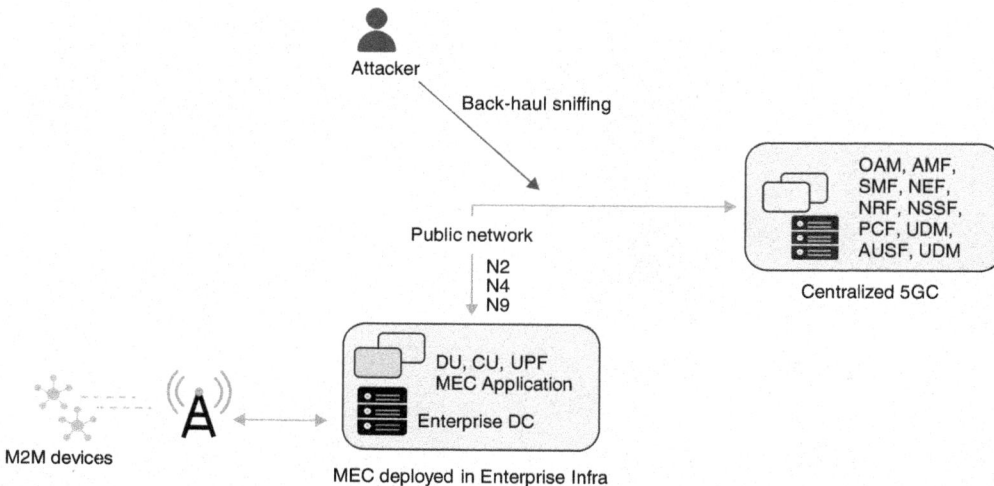

FIGURE 5-57 Back-haul Sniffing Due to Untrusted Public Networks

In the example illustrated in Figure 5-57, the transport from the Enterprise MEC toward the centralized 5GC data center is based on the public network. The information carried in this transport from the enterprise MEC to the centralized 5GC involves the N2, N4, and N9 interfaces:

- **N2:** Interface between gNB and AMF

- **N4:** Interface between UPF and SMF

- **N9:** Interface between UPFs (UPF from MEC and UPF in centralized 5GC)

The N2 and N4 interfaces need to be protected for integrity, confidentiality, and replay when using insecure public networks. They do not have these built-in security layers.

Mitigation Examples

This section takes you through the mitigation steps to secure the MEC deployment from the threats shown in Figure 5-53. This section will be of particular interest to readers trying to understand the exact scope of the threats and what specific solutions could be used as a tactical method to mitigate the four threat vectors, as shown in Figure 5-53 earlier in the chapter.

Applying DDOS Protection and Securing API: Threats Vectors A and B

Figure 5-58 shows the implementation of an API gateway to secure the API as well as DDoS protection techniques to protect the MEC deployment. For this example, the requirement of the DDoS protection is shown only for the MEC applications hosted in the public cloud as a defense mechanism against the device-based volumetric attacks on the MEC applications. An API gateway/API firewall is recommended in both types of MEC deployments.

As shown in Figure 5-58, the API gateway can be used as one method to mitigate the API attacks. In this example, the API gateway/API firewall is deployed at the MEC locations as well as the centralized 5GC to protect the APIs from API vulnerable threats. The API gateway should have the following capabilities to protect the 5G applications and network functions from API-based attacks. The rate limiting and throttling features in the API gateway will provide protection against excessive request attacks from the IoT devices in the network. They will also secure the MEC applications from attackers trying to exploit any API vulnerabilities, such as Broken Functional Level authentication and authorization checks. This is done by having strict authentication and authorization checks carried out by the API gateway/API firewall against the identity, authentication, and authorization solutions.

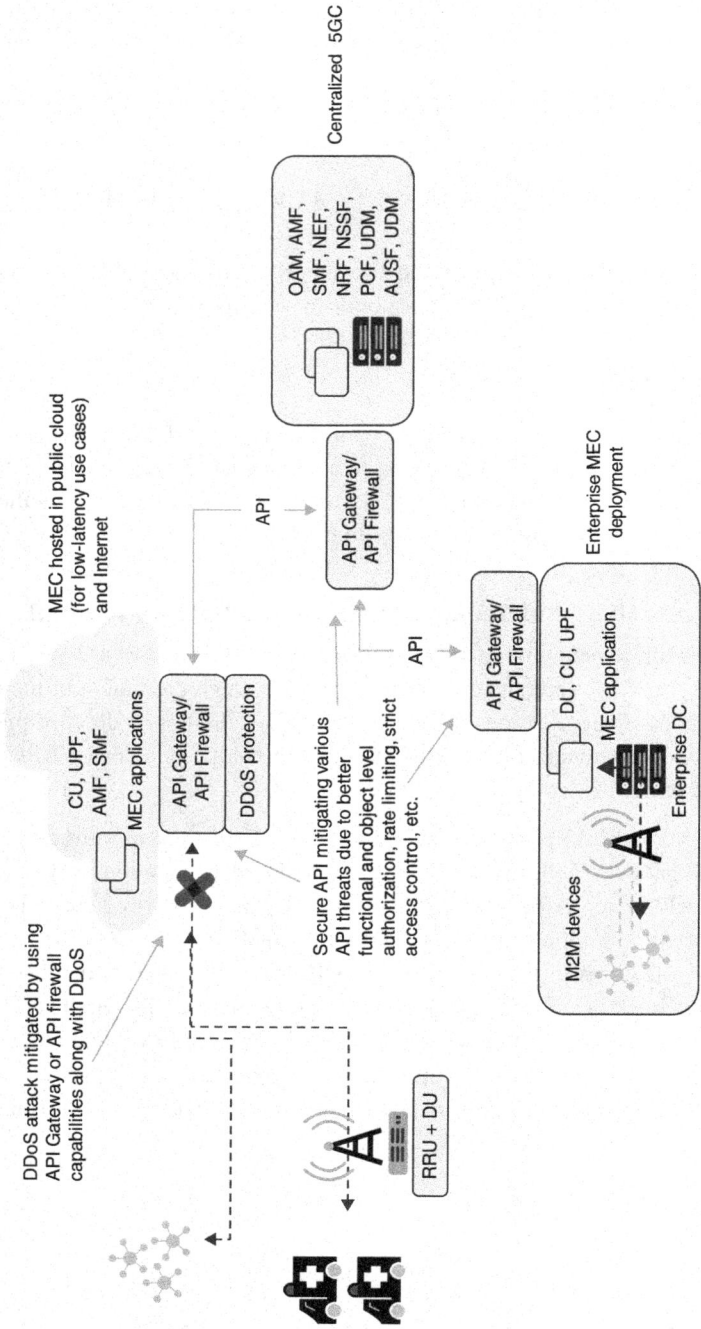

FIGURE 5-58   API Gateway/API Firewall Deployments in MEC

Depending on the MEC deployment model, an option of on-premises, cloud, or hybrid DDoS protection could be used. The DDoS protection is recommended to be deployed at the MEC implemented in the public cloud to thwart the DDoS attacks from the devices being catered to by the network. As this mode of MEC deployment is primarily to support URLLC use cases, the latency impact should be kept to a minimum and provide near-real-time response to requests. The DDoS solution implemented for the public cloud–hosted MEC applications should have features for scanning the HTTP request, and checks should be carried out on TLS version, HTTP version, query string, and user agent, among others.

Securing MEC Deployments Against APT Threats: Threat Vector C

The threat vector in this example is based on the attacker inserting malicious code in the software updates or patches. The software update and the patch are then installed by the network device or virtualized network function (container/virtual machine). The malicious code can have a variety of malicious actions. The action in this example is for extracting the sensitive data from the 5G MEC application hosted in the enterprise data center. Once the data is extracted, the attacker can analyze it for any sensitive data, such as PII and network information. The attacker can then use the gathered PII to exploit the user data and sell the details on the web. The other method could be for the attacker to use the gathered network information to understand the type of network devices and software being used by the MEC data center in the enterprise network.

This section explains some methods for mitigating this threat vector.

Figure 5-59 shows one of the methods using an anomaly detection system to mitigate the APT-based threat vector.

As shown in Figure 5-59, the attacker inserts a malicious code in the software update or patch files in the original update package from the MEC application vendor. The MEC application then downloads the update file for installing the software update. This could also impact the network devices if the update is for a firmware update for the MEC data center server or switches. The following steps would explain the method used by the anomaly detection system to mitigate this APT threat vector:

1. The anomaly detection system is basically an intelligent network visibility and monitoring solution with enhanced machine learning and artificial intelligence algorithms. The anomaly detection solution will collect network telemetry such as NetFlow, jFlow, sFlow, IPFIX, syslog, events, and so on from network devices such as switches, routers, virtual switches, and so on. It would also collect telemetry from virtual instances, such as containers, and for virtual machines by using hypervisor telemetry. Cloud telemetry such as VPC logs can also be collected to have end-to-end visibility. Features like VXLAN stripping and packet de-duplication are required to make the information gathering more efficient.

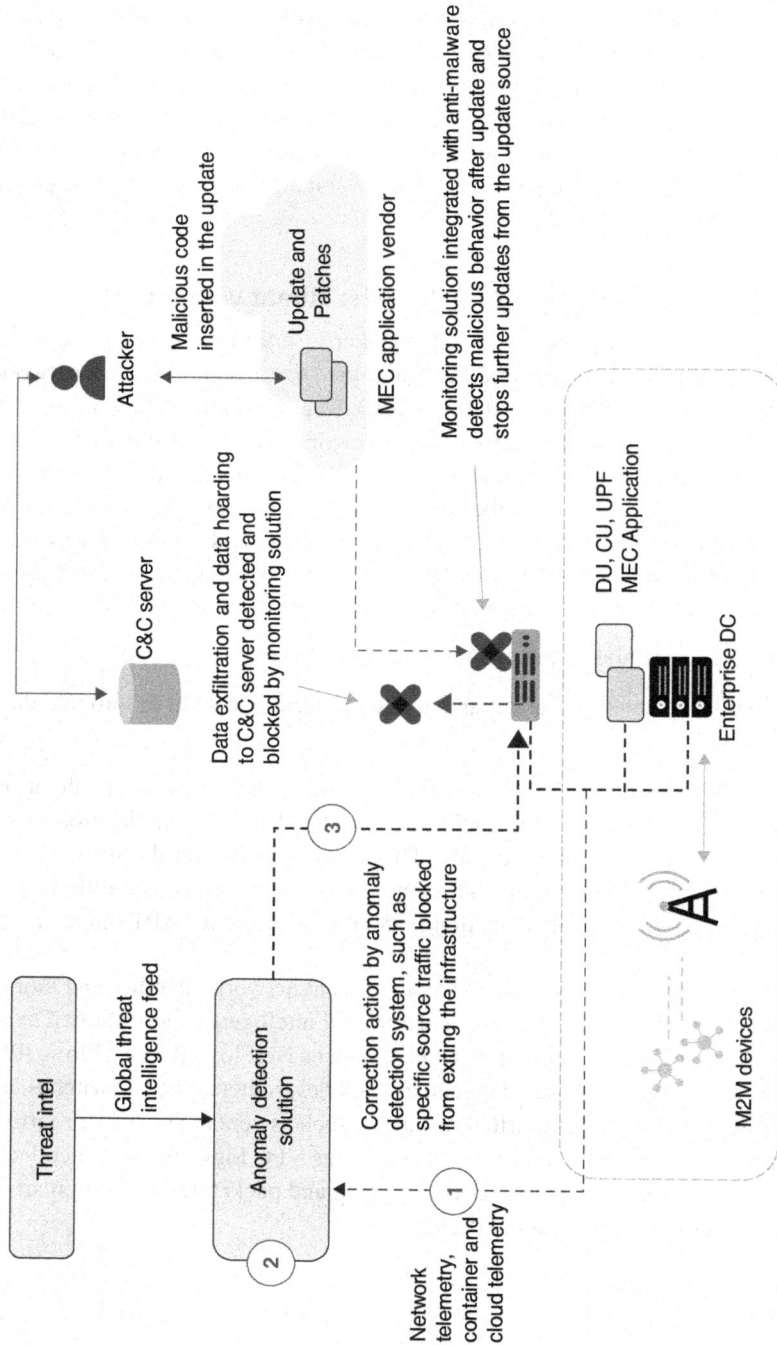

FIGURE 5-59   Anomaly Detection System to Mitigate APT-Based Attacks

2. Analyze the data gathered and find any anomalies in it. For this to happen, it's important that the normal network behavior is known. The normal network behavior, also known as baseline network, is identified by gathering the network information for 3–4 weeks and just feeding it into the learning database. This database is used by the anomaly detection system to learn what is normal behavior of the network and uses this as the baseline. Any variation from this threshold used for baseline will be considered abnormal behavior, otherwise known as anomalous behavior of the network. The anomaly detection solution should also have a very robust threat intelligence feed for global threat awareness. Having global threat intelligence would allow the advantage of the threats seen in any part of the world to be blocked if seen in part of the network. Threat intelligence is used to inform the threat intel consumers such as service providers and enterprises what threat actors are active, what attack model they use, the attacker's intentions and motivations, and which malware arsenal they deploy. Having this information fed into the intrusion detection and prevention systems would give the systems' algorithms access to a wide array of data as active threats emerge, thus providing effective mitigation actions.

3. The anomaly detection systems will also indicate any actions of compromise, such as data hoarding and data exfiltration. This would need robust machine learning algorithms that are trained using data from such specific and targeted attacks. Once the ports through which the data is being exfiltrated are detected by the anomaly system, it will create alerts and necessary actions such as blocking the internal traffic from being exfiltrated.

4. One of the key techniques used by the APT is to use the encrypted traffic to prevent detection by IPS deployed in network. So, when a new device is infected by the malicious code within the software update, the device would then initiate an encrypted connection to the next connected host using the existing ports, thereby attempting to infect the neighboring host without detection. Here is where we need an enhanced visibility layer and mechanisms of detecting any malicious traffic without decrypting the encrypted traffic. This is provided by some vendors in the market today and enables the service provider/enterprise to detect malware in encrypted traffic. Once the anomaly detection solution detects the new behavior from the network devices and network functions after the software update/patch update, corrective actions can be used to prevent any more solutions being updated. As an added measure to counteract the malware propagation, advanced anti-malware solutions should also be used. The chosen advanced anti-malware solution should inspect the memory, CPU, firmware, software OS, and the application layer to identify malicious scripts and behavior using methods such as sandboxing. Once the source of the malware is found, it is recommended to quarantine the hosts and take preventive actions, such as halting the specific action that caused the malware intrusion.

Securing MEC Against Sniffing Attacks: Threat Vector D

The MEC deployment within the enterprise in this example requires back haul toward the centralized 5GC data center of the service provider. This back haul is vulnerable to a back-haul sniffing threat vector, as explained in threat vector D.

Figure 5-60 shows one of the mechanisms to mitigate back-haul sniffing at the transport layer.

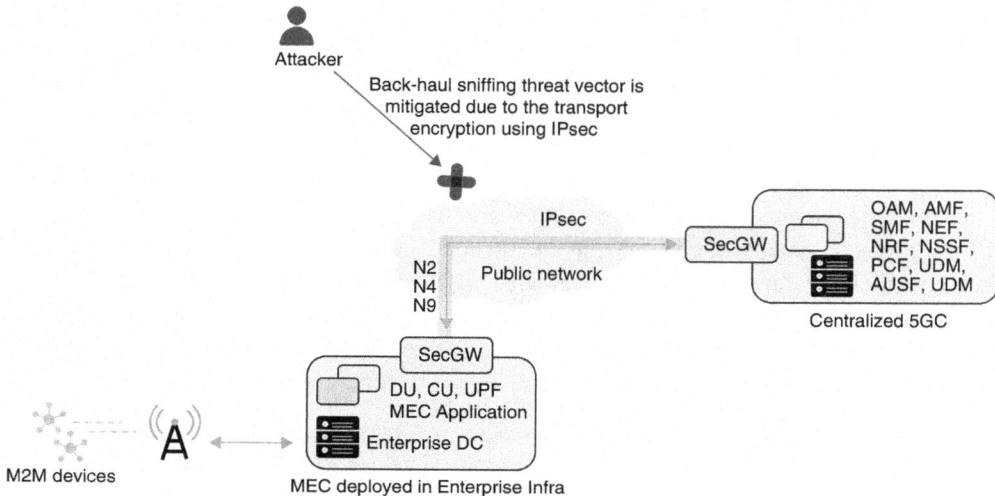

FIGURE 5-60   Mitigation of Back-haul Sniffing Attacks

Figure 5-60 illustrates the security control between the enterprise MEC and the centralized 5GC by implementing IPsec using a security gateway (SecGW). In the enterprise MEC deployment in this example, the transport for the N2, N4, and N9 interfaces uses the back hauling on the public network, which is insecure. To prevent any eavesdropping in the back haul, IPsec is recommended. IPsec is a collection of protocols for security defined by the IETF RFC 2401 and provides data authentication, integrity, and confidentiality.

Summary

As discussed in this chapter, an MEC implementation will enable you to deploy low-latency use cases. Based on your chosen deployment method, you can implement MEC in the public cloud, on premises, on the enterprise data center, or you can choose a hybrid implementation. The "Threat Surfaces in 5G MEC Deployments" section took you through the different threats in the MEC deployment, the mitigations of which were explained in the "Securing 5G MEC" section. The "Real Scenario Case Study: MEC Threats and Their Mitigation" section took you through the multidomain threat examples and their mitigations.

Figure 5-61 shows you the multilayered security controls to secure the MEC deployments.

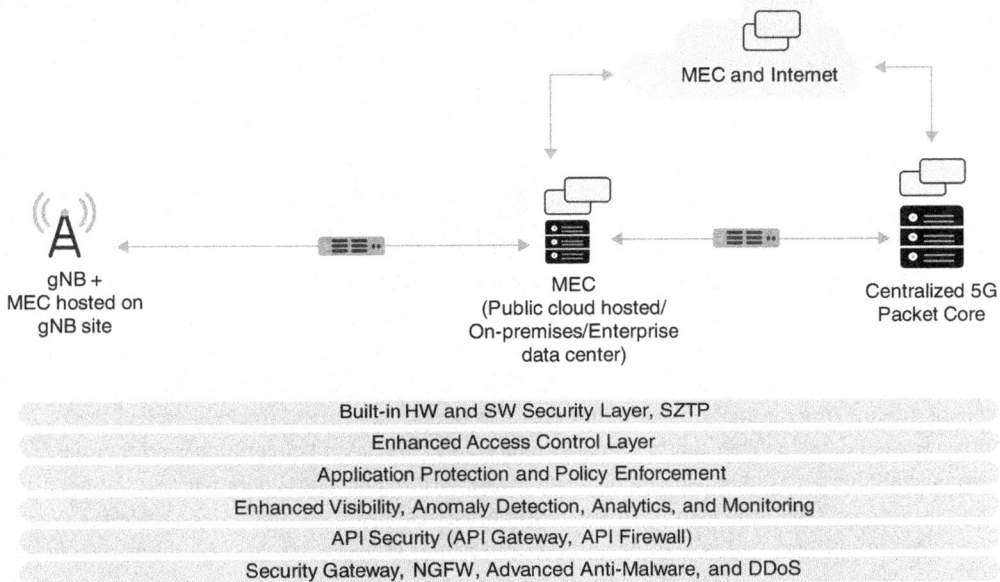

MEC and Internet

gNB +
MEC hosted on
gNB site

MEC
(Public cloud hosted/
On-premises/Enterprise
data center)

Centralized 5G
Packet Core

Built-in HW and SW Security Layer, SZTP

Enhanced Access Control Layer

Application Protection and Policy Enforcement

Enhanced Visibility, Anomaly Detection, Analytics, and Monitoring

API Security (API Gateway, API Firewall)

Security Gateway, NGFW, Advanced Anti-Malware, and DDoS

FIGURE 5-61 Multilayered Security Controls to Secure the MEC Deployments

Figure 5-61 shows the key security controls required to secure the various modes of MEC deployment. It has to be noted that you can share the security controls for other parts of the infrastructure if your 5G network deployment has other parts of the 5G infrastructure such as 5G RAN and 5GC.

The security control layers to secure 5G MEC applications and deployments are summarized as follows:

- **Built-in hardening (hardware and software layers):** The built-in security capabilities, such as hardened hardware and software, including supply chain security and secure provisioning of the network components. This layer is usually checked or verified by a handful of service providers and enterprise customers. It is very important that you ensure that the built-in hardening is part of the checklist you include while selecting the vendors of your choice. The list provided in the "Hardening Hardware and Software" section in this chapter is a good place to start.

- **Enhanced access control layer:** Having application access control based on the Zero Trust principles is a good way to have the security controls to focus on users, assets, and resources. The aim is to protect the 5G MEC resources such as MEC applications, servers, and network devices, irrespective of the MEC location (cloud/on premises/enterprise). For the enterprise MEC use cases, you should ensure that the access requests from assets located in the enterprise-owned network infrastructure meet the same security requirements as access requests and communication from any other non-enterprise-owned network. Due to the cloud-native mode of deployment in 5G, the MEC virtual instances are dynamic in nature, and the appropriate security controls should also evolve to secure the access to these dynamic components.

- **Application protection and policy enforcement:** To secure the MEC applications and provide application-level visibility and policy enforcement, it is important to use a solution that can provide visibility at the application layer, including user-to-application mapping, enabling you with the option of strict enforcement on the applications. The application protection and policy enforcement layer should also scan all images before they run and enforce policy checking to ensure that they are allowed to be executed in your environment. The application protection and policy enforcement layer should also provide vulnerability and forensics capabilities. This will enable you to investigate any abnormal behavior of the user and applications.

- **Enhanced visibility, anomaly detection, analytics, and monitoring:** Decrypting the packets, analyzing them, and then encrypting the packets again increases latency in the MEC applications. This might be an issue for URLLC use cases, some of which have stringent latency requirement of sub-1ms. An alternative would be to use intelligent solutions to cater for such attacks by identifying anomalies in the cipher suites being chosen and the behavior of the client/server handshake. This can be done by analyzing the packet header data, and there are solutions available today that can detect the malware within encrypted data without the need for decryption. The analytics part of this solution must also use the threat intelligence feed and combine it with the analysis of the gathered information using ML and AI algorithms to proactively detect any change in behavior of the network.

- **API security:** In 5G, APIs will play a much larger role of supporting intelligent flow of communications between the 5G network functions and applications for use cases such as IoT, M2M, and others. Having a strong, secure API strategy such as conducting penetration testing and audits on the APIs and applications using APIs will ensure that your API is secured and that any security holes/gaps are understood. The gaps can then be filled by ensuring the right feature or capability is enabled within the relevant security control. An example is a high value of rate limiting, which may allow DDoS/DoS attacks to be possible. A corrective measure such as optimizing the rate-limiting threshold can be performed to minimize the attacks.

- **SecGW, NGFW, advanced anti-malware, and DDoS security layer:** MEC deployments such as enterprise deployments to fulfill enterprise use cases would need proper segmentation and encryption of the data being transferred between the enterprise sites and the centralized 5GC. Segmentation of traffic should be provided by an application-aware next-generation firewall (NGFW), which can apply policies based on the defined policy as well as protocols being used by the application; this will ensure that any malicious application within the enterprise network is not migrated to the service provider network and that any malicious instance from within the service provider infrastructure is not impacting the enterprise network. The NGFW should also provide audit and compliancy reports. A security gateway (SecGW) should be used when the 5G traffic traverses any public network or any untrusted network. IPsec uses a SecGW to mitigate the possibility of eavesdropping and man-in-the-middle (MitM) attacks. 5G MEC should also be protected by advanced anti-malware to protect the MEC network from malicious code that makes its way to the internal network. DDoS protection is an important part of the security control to mitigate attacks from internal devices within the MEC infrastructure and any inbound attacks from the Internet.

Acronym Key

Acronym	Expansion
5GC	5G Core
AEF	API Exposure Function
AMF	Access and Mobility Management Function
API	Application programming interface
AS	Access Stratum
AUSF	Authentication Server Function
C-RAN	Cloud RAN or Centralized RAN or Central RAN
C&C	Command and Control
CA	Certificate Authority
CAPIF	Common API Framework
CCF	CAPIF Core Function
CD	Continuous Delivery
CI	Continuous Integration
CNF	Cloud-native function
CP	Control plane
CU	Centralized Unit
CUPS	Control Plane User Plane Separation
DCT	Docker Content Trust
DDoS	Distributed denial of service
DoS	Denial of service
DU	Distributed Unit
EAP	Extensible Authentication Protocol
ESP	Encapsulation Security Payload
FH	Front haul
FIPS	Federal Information Processing Standard
HSM	Hardware security module
IPsec	IP Security
ISAKMP	Internet Security Association and Key Management Protocol
M2M	Machine-to-machine
MAC	Medium Access Control
MEC	Multi-access edge computing
MitM	Man in the middle
mMTC	Massive machine-type communication
NAS	Non–Access Stratum
NEF	Network Exposure Function

Acronym	Expansion
NR	New Radio
NRF	Network Resource Function
NSA	Non-standalone
NSSF	Network Slice Selection Function
O-CU	O-RAN Compliant Centralized Unit
O-DU	O-RAN Compliant Distributed Unit
O-RAN	Open RAN
O-RU	O-RAN Compliant Radio Unit
OAM	Operations Administration Management
PCB	Printed circuit board
PCF	Policy Control Function
PFCP	Packet Forwarding Control Protocol
PGW-C	Packet Data Network Gateway – control plane
PGW-U	Packet Data Network Gateway – user plane
PKI	Public key infrastructure
RAN	Radio access network
RAT	Radio access technology
RBAC	Role-based access control
RIC	RAN Intelligent Controller
SA	Standalone
SBA	Services-Based Architecture
SCTP	Stream Control Transmission Protocol
SDL	Secure development lifecycle
SDN	Software-defined network
SecGW	Security gateway
SEG	Security gateway
SGW-C	Serving gateway – control plane
SGW-U	Serving gateway – user plane
SMF	Session Management Function
SSL	Secure Sockets Layer
SUCI	Subscription Concealed Identifier
SUPI	Subscription Permanent Identifier
SZTP	Secure Zero-Touch Provisioning
TLS	Transport Layer Security
TPM	Trusted Platform Module
UDM	Unified Data Management
UE	User equipment

Acronym	Expansion
UP	User plane
URLLC	Ultra-reliable low-latency communication
VRAN	Virtual RAN
ZTP	Zero-Touch Provisioning

References

3GPP TS 33.501, "Security Architecture and Procedures for 5G system," Release 16, v 16.5.0, December 2020

Pramod Nair, "Securing 5G and Evolving Architectures," Cisco Knowledge Networks, Cisco Public, September 17, 2020

Pramod Nair, "Why 5G is changing our approach to security?", Cisco Public, June 10, 2020

Pramod Nair, "Securing 5G cloud native deployments," Open5GCon, September 30, 2020

Cisco Service Provider VNI Forecast 2019

CISCO CYBERSECURITY SERIES 2019

https://github.com/kubernetes/kube-state-metrics

https://owasp.org/www-project-top-ten/

https://digital.nhs.uk/cyber-alerts

https://blog.talosintelligence.com/2020/12/xanthe-docker-aware-miner.html

https://docs.docker.com/engine/security/trust/

Chapter | **6**

Securing Virtualized 5G Core Deployments

After reading this chapter, you should have a better understanding of the following topics:

- Threats in virtualized 5G packet core deployments
- Securing virtualized 5G packet core deployments
- Real scenario case study examples of virtualized 5G core threat surfaces and threat mitigation techniques

5G introduces service-based architecture (SBA) for 5G Core (5GC), which is designed for cloud-native deployment. This chapter will take you through the threat surfaces in virtualized 5G packet core Cloud-Native Function (CNF) deployments and mechanisms to mitigate the threats. The same threats and mitigation techniques would apply to the virtualized RAN CNF deployments as well, but as 5GC SBA architecture based on cloud-native architecture is mandated for 5G deployments by 3GPP, the focus of this chapter is primarily aimed toward the 5GC use cases.

This chapter will be of particular interest to the following teams from enterprise and 5G service providers deploying 5GC, 5G vendors developing 5GC products, and cybersecurity vendors planning new product developments and new functionalities to secure the deployment of 5GC:

- Mobile infrastructure strategy teams of service provider deploying 5GC
- Security strategy teams within service provider planning on deploying 5GC
- Security strategy teams within enterprise verticals planning on deploying 5GC
- Cloud computing and data center teams involved with 5G strategy and deployment
- Security architects and design teams looking at securing the mobile infrastructure
- Enterprise solution and security architects deploying 5GC on enterprise premises

- Enterprise solution architects and enterprise security architects working with enterprises integrated with service provider 5GC

- Government departments looking at regulating security for service providers

- Cybersecurity vendor security architects looking to secure 5GC deployments for their customers

- Cybersecurity vendor product managers looking to enhance security products to cater for 5GC security use cases

A Brief Evolution of Virtualization in Telecommunications

Virtualization has been around since the 1960s, thanks to IBM CP-40, which was used to logically split the system resources in mainframes between different applications. But virtualization was primarily limited to mainframes, which made proprietary monolith hardware the norm for non-mainframe deployments. This followed into the data center (DC) world, where you had proprietary hardware from multiple vendors performing specific functions. In the telecom world, there was vendor-specific hardware such as Base Transceiver Station (BTS), Base Station Controller (BSC), transcoders, Mobile Switching Center (MSC) for 2G, NodeB, Radio Network Controller (RNC), Serving GPRS Support Node (SGSN), GPRS Support Node (GGSN), Home Location Register (HLR) for 3G and eNodeB, Mobility Management Entity (MME), Serving Gateway (SGW), and Packet Data Network Gateway (PGW) for 4G. These were deployed in the telco data centers as monoliths.

Then came the first stages of virtualization, where you had physical servers racked in your data center with an operating system deployed on top of them, and a specific application could now be deployed on top of this operating system.

Telco vendors saw this as an opportunity to deploy the network functions, but the service providers wanted a data center with a single operating system that could be used to deploy the network functions. This led to the evolution of the hypervisor, which existed since the 1960s. This required an evolution in the hardware as well, and x86 server chipset vendors started to add hardware virtualization in their chips, which happened around 2005.

You were able to reuse your data center servers, but rather than installing a single operating system and specific custom application on that server, you could install a hypervisor operating system and use it to deploy multiple virtual machines (VMs) that can run many different network functions all at the same time in a single x86 server. This gave you the advantage of optimizing your data center resources for telco deployments and the flexibility to deploy the network functions wherever you had a data center footprint.

Then came more innovations in the computing world in the form of containers and cloud-native deployment models. The cloud-native approach is a method of building and deploying applications enabling you to fully consume the advantages provided by cloud deployments. You can fulfill the cloud-native approach using both virtual machines (VMs) and containers.

Containers are very lightweight and provide a standalone executable software package that is self-sufficient and includes everything to run on, including executables, binary code, libraries, system tools, and so on. These containers will always run the same, regardless of the infrastructure where you plan to deploy them. This allows you to create a network function in any program of your choice, deploy it in any infrastructure of your choice, use any hypervisor of your choice, deploy it on any public cloud infrastructure, and also deploy it on bare metal. The telecom industry saw this as the right opportunity to use a new cellular technology to take advantage of these innovations.

Enter 5G. 5G is the first technology to make use of open source and Cloud-Native Functions to deploy network functions. The key idea is to make it extremely flexible to deploy so that you don't have to worry about proprietary hardware, software stacks, or interfaces for interoperability. You just need to think about creating new use cases and business models. When we look at the container deployments today, the most widely used container platform is Docker, and Kubernetes is the most widely used orchestration layer.

Figure 6-1 illustrates the evolution in telecom infrastructure from physical monolith equipment in 2G and 3G to the virtualized cloud-native deployments in 5G.

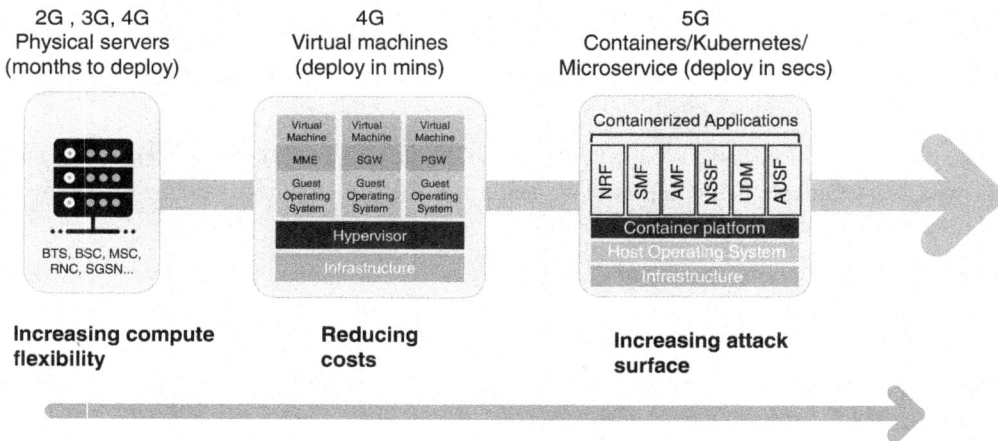

FIGURE 6-1   Evolution of Telecom Infrastructure from Monolithic to Cloud Native

As shown in Figure 6-1, while virtualized cloud-native deployment of 5G network functions such as 5GC and 5G RAN allows flexibility and scalability, it brings in new vulnerabilities. Although the system resource utilization in proprietary physical servers used in legacy cellular technologies like 2G and 3G is poor, it allows good isolation by providing physical separation of management and the user plane, including options of separate ports, line cards, and CPUs. Deployment of 4G network functions using virtual machines allowed multiple workloads to be deployed on a single x86 server. This provided a more efficient use of system resources and also provided isolation by permitting applications to run in a guest operating system. As compared to the proprietary monolith hardware, VM deployments offer less isolation because the same CPUs and interfaces can be used for control plane and user plane. When VMs are being installed, you have to allocate RAM and CPU resources that are earmarked for the VM instance regardless of whether usage leads to wasting resources. The VMs copy not only the operating system instances but also the libraries, binaries, and copies of the virtual hardware needed by the OS. Repetitive files suck up a large part of the RAM and CPU resources of the servers.

5G Cloud-Native Functions (CNFs) deployed using containers or VMs provide a very flexible deployment model because they are lightweight and it's easy to deploy within an on-premises data center or public cloud. Container deployment in 5GC provides effective resource usage, as the container is packaged with dependencies from the operating system, binary code, libraries, and so on.

Although virtualization provides better resource optimization, flexibility in deployment, and reduced cost, the attack surface has also increased due to the dissolving perimeter. This is because 5G RAN and 5G Core (5GC) CNFs can now be deployed in any public cloud provider's infrastructure and on-premises DC along with other existing applications, and the 5GC NFs can be developed using open source software stack that might have existing vulnerabilities.

Figure 6-2 illustrates the key 5GC control plane and user plane CNFs, further defined in the list that follows:

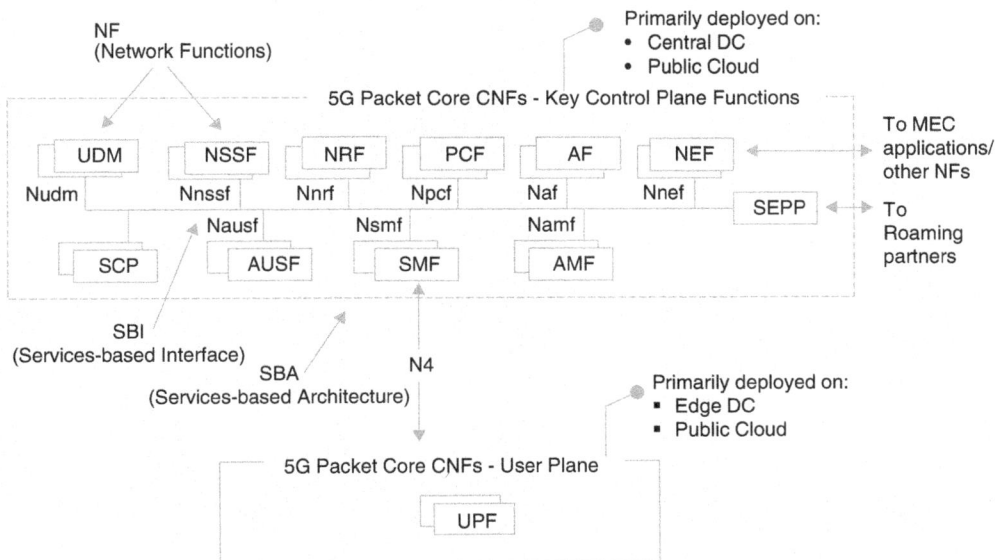

FIGURE 6-2   Key 5G Control Plane and User Plane Network Functions

- **Network function (NF):** 5GC is composed of cloud-native, virtualized functions to support a wide array of use cases. 5GC CNFs are basically 3GPP 5GC control plane and user plane functions that are deployed to fulfill the 5G use cases. The control plane messages are mainly between the 5GC NFs. The user plane messages are mainly between the user equipment and Internet, with the User Plane Function (UPF) providing the interconnect, encapsulation, and decapsulation of GPRS Tunneling Protocol User Plane (GTP-U) and other functions related to the user plane traffic.

■ **Service-based architecture (SBA):** SBA is a framework for 5GC CP communications with the aim of replacing signaling messages by API calls. It also replaces the point-to-point signaling used in legacy technologies such as 2G, 3G, and 4G with a service bus everyone-to-everyone communication. As illustrated in Figure 6-2, the principles of SBA apply only to the control plane functions of the 5GC. Using the SBA framework, you have the option of using any vendor for any network function. For example, you could have AMF/SMF from Vendor A and NSSF/AUSF from Vendor B. As SBA uses API calls for communications, you shouldn't really have any issues with interoperability. Having the SBA framework, you can now also deploy the network functions in the public cloud. Another advantage of SBA is that it enables developers to develop applications that interact with the 5GC components using API calls using Network Exposure Functions (NEF).

■ **Service-based interface (SBI):** SBI is the term used for the API calls between the 5GC network functions in the SBA architecture. The design principles and documentation guidelines for 5GC SBI APIs are specified in ETSI TS 29.501. All 3GPP NFs have a mandatory requirement to support TLS, which should be used if other means of network security, such as Network Domain Security (NDS) using security gateways (SecGW/SEG), are not implemented in the 5GC PLMN network.

Figure 6-3 provides the view of the deployment locations for the 5G cloud-native control plane and user plane components.

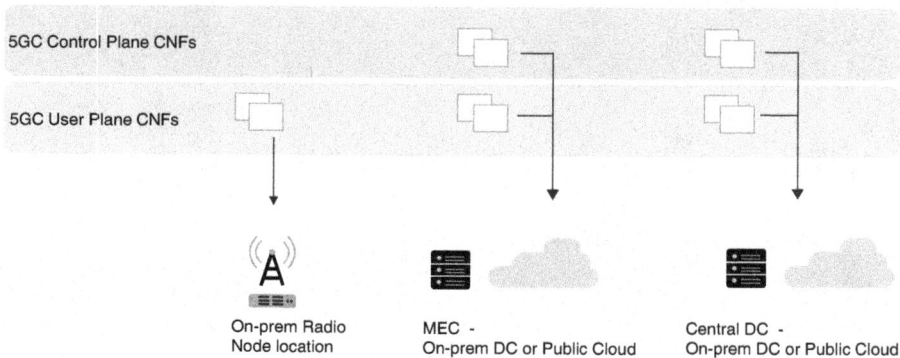

FIGURE 6-3 Deployment Location Options for 5GC Control Plane and User Plane Network Functions

As shown in Figure 6-3, the 5GC user plane (UP) components can be deployed even at the radio node location to enable ultra-reliable low-latency communication (URLLC) use cases. The 5GC control plane (CP) functions can also be deployed in the radio node location, but because the latency requirement is specifically for user plane and because the server at the radio node has an extremely low footprint, there is no need to have the 5GC CP functions deployed at that location.

Although the 5GC CP CNFs can be deployed in the edge and central DCs, most of the use cases today do not need the CP to be deployed at the edge.

Figure 6-4 illustrates the different modes of 5GC NF deployment.

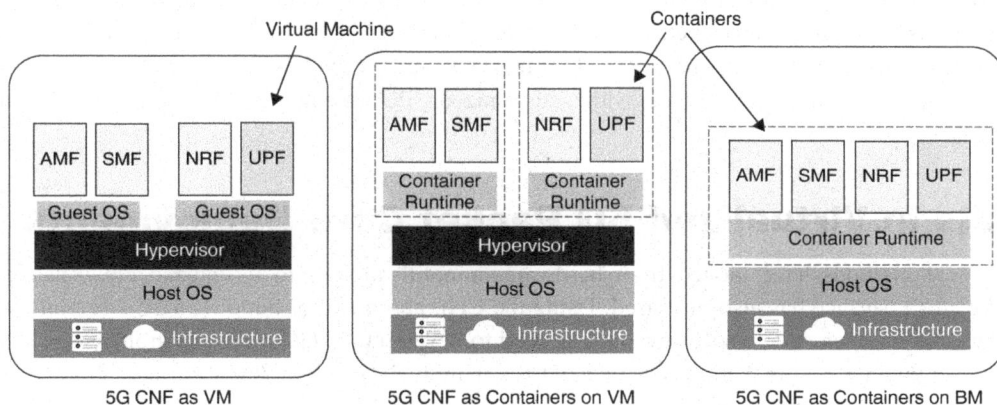

FIGURE 6-4   Deployment Modes for Cloud Native 5G Packet Core Functions

Service providers and enterprise verticals want a choice of deployment models to enable use cases. The reasons could include whether you want to support multiple modes of deployment based on use case and workload type or to accommodate an existing single-cloud deployment or due to a shortage of resources for optimal deployment of a specific method. You want the flexibility to be able to run containerized applications on bare metal (BM) or virtual machines (VMs) in your on-premises data centers, whether on a RAN node location, an MEC or centralized DC, a multistack public cloud, or a hybrid deployment model. As shown in Figure 6-4, there are three options for deploying 5GC CNFs that can be deployed on premises or in the multicloud public cloud stack. One option is based on a virtual machine and the other two options are based on containers:

- **5GC CNF as VM:** VM deployment is not a new deployment model. Many of the test beds used initially for the 5G non-standalone model (NSA) used the virtual network functions of 4G and upgraded the software to provide initial 5G NSA services. Many vendors also have parts of their 5GC network functions in VM format to make full use of the service provider's resources. All the major vendors are moving toward the container in VM and container in BM models to make full use of the cloud-native approach.

- **5GC CNF as containers on virtual machine (VM):** Having 5GC NFs deployed as a container on VMs is the option many service providers are choosing to deploy, mainly because of infrastructure readiness and lack of readiness from 5GC vendors in providing a true container stack. Many of the service providers are also heavily invested in a single private cloud platform, and a step in the direction of deploying containers on VMs is in the right direction of moving toward the hybrid cloud environment. Deploying containers on VMs will provide you with better isolation as compared to deploying containers on bare metal. Containers deployed on VMs are also easier to move between hosts using guest images.

- **5GC CNF as containers on bare metal (BM):** Containers being deployed on bare metal will actually provide you with better performance for 5GC use cases. Containers deployed on bare metal utilize system resources better than containers deployed on VMs. One of the key reasons is the direct access to physical hardware. This makes it suitable for low-footprint server deployments, such as in the gNB locations, with a customized CPU for better performance in MEC use cases.

Threats in Virtualized 5G Packet Core Deployments

As deployment methods have moved from hardware monolith models to virtualized cloud-native models, the attack surface has also increased. Using the CNF approach for building 5GC NF components means that an open source software stack is used to develop the 5GC NFs, which increases the risk in 5GC deployments. This section is dedicated to explaining the different threat surfaces in a virtualized 5GC CNF deployment.

Figure 6-5 illustrates the threat surfaces for different 5G RAN and core CNF deployments. This section will take you through the 5GC container, internal, and external 5GC NF vulnerabilities. The hardware vulnerabilities are covered in Chapter 5, "Securing MEC Deployments in 5G" in the section "5G MEC Infrastructure and Transport Vulnerabilities."

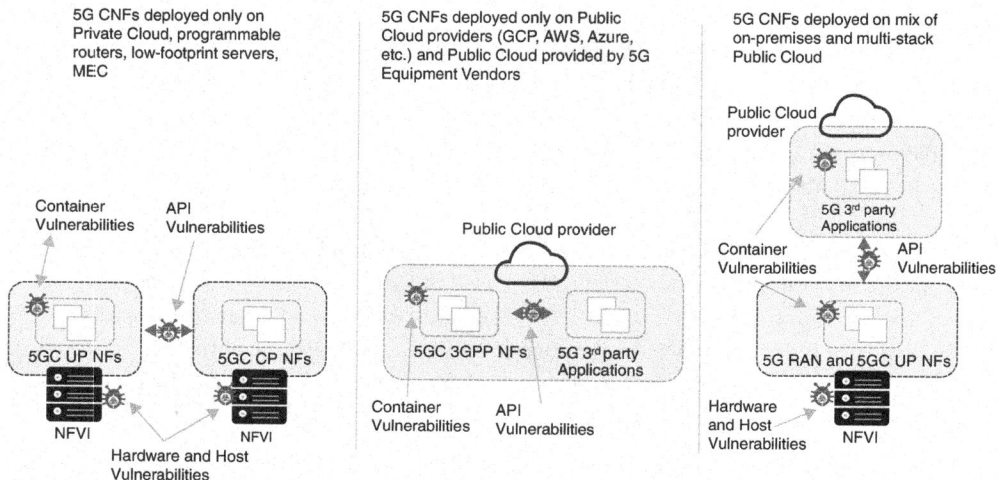

FIGURE 6-5   Threat Surfaces for 5G CNF Deployments

Figure 6-5 shows the two models of deploying the 3GPP 5GC network functions: on the public cloud and the on-premises data center. These new deployment models require a good understanding of the multiple cloud-native component integrations and the different layers of testing required to achieve secure cloud-native deployments.

The threats within the 5GC container deployments can be broadly split into three categories:

- **5GC container vulnerabilities:**
  - Insecure container build
  - Image vulnerabilities
  - Container runtime vulnerabilities
  - Insecure container host
  - Malicious container network traffic
  - Malicious 5GC CNFs
  - Insecure container management and orchestration
  - Improper access control
  - Improper configuration
  - Insufficient isolation
- **Insecure container networking:** This can be split into two areas—internal container communication within a pod and communication with external pods or servers:
  - **Internal interfaces**:
    - API vulnerabilities within SBI communication
    - Insecure network leading to eavesdropping
  - **External interfaces**:
    - Improper isolation and segmentation with enterprise network
    - API vulnerabilities
    - Vulnerabilities related to non-3GPP MEC NFs
    - Insecure roaming interface and misconfigurations
- **Hardware and host OS vulnerabilities:**
  - NFVi hardware and software vulnerabilities
  - Improper access control

5GC Container Vulnerabilities

In previous cellular technologies, there was very little to no deployment related to container technologies. Use of container technology for 5GC NFs necessitates the Continuous Integration/Continuous Development (CI/CD) process, which helps you achieve quicker delivery of applications. If you are a 5GC vendor, you will use the CI/CD process to build your 3GPP 5GC NFs. If you are an enterprise or service provider, you will use the CI/CD process to build non-3GPP 5GC NFs to interact with 3GPP 5GC NFs for various use cases such as enhancing the functionalities of the 5GC NFs or allowing integration for enterprise use cases and monetizing 5G.

Simply following the CI/CD process is not enough to build the 5GC NFs. CI/CD, while ensuring quicker development of application, also brings in new threat surfaces related to insecure 5GC NF images and their distribution. 5GC 3GPP and non-3GPP NFs are susceptible to vulnerabilities and malicious software, which pose a threat during runtime. The malicious images could be used by an attacker to exfiltrate sensitive information or cause DoS attacks.

Figure 6-6 illustrates the key threat surfaces in the container runtime environment.

The text that follows describes the threat surfaces illustrated in Figure 6-6.

5G provides the flexibility to build or create 5GC network functions with open source components to make them truly deployable in a cloud-native environment. *Cloud native*, a term used to describe container-based deployment, is a new concept in 5GC deployments. The network abstraction brought along by 5G addresses not only the need for portability across cloud environments but also the ability for developers to take advantage of emerging patterns to build and deploy applications. Some of the key threat surfaces illustrated in Figure 6-6 are as follows.

1. The developer for the 5GC vendor uses open source software to build the 5GC network functions. One of the threat surfaces here is the existing vulnerability of the program being used to build the image.

2. In the example shown in Figure 6-6, the developer uses software components with an unpatched vulnerability, images missing critical security updates, or outdated patches. Once the developer builds the image with an existing vulnerability in the software components, it leads to the image having built-in vulnerabilities that can be exploited by attackers.

3. Once the image is built, it is stored in a container repository. You will then download the containers from the container repository and deploy them in your network. If you download an image that has an embedded vulnerability, it might make your network vulnerable to attacks due to attackers exploiting the known vulnerability.

4. Another attack vector at the container repository is the attackers scanning the images in the container repositories for any unprotected containers and then exploiting them. This attack vector is explained in detail in the section "Virtualization Threat Vectors" in Chapter 5.

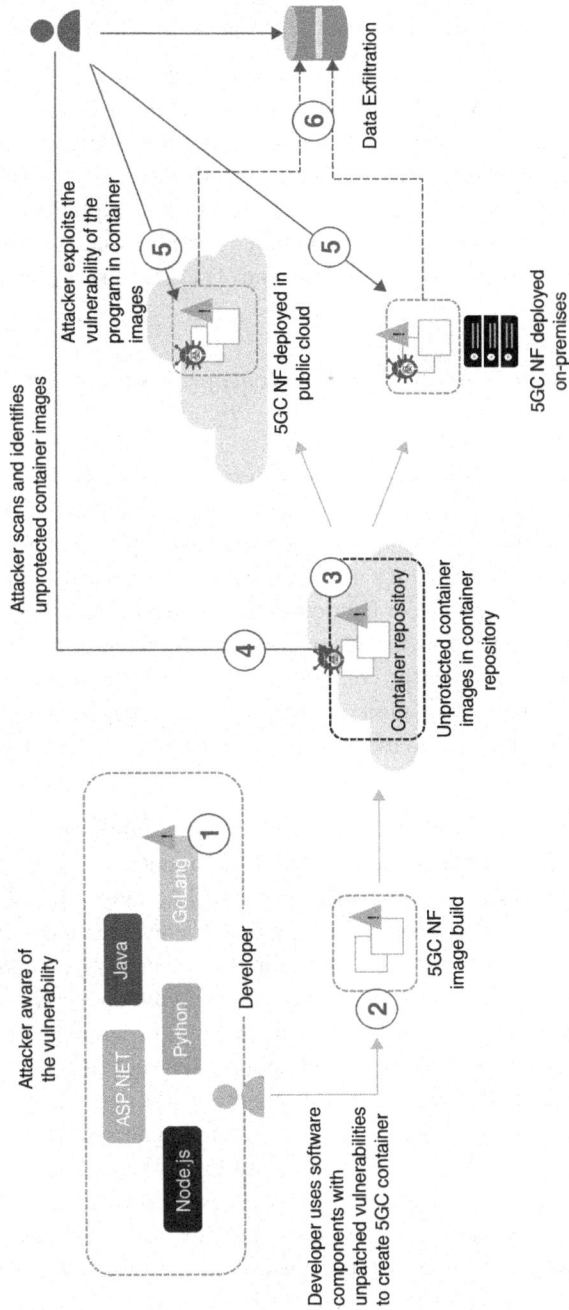

FIGURE 6-6 Key Threat Surfaces in Containerized Runtime Environment

5. Once the vulnerability in the deployed 5GC network function is exploited, the attacker might be able to map other systems in the environment, attempt to elevate privileges within the compromised container, or abuse the container for use in attacks on other systems.

6. Due to the portability and the modular characteristics of containers, the packages can be configured once and deployed on multiple servers or multiple cloud instances. An example would be a user plane component like User Plane Function (UPF) or a control plane component like Session Management Function (SMF), which can be deployed in multiple edge on-premises servers or public cloud with a predefined template. If the attacker exploits the UPF image within the container repository, all the UPFs deployed within your infrastructure are now susceptible to attacks such as exfiltration to a command and control (C&C) server defined by the attacker in the exploited image.

Apart from the aforementioned vulnerabilities in the container build and runtime, other key container vulnerabilities are as follows:

■ The container registry is a location where you will store and distribute images. The registry will be used by vendors to upload signed images that you can download and deploy in your network. The registry will also be used by you if you are developing a 5GC application and would like it to be deployed across different parts of the network. Due to the container environment where an application can have tens or hundreds of containers, depending on the use cases, you might have thousands or tens of thousands of containers in the registry. An insecure container registry or API implementation of the containers can lead to attackers having access to the containers, and the attackers can exploit any weakness to deploy malicious code within the container that you have uploaded to the registry or replace the container with a malicious version.

■ Patch management is another area that is impacted by the deployment of container-based network functions. In existing technologies where the deployments are based on proprietary hardware or virtual machine instances, the patches are usually provided by the vendors a couple of times per year. The real issue for you in relation to 5GC containerized network function deployment is the time taken to patch the vulnerability in the open source component of the 5GC CNF. An example would be one or two major patches coupled with one or two maintenance patches per year. Container deployment is basically deployment of disaggregated software, and this will allow increased updates, basically destroying the existing 5GC container and deploying a new image with the updated patch. There are a couple of risks attached to this new procedure. One of them is having gaps in visibility and monitoring in understanding which of the deployed 5GC network functions are up to date. The other key risk is an improper patch update procedure ensuring right configurations and security policies are applied to the latest 5GC container image. Normally when a vulnerability is disclosed in open source software, the person who finds the vulnerability reports it through the open source community. The 5GC vendor should now create a customized kernel with the patch and follow the procedures for signing and uploading it to the registry. This should then be downloaded, the old container image destroyed, and new the container image with the vulnerability patched should be deployed. The key issue here is the time taken between the vulnerability disclosure and deployment of the new container in your 5GC environment. This gives enough time for an attacker to exploit the vulnerability and cause harm to your network or exfiltrate critical information. One other key issue is the customization of the Linux kernel used in a 5GC containerized deployment. This will further extend the time taken to provide the patched container image, as you cannot just patch it from the upstream kernel. In short, patch management for the 5GC components should be one of the key criteria that's thoroughly discussed with the 5GC vendor.

- The very dynamic nature of 5GC container-based network functions during runtime can make it very difficult to monitor them. The key risk here is the lack of visibility in the dynamic container environment to detect any anomalies.

- Improper configuration of a 5GC container-based network function such as the 5GC components running with a higher privilege is another area that can lead to your network being compromised.

- One of the reasons for the increase in threat surfaces in virtualized 5GC NFs is due to the polyglot nature of microservices, which is the result of the flexibility in using multiple open source components in 5GC. This threat surface is explained in Chapter 5 in the section "Virtualization Threat Vectors."

Insecure Container Networking

In 5GC container-based network function deployments, the communications are based on the service-based interface (SBI), which employs REST interfaces using HTTP/2.

As illustrated in Figure 6-7, there are multiple interfaces in 5GC where the REST API calls would be used, such as the SBI message bus for control plane communications within the 5GC network functions, by the Network Exposure Function (NEF), which exposes capabilities and events to third-party Application Functions (AFs) and the orchestration layer.

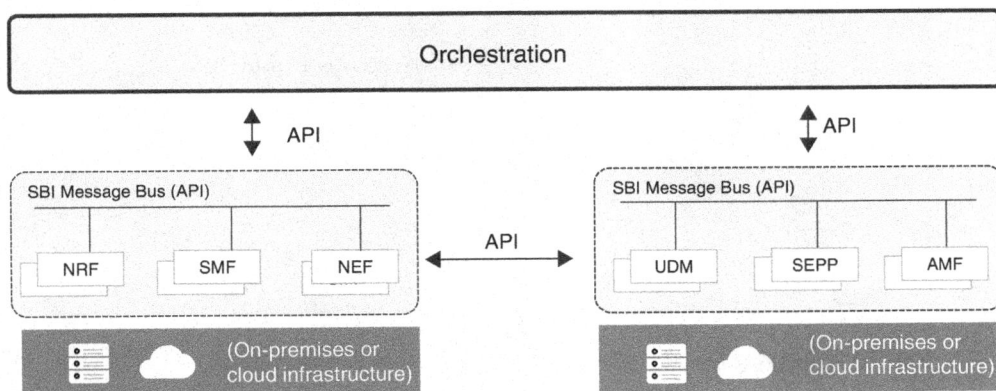

FIGURE 6-7 API Communication in the 5GC Environment

The deployment of 5GC containers also requires orchestration for creating, operating, and managing the containers. One of the most popular container orchestration tools used today is Kubernetes. While deploying 5GC CNFs, the majority of the service providers and enterprise verticals plan to use Kubernetes as an orchestration tool. One of the main reasons for this is the existing skillset within the in-house development teams and the Docker and Kubernetes options provided by the vendors.

The Kubernetes API is the main mode of communication between the Kublet on the Kubernetes worker node and the Kubernetes API server. APIs are used by Kubernetes to create, configure, and manage the clusters.

Internal Interfaces

Let's first have a look at the internal container communication threat surfaces.

Although the 5GC network functions are containerized and can be ideally deployed in a single server, that is almost never the case. The 5GC network functions would need to serve millions of users, and this requires multiple instances to be instantiated to serve the users. There are also multiple considerations such as the vendor you are choosing to use for user plane functions and control plane functions. Looking at the scenario in a couple of years after this book is published, you will have so many different vendors to choose from, especially in MEC use cases catering for ultra-reliable low-latency communication (URLLC).

Figures 6-8 and Figure 6-9 illustrate some of the different threat surfaces due to the deployment types. In this example, part of the 5GC CNFs are deployed on the public cloud and parts of them are deployed on premises. 5GC CNFs deployed on premises have separate host servers for SMF, AMF, and NRF that can be from multiple vendors. PCF, UDM, and AUSF components are deployed in the public cloud.

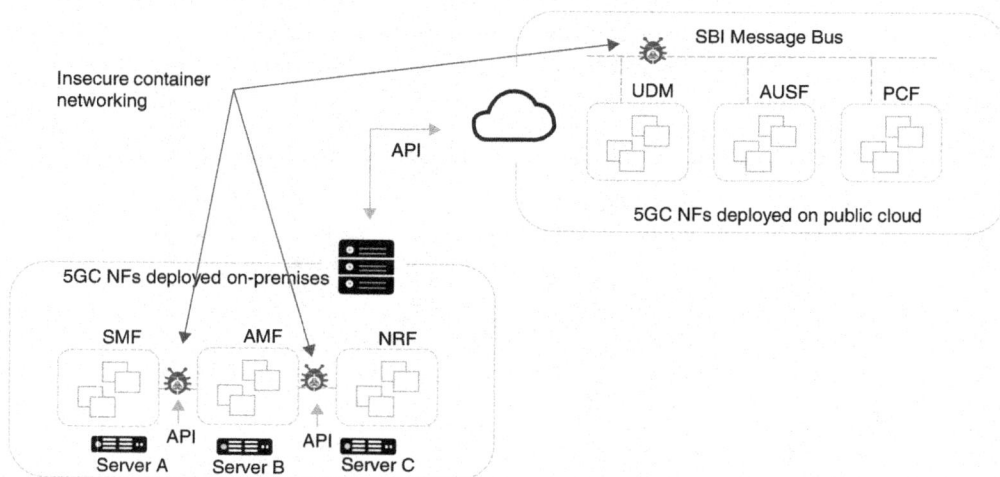

FIGURE 6-8   Insecure Inter-5GC NF API Communication

Figure 6-8 shows the threats in the SBI-based communications used for internal communications between the 5GC network functions due to insecure container networking both on the premises and in cloud-deployed CNFs. As can be seen from Figure 6-8, due to the multiple-server deployment on the premises to cater for the higher capacity, the API calls between the 5GC CNFs have to be handled through switches and virtual switches. Insecure interfaces between the switches and the unprotected API calls increase the risk of man in the middle (MitM) attacks. When deployed in the public cloud, the same SBI message bus using REST HTTP/2 API calls is used for communication between the 5GC control plane CNFs. The unprotected API communication between the containers is prone to MitM attacks. Figure 6-9 illustrates threats due to the hybrid deployment model of 5GC containers.

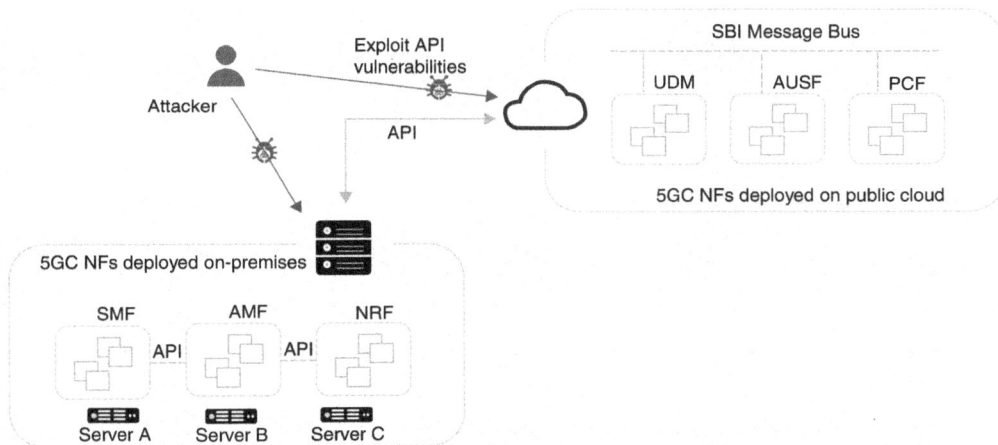

FIGURE 6-9   Insecure External 5GC NF API Communication

Apart from the internal communication threats within the on-premises and public cloud deployment models, deploying containers in a hybrid model (that is, having some 5GC containers on premises and some on the public cloud) brings in new threat surfaces. One of the key threat surfaces is unprotected communication between the on-premises and cloud-deployed 5GC containers, as shown in Figure 6-10. The other threat surface is improper configuration of the access controls, which might lead to improper access privilege for users accessing the API calls.

Another threat surface for the 5GC control plane CNFs is the implementation of the new network function called Network Exposure Function (NEF), which facilitates the access of 5GC network functions to external applications created by third-party developers, as shown in Figure 6-10. The concept of an exposure function is not new to cellular technologies. In 4G technology, Service Capability Exposure Function (SCEF) was used to expose the 4G packet core capabilities for IoT use cases, event reporting, and so on. The key differentiator in 5G for the exposure function is the capabilities of the service-based architecture (SBA) of the 5GC control plane containers, which use RESTful API calls. This makes the exchange of information easier and enables innovation in the development of applications for 5G use cases. In other words, NEF can be seen as an evolution of SCEF. The key threat surfaces arising in implementations of NEF are mainly due to improper access control and improper implementation of APIs.

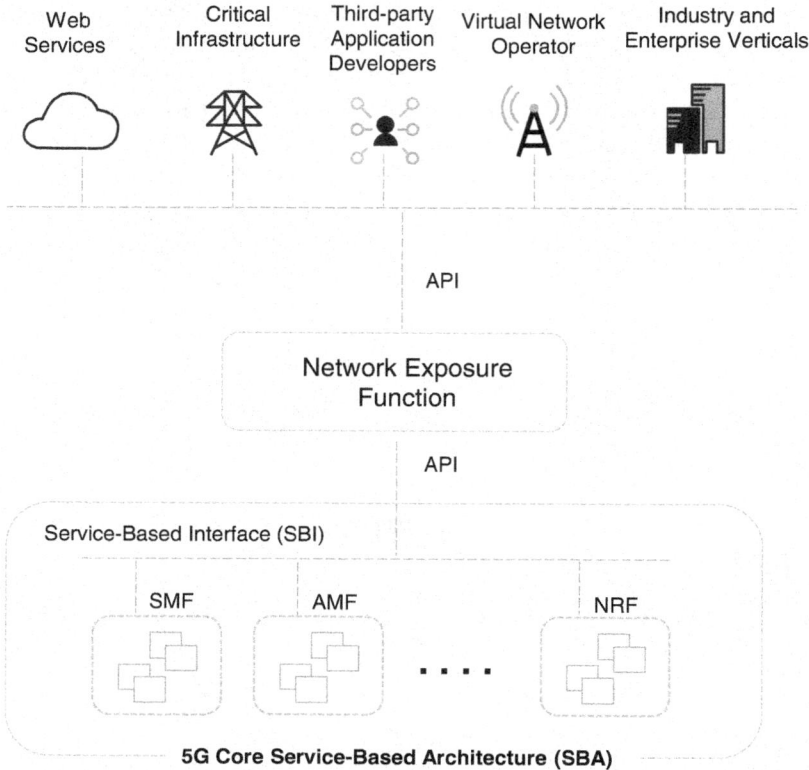

FIGURE 6-10   5GC Network Exposure Function

External Interfaces

Apart from the API-based calls in the 5GC control plane SBA architecture, there are other non–control plane communications required for the 5G network to operate and fulfill use cases seamlessly. Figure 6-11 illustrates the external components for the 5GC control plane, and Figure 6-12 illustrates external components for the 5GC user plane, which requires integration with external components.

Traffic between the containerized 5GC network functions is normally routed over a virtual overlay network that is managed by the container orchestration platform and usually bypasses the network security and management layer within your organization. This becomes an issue when the traffic between the containers is encrypted. Existing network security controls in your network will not be able to provide visibility into the actual container communications, and it's not ideal technically and commercially to deploy a decryption engine between each and every container. This causes a lack of visibility in your container environment and ineffective monitoring of your network traffic.

5G Packet Core CNFs - Key Control Plane Functions

| UDM | NSSF | NRF | PCF | AF | NEF |

Nudm   Nnssf   Nnrf   Npcf   Naf   Nnef

Nausf   Nsmf   Namf   SEPP

| SCP | AUSF | SMF | AMF |

- UE NAS messages
- Access Network
- 5GC UP NFs deployed on MEC (public cloud/on-premises)
- Non-5GC NFs deployed on MEC (public cloud /on-premises)
- Roaming Partners
- Integration with enterprise non-3GPP NFs

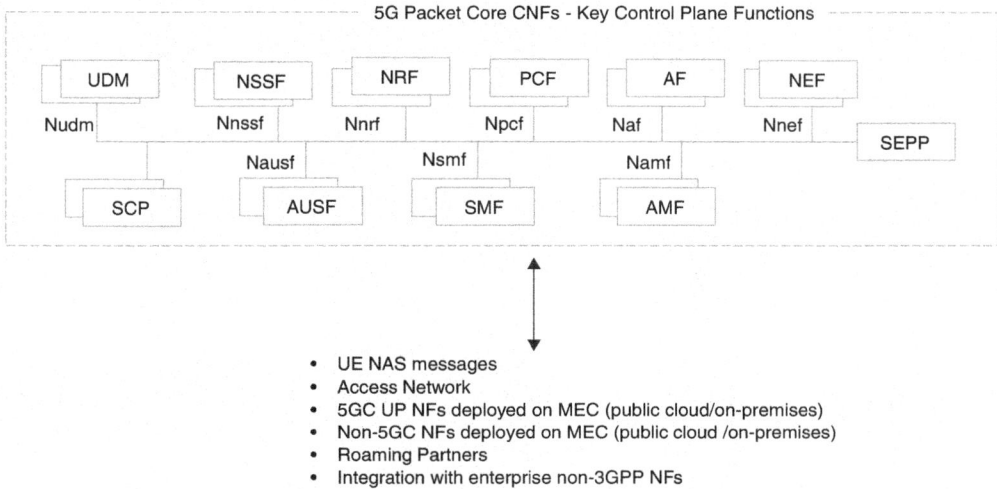

FIGURE 6-11 5GC Control Plane Integrations with External Network Components

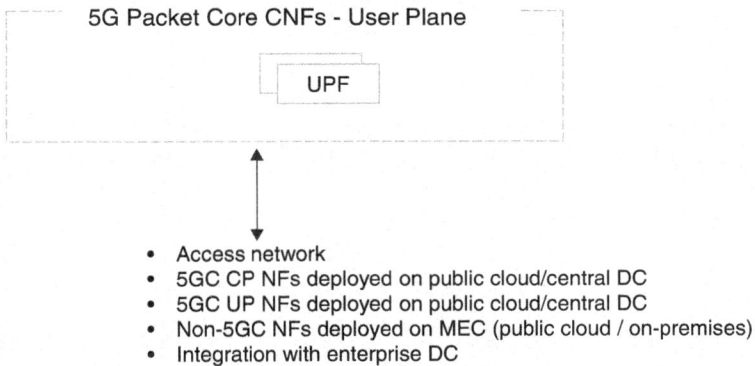

5G Packet Core CNFs - User Plane

| UPF |

- Access network
- 5GC CP NFs deployed on public cloud/central DC
- 5GC UP NFs deployed on public cloud/central DC
- Non-5GC NFs deployed on MEC (public cloud / on-premises)
- Integration with enterprise DC

FIGURE 6-12 5GC User Plane Integrations with External Network Components

This opaque network layer caused by encrypted inter-container traffic becomes critical if you have a mix of critical applications and Internet-facing web applications share the same virtual infrastructure. You will face this issue when containerized 5GC functions are deployed along with non-3GPP 5G applications in the MEC layer, as shown in Figure 6-13.

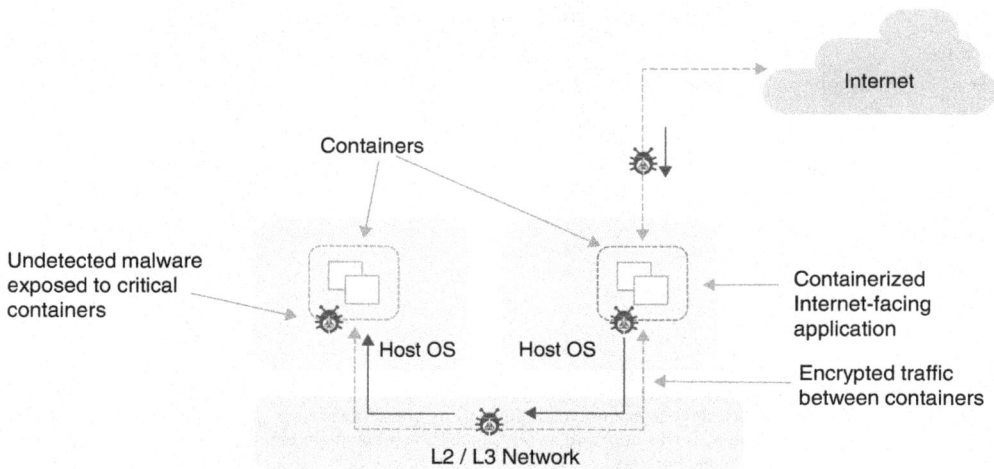

FIGURE 6-13   Impact of Compromised Web-Facing Application

Figure 6-13 illustrates how a compromised containerized web-facing application could impact the critical container sharing the same virtual infrastructure. The communication between the containers is encrypted with methods such as Transport Layer Security (TLS). While the communication is passing through the L2/L3 network, which could be a virtual switch deployed on the infrastructure, the traditional network filters would see only the encrypted packets between the two hosts, without any visibility into the traffic between the container endpoints. If the web-facing application is infected, any malware being sent using the encrypted path would be invisible to you.

Figure 6-14 illustrates the roaming interfaces of the 5G non-standalone architecture (NSA) and different integrations required between the home public land mobile network (PLMN) and visitor public land mobile network (VPLMN).

As shown in Figure 6-14, NSA deployments have to cater for the integrations between the 4G Evolved Packet Core (EPC) and 5GC network functions for roaming scenarios. The control plane vulnerabilities of GTP (control plane and user plane) and Diameter protocols are well-researched areas, and the attack vectors are well documented. For the 5GC integrations, the following reference points would use API calls in a roaming scenario:

- **N8:** Reference point between VPLMN AMF and HPLMN UDM

- **N16:** Reference point between VPLMN SMF and HPLMN SMF

- **N24:** Reference point between VPLMN PCF and HPLMN PCF

These interfaces (N8, N16, and N24) are susceptible to attacks exploiting the API vulnerabilities due to improper API implementation.

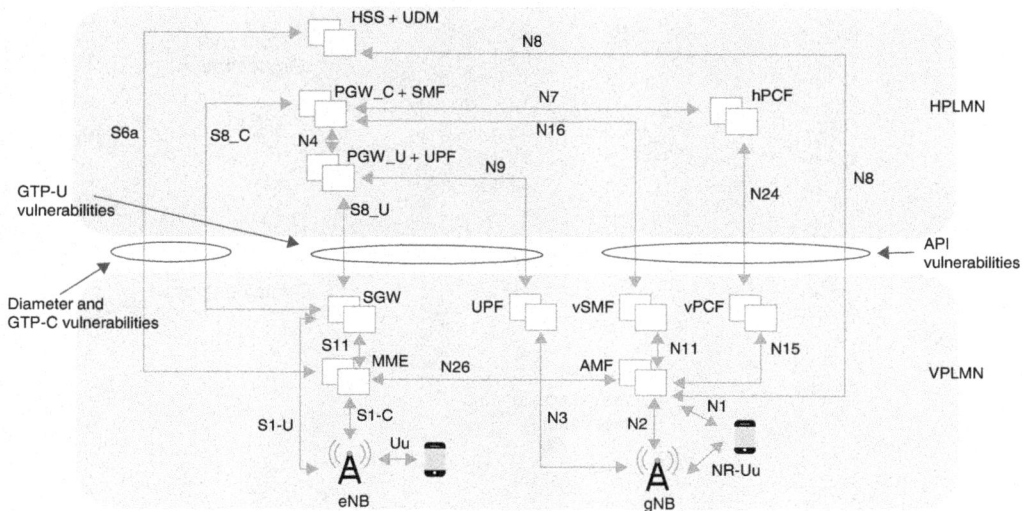

FIGURE 6-14   Roaming Interface for NSA and SA Deployments

Orchestration

Containerized 5GC network functions, whether on the premises or in the public cloud, will be dynamic in nature and should have the capability to maintain state, which is very important to ensure high availability and provide services to your 5G subscribers and use cases. This capability is provided by the orchestration layer.

This makes the orchestration layer one of the key targets for attacks. The orchestration layer usually manages thousands of containers and is designed to handle multiple users. This brings in one of the key threat surfaces of improper authentication and authorization of users accessing the orchestration platform. Figure 6-15 illustrates the risk of an attacker having unauthorized access to the orchestration platform.

Container orchestration platforms normally allow every application access to query the API. Container orchestration platforms use a role-base access control (RBAC) that integrates with an external authentication directory such as Lightweight Directory Access Protocol (LDAP) or an existing authentication solution being used by the specific teams operating the packet core, which is separate from the user directories used by your IT department or organization. Having multiple authentication solutions is quite common with service providers where one or more authentication solutions are implemented per department. This leads to an isolated directory system that is prone to weaker management and improper access control. The container orchestration platform usually caters for managing containers with high privileges. If the authentication directory within such orchestration platforms is misconfigured or not properly audited and managed, the permissions will be escalated quickly, consequently allowing the attacker to inject malicious code into the 5GC network functions or deploying a container with malicious code in it. Once the attacker performs privilege escalation, they could destroy the legitimate 5GC containers and cause disruptions or an outage of your 5G network.

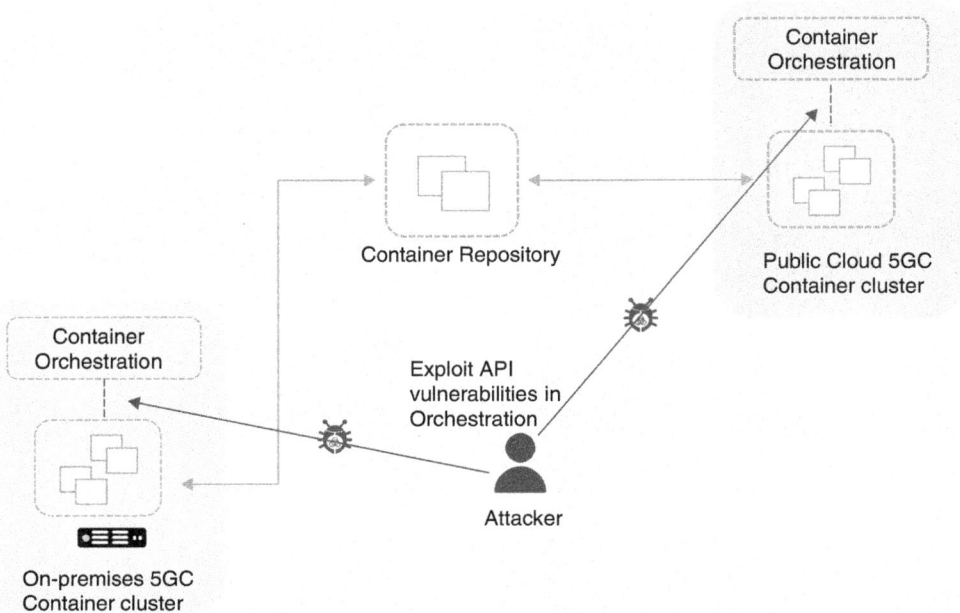

FIGURE 6-15   Attacker Having Unauthorized Access to Orchestration Platform

Another area of risk in the container orchestration layer is unauthorized hosts joining the container cluster. Improper configurations on the orchestration platform can lead to unauthorized access and unauthorized host configurations, which can expose the container clusters to increased risks.

Container Host and HW Vulnerabilities

The previous sections covered the vulnerabilities and risks in containers, container networking, and orchestration. The underlying host introduces new attack vectors, which are further discussed in this section. One of the key attack vectors is exploiting the CPU optimization methods used for code execution, and the range of attacks is aimed at applications running on the same core. There are also other attack vectors due to insecure supply chain, such as changing the design of the circuit board to include malicious chipsets or shadow chipsets. Apart from malicious chipsets, the legitimate off-the-shelf chipsets also have risks such as unauthorized read/write access to their virtual memory, which can lead to leakage of cryptographic keys. These malicious chipsets and vulnerabilities within off-the-shelf chipsets could be used to execute the load instructions before the legitimate CPU gets them. Many chipsets available today have micro-codes embedded into the firmware for management. These micro-codes are found to be vulnerable, enabling attackers to have root access leading to data from encrypted hard disks being decrypted, device IDs being forged, and even the ability to extract digital content. Other attacks, such as Spectre and Meltdown, exploit the read/write implementations of the chipset manufacturer, leading to exfiltration of data. Critical vulnerabilities and threat related to the hardware vulnerabilities are explained in the "Hardware and Software Vulnerabilities" section in Chapter 5.

The main issue in deploying 5GC Cloud-Native Functions in the public cloud is the hardware used by the public cloud providers for hosting your applications. If the hardware used by the cloud vendor uses chipsets that are insecure, the cloud provider's DCs are already exploited, and your data is not safe from the time it is deployed in the compromised cloud provider's environment.

One of the key areas of concern while deploying the 5GC containerized network function in the public cloud or on-premises data center is the level of isolation provided during deployment. Containerized 5GC functions can be deployed on VMs or bare metal. Many vendors today are offering bare-metal container deployment options, but they need proper validation to ensure that the container environment provides proper isolation. Let's discuss some of the isolation issues related to container deployments.

When 5GC network functions are deployed as containers on virtual machines, they use the VMs as the host operating system rather than running directly on bare metal. VMs provide inherent isolation between the VMs and host operating system, as illustrated in Figure 6-16. A container-on-VM deployment is one of the key methods used by public cloud providers to enhance the isolation between your 5GC network functions deployed on the cloud and the other customers' applications deployed on VMs or containers.

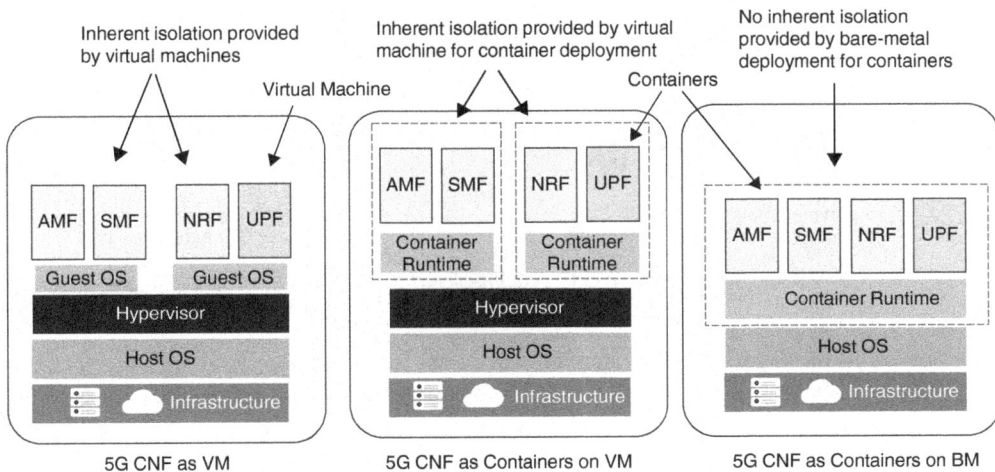

FIGURE 6-16 Isolation Provided by Different 5GC CNF Deployments

Isolation is not the only consideration you will have before choosing the deployment option. If you have a colleague from the data center compute team working with you for the dimensioning part, the criteria would be consumption of computing resources, such as memory, storage, and CPU, which is very efficient in containers deployed on bare metal. Another aspect could be ease of operations and maintenance, which are again better handled by deploying containers on bare metal. Containers on bare metal also help with the ease of deployment and scaling, as they are deployed on a single oper-ating system (OS), which reduces the number of OSs that need patching and so on. All these benefits of deploying containers on bare metal will lead to you choosing this option over containers being deployed on VMs. But in pragmatic network deployments, you might actually end up using containers on VMs and BMs, as shown in Figure 6-16.

A containers-on-BM deployment is chosen where latency is critical to enable ultra-reliable low-latency communication (URLLC). These deployments would be typical at the RAN site location or at the far-edge DC, as shown in Figure 6-17. In other locations, such as the MEC and central DC, you can deploy containers on BM or VMs. Let's therefore take a deeper look into the insufficient isolation in bare-metal deployments and their vulnerabilities.

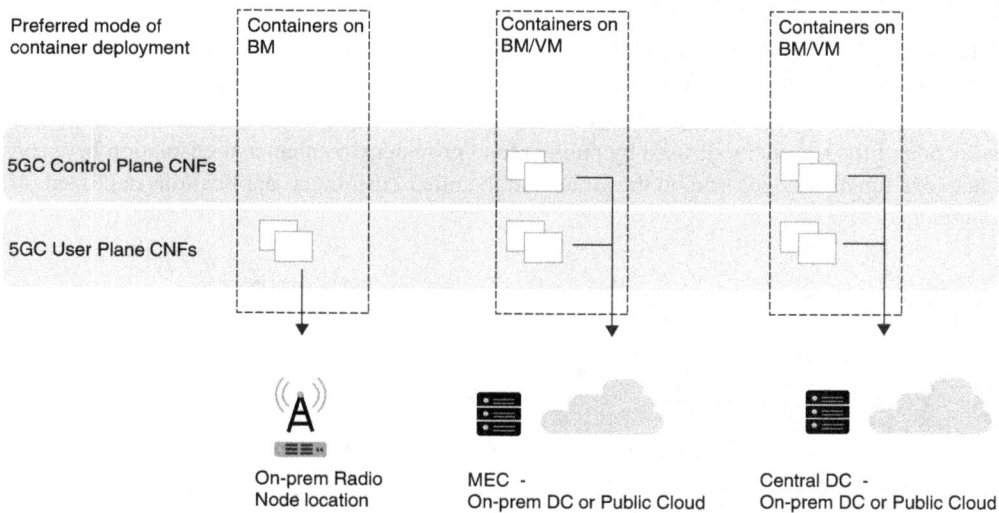

FIGURE 6-17   Preferred Mode of Containerized 5GC Deployment

Containers allow you to package your 5GC network functions and their dependencies together in a single object. The isolation for containers is provided by namespaces, which define the boundary by partitioning kernel resources for a process. For some use cases, such as running processes that require direct hardware access such as read/write access to the root partition, some containers can be run as privileged. For example, in Docker, "-privilege" is a special flag that you can set at runtime specifically to allow a Docker container to break free from its namespaces and access the entire system directly. It is just like you having access to the host machine. This can be exploited by the attacker, as shown in Figure 6-18.

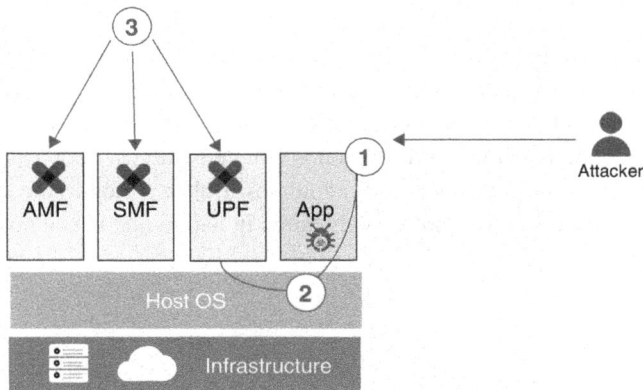

FIGURE 6-18   5GC NF Being Impacted Due to Privilege Escalation Type Attack

The attack vector shown in Figure 6-18 follows these steps:

1. The malicious application deployed in your container environment is run as privileged.

2. Once the malicious container of the attacker is run as privileged, it can allow malicious code from the attacker's container to overwrite or modify the SSH keys to allow remote users to log in to your container environment.

3. The malicious code then proceeds to destroy the AMF, SMF, and the UPF 5GC CNFs, thereby causing outage and service disruption in the network.

In another version of the attack, the attacker can use the container escape using the "-privilege" flag. This is an attack seen when the containerized network function that needs to have host resources from the container environment is given a "-privilege" flag. If an attacker can gain control over this containerized network function, they can escape the container environment and access the root privileges.

The scenarios illustrated in Figure 6-18 should be carefully considered when deploying 5GC containerized network functions in an enterprise environment in MEC use cases. The enterprise environment will have many built-in container applications and third-party applications already deployed in the enterprise data center. If the 5GC containerized network functions are deployed in the same virtual environment as that of the third-party applications, some with the "-privilege" flag, the container escape attack by an attacker can destroy the critical applications, causing data exfiltration and denial of service (DoS) attacks.

Improper Access Control

Figure 6-19 illustrates the access control provided by traditional segmentation methods used in legacy networks like 4G and 3G. The traditional network segmentation methods are based on IP address and ports. IP addresses are very important to traditional network security controls, as the network is segmented based on the IP and port information. When this information is coupled with user context, you can have policies that map users to an IP address or pool of IP addresses. But in a cloud-native deployment, IP addresses are not very helpful because they are very dynamic in nature and hence not considered the best way to apply security policies.

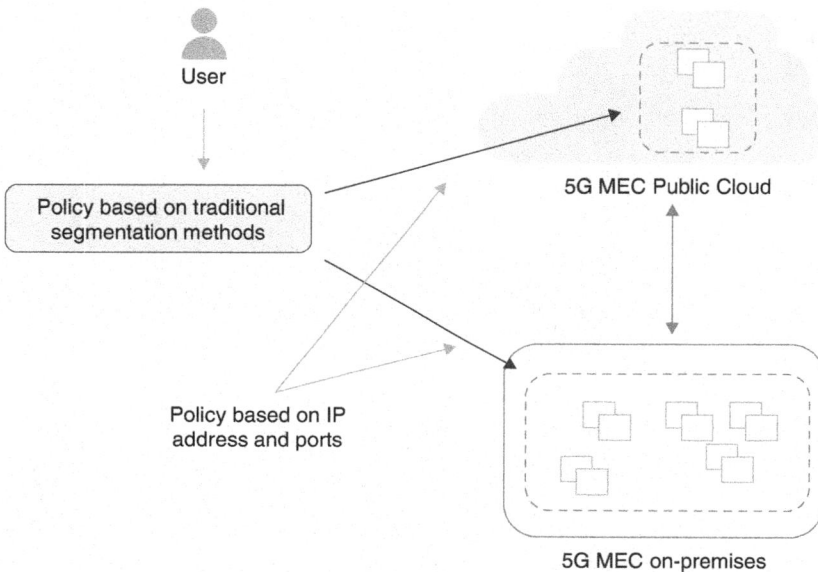

FIGURE 6-19 Access Control Based on Traditional Segmentation Policies

Another issue with the traditional access control mechanism is that there is no stickiness to the policy, as illustrated in Figure 6-20. It means that there is no flexibility in moving the policy along with the 5GC NF workload when you migrate it from on-premises to public cloud. Although methods like role-based access control (RBAC) can be used to restrict system access to only authorized users, the main aim of RBAC is to provide access to a wide range of users.

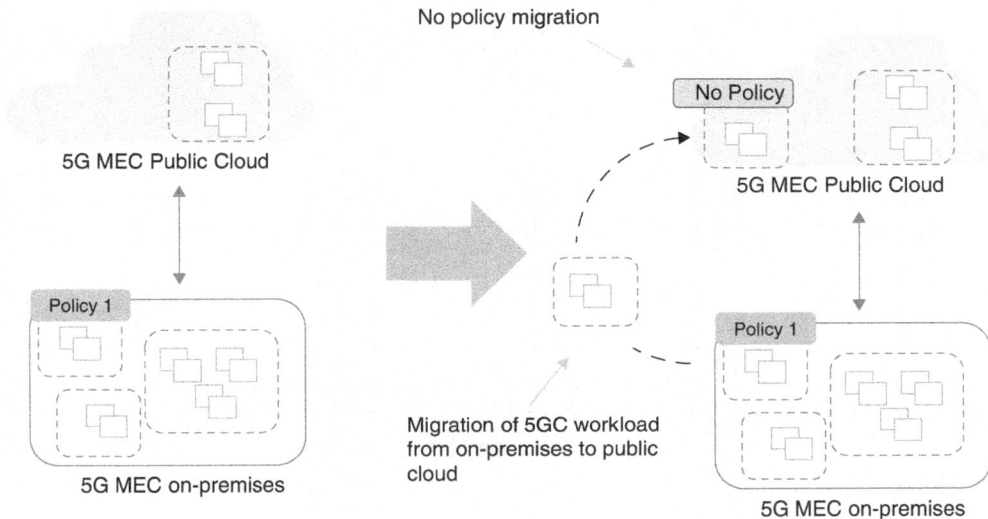

FIGURE 6-20   Lack of Policy Stickiness Using Traditional Segmentation Methods

Securing Virtualized 5G Packet Core Deployments

Securing virtualized 5GC NFs that follow the cloud-native approach requires a new mindset toward securing your 5GC environment. The key areas of discussion in this section are as follows:

- Secure CI/CD

- Securing 5GC NF runtime

- Securing 5GC NFs and 5GC NF traffic

- Securing 5GC NF orchestration and access controls

- Securing 5GC NFs in roaming scenarios

- Securing host OS and hardware

A majority of service providers are planning to deploy container-based virtual 5GC NFs. This section will discuss the aforementioned key discussion areas in much more detail.

Secure CI/CD

The Continuous Integration/Continuous Delivery (CI/CD) process, as covered earlier in the chapter, is very crucial to the software development lifecycle. It includes code and image repositories, build servers, containers, and third-party tools. CI is a process where software developers merge software code. CD is a process that includes provisioning the infrastructure for container deployment. Figure 6-21 illustrates the high-level stages of a CI/CD process.

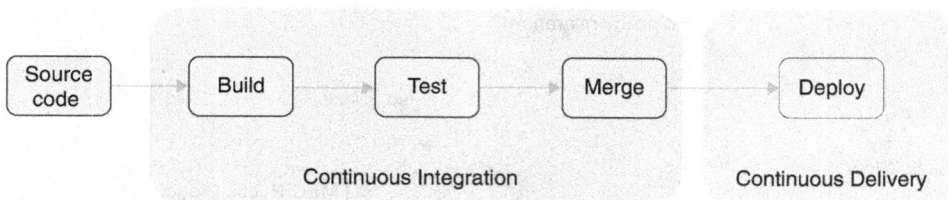

FIGURE 6-21   Stages of CI/CD Process

Securing CI/CD would require close integration between the DevOps process and the container security controls. Such integrations are also called DevSecOps. Using DevSecOps will help you to automate the container security checks, thus ensuring trusted images are deployed to runtime and are monitored for anomalies once deployed in production. This should be followed by the vendors' development teams developing containerized 5GC network functions or by teams developing applications that would interact with the 3GPP 5GC containerized network functions, as illustrated in Figure 6-22.

Some of the key integration points illustrated in Figure 6-22 are as follows:

- **Continuous scanning**: The scanning should include container build and deployment. Continuous scanning should be implemented throughout the CI/CD process to identify any issues related to misconfiguration (intentional/unintentional) in the deployment scripts. The scripts should be scanned for any keys, credentials, and secrets. Any such details found should be removed and be protected by trusted secret managers. 5GC CNF will include open source components for the images. These images should be scanned continuously for any vulnerabilities—during development, while being moved into production, and during production.

- **Continuous monitoring**: The expected behavior of the 5GC container should be modeled and any identified anomalies from the expected behavior should be identified and an alert created for the security teams to ensure deeper investigation, if required. There are many options available today to have automated monitoring of container behavior.

- **Image configuration**: Improper configuration is one of the key challenges in configuring containers apart from compliancy. You should ensure that the images are configured to run as non-privileged users. Many of the 5GC containerized network functions come with vendor-specified recommendations. Any methods for remote users, such as SSH, should never be enabled. Only API access should be allowed for any remote management.

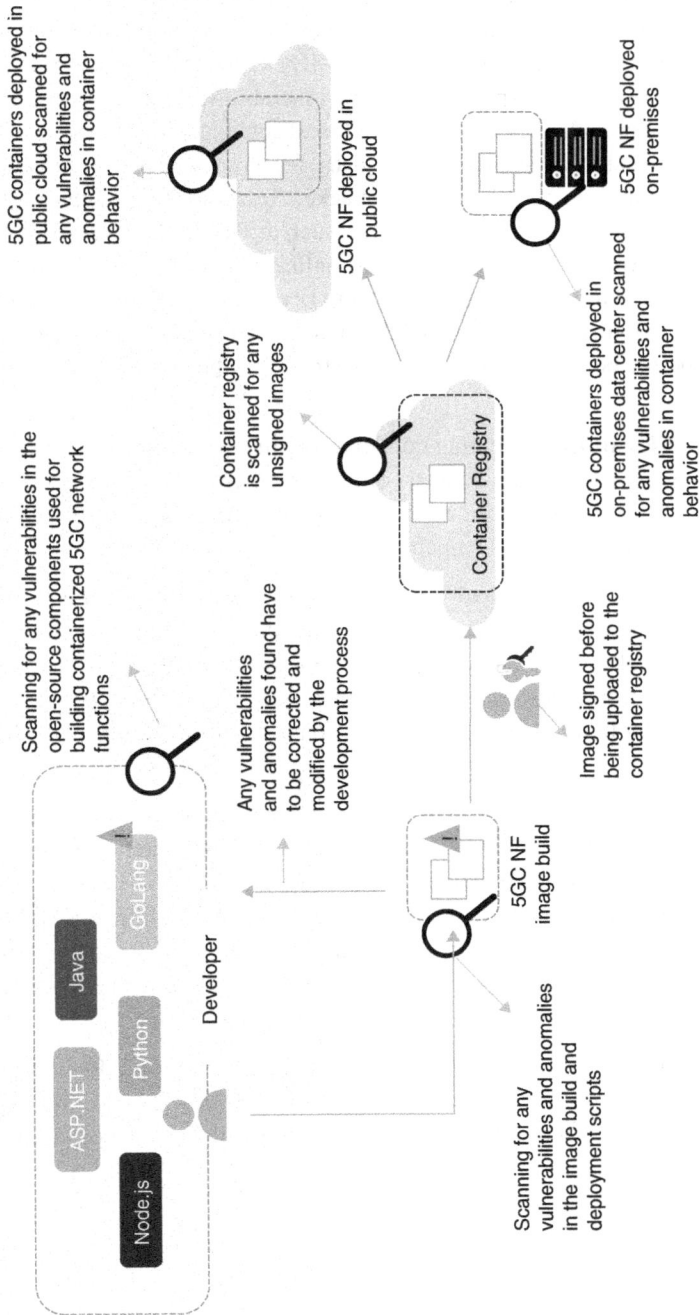

FIGURE 6-22 Secure CI/CD Process

5GC containers deployed in public cloud scanned for any vulnerabilities and anomalies in container behavior

5GC NF deployed in public cloud

5GC NF deployed on-premises

Container registry is scanned for any unsigned images

Container Registry

5GC containers deployed in on-premises data center scanned for any vulnerabilities and anomalies in container behavior

Scanning for any vulnerabilities in the open-source components used for building containerized 5GC network functions

Any vulnerabilities and anomalies found have to be corrected and modified by the development process

Image signed before being uploaded to the container registry

Developer

GoLang

Java

Python

ASP.NET

Node.js

5GC NF image build

Scanning for any vulnerabilities and anomalies in the image build and deployment scripts

- **Secure registry and include image signing**: A container registry such as a Docker registry is a kind of repository for developers uploading container images. This is a very useful concept because it enables providing the most up-to-date images with all the vulnerabilities patched in one location. The key issue, as explained in the "5GC Container Vulnerabilities" section earlier in the chapter, is the attacker exploiting known vulnerabilities and modifying the images with obsolete code or malicious code that can be executed in your production environment. The other risk is the attacker scanning the registry for any vulnerable or insecure API implementation that can then be exploited by the attacker. To offset these issues, container registries these days use the concept of a *trusted registry*, such as Docker Content Trust, that implements The Update Framework (TUF), which uses the online/offline key concept, where the offline key is stored securely and used to sign the timestamp key and tagging key. To secure the image, you should have the visibility of the entire 5GC container lifecycle, from the build process to signing the image before being uploaded to the registry, to visibility during runtime to detect any anomalies.

- **Securing 5GC NF runtime:** Securing container runtime means securing the container application environment, which includes analysis of the container and host activity as well as monitoring the protocols and payloads of the network connections. Having a secure runtime would require integration with the CI/CD pipeline.

- **Unique identity for each workload**: This method involves creating a unique identity for each workload early in the CI/CD pipeline and then using these identities to enforce security policies on the clusters, thereby preventing unauthorized hosts from being deployed and communicating with legitimate workloads, as shown in Figure 6-23. This method would require interaction with the 5GC vendor building the network functions and the service provider deploying the container.

5GC CNFs being deployed as containers will require authentication and encryption of the inter-5GC container communications as well as external communications. TLS should be implemented to ensure that inter-5GC control plane network functions are secured. 5GC control plane network functions use service-based architecture (SBA), which implements a service-based interface (SBI) using RESTful APIs based on HTTP/2. RESTful APIs are also used by the Network Exposure Function (NEF) for capability exposure to third-party application providers and integration with industry verticals enabling 5G use cases.

Although API communication facilitates integration between 5GC NFs, API vulnerabilities are well known by the attackers, consequently increasing the vulnerability of your network. The security controls illustrated in Figure 6-24 and described in the list that follows will help secure your 5GC NFs' communication at runtime.

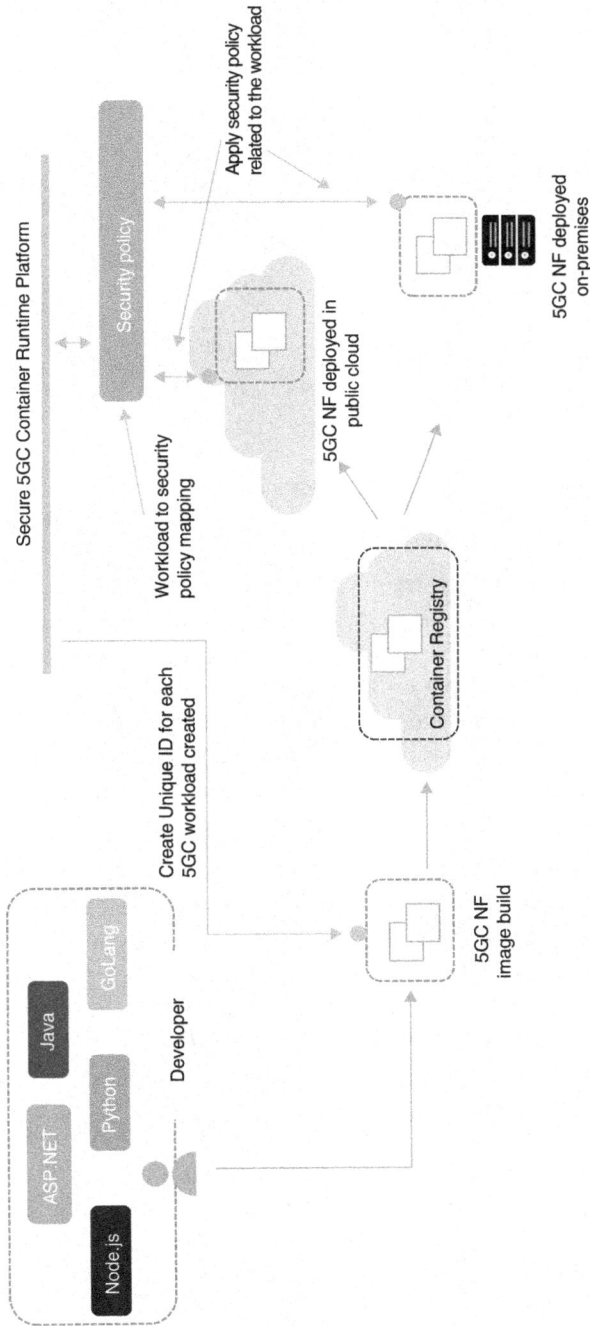

FIGURE 6-23 Secure 5GC Container Runtime

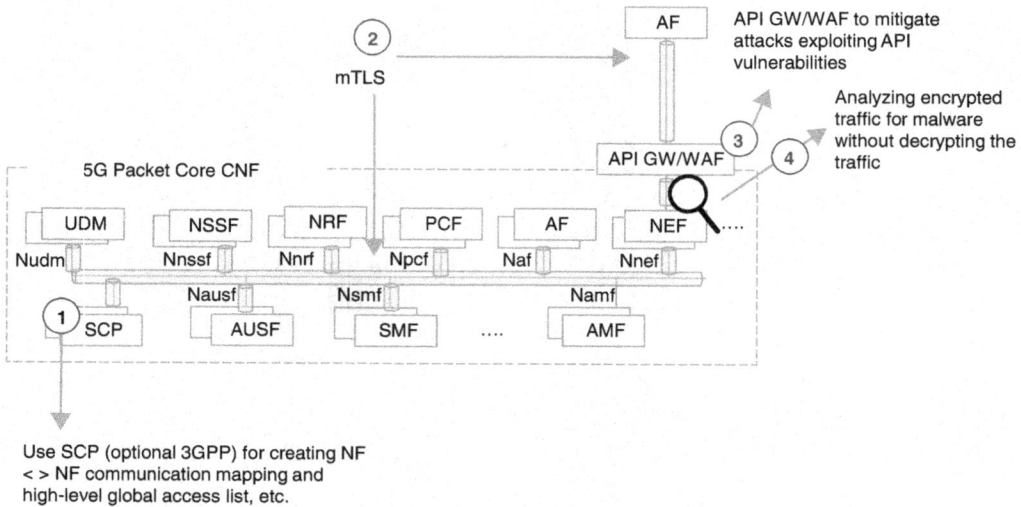

FIGURE 6-24   Security Controls to Secure 5GC NF Runtime

1. **Service communication proxy (SCP):** SCP is defined in the 3GPP TS 23.50, Release 16 spec-ification for system architecture for the 5G system. It is not a mandatory feature/component of the 5GC. SCP will enhance the 5G network function service discovery and selection process by enabling and managing the 5G NF-to-NF mapping. It can also implement access control lists to prevent unauthorized communications between NFs. SCP also provides cryptographic verifica-tion using authentication mechanisms like mutual Transport Layer Security (mTLS), which is an extension of TLS, between the SCPs and the NFs and between SCPs within the PLMN.

2. **TLS/mTLS:** All 3GPP NFs have a mandatory requirement to support TLS, which should be used if other means of network security such as Network Domain Security (NDS) using secu-rity gateways (SecGWs/SEGs) are not implemented in the 5GC PLMN network. The protocol stack for the SBI is illustrated in Figure 6-25. Any vendor you choose for the 5GC deployment should support TLS/mTLS within the 5GC NFs. This will ensure that you can enable TLS to secure the 5GC NFs when you want to implement it later in the deployment stage. It is recom-mended to use TLS 1.3 because it provides a faster, simpler, and more secure cipher suite, thereby providing better performance and enhanced security as compared to its predecessor, TLS 1.2.

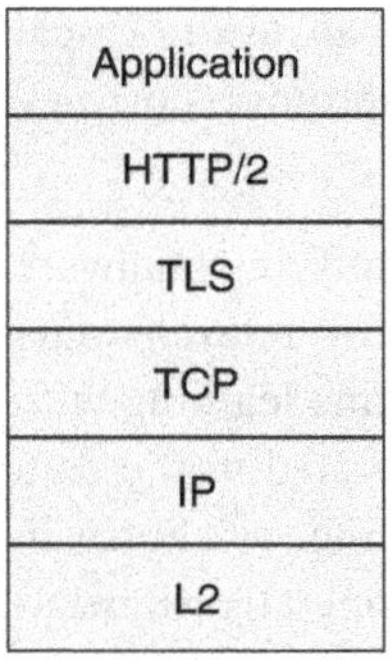

FIGURE 6-25   Service-Based Interface Protocol Stack

3. **API GW/WAF:** TLS is used to provide integrity, replay, and confidentiality protection for the interface between NEF and AF network functions. Once authenticated, the NEF will determine if the AF is authorized to send requests toward the 3GPP 5GC network function. This authorization is performed by using the OAuth 2.0 authorization framework, as specified in the IETF RFC 6749. The Common API Framework (CAPIF) can also be used for interactions between NEF and APF. CAPIF is discussed in the section "Securing API" in Chapter 5. Due to improper implementations of the API, most of the API calls from the third-party Application Function (AF) and any MEC application to the Network Exposure Function (NEF) are vulnerable to API risks. To mitigate any risks from vulnerable APIs or improper implementation of the APIs, it is recommended to use the API gateway (API GW) or a web application firewall (WAF). An API GW or WAF would protect the 5GC components from any attacks, such as cross-site scripting (XSS), SQL injection, and so on.

4. **Analyzing encrypted traffic:** TLS communication between the 5G NFs and between 5G NFs and AFs enables enhanced privacy and security for use cases by using keys and certificates to ensure security and trust. However, service providers and enterprises are not the only ones to benefit from encryption. Threat actors have leveraged these same benefits to evade detection and to secure their malicious activities. Some categories of threats that use encrypted traffic are as follows:

   - Illicit crypto mining
   - Android OS Trojans
   - Ad injectors
   - Sality malware
   - Malware using SMB (Server Message Block) service discovery
   - Potentially unwanted applications such as Tor and BitTorrent

Traditional threat inspection with bulk decryption, analysis, and re-encryption is not always practical or feasible for 5G use cases due to performance and resource reasons. Also, it compromises privacy and data integrity.

Today, a couple of solutions providers from cybersecurity vendors focus on identifying malware communications in encrypted traffic through passive monitoring, the extraction of relevant data elements, and a combination of behavioral modeling, artificial intelligence, and machine learning with cloud-based global visibility. The solution you choose should analyze the new encrypted traffic data elements in enhanced NetFlow by applying machine learning and statistical modeling, as shown in Figure 6-26. Rather than decrypting the traffic, the solution should use machine learning algorithms to pinpoint malicious patterns in encrypted traffic to help identify threats and improve incident response. Upon discovery, the malicious encrypted traffic should be blocked and quarantined to an isolated and segmented network.

The solution used for analyzing encrypted traffic should identify encryption quality instantly from every network conversation, providing the visibility to ensure enterprise compliance with cryptographic protocols, as shown in Figure 6-26. This will enable you to identify any noncompliant cryptographic protocol implementations in the network and take corrective actions.

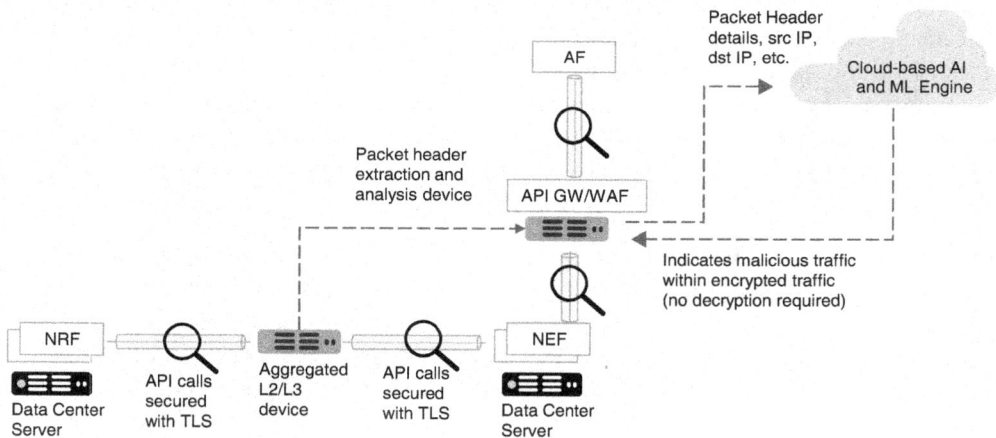

FIGURE 6-26 Detection of Malicious Traffic in Encrypted API Calls Without Decryption

Figure 6-26 shows one of the methods to detect malicious traffic within encrypted traffic without decrypting it. This method shows additional telemetry from routers and switches specifically related to encrypted data. The AI and ML engine in the cloud then analyzes specific details collected and determines whether the encrypted traffic between the 5GC NFs or between the 5GC NEF and the Application Function contains any malware. This would enhance the security in the 5GC virtualized deployment.

Securing 5GC NFs and 5GC NF Traffic

Most vendors are proposing container-based deployment (VM based and BM based) to make full use of virtualization technology. To secure 5GC NFs, security controls should be closely mapped to the workloads themselves, moving with the instances and data anywhere they might run (in data centers, virtual infrastructure, or cloud environments). Using this method, you will have a much stronger chance of protecting data regardless of where the system runs. The actual behavior of the applications and services as well as the mapping of applications and services also needs to be catered for to enable the right security controls to the right applications, which requires granular isolation between specific workloads, regardless of the environment in which they are deployed. This requires a new mindset for securing the network, as illustrated in Figure 6-27.

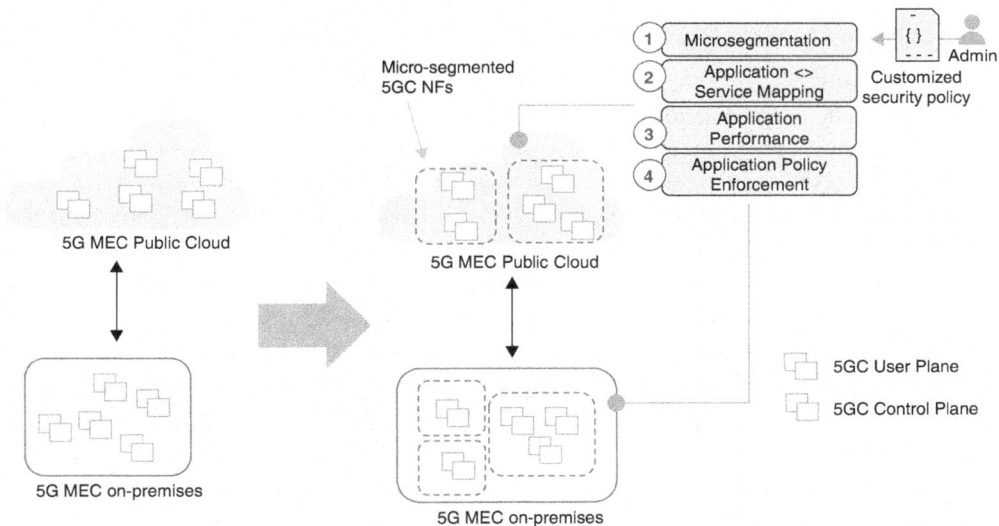

FIGURE 6-27   Security Controls to Secure 5GC NFs and 5GC NF Traffic

Microsegmentation

Deployment of virtualized 5GC NFs requires a new approach to network segmentation and access control. Microsegmentation is one of the primary methods by which this is achieved by following the principles of zero trust. The idea here is to ensure that specific policies can be applied to specific workloads without worrying about the nature of deployment of the 5GC NFs (VMs, containers, and so on) or the environment where the 5GC NFs are deployed (on-premises or public cloud), as shown in Figure 6-28.

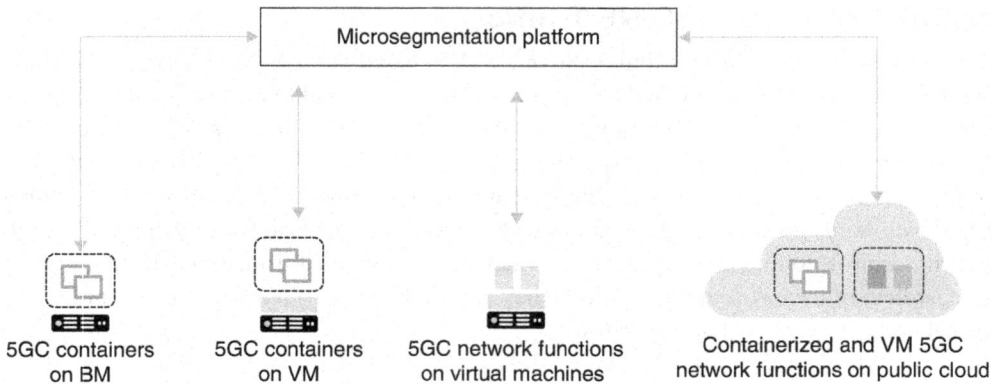

FIGURE 6-28   Microsegmentation Applied to All 5GC Deployment Types

To have an effective microsegmentation, it is recommended to start with monitoring of the infrastructure and applications, understanding the behavior of various NFs, and then creating a baseline of behavior for all 5GC NFs and the relationship between the 5GC NFs and services like DNS, NTP, and so on. Once the data is collected and analyzed, you will have a better idea of how to segment it and isolate the workloads better. Various vendors have refined the algorithms to provide you with an automated segmentation. Microsegmentation in a 5GC container environment would need to include application context, such as a specific non-3GPP network function where integration or communication is required for specific use cases. This would allow you to reduce risk in case a workload has been compromised. Essentially, it means that a hacker who has breached a workload serving an application cannot move laterally to attack another application host. In this model, the hacker is isolated to the workload servicing the specific tier of the compromised host of the application. In real-life 5GC deployments, you would need to cater for virtual machines, containers on VMs, containers on BMs, and containerized 5GC network functions deployed on the cloud.

The microsegmentation layer should be able to cater for providing isolation and applying segmentation policies for all modes of 5GC deployment. The microsegmentation layer should automatically tag security policies on all enforcement points in your 5G network, such as SDN, multivendor firewalls, or the workload level (virtual machine, bare metal, or container) by using software agents on servers (virtual machines, bare metal, or containers) to enforce a consistent, distributed zero-trust policy at scale. In some cases, there is no need for software agents because the microsegmentation policy can be enforced using the data processing unit (DPU) of your data center infrastructure. Although software agents are not required, closer integration between the microsegmentation solution and the DPU chipset vendor will be required.

Application Service Mapping

Application service mapping is sometimes referred to as application dependency mapping by vendors. Application dependency mapping is the process of figuring out which applications are dependent on what other applications and services. In 5G networks, application dependency mapping or application service mapping can be achieved by observing and collecting all interactions between applications and all systems and devices. Analytics is then used on this data to determine what 3GPP and non-3GPP 5G applications exist in your infrastructure and how these applications interact with other applications, systems, and devices.

Implementing application service mapping in your 5G network will provide you with the following key capabilities:

■ Understanding the services linked to the 5GC workloads, including 5G MEC applications. Figure 6-29 illustrates an example that shows the dependencies and services linked to an MEC application client.

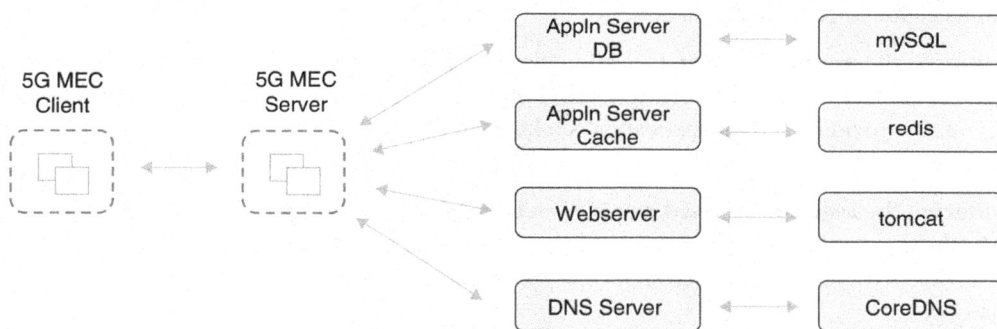

FIGURE 6-29   Example of Application Dependency Mapping for a 5G MEC Application

■ Discovering and mapping 5GC workloads such as 5GC 3GPP and non-3GPP NFs across multiple environments—hardware, on-premises virtual deployment, and public cloud deployment.

■ Discovering all types of 5GC workloads, including monolith, virtual machines, containers on VMs, and containers on BMs.

■ Discovering dynamic workloads (such as 5GC UPF network functions) and creating a topology map for the most recent instantiated workload.

■ Verifying the 5GC traffic flows against your customized segmentation policy for 5GC workloads.

■ Indicating and creating alerts for any unauthorized traffic flows between 5GC workloads and between 5GC and non-5GC workloads.

■ Indicating and creating alerts for workloads not properly protected.

■ Simulating the 5GC workload impacts of adding or removing a service or security policy. An example would be showing you the impacts of adding a virtual firewall with a customized security policy between 5GC and non-5GC workloads and the impacts it would have on the topology, such as breaking an existing active traffic flow that is required for certain 5G use cases.

Application Performance Monitoring (APM)

In legacy technologies like 4G, key performance indicators (KPIs) such as availability rate, downtime, and so on were used to monitor the health of the infrastructure. In 5G, all the 5GC components are virtualized and deployed in different form factors and in different environments (on-premises/public cloud), so you need to look at performance monitoring a bit differently. The 5GC vendors will of course provide the KPIs related to subscribers, such as drop rates, inter-RAT and intra-RAT handover success rates, and so on, but you will still need deeper performance monitoring of the 5GC workload itself. You will need a solution to monitor the elastic 5G network function workloads, which are dynamic in nature, across your multiple virtual environments. This will be fulfilled by APM. The APM vendor you choose should provide you with the following capabilities for both 3GPP and non-3GPP workloads:

- **Average response times of 5GC workloads**: This is critical for ultra-reliable low-latency communication (URLLC) use cases.

- **Monitoring the uptime of the 5GC workloads**: This is critical for multiple use cases, such as mitigating denial of service or interruptions for legitimate consumers accessing the applications or if you are providing certain applications to the 5G industry verticals and you need to maintain an SLA.

- **Monitoring the requests received per 5GC network function and detecting any anomalies**: APM solutions need to create a baseline and a threshold for all your applications. Any change of behavior from the measured threshold can be indicated as an anomaly and lead to creating alerts.

- **Continuous profiling of applications**: Continuous monitoring the CPU usage, memory, and disk read/write speeds of the 5GC workload that are mapped to specific requests would enable you to determine if any requests are impacting the performance of 5GC workloads and give you code-level performance details. Figure 6-30 illustrates how this could be achieved.

- **API validation**: This enables your 5GC development team to validate and test API calls and 5G network function workload impacts by mapping the performance impact for each API call. The changes can then be made by the development team before deploying them to production.

- **Monitoring user access:** This provides you visibility into the users accessing the 5GC workloads deployed in your infrastructure, such as MEC applications deployed across multiple virtual environments.

- **Openness and third-party integration:** The APM you choose should also provide or allow you to integrate with the vulnerability detection system to ensure that the untested libraries with vulnerabilities are detected and flagged during the development stage before being moved to production.

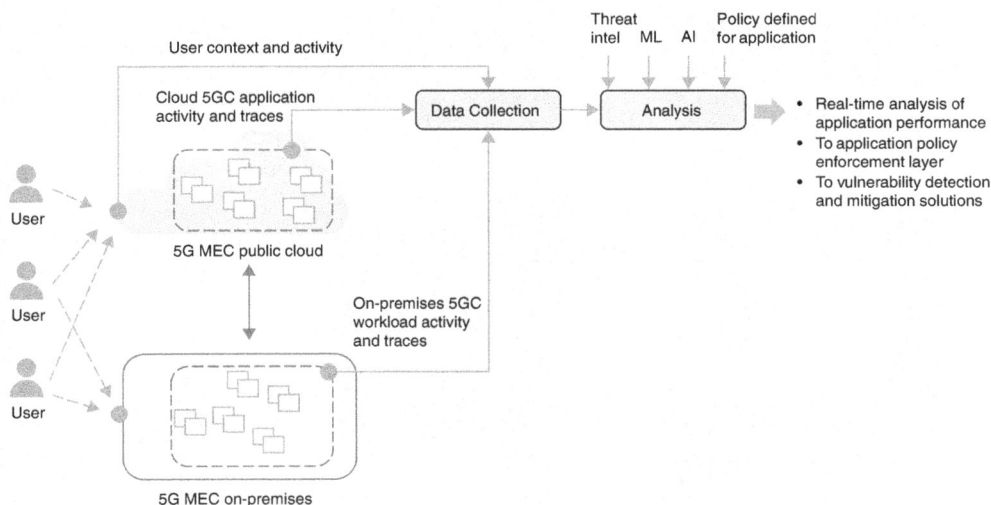

FIGURE 6-30   Application Performance Monitoring for 5GC CNF Deployments

Application Policy Enforcement

Traditionally you would enforce the policies for proprietary hardware using a firewall appliance. 5GC NFs being dynamic assets that can scale in and out depending on the traffic and use case requirements would need a smarter enforcement mechanism. 5GC NF application policy enforcement would need to stick to the 5GC NFs wherever they are deployed. This requires you to look beyond fixed network enforcement points. To enable the evolution of policy enforcement, you need the aforementioned layers of microsegmentation, application-to-service mapping, and application performance monitoring. Once you have implemented all these layers, the next step is to enforce the policies to the application/5GC NFs. The security policy for runtime could be, for example, what workloads can be deployed in what clusters and the connection rules that determine what connections can be set up between workloads. These connections need to be set at the workload level and not at the IP address or port level, which is usually done in firewalls, as shown in Figure 6-31. It could also contain deployment rules that control which workloads can be deployed in which environments.

As illustrated in Figure 6-31, the policy enforcement could be done directly to 5GC network functions instead of the firewall using the orchestration system. Policies could be pre-fed into the orchestration system, such as pod security policies in Kubernetes, which is a cluster-level resource that specifies security aspects for the pod. The fine-grained policies could also be enforced by agents that can be installed by the policy enforcement solution on the VM, bare metal, container host, Linux, and so on. These sensors/agents would collect data from the workload (including process data, software package details, and so on) and inter-5GC NF communication and send it to the policy enforcement solution. This allows the policy to be attached to the 5GC NF and would be enforced wherever it is deployed, as illustrated in Figure 6-32.

Legacy method of policy enforcement | Policy enforcement for 5GC virtual deployments

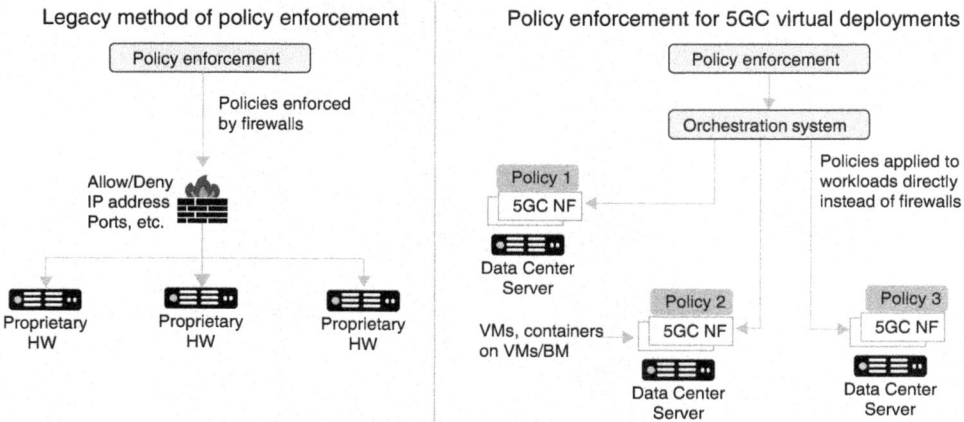

FIGURE 6-31   Policy Enforcement at 5GC NF Level

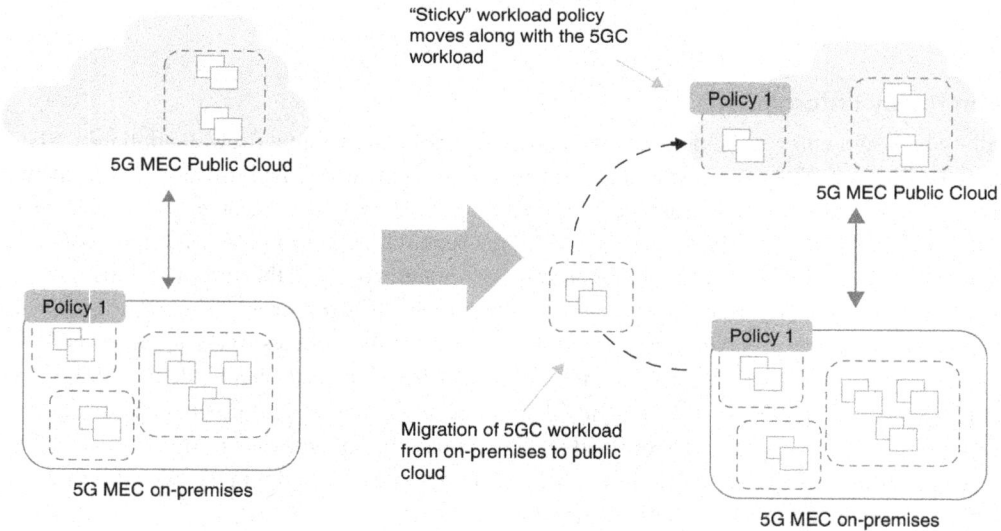

FIGURE 6-32   Policy Stickiness Due to Policy Being Applied at NF Level

Figure 6-32 illustrates the recommended method of policy enforcement, which is linked to the 5GC NF or application rather than based on the traditional method of segmentation based on IP address, port, and so on. The traditional network layer segmentation method will not be efficient for you because it cannot provide a granular security policy for mapping applications to users. There are solutions available today that enable you to have security policies based on the workloads.

These solutions, with both software agents and agentless options, allow you to scan container images and enforce policies as part of Continuous Integration and Continuous Delivery (CI/CD) workflows, continuously monitor code in repositories and registries, and secure both managed and unmanaged runtime environments. Some of these solutions also leverage the Istio service mesh for protecting network communications, multicluster communications, and secure communications with external resources as well as enforcing segmentation and workload isolation based on Layer 7 protocols mitigating pod communication with service mesh controls. These solutions also integrate with remote access virtual private network (RA-VPN) clients and Network Access Control (NAC) for users and devices to bring user and endpoint context into the segmentation policy. This helps you to define granular policies to restrict application access based on the user, user group, user location, or other user-related attributes, as compared to enforcement of security policy based on just IP addresses and ports provided by traditional firewalls.

Securing 5GC NF Orchestration and Access Controls

Although you will have both VM and container deployments for 5GC NFs, a majority of the 5GC deployments would be based on containers (containers on VMs and containers on BM).

You will also need the deployments to be automated, scalable, maintain the state for the number of 5GC NFs, and provide secure networking for the containers. This should be catered for by the container orchestration solutions you choose. Due to the criticality of the orchestration function, it should support the security capabilities described in the sections that follow.

Role-Based Access Control (RBAC)

RBAC will enable you to regulate access to the orchestration solution. The solution you choose should be able to work with on-premises and cloud-deployed access control systems. RBAC should also be used at the API access control layer to ensure you are authenticated before your requests are authorized. Kubernetes, which is a widely used orchestration system, authorizes API requests using the API server. It evaluates all of the request's attributes against all policies and allows or denies the request; however, your deployment might have NFs that are not containers, and you need to ensure that RBAC is catered for by all the deployment nodes. Try to create policy templates that can be reused for a group of similar pods or applications; this will help you when you redeploy the applications or are deploying a similar application.

Security Policies

For the containerized 5GC NFs orchestrated by well-known orchestration systems like Kubernetes, network policies are used to enforce segmentation policies. In Kubernetes, the 5GC NFs will be run in pods, which consists of one or more NF containers that are deployed together. Each pod is assigned an IP address and is routable with other pods. Network policies will determine the access permissions for those pods. In pragmatic deployments, however, you will have to manage large clusters with multiple orchestration platforms that contain several services developed by multiple teams. This includes third-party 5G MEC application services interacting with the 3GPP 5G NFs using the Network Exposure Function (NEF), which leads to challenges in having consistent security policies across different clusters. To solve this issue, you should implement multiple layers of security policy enforcement and look at deploying a service mesh, as discussed further in this section.

The orchestration system you choose should offer the capacity or integration options with a solution that provides a security policy per workload within the pod, which is more effective, as illustrated in Figure 6-33. The chosen solution should also allow application-to-service mapping and visibility on 5GC NF-to-NF-level communication. The default policy should deny all requests so that you can select the right security policy for each and every workload or at least define a policy on a set of similar workloads. This can be planned well in advance, and methods such as the aforementioned Application Policy Enforcement will help you plan and enforce the policies across 5GC workloads.

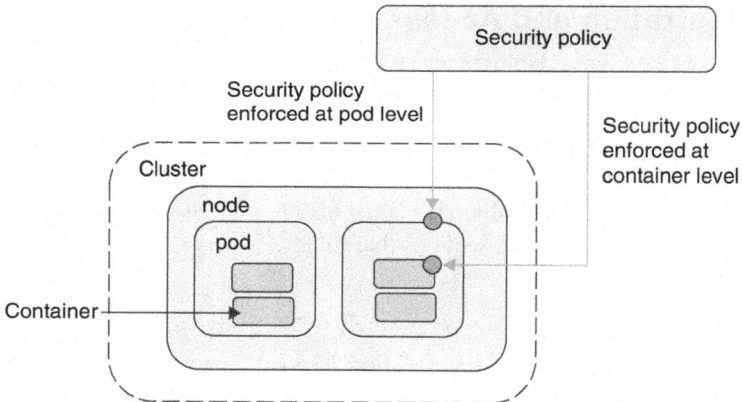

FIGURE 6-33   Security Policy Enforcement Levels

Secure Communication

One of the key challenges of managing virtual deployments of 5GC (specifically container-based NF deployments of 5GC), other than securing the container itself, is to understand how to properly secure and encrypt the traffic flow and communication between NFs. mTLS and TLS are recommended by 3GPP TS 33.501 to be used for inter-NF traffic flows. For 5GC NF-based deployments, vendors are aiming to provide TLS and mTLS options. The API calls from the container orchestration system to any internal or external container nodes should be protected by TLS, as illustrated in Figure 6-34.

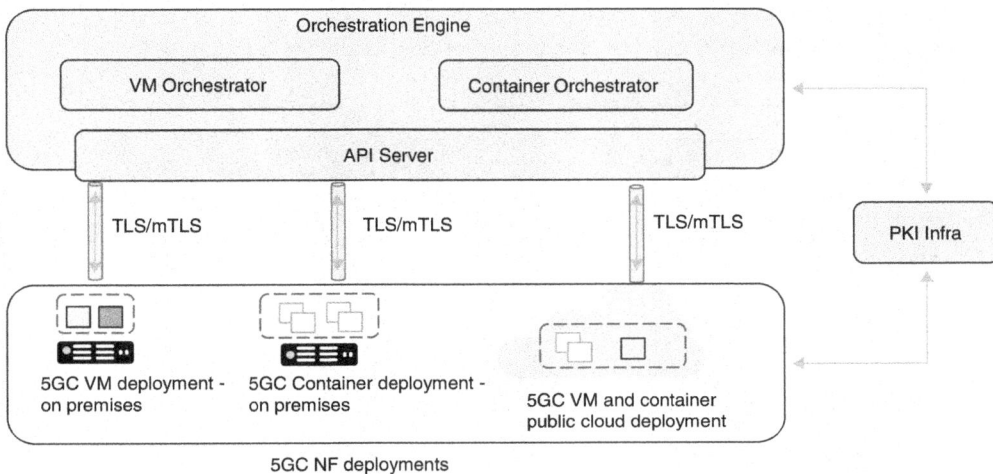

FIGURE 6-34 Secure Communication Between Orchestration and 5GC NFs

As illustrated in Figure 6-34, the communication between the orchestration system and the 5GC NFs should be encrypted using TLS/mTLS encryption mechanisms. For the VM orchestration, you would normally use a virtual network function manager (VNFM) and an orchestrator. For the container deployment, Kubernetes is widely used. It is recommended to use a dedicated PKI infrastructure for the 5GC components.

One of the other options could be to leverage a service mesh and use a method to add a layer of security capabilities, such as encryption and authentication, between services inside your 5GC NF cluster and outside your 5GC NF cluster. A service mesh also provides more visibility for microservice inter-actions to the operations team, especially when observability and visualization of east-west cluster traffic are included as part of the service mesh solution. In deployments that have large clusters with multiple teams developing different 5G-related services, a service mesh will help you create a network encryption layer that allows all 5G-related services to communicate with each other securely. Basically, a service mesh consists of a control plane and data plane. In actual deployments, the data plane is typi-cally implemented as a proxy such as Envoy that is run "out of process" as a sidecar component. The control plane caters for policy and configuration for the data planes running in the mesh. The control plane also allows telemetry collection, which can then be consumed by your network analytics and monitoring solution. In a service mesh, a sidecar agent is attached to each pod and provides commu-nication between services by fine-tuning the set of ports and protocols that the proxy will accept when forwarding traffic to and from the workload. In addition, it is possible to restrict the set of services that the proxy can reach when forwarding outbound traffic from workload instances. Apart from providing secure communication, service meshes also enable service discovery as well as monitoring/tracing/logging, and they avoid service interruption because they allow you to write applications that limit the impact of failures, latency spikes, and other undesirable effects of network peculiarities. A service mesh could be designed using open source services like Istio. It is basically a mesh of Layer 7 proxies that aims to establish and maintain service-to-service connections, as shown in Figure 6-35.

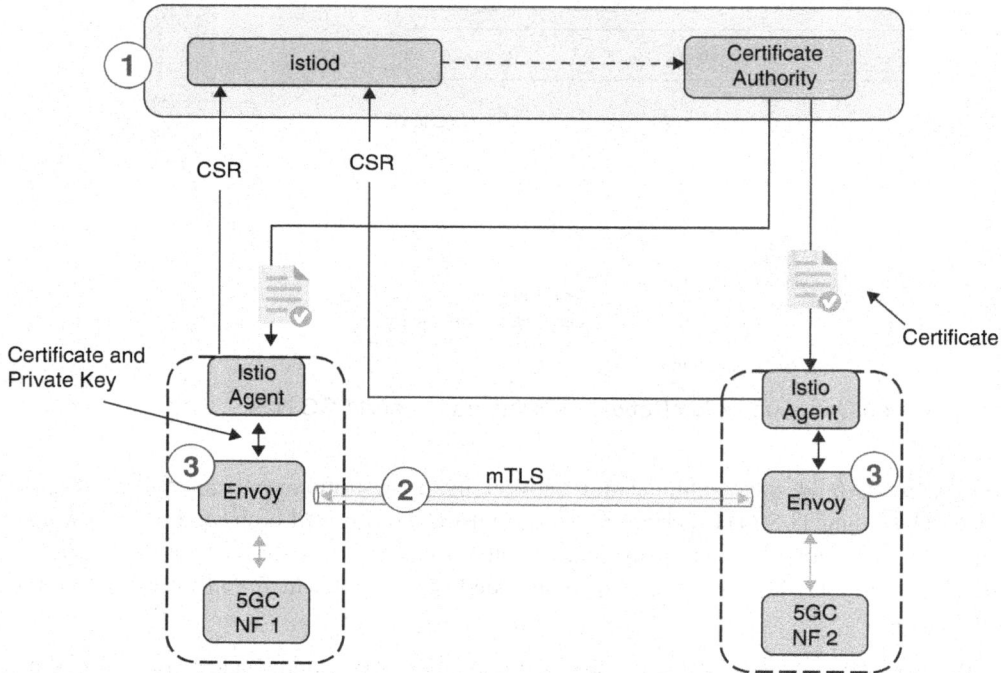

FIGURE 6-35   Secure Service to Service Communications in Service Mesh

Secure Production Identity Framework for Everyone (SPIFFE) uses a form of X.509 certificate to provide a secure identification process to each and every workload. The framework, once completed, will provide an enhanced method of identification and will remove the need for passwords or API keys. Figure 6-35 illustrates one of the methods for NF-to-NF/orchestration-to-NF authentication based on Istio's service, which is explained as follows:

1. In a container-based 5GC NF deployment, all the 5GC NFs can be securely provisioned using an Istio service mesh with X.509 certificates. The Istio agent placed next to the Envoy proxy creates the private key and certificate signing request (CSR) toward the CA. The Istio service mesh uses Remote Procedure Call (gRPC), which is a protocol built on HTTP/2, to take the CSR. The CA in the Istio daemon (istiod) validates the CSR. Once the CSR is successfully validated, the CA signs the CSR to generate the certificate. When the 5GC NF is started, the Envoy proxy will request the certificate and key from the Istio agent using the Envoy Secret Discovery Service (SDS) API. The Istio agent then sends the certificate received from istiod and the private key to Envoy using the Envoy SDS API. The workload certificate is then continuously monitored by the Istio agent.

2. NF-to-NF security in Figure 6-35 is provided by Envoy proxies, which enable tunneling of traffic between 5GC NF 1 and 5GC NF 2, and is secured using mTLS.

3. When the client-side Envoy starts a mutual TLS handshake with the server-side Envoy, the client-side Envoy performs a secure naming check to verify that the service account presented in the server certificate is authorized to run the NF service.

Secure Access Control Based on Zero-Trust Principles

While you're designing the access control for the virtualized 5GC NF deployment, it is important to follow the zero-trust principles. This would ensure that each and every request has to go through multiple security controls, which reduces the risk of unauthorized access to 5GC NFs.

For the communication between the NFs, if you don't use the Service Communication Proxy or any network security controls, it is recommended to have security controls that can cater for authentication, authorization, and validation as well as compliancy of the inter-NF API calls, as shown in Figure 6-36.

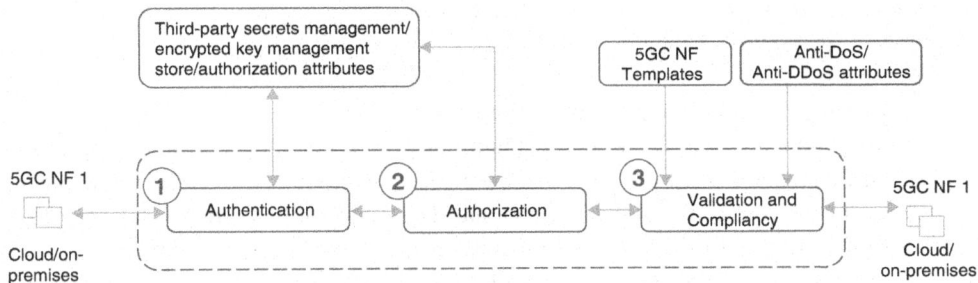

FIGURE 6-36   Access Control for Inter-NF API Calls

The key features of the recommended security controls illustrated in Figure 6-36 for 5GC NFs can be deployed as a part of the API GW/WAF solution or can be deployed separately, as detailed in the following list:

1. In the initial deployment of 5GC NFs—be it based on virtual machines or containers—the authentication would be based on a TLS/mTLS mechanism, as that is the method most of the 5GC vendors are focused on today. But once you have a widely deployed multivendor environment in a couple of years' time, it would be important to have on-demand ephemeral secrets and access permissions to ensure that all the 5GC NFs, including dependent services like orchestration, are secured using zero-trust principles. This can be achieved using a secrets management solution that is provided today by multiple vendors and integrating that with your authentication management/authentication solution. The 5GC vendors you choose today should at least have a plan—not necessarily a roadmap, as many product teams would not provide their future product "secrets"—to ensure that the 5GC NF authentication can be performed by any third-party

authentication/secret management solution you choose. This is particularly important for 5GC NFs that are deployed external to your DC. While choosing the vendor for secrets management, it is recommended to validate the encryption for data at rest and authorization capabilities of the provided solution. These aforementioned controls will bring a truly robust authentication strategy to your virtual 5GC environment.

2. The authorization capabilities should also be capable of integrating with the third-party secrets management solution. You should also be capable of integrating the authorization subsystems with the third-party authorization solution to provide enhanced capabilities such as centralized authorization for all API calls originating outside your network toward your internal 5GC NFs. As these API calls could be from your 5GC NFs, secure exposure NFs like NEF might be bypassed; this makes the integration of the authorization controls to the external third-party secrets and authorization management very important in hybrid 5GC deployments.

3. Once the API calls are authenticated and authorized, they should be inspected for any anomalies or any invalid attributes. It is important to have the API validation at both the client side and server side. Many of the API validations today are server side due to developers not understanding the implications of improper API security design. This topic is explained in more detail in the "Secure API" section of Chapter 5.

For seamless deployment, operations, and maintenance of virtualized 5GC NFs, the security design following the zero-trust principles, as illustrated in Figure 6-37, is recommended to be deployed to mitigate any risks due to unauthorized users accessing the 5GC NF environments. The key components of this design are the integration of the authentication and the authorization controls with the third-party Identity and Access Management (IAM), secrets management, and encryption key management (EKM) solutions with the multifactor authentication (MFA) solution, which is then integrated with application policy management, which includes the application-to-user mapping and application policy enforcement.

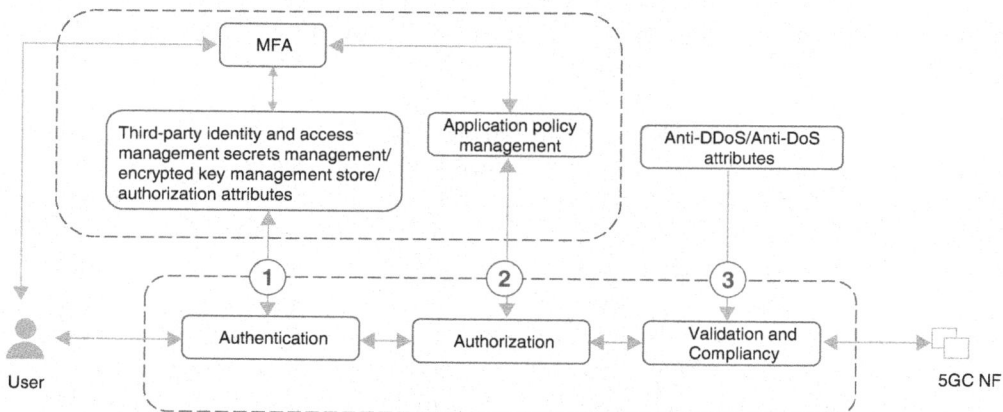

FIGURE 6-37 Access Control for Users and User Devices

The security design illustrated in Figure 6-37 is further explained as follows:

1. Access control for 5GC users should be carefully planned to reduce any attack surfaces due to unauthorized access and privileged mode escalations. The multivendor 5GC NF deployment in hybrid virtualized environments should have centralized authentication for users accessing the 5GC NFs. The authentication solution should be able to provide multiple authentication methods such as JSON Web Tokens (JWT), Open ID Connect (OIDC), LDAP, RADIUS, and so on. Adoption of containerized 5GC will see many different options of how the policy is attached to the user. This is because there are many standards allowing more compact and efficient methods of securely transmitting information between a client and a server using URL, POST attributes, or an HTTP header that are self-contained; there needs to be less to and from queries. One such method is JWT. When a user successfully logs in to the identity authentication server, it responds with JWT, which consists of what routes, resources, and protected services or 5GC NFs the user can access. This method is quite useful in API calls, which the user will have to use to access the 5GC NFs for any configuration or operational reasons. You should ensure that the 5GC vendor's RBAC functions can integrate with the third-party IAM and secrets management solution. The MFA solution will enable establishing trust of the user and device by establishing user and device identity, checking the device posture and vulnerabilities, and ensuring access only if the device meets certain preset minimum criteria for access. All these security controls will provide a robust security posture for users accessing the 5GC NFs and allow you to future-proof your 5GC user access control mechanism.

2. Application-to-user mapping (explained in detail in the section "Application Performance Monitoring" in this chapter) will enable you to further enhance the user-to-application authorization policies by providing the visibility of any users violating the application-to-user mapping. For example, if User A is mapped to 5GC NF 1 only and tries accessing 5GC NF 2, an alert should be created and should be able to be integrated with your existing logging and monitoring solution.

3. Once the user and device are authenticated and authorized, they should be validated for any unallowed URL or POST attributes and should undergo rate-limiting checks to prevent distributed denial-of-service (DDoS) or denial-of-service (DoS) attacks.

Securing 5GC CNF in Roaming Scenarios

Apart from the internal traffic flow in 5GC between the 3GPP and non-3GPP 5G network functions, the virtual 5GC deployments need to be secured against threats from external connections, such as roaming partners. Figure 6-38 illustrates the security controls for different 5G NSA and SA interfaces during a roaming scenario. In a pragmatic scenario, you will have 4G and 5G networks co-existing with each other. While you're considering the roaming interface, remember that a majority of your roaming partners for the next 2 to 3 years will still be based on 4G or 5G NSA deployments, which will require Diameter and GTP-C protection for the control plane and GTP-U protection for the user plane.

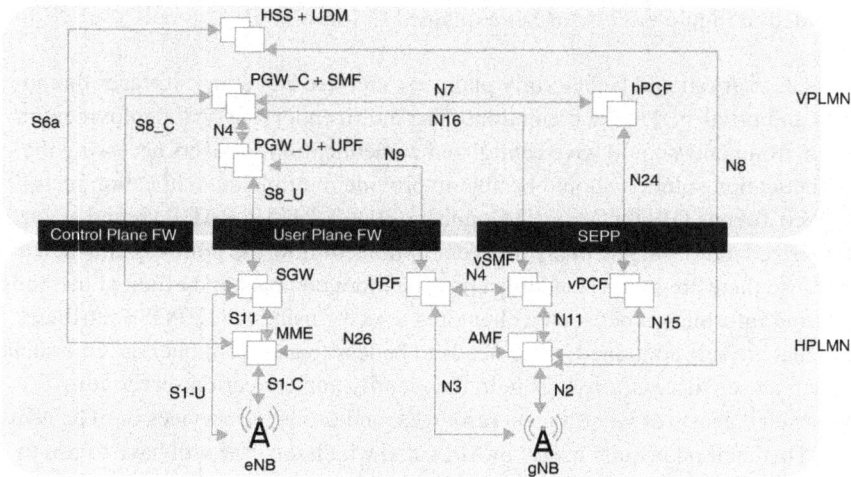

FIGURE 6-38   Securing 5GC NSA and SA Roaming Scenario

As shown in Figure 6-38, if your roaming partner is using a 5G NSA/4G deployment, control plane firewalls such as Diameter firewalls and GTP-C firewalls can be used to secure the roaming interfaces. For 5G roaming scenarios, the legacy protocols on the control plane, such as Diameter and GTP-C, have been replaced by APIs and a new NF called Security Edge Protection Proxy (SEPP), which is defined to secure the API communication between the home network and visitor 5G network. The SEPP function is defined by 3GPP to enable secure roaming interconnect between 5G networks. SEPP protects the control plane by performing mutual authentication and negotiation of the cipher suites. It also handles the key management aspects that include setting up the required cryptographic keys needed for securing messages between the home PLMN (HPLMN) SEPP and the visitor PLMN (VPLMN) SEPP apart from performing topology hiding by limiting the internal topology visible to the external roaming partners. As compared to the 4G roaming interconnect method, the SEPP function in 5G allows simpler configurations required to secure the roaming control plane. Figure 6-39 illustrates a simplified view of the SEPP function.

The SEPP function, as illustrated in Figure 6-39, secures the 5GC NF function from roaming threats by validating requests from 5G roaming entities. For securing the user plane communications, the 5G user plane function uses the Inter-PLMN UP Security (IPUPS), which enforces GTP-U security on the N9 interface between the UPFs of the visitor and home PLMNs.

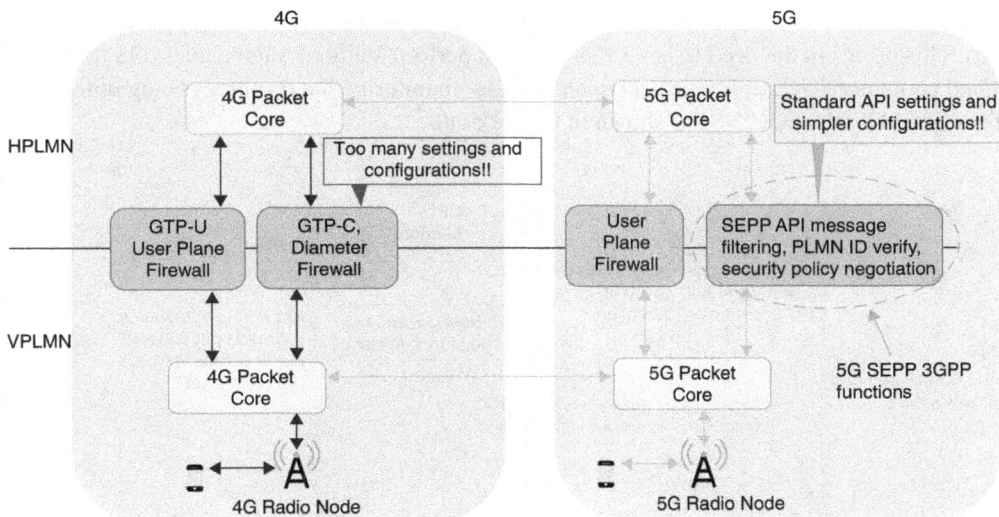

FIGURE 6-39 Simplified View of 5GC SEPP Network Function

Securing the Host OS and Hardware

The security of the hardware and software on which the 5G NFs would be deployed is very important for the overall security of the 5GC CNFs. Here are some key security controls and best practices related to securing the host OS and hardware for 5GC CNF deployments:

- The host hardware for containerized deployment should be hardened as covered in detail in Chapter 5 in the section, "Hardening Hardware and Software." The hardware should have the hardware root of trust using Trusted Platform Module (TPM) and Trust Anchor modules (TAMs). Host hardware should also support Unified Extensible Firmware Interface (UEFI). Methods such as UEFI secure boot detect tampering with boot loaders, operating system files, and unauthorized ROM by validating their digital signatures. This would help you verify if the code launched by the firmware is trusted or not, and any detections are blocked by running before they can infect the system.

- It is recommended to validate the components for base OS and functionality, which is less of an issue for container-based 5GC NFs because most of the dependencies are packaged within the container. For any VMs being deployed, having the validation of the base OS and functionality is a critical step.

- The dependencies on the host OS for the deployed 5GC NFs should be kept to a minimum. This is one of the key reasons why many vendors and service providers are aiming for container deployment for 5GC NFs. When you deploy a containerized 5GC NF, the dependency on the host is bare minimum, as most of the dependencies are packaged within the container itself.

This brings in a lot of flexibility to deploy the 5GC NF wherever you want (on-premises or cloud). This brings in the need to have visibility and perform auditing on the guest OS for both VM and container 5GC deployments, which includes monitoring the resources being utilized and consumed by the 5GC NFs as shown in Figure 6-40.

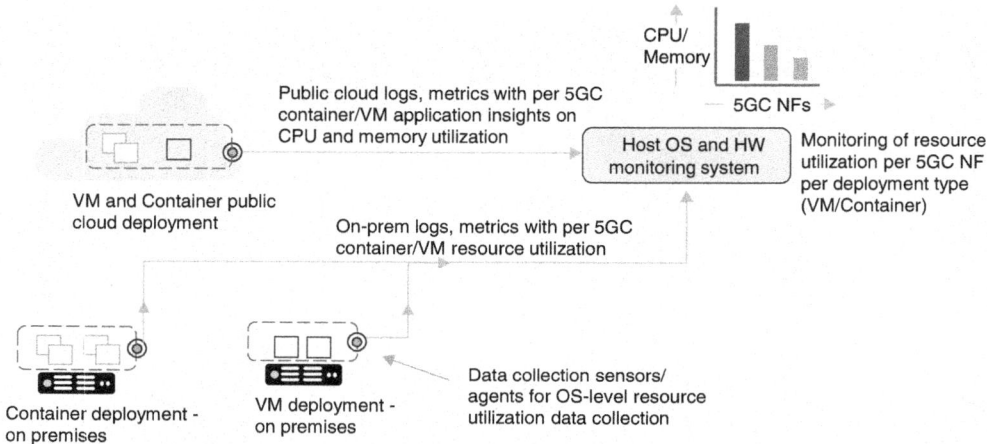

FIGURE 6-40   Monitoring System for Host OS and HW

- It is recommended to keep the host OS current with security updates and any other component updates recommended by the vendor providing you with 5GC NFs.

- When 5GC NFs are deployed on VMs, hypervisors provide isolation and allocate resources such as CPU, memory, and storage to the guest OSs. This ensures that the guest OSs do not overlap on each other's resources and mitigates the risk of DoS attacks caused by excess resource consumption by malicious code in a guest OS deployed in the same server/hypervisor. In real-life deployments, however, virtual machine deployments use guest tools, wherein the guest OS can access files, directories, and so on, on the other guest OS, or sometimes even the host OS. This could be used as an attack vector. The deployment of any such guest tools should be avoided if VMs are used for 5GC NF deployments.

- Configure control groups (cgroups) to limit and isolate the resource usage for a group of processes. This will ensure that containers on the same host do not impact the performance of each other. Specifically, for this purpose, it is recommended to have the critical 5GC NFs placed on their own dedicated host so that the host can be secured specifically for the services offered by the critical 5GC NFs and so that an attack on any other service would not impact the critical services. This strategy should be applied to any format of virtualization (VMs/containers).

Real Scenario Case Study: Virtualized 5GC Threats and Mitigation

This section focuses on multiple attack surfaces on virtualized 5GC environments and discusses methods to mitigate the mentioned attacks. At the time of writing this book, 5G was not yet widely deployed, and the scenarios mentioned in this section are based on discussions with customers and experiences with lab and pilot deployments. Figure 6-41 illustrates the virtualized 5GC environment being considered for the case study.

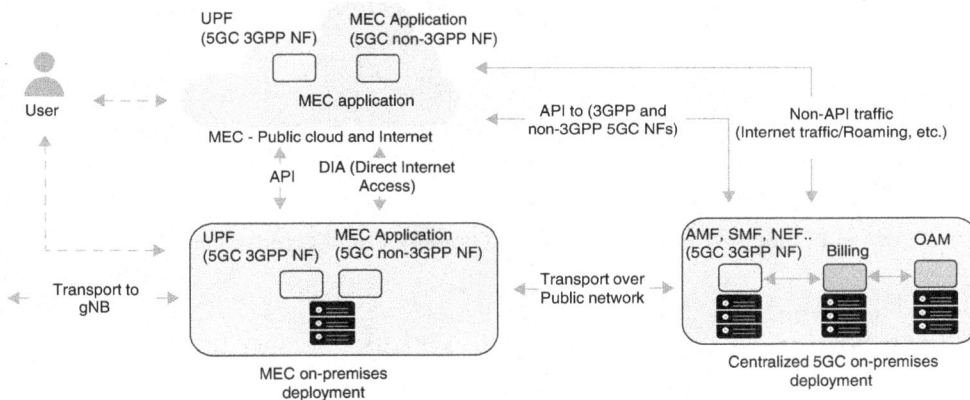

FIGURE 6-41   5GC Deployment Example for the Case Study

As shown in Figure 6-41, the 5GC NFs are deployed both on premises and on the public cloud. The architecture for this deployment is explained in the following list:

■ The on-premises MEC DC has a 5GC 3GPP NF called intermediate UPF (I-UPF), which redirects the traffic flows to the designated UPF in the public cloud. I-UPF acts as a classifier that redirects the flows using traffic-matching filters. N9 is the reference point/interface between I-UPF and UPF session anchor. This design reduces the latency and will help you to validate the design and performance for ultra-reliable low-latency communication (URLLC) use cases.

■ The MEC application is a third-party non-3GPP application that is deployed on the same DC where the I-UPF is deployed. It is located in the edge DC to terminate MEC application-related queries from IoT devices and sessions that are redirected from the I-UPF 5GC NF.

■ The MEC application and the I-UPF are deployed on the same server.

■ The MEC DC provides API external/northbound interfaces for orchestration and API calls between MEC applications. Non-API interfaces are provided to allow Direct Internet Access (DIA) for use cases that require Internet access from the MEC.

- MEC layer using public cloud deployment is also considered to enable deployment of UPF and third-party MEC applications, similar to the on-premises MEC, with the only exception of I-UPF, which is dedicated in the on-premises MEC.

- The centralized 5GC on-premises DC hosts the 5GC control plane 3GPP NFs such as AMF, SMF, PCF, and so on, 5GC user plane NFs such as UPF for non-URLLC use cases, and non-3GPP NFs such as dedicated third-party billing functions from vendors providing a telecom billing infrastructure.

- The transport infrastructure between the MEC and centralized 5GC deployment carries traffic between the 5G 3GPP NFs such as N9 (I-UPF < > UPF), N2 (gNB < > AMF), and N4 (I-UPF < >SMF).

- The centralized DC hosting 5GC NFs also provides API interfaces toward the MEC and orchestration layers. The northbound API from the centralized DC enables integration between the 5GC control plane NF deployed in the centralized DC with the 5GC control plane NFs deployed in the public cloud.

- The user shown in Figure 6-41 is allowed access to both the cloud-deployed and on-prem NFs.

Threats Case Study

This section takes you through the threat examples in the virtual 5GC environment illustrated in Figure 6-42.

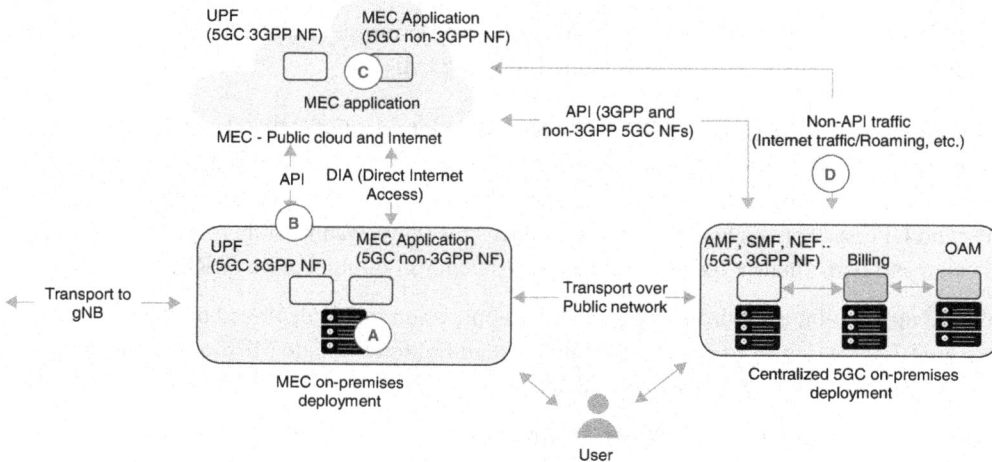

FIGURE 6-42   Threats Scenarios

Figure 6-42 shows multiple attack vectors in the deployed 5GC network:

A. A single server is used to host both the 5GC NF and MEC application. This increases the risk for high-sensitivity workloads such as 5GC 3GPP UPF workloads being impacted due to lower security measures undertaken in the lower-sensitivity workloads such as MEC applications. MEC applications created by third-party developers usually don't apply robust security policies, which leads to attack vectors such as unauthorized access to secure applications by using privilege escalations. In Figure 6-43, attackers use a weak API implementation on the MEC to gain access to the host OS. Once attackers gain access to the host OS and hardware (HW), they can access and exploit the 5GC 3GPP NFs and perform actions such as destroying the containerized 5GC NFs.

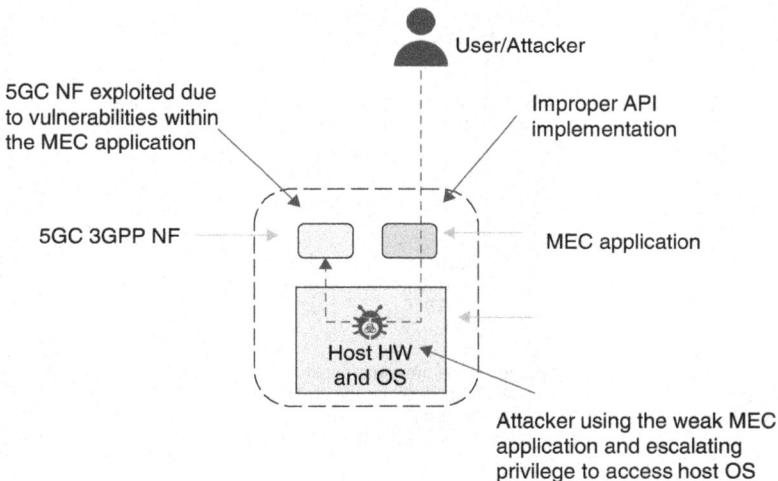

FIGURE 6-43   5GC NF Impacted Due to Weak Security in MEC Application

B. The MEC DC provides an API interface to enable inter-NF SBI-based communications that are prone to API vulnerabilities. The 5GC vendor provides 5GC NFs based on service-based architecture (SBA) implementing a service-based interface (SBI) that employs a RESTful API interface using HTTP/2. Secure implementation of API purely depends on service providers and vendors providing 5GC. Attackers can steal or modify such weakly protected data to conduct fraud, identity theft, or other crimes. Sensitive data might be compromised without extra protection, such as encryption at rest or in transit, and requires special precautions. But the most seen issue during proof of concepts or during conversations with different customers is security misconfiguration. This is due to improper deployment methods such as detailed error messages with component details being sent to the user, using insecure default configurations, incomplete configurations, misconfigured HTTP headers and unpatched vulnerabilities, and improper security configurations related to rate limiting for API calls, as shown in Figure 6-44. Improper security configurations allow the attacker to do the following:

- Understand the APIs being used internally

- Map the various services integrating with each other by gaining insight on NEF/SCP/NRF NFs, which opens up doors to attack chaining

- Gather the versions and types of applications and NFs being used

- Deny service to the 5GC system by forcing the system into a deadlock or an unhandled exception

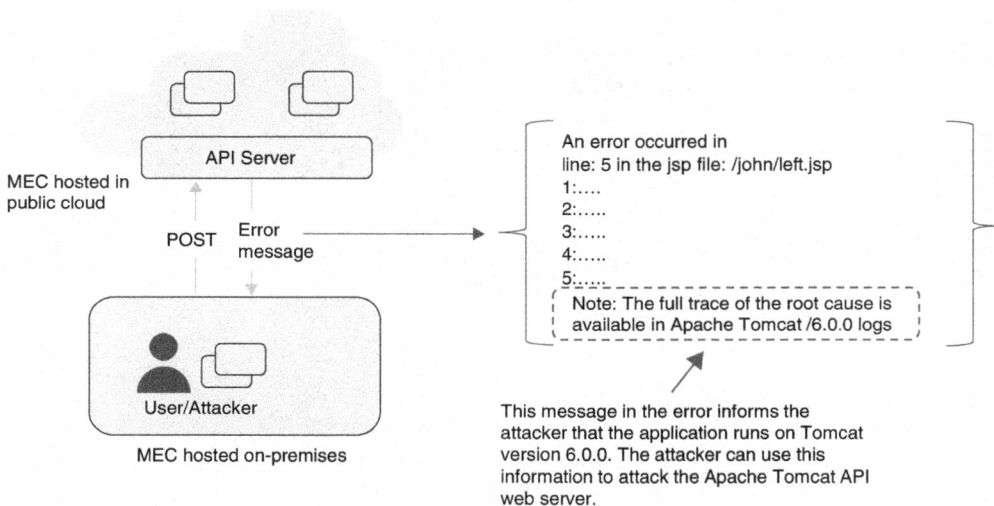

FIGURE 6-44   Security Misconfiguration Example

As illustrated in Figure 6-44, improper API security configurations such as passing more information in the error message to the API user would result in the user/attacker having information such as the API server being used and where the log is being stored. The attacker can then check the web server vendor's website for security vulnerabilities and can attack the web server. In many instances, the security vulnerabilities in the web server are overlooked by the packet core team, as it is not a part of the main 5GC subsystem. This vulnerability, if left unpatched, would increase the risk for the virtual 5GC environment.

C. Using cloud-native methodology to develop 5GC opens up an entirely new way of deploying 5GC NFs. This includes VM-based and container-based NFs that are constructed using open source and polyglot programs the developer is comfortable and efficient with. This new development of 5GC NFs also brings in new attack vectors due to vulnerabilities existing in the open source programs used in 5GC NFs. This vulnerability can be exploited by the attacker using

API calls with escalated privileges to exfiltrate sensitive information or to change attributes within the NF to cause a DoS, as illustrated in Figure 6-45. As the 5GC CNF development is in its infancy, these threat surfaces are quite frequently found in the 5GC NFs from various vendors.

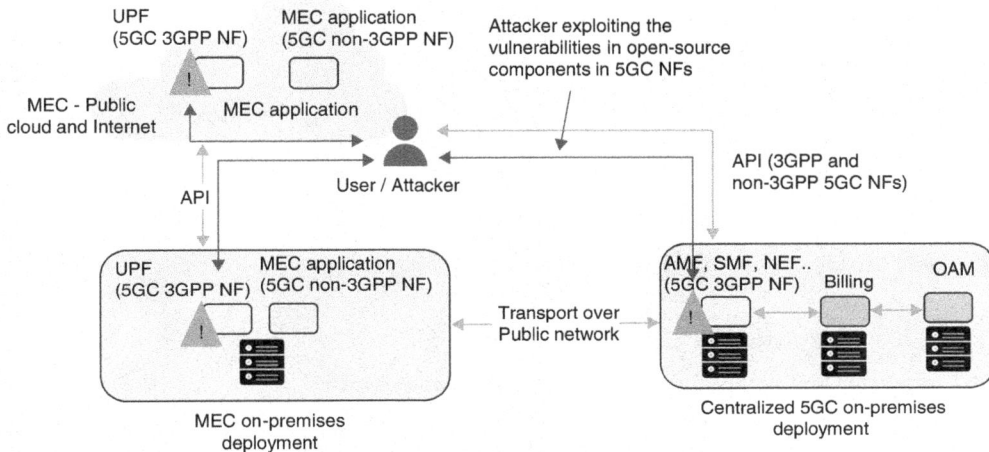

FIGURE 6-45   Attacker Exploiting Weak Security in 5GC NFs

D. One of the key risks in virtualized 5GC deployments is related to improper access control for users trying to access 5GC NFs and 5G related MEC applications. Due to the dynamic scale-in and scale-out nature of 5GC NFs, it is quite complex to maintain the user-to-application mapping using traditional segmentation methods such as network layer controls. The other key risk is the inconsistency of access management policies between multiple identity and management (IAM) solutions. The primary mode of access control provided by 5GC equipment and software vendors is based on role-based access control (RBAC). This RBAC is then integrated with an identity and management (IAM) solution with customized policy for access control. Due to the nature of commercial offers from 5G vendors, a majority of the vendors usually provide an option of their own IAM solution or partner IAM solution as a bundled offer. This leads to multiple IAM solutions with inconsistent policy implemented in your network, making the access control inefficient and inconsistent as shown in Figure 6-46.

Mitigation Examples

This section takes you through the techniques to mitigate the threats described under the "Threats Case Study" section based on the deployment illustrated in Figure 6-41. This section will be of particular interest to readers who are involved in a proof of concept or in early stages of virtualized 5GC deployments. Figure 6-47 illustrates the method used to deploy the virtualized 5GC 3GPP and non-3GPP NFs.

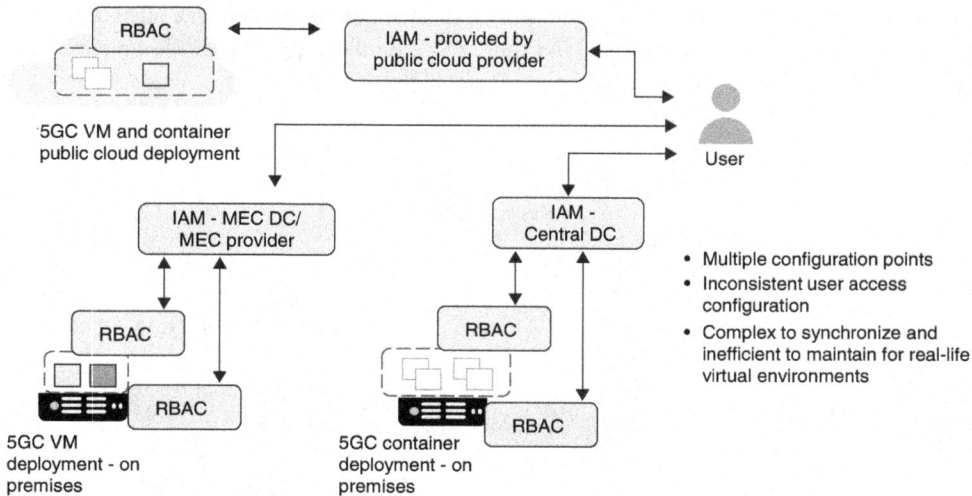

FIGURE 6-46   Improper Implementation of IAM for 5GC Deployments

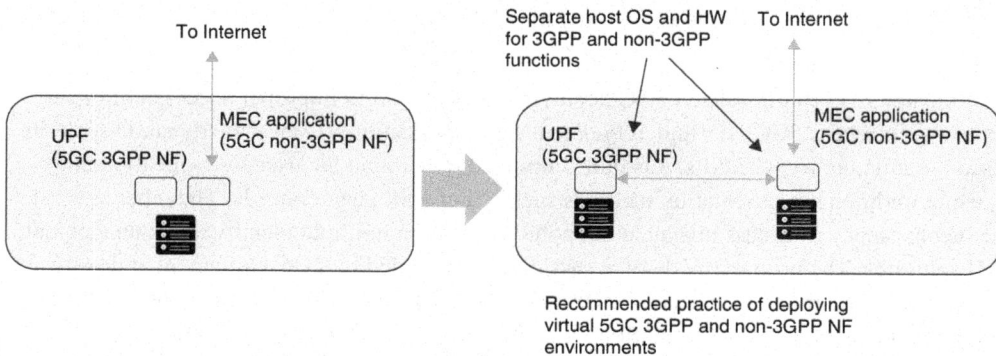

FIGURE 6-47   Separation of Critical and Non-Critical 5GC Workloads

As shown in Figure 6-47, it is considered best practice to deploy the non-3GPP 5G NFs, such as MEC applications, in a separate NFVi infrastructure from the 5GC 3GPP NFs, specifically when the non-3GPP NFs are exposed to the Internet. In the containerized deployment of 5GC NFs, it is good to verify with the 5GC vendor if any data or state is stored on the host or if any application-level dependency is required to be provided by the host. Any dependency of a containerized 5GC NF toward the host OS and HW should be prevented or kept to a minimum. This will reduce the attack surface to a great extent, provide you with an easier way to identify any anomalies, and allow you to air-gap NFs that do not need exposure to the Internet. All the dependencies should be packaged within the 5GC NF image itself. 5GC vendors are working in this direction but are not yet there completely.

Enhance Security Configurations and Handling or Error Messages/API Exceptions

5GC NF deployment is based on SBA and uses API calls to communicate between the NFs and between the NFs and non-3GPP functions. To mitigate the risks due to improper error handling and exposing more information within the error message, the following steps should be taken by the 5GC vendor and service provider.

Here are some key checks to be done by the 5GC vendor:

- The testing team of the 5GC vendor should ensure that error messages don't provide the user with extra information, such as the version and vendor type of the API web server being used in the infrastructure, as illustrated in Figure 6-44.

- The testing process by the 5GC vendor should include checks for Top 10 OWASP vulnerabilities.

- The test and validation team of the 5GC vendor should ensure that default and known files have been removed before moving the API server code to a production image that can be downloaded by customers.

- The test and validation team of the 5GC vendor should ensure that debugging code or extensions have been removed before moving the API server code to a production image that can be downloaded by customers.

- Ensure that the logging mechanism of the API server is customizable and can be set by the customer.

- The testing team should validate the expected input type (JSON, XML, and so on).

Here are some key checks to be done by the service provider:

- It is recommended that you understand the key points (east-west, north-south) of API calls within your deployment. This includes inter-NF SBI API calls, NF-to-third-party API calls, API calls to and from the orchestration solution, and API calls to and from the OAM solution. Having this understanding would allow you to map the weaker parts of your infrastructure that need security control enhancement.

- Ensure that the default password for the API server is changed and updated.

- Ensure that the logging mechanism is secure and, if required, can be chosen to be encrypted at rest.

- Ensure that the API server is configured to handle overloads and that you have options to rate-limit the API calls to prevent any DoS attacks.

- Use an API GW/WAF for securing the external API calls and use TLS/mTLS to secure 5GC SBI traffic flow.

Vulnerability Assessment and Patch Management

Figure 6-48 illustrates the mechanism to mitigate the vulnerabilities of the open source components used in the 5GC NFs. The security controls illustrated in Figure 6-48 and the best practices are explained in the list that follows:

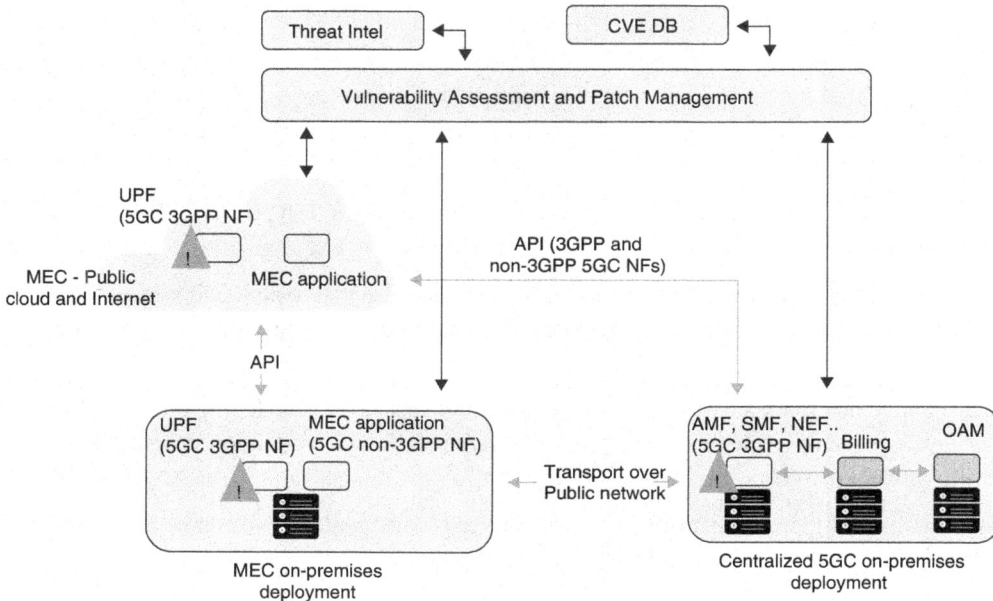

FIGURE 6-48   Vulnerability and Patch Management for 5GC NFs

- Although the 5GC NFs are now based on open source components, some vendors are still planning to provide patched images a couple of times per year. Although this might be an increased number of patches of one to two times a year from previous deployment methods, it is not efficient enough. This is bound to improve once the 5GC vendors undergo more rigorous testing procedures and secure CI/CD procedures using a DevSecOps process where security is embedded into every stage of the software cycle, reducing the vulnerabilities and thereby reducing the patch cycles. Ideally, patches should be deployed as frequently as required, as soon as a vulnerability has been identified and patched open source software released. But in real-life scenarios, it would be good to have more frequently patched images with a possible frequency of once per month and then based on the severity of the exposed vulnerability. The reason a patching frequency of at least once per month needs to be checked is not for ensuring that each month the vendor has to release a patch, but to ensure that if there are multiple severities detected, the vendor has a process in place to provide expedited patched images that are signed and uploaded in a trusted registry for you to download and deploy.

- 5GC vendors should use semi-structured inputs while fuzz-testing the 5GC NFs to validate the input verifications done by the 5GC NF and check whether it throws any exceptions. If the fuzz testing is not being carried out on the 5GC NFs, it's time that you start the fuzz testing for the 5GC NFs—at least the NEF and SCP NFs, which will be exposed to the Internet.

- The 5GC NFs you deploy should be capable of being integrated with a vulnerability solution that can cater for all modes of virtual deployments, such as virtual machines, containers on VMs, containers on BM, and virtual instances deployed on the public cloud.

- The vulnerability assessment solution you choose should have integrations with the threat intelligence and Common Vulnerabilities and Exposures (CVE) database, which will enable you to identify the vulnerabilities in your entire virtual environment.

- The vulnerability assessment solution should also allow you to mandate policies such as isolating web-facing workloads with a certain condition, such as having a CVE severity score of greater than 9 after giving the developer a certain period of notice.

Granular Access Control Based on Zero-Trust Principles

Figure 6-49 illustrates the security controls and mechanisms to mitigate the risk of improper access control for the 5GC NFs. The list that follows explains this in more detail. This method can be used for access control for all CNFs in 5G deployments, including 5G RAN, core, third-party MEC applications, and orchestration functions.

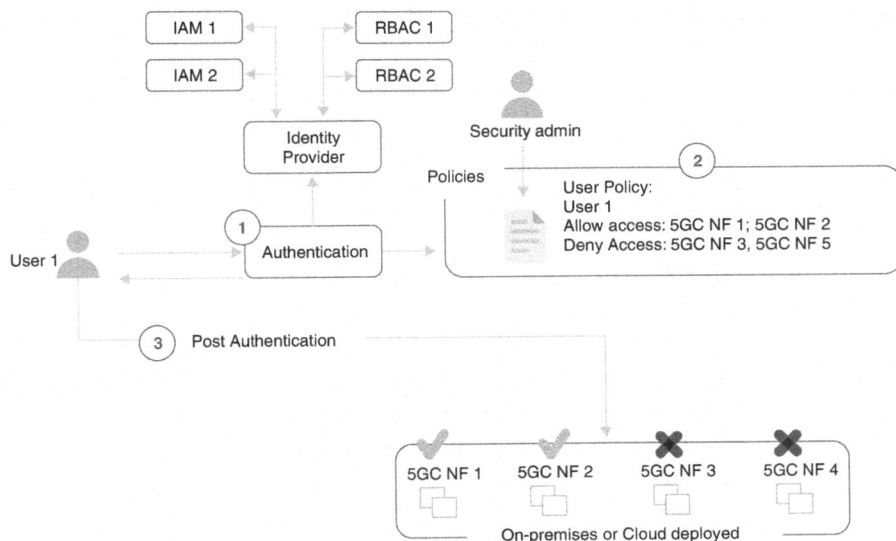

FIGURE 6-49   Access Control Policy Deployment Example for Users Accessing 5GC NFs

1. The 5GC vendor you choose should allow the integration of their RBAC with an external authentication server. The authentication server you choose should be able to be integrated with multiple IAM solutions. This is actually one of the most overlooked issues in identity and access management for 5GC deployments. Many miss the fact that each 5GC vendor might have a preferred IAM vendor or an integrated IAM (part-IAM) solution for access control. If you choose around four to five different vendors only for the 5GC solution deployment, you will have an extra four to five IAM solutions to operate and maintain. We are seeing this in pilot deployments already, and the production systems will have the same issues. The primary point here is not about cost, as the IAM solutions can be deployed for a few thousand dollars/euros, but about efficient identity and access management for your end-to-end multivendor 5G solution, including the 5GC domain.

2. The solution you choose for the IAM should provide you with the option of configuring custom policies for users and should be able to map users to applications.

3. The authentication solution should be able to provide multiple authentication methods such as JSON Web Tokens (JWT), Open ID Connect (OIDC), LDAP, RADIUS, and so on. Using a method such as JWT, claims can be transferred between the user and authentication server. These claims are encoded as JSON objects, which are digitally signed using JSON Web Signature (JWS) and, if required, encrypted using JSON Web Encryption (JWE). Based on these claims, the user will have access (or not have access) to the 5G NFs. For example, the authentication solution can set a claim saying 'isAdmin:true' and issue it to the admin user trying to access the 5G NF or third-party MEC application to make configuration changes upon successfully logging in to the application. The admin user can now send this token in every consequent request to the authentication solution to prove their identity.

This method can provide granular access control by ensuring mapping between users and applications, including the role of the user or application, to determine if the user can access and modify the configuration of the 5G NFs.

Summary

As discussed in this chapter, 5G brings the cloud-native approach to developing 5GC network functions using an open source software stack, deploying and maintaining the 5GC network functions dynamically enabling elasticity, optimizing the use of resources, improving overall agility, and bringing together a whole new software ecosystem that can integrate with your 5GC. It also brings in new threat surfaces not present in previous technologies, requiring a mindset change in the way you perceive security for packet core and ecosystems, depending on your packet core. The section "Threats in Virtualized 5G Packet Core Deployments" took you through the key threat surfaces in virtualized 5GC deployments, where different threat surfaces are explained in detail. The section "Securing Virtualized 5G Packet Core Deployments" then took you through the mitigation techniques in detail. You also saw the real-life scenarios where customers are concerned about the attack vectors and risks during the initial deployment of 5GC as well as some of the threats that have been seen in present deployment phases.

Figure 6-50 illustrates multilayered security controls to secure virtualized 5GC deployments.

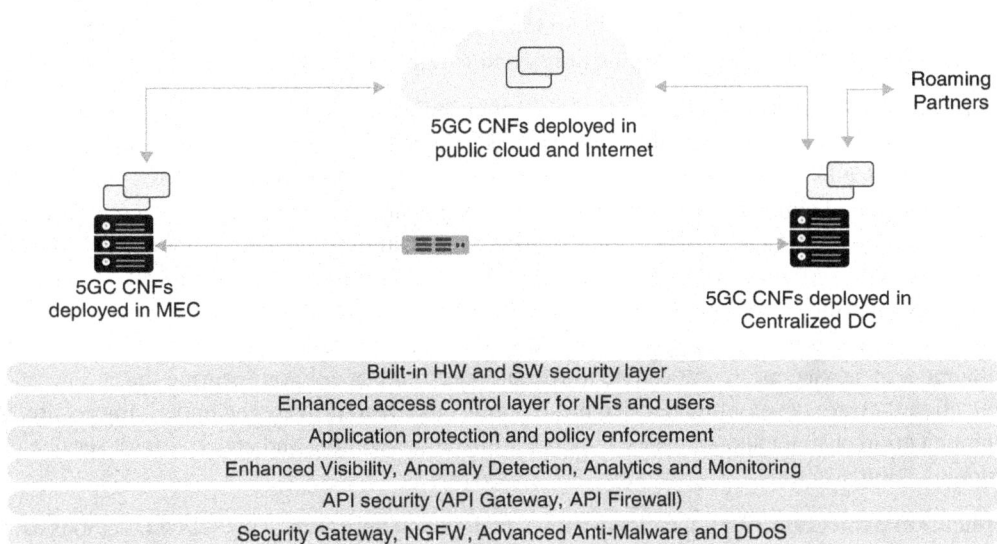

Roaming Partners

5GC CNFs deployed in public cloud and Internet

5GC CNFs deployed in MEC

5GC CNFs deployed in Centralized DC

Built-in HW and SW security layer

Enhanced access control layer for NFs and users

Application protection and policy enforcement

Enhanced Visibility, Anomaly Detection, Analytics and Monitoring

API security (API Gateway, API Firewall)

Security Gateway, NGFW, Advanced Anti-Malware and DDoS

FIGURE 6-50 Multilayered Security Controls to Secure Virtualized 5GC Deployments

Figure 6-50 illustrates the key security controls needed to secure the various deployment modes of virtualized 5GC NFs. The security controls can be reused for other parts of the infrastructure but are mapped here specific to the 5GC so that the reader interested in only this section can identify the key multi-security layers required to protect the virtualized 5GC deployment.

The security layers required to secure virtualized 5GC deployments are summarized in the following list:

- **Built-in hardening:** Ensure that the host OS and the firmware of the host HW are patched regularly to prevent attackers from exploiting the host OS for attacks based on privilege escalation or, in the case of HW, to narrow the possibilities of attacks based on side channel. It is recommended to use server manufacturer–supplied patch management and secure boot visibility solutions, which will help you to identify the patch history of the firmware and also alert you whenever a vulnerability is exposed and when the corresponding patch is made available by the vendor.

  The 5GC deployment should follow the true immutable manner where critical data is not stored on the host OS. The containerized 5GC NFs should have all the necessary application dependencies within the container itself, making it truly stateless.

The hardware should have the hardware root of trust using Trusted Platform Module (TPM) and Trust Anchor Modules (TAMs). Host hardware should also support Unified Extensible Firmware Interface (UEFI). This will help you verify whether or not the code launched by the firmware is trusted, and any untrusted code executions are stopped before they can infect the system. The hardware vendor should be able to provide you with the supply chain security measures it has undertaken. This is very important if you plan to use the hardware for 5G use cases for critical verticals such as public utility and national security verticals.

- **Enhanced access control layer:** Due to the flexible deployment nature of 5GC workloads and API being the mode of communication, access control needs to be carefully planned while deploying 5G CNFs. 5GC vendor should provide role-based access control (RBAC) capable of being integrated with third-party access management solutions which support secrets management and encrypted key management. Network administrators and teams from vendor or contractors who configure and maintain the multi-cloud deployed 5G NFs should be allowed access only after they go through the Multi-Factor Authentication (MFA) process and after verification of devices used to configure the network. For containerized deployments of 5GC NFs, deployment of service mesh will enable the developers to focus only on securing the application they are developing. Service mesh options like Istio provide the underlying secure communication channel and manages authentication, authorization, and encryption of service communication at scale.

- **Application protection:** With the cloud-native deployment approach using CI/CD methods as well as options for deploying containerized instances from vendors, the whole security ecosystem around having network security first has now changed to having application first security, which means you will now think about securing applications and their dependencies first, be it 5GC 3GPP NFs (VMs/containers) or non-3GPP 5GC NFs such as third-party MEC applications. Having a secure 3GPP 5GC NF starts with the vendor having robust CI/CD solutions to ensure any vulnerabilities are captured early in the development lifecycle and rechecked and audited again before the images are signed and uploaded into a container registry for customers to download and deploy.

You should deploy vulnerability scanning for all the virtualized 5GC components deployed in all types of virtual environments, such as on-premises and public cloud. You should plan to deploy solutions for vulnerability management that make use of artificial intelligence (AI) and machine learning (ML). They will improve efficiency and provide quicker time to detect (TTD). You will need to look at solutions that provide an application-level policy; in other words, not only network-level policy creation but container-level policies that can be mapped to users and other applications. Mechanisms like microsegmentation and providing sticky policy mappings between users and applications will have to be planned and implemented once your network starts to scale using multiple deployment models.

- **Enhanced visibility, anomaly detection, analytics, and monitoring:** To mitigate risks related to poorly implemented security controls within third-party applications and any vulnerabilities within the 5G CNFs, it becomes very important to have an enhanced visibility layer. For this to happen in virtualized 5G CNF deployments, the 5G vendor should provide the integration options with external monitoring solutions. This will allow the service providers to have a centralized monitoring system for a multivendor 5G ecosystem.

  You should also ensure that the 5G vendors allow integrations with external monitoring solutions, even if the 5G vendor provides its own vendor-specific inter-NF-monitoring capability. This is due to the fact that you will have multiple 5G vendors in your network due to the new open 5G ecosystem. You should also look at having enhanced visibility at the encrypted layer without having to decrypt the encrypted traffic. There are vendors today who provide these capabilities using AI and ML techniques.

- **API security:** The starting point of having a secure API is the definition of the API itself, which should undergo robust testing to fill any gaps or identify any security misconfigurations. You should also have detailed API security layers to ensure the validation of API calls, verifying the API calls for any malformed requests or packets, ensuring that the API calls are authenticated, and ensuring that policies are applied to the API request using mechanisms such as JWT Tokens and ensuring that critical information is not passed on to the users. Orchestration API calls should be checked to ensure they are valid orchestration API calls and that any critical attribute changes should have multilayered authentication before they are executed. Auditing of the entire API ecosystem (SBI as well as external API) should be carried out at least every 3 months, and the encrypted API calls should be continuously monitored for any anomalies without having to decrypt them.

- **SecGW, NGFW, advanced anti-malware, and DDoS security layer:** The initial deployment of 5G CNFs is aimed at on-premises DC and limited deployment on the public cloud. 5G NFs being deployed in the public cloud require integration with the on-premises 5GC NFs, which can be protected using encrypted API calls. Any non-API calls can be protected with IPsec by using a security gateway (SecGW). The network layer segmentation of the NFs and inspection of the packets can be achieved by using next-generation firewalls (NGFWs). The operation and management (OAM) systems should also be inspected for any malware, which can be achieved by using an anti-malware solution that can also validate user devices to mitigate any risks from user devices impacting the 5GC OAM network.

Acronym Key

Acronym	Expansion
3GPP	3rd Generation Partnership Program
5GC	5G core
AF	Application Function
API	Application programming interface
APIGW	Application programming interface gateway
APM	Application performance monitoring
AUSF	Authentication Server Function
BM	Bare metal
C&C	Command and control
C2C	Command and control
CD	Continuous Delivery
CI	Continuous Integration
CNF	Cloud-native function
CPU	Central processing unit
CSR	Certificate signing request
DC	Data center
DDoS	Distributed denial of service
DevOps	Development and operations
DNS	Domain Name System
DoS	Denial of service
HPLMN	Home public land mobile network
HTTP	Hypertext Transfer Protocol
IAM	Identity and Access Management
JSON	JavaScript Object Notation
JWE	JSON Web Encryption
JWT	JavaScript Object Notation Web Tokens
LDAP	Lightweight Directory Access Protocol
MEC	Multi-access edge compute
MFA	Multifactor authentication
MitM	Man in the middle
mTLS	Mutual Transport Layer Security
NAC	Network Access Control
NAS	Non–Access Stratum
NEF	Network Exposure Function
NF	Network Function

Acronym	Expansion
NSA	Non-standalone
NTP	Network Time Protocol
OIDC	Open ID Connect
PCF	Policy Control Function
RA-VPN	Remote access virtual private network
RAN	Radio access network
RBAC	Role-based access control
REST	Representational State Transfer
RPC	Remote procedure call
SA	Standalone
SBA	Service-based architecture
SBI	Service-based interface
SCEF	Service Capabilities Exposure Function
SCP	Service Communication Proxy
SDS	Secret Discovery Service
SEPP	Security Edge Protection Proxy
SMF	Session Management Functions
SPIFFE	Secure Production Identity Framework for Everyone
TAM	Trust Anchor Modules
TLS	Transport Layer Security
TPM	Trusted Platform Modules
TTD	Time to detect
UDM	Unified Data Management
UE	User equipment
UEFI	Unified Extensible Firmware Interface
UP	Uplink
UPF	User Plane Function
URLLC	Ultra-reliable low-latency communication
VM	Virtual machine
VNFM	Virtual Network Function Manager
VPLMN	Visited public land mobile network
WAF	Web application firewall
XSS	Cross-site scripting

References

3GPP TS 23.222, "Common API Framework for 3GPP Northbound APIs," Release 16, v16.10.0, June 2021

3GPP TS 33.501, "Security architecture and procedures for 5G system," Release 16, v16.7.0, June 2021

NIST Special Publication 800-190

https://docs.docker.com/engine/security/trust/

https://www.appdynamics.com/

https://istio.io

Pramod Nair, "Securing 5G and Evolving Architectures," Cisco Knowledge Networks, Cisco Public, September 17, 2020

Pramod Nair, "Why 5G Is Changing Our Approach to Security," Cisco Public, June 10, 2020

Pramod Nair, "Securing 5G Cloud-Native Deployments," Open5GCon, September 30, 2020

https://kubernetes.io

https://www.portshift.io

https://www.kennasecurity.com

Chapter 7

Securing Network Slice, SDN, and Orchestration in 5G

After reading this chapter, you should have a better understanding of the following topics:

- Threats in SDN, orchestration, and network slice deployments in 5G
- Securing SDN, orchestration, and network slice deployments
- Real scenario case study examples of SDN, orchestration, and network slice threat surfaces and threat mitigation techniques

This chapter will take you through the threat surface and mechanisms to mitigate threats in the 5G network slice and its enablers—software-defined networks (SDNs) and orchestration.

This chapter will be of particular interest to the following teams from enterprise and 5G service providers deploying 5GC and cybersecurity vendors planning product developments and new functionalities to secure the deployment of 5GC:

- SDN and orchestration teams within enterprises, service providers, and vendors
- Automation strategy teams of service providers deploying 5G
- Security strategy teams within service providers deploying 5G network slice
- Service providers planning to offer network slice as a service (NSaaS)
- Enterprises planning to monetize the network slice provider marketspace
- Security strategy teams within enterprise verticals planning on deploying 5G
- RAN, transport, packet core, and data center teams involved with 5G strategy
- Security architects and design teams looking at secure SDN and automation deployments for 5G

- Enterprise solution and security architects deploying 5G network slicing on enterprise premises

- Enterprise solution architects and enterprise security architects working with enterprises integrated with service provider 5G networks

- Government departments looking at regulating security for service providers

- Cybersecurity vendor security architects looking to secure 5GC deployments for their end customers

- Cybersecurity vendor product managers looking for use cases or features to enhance security products to cater for secure 5G deployments

Network Slicing and Its Enablers—SDN and Orchestration

Network slicing is a key concept in 5G. It is the ability to build what looks like discrete end-to-end networks for different 5G use cases or customers. However, one requirement of network slicing is that network functions might need to be shared between slices. For example, the multiple slices might share the control plane, but each individual slice might have its own user plane function. One of the other key requirements of 5G is that the network needs to be designed to be able to offer a different mix of capabilities to meet all diverse requirements, such as high-bandwidth high throughput and low-latency low throughput, at the same time.

In legacy networks such as 4G, an Evolved Packet System (EPS) bearer is the level of granularity for bearer-level quality of service (QoS) control. That is, all traffic mapped to the same EPS bearer receives the same bearer-level packet-forwarding treatment, such as scheduling policy, queue management policy, rate-shaping policy, Radio Link Control (RLC) configuration, and so on. One EPS bearer is established when the user equipment (UE) connects to a Packet Data Network (PDN), and that remains established throughout the lifetime of the PDN connection to provide the UE with always-on IP connectivity to that PDN. That bearer is referred to as the *default bearer*. Any additional EPS bearer that is established for the same PDN connection is referred to as a *dedicated bearer*. An EPS bearer uniquely identifies traffic flows that receive a common QoS treatment between the UE and a PDN GW for GPRS Tunneling Protocol–based S5/S8. An EPS bearer identity uniquely identifies an EPS bearer for one UE accessing via LTE eNB.

In 5G, the QoS model is based on QoS flows, which are controlled by the Session Management Function (SMF). The SMF performs the binding of Policy and Charging Control (PCC) rules to QoS flows based on the QoS and service requirements. The QoS flow is the finest granularity of QoS differentiation in the PDU session. A QoS flow ID (QFI) is used to identify a QoS flow in the 5G system. User plane traffic with the same QFI within a PDU session receives the same traffic-forwarding treatment (for example, scheduling and admission threshold). The QFI is carried in an encapsulation header on N3, as well as N9 (that is, without any changes to the end-to-end packet header). The 5G QoS model supports both QoS flows that require guaranteed flow bit rate (GBR QoS flows) and QoS flows that do not require guaranteed flow bit rate (non-GBR QoS flows). The 5G QoS model also supports reflective QoS, which enables the UE to map uplink (UL) user plane traffic to QoS flows without Session Management Function–provided QoS rules.

Figure 7-1 illustrates 4G and 5G QoS.

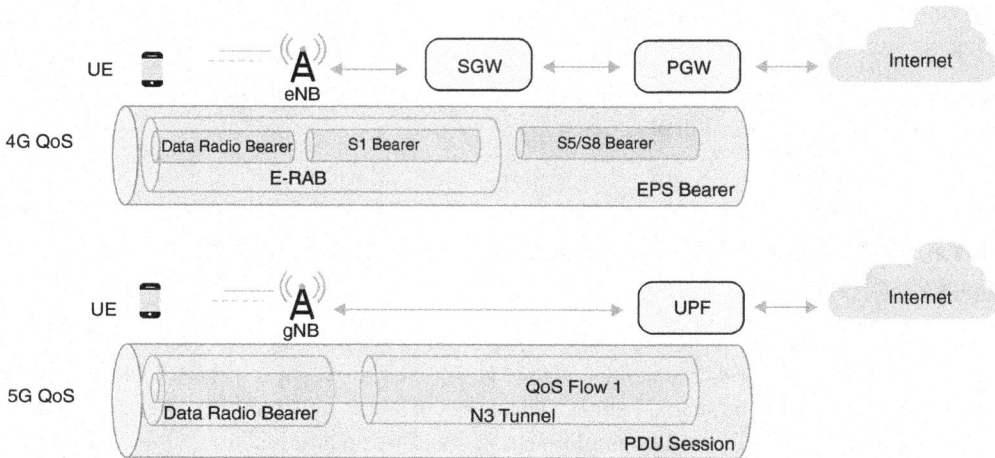

FIGURE 7-1   4G and 5G QoS

As shown in Figure 7-1, the traditional network deployment options did not allow the logical splitting of the network to cater for different use cases. It used the concept of Access Point Names (APN), which is basically the name of a gateway between the mobile network and public Internet. Some examples are provided as follows:

2G / 3G APN example: internet.mnc011.mcc123.gprs

4G APN example: internet.apn.epc.mnc011.mcc123.3gppnetwork.org

The key difference between 4G APN and 2G/3G network APN is the use of "apn.epc" before the mobile network code (mnc) and use of "3gppnetwork.org" instead of "gprs" after the mobile country code (mcc).

Different APNs are configured in the Evolved Packet Core (EPC), which is then used to identify the QoS or access restrictions on roaming assigned to the user. APN names are chosen by the service providers and some examples of APN names are as follows:

APN 1: internet.*unlimtedthroughputapn*.epc.mnc011.mnc123.3gppnetwork.org

APN 2: internet.*restrictedthroughputapn*.epc.mnc011.mnc123.3gppnetwork.org

In the aforementioned example, the configurations for APN 1 in the PGW context allows the users associated with the APN 1 to have unlimited throughput and the configurations for APN 2 in the PGW context restricts the throughput for the users associated with APN 2.

The PGW will use the APN configuration to select the PDN context and IP addresses are assigned from the IP pool configured in the selected PDN context. The QoS policies are assigned by the Policy Control and Charging Rules Function (PCRF) within the EPC. The key issue in legacy deployments such as 4G was the inability to have a use case–specific dedicated resource. 5G brings significant improvement and flexibility in the QoS, which helps you to deliver ultra-reliable low-latency communication (URLLC) for mission-critical communication services. It also provides you different prioritization levels for different applications, as per preconfigured requirements for specific subscribers or industry/enterprise verticals.

The successful delivery of QoS along with the realization of new use cases, such as URLLC, enhanced mobile broadband (eMBB), and massive machine-type communications (mMTC), requires a tailored logical network using techniques such as network slicing, enabled by multidomain orchestration and SDN architectures to allow the common infrastructure to efficiently support multiple network instances with tailored services. To achieve end-to-end (E2E) automation of the 5G network, an orchestrator should be able to receive the "intent" of the service you require and translate that to real change in the 5G network. The orchestration solution you choose should fully automate the creation, deletion, and runtime modification of network services. It should map design-time service definitions to runtime network operations through a single, flexible data model for a service. The orchestration solution should then derive the minimum network changes required and execute them.

Orchestration and automation tools have typically been bound to a technology domain: a tool for the packet core network, a tool for the transport network, a tool for the data center network, a tool for the WAN, a tool for the optical network, and perhaps tools to manage firewalls and other L4–L7 devices. This creates inflexibility and higher complexity for 5G orchestration. The orchestration solution you choose for 5G should span multiple technology domains, allowing you to create and automate cross-domain service chains much more easily and dependably for successful deployment of network slices. Furthermore, orchestration should be done in a loosely coupled manner so that the operations teams responsible for each technology domain can work cooperatively with the teams building the service chains.

The actual 5G deployments will require high levels of programmability to ensure that the underlying 5G infrastructure can be abstracted for applications and service. This is enabled by using SDN. Its key concepts (resource virtualization, dynamic initialization, and so on) support the envisioned high degree of flexibility through network slicing. Common abstractions allowed by SDN enable slice resources (networking, processing, and storage types) included in a slice to fulfill a business purpose, while open and programmable interfaces in the SDN and orchestration layers allow dynamic control and automation of network slice creation and operation. This is further realized using new specifications in 5G and new cloud-native deployment methods for 5G Core (5GC). This will help you with flexible configurations and allows multiple networks to be logically deployed on the underlying infrastructure, such as radio, transport, and packet core network, as shown in Figure 7-2.

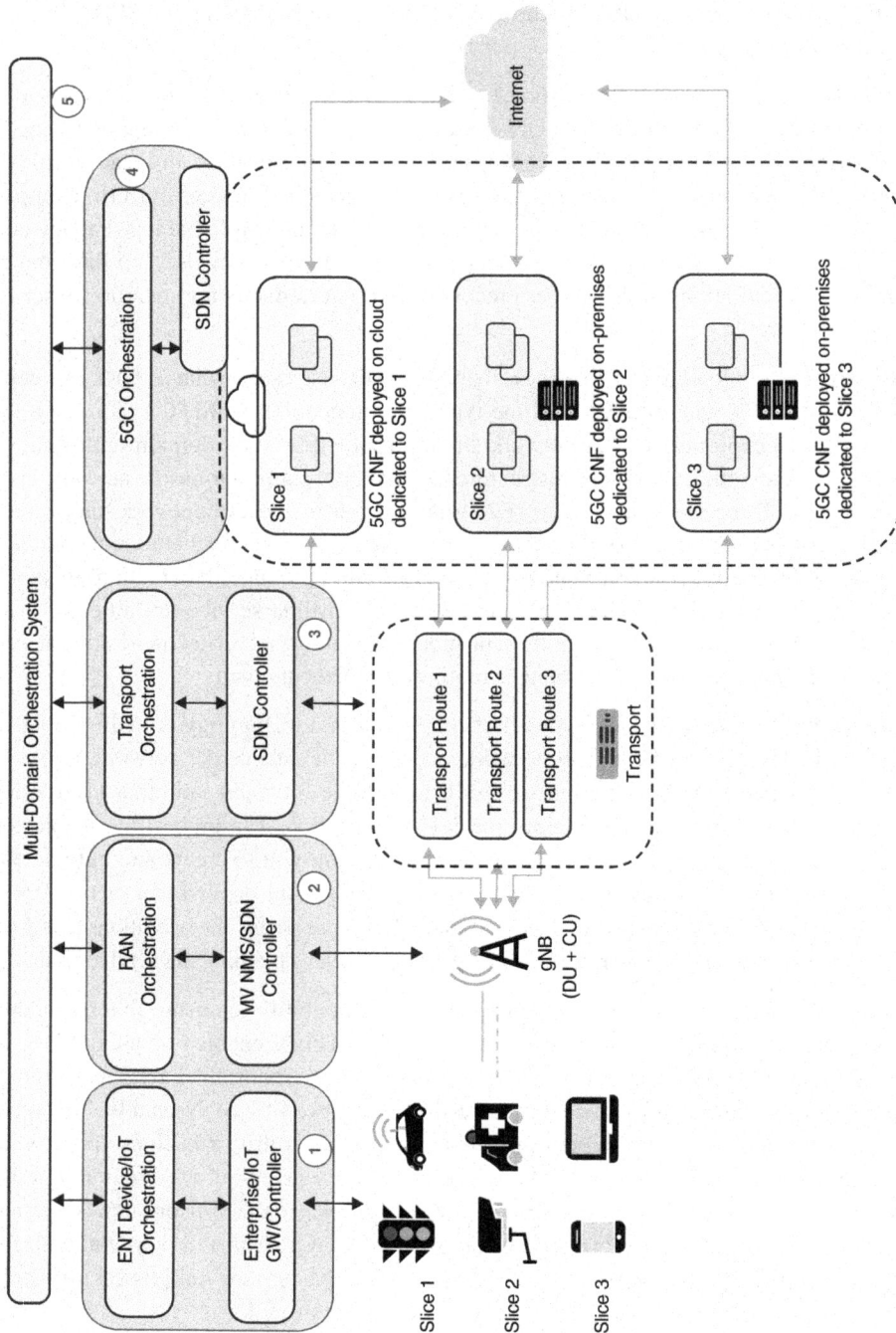

FIGURE 7-2   Network Slice Deployments Enabled by Orchestration and SDN

Figure 7-2 illustrates the components required for efficient network slice 5G deployments. The key functions are explained in the list that follows.

1. **Devices:** Use cases such as vehicle to everything (V2X), ultra-reliable low-latency communication, and enhanced mobile broadband (eMBB) are allocated the following slices:

   - **Slice 1:** V2X based on machine-to-machine (M2M) use cases

   - **Slice 2:** URLLC use cases

   - **Slice 3:** eMBB use cases

   The IoT/IIoT devices used in industry verticals such as smart factories enabling M2M and URLLC use cases will be connected to IoT gateways, which provide simple, essential connectivity for assets at mass scale. The gateways will provide connectivity to the 5G network if the devices themselves are not 5G capable. IoT data orchestration software extracts, transforms, and delivers data of connected assets from edge to multicloud destinations with granular data control.

2. **RAN:** You would typically have a multivendor radio access network (MV-RAN) in your 5G network that is managed using Network Management System (NMS), vendor-specific Element Management System (EMS), or configured by an SDN controller. This is indicated in Figure 7-2 as multivendor NMS (MV NMS/SDN controller). A RAN orchestration solution will orchestrate the deployments of virtual RAN components such as Centralized Unit (CU) and Distributed Unit (DU) by integrating with the respective RAN vendor's NMS/EMS system or third-party SDN solution, as shown in Figure 7-3.

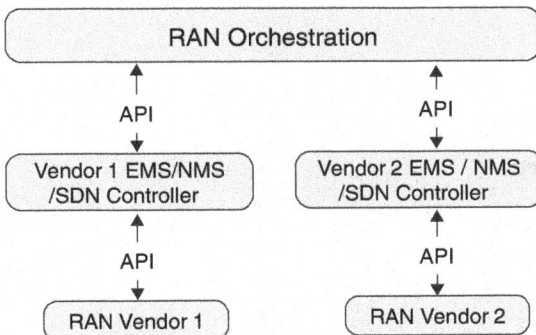

FIGURE 7-3 RAN Orchestration

3. **Transport and SDN:** In 5G, the transport layer is recommended to be programmable and would require a software-defined networking (SDN) approach. Transport devices being deployed in 5G would need to cater for different requirements on the traffic types. In some cases, the transport network would need to handle high-throughput data; in other cases, it would require traffic with a demand of low latency with very strict multidomain service level

agreements (SLAs) required for a specific traffic profile. This will require dynamic traffic path changes to ensure that the transport network meets the SLAs and converged, end-to-end packet infrastructure, beginning in the access layer and stretching via the network data center all the way to the core, based on segment routing and packet-based QoS. The SDN controller will be integrated using API in the northbound interface with the transport orchestration solution, as shown in Figure 7-4.

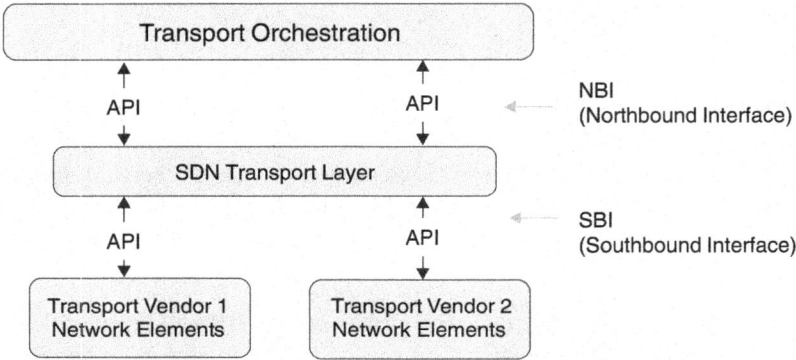

FIGURE 7-4   Transport Orchestration

4. **5G Core (5GC):** To meet SLAs and to optimize resources per traffic profile and use cases, the cloud-native-based 5GC control plane (CP) and user plane (UP) network functions (NFs) can be dedicated for specific use cases. In some deployments, you will see 5GC CP NFs being shared to further optimize the number of CP instances required to fulfill the use cases, as shown in Figure 7-5.

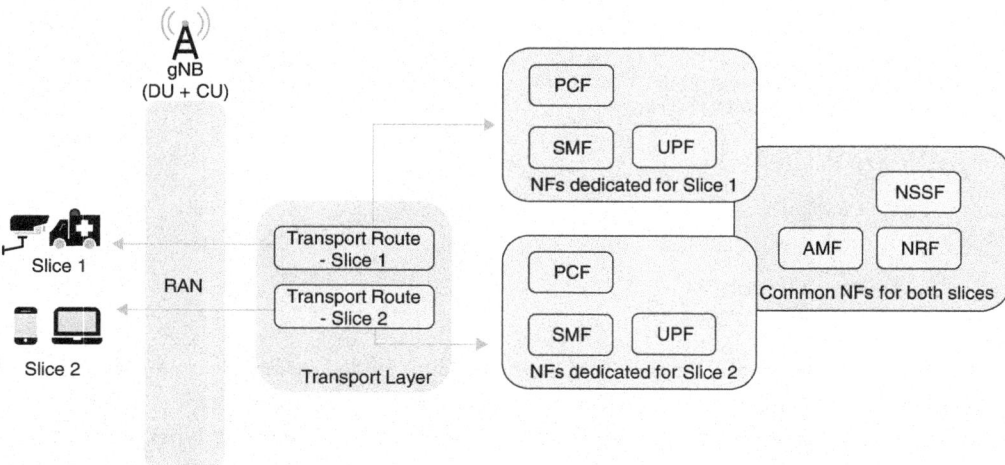

FIGURE 7-5   Shared Network Slice Deployment

The sample shared slice deployment in Figure 7-5 shows where AMF, NRF, and NSSF are shared for separate slices.

The key features enabling network slice is explained in the list that follows:

- **Network Slice Selection Function (NSSF):** Supports the following functionality:
  - Selecting the set of network slice instances serving the UE
  - Determining the allowed NSSAI and, if needed, the mapping to the subscribed S-NSSAIs
  - Determining the configured NSSAI and, if needed, the mapping to the subscribed S-NSSAIs
  - Determining the AMF set to be used to serve the UE, or, based on configuration, a list of candidate AMF(s), possibly by querying the NRF

- **Network Slice Selection Assistance Information (NSSAI):** Used for uniquely identifying a network slice. Single NSSAI (S-NSSAI) contains Slice Service Type (SST) and an optional Slice Differentiator (SD) that's used as an additional differentiator if multiple network slices carry the same SST value.

- **UE Route Selection Policy (URSP):** Feature enabled by the Policy Control Function (PCF). Although not a network slice function, it helps inform the network slice status to the UE via the AMF, enabling you to configure a new service to the UE.

- **Multidomain orchestrator:** The multidomain orchestrator has the task of interacting with multiple components to orchestrate the necessary changes in the network. The orchestration solution you choose provides a single, network-wide interface to all network devices and services—both physical and virtual—using a common modeling language and data store. It should also let development teams define services and hand them off to operations teams that can then automate service activations and changes quickly and simply, moving from high-level intent to granular device configurations in a single transaction. It should also support the following access options:
  - **CLI:** For network engineers who prefer the command-line interface
  - **Web interface:** For network engineers who prefer a graphical interface
  - **REST:** For programmatic access (exactly the same feature set as the CLI and web interface)
  - **Java/C:** For building custom applications and service provisioning logic
  - **JavaScript:** For embedding orchestration functions in portals
  - **NETCONF:** For importing and exporting XML configurations
  - **SNMP:** For reading status and receiving orchestration alarms
  - **Python:** At least support Python for scripting network-wide configuration changes

Automation, which is very closely tied with orchestration, becomes very important for you because it helps to manage your services and 5GC NFs across their entire lifecycle: from onboarding and deploying new NFs to monitoring health, to scaling up and down instances in response to demand, to automatically remediating issues, to finally spinning down instances and freeing resources for other apps and services. Figure 7-6 illustrates the multidomain orchestration solution interacting with the 5G infrastructure in the southbound interfaces (SBI) covering both SDN-capable and non-capable 5G infrastructure components. 5G CNFs could be deployed on virtual machines (VMs), containers on VMs/bare metal, or on the public cloud, where the 5G CNFs could run as either VMs or containers on top of the compute and storage infrastructure. The northbound interface (NBI) receives the "intent" of the service you require and translates that to real change in the 5G network. It will also be used by third-party application developers to integrate certain software modules within the orchestration layer to enable 5G use cases.

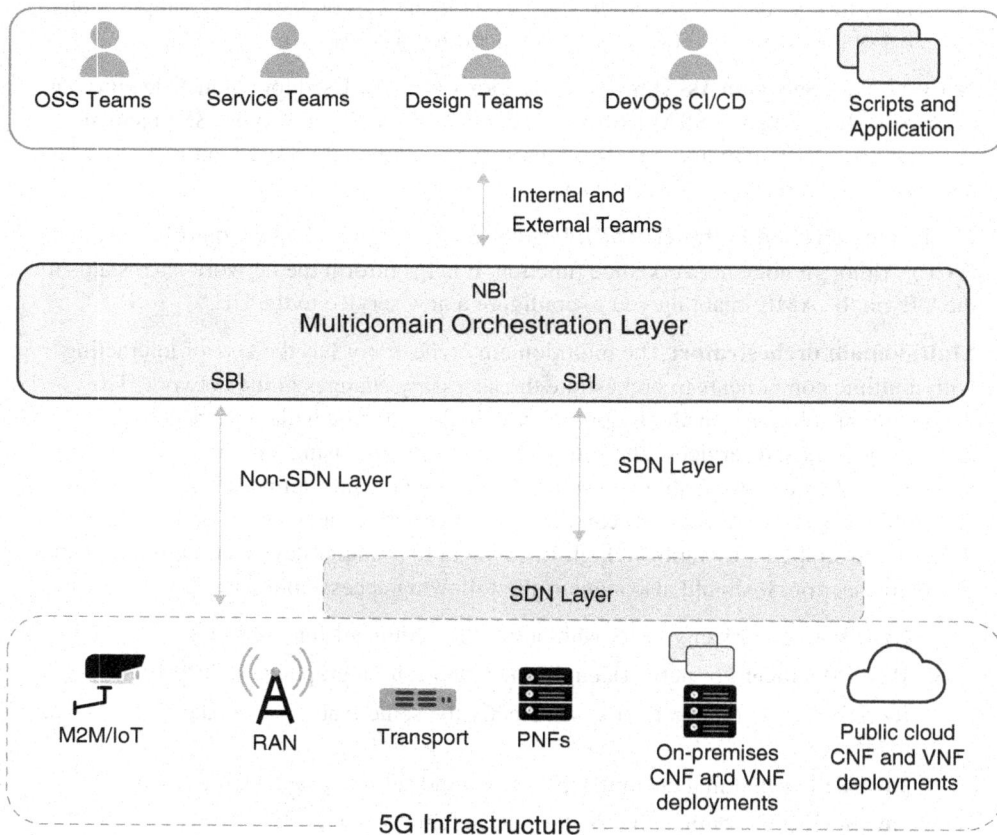

FIGURE 7-6   Multidomain Orchestration for Multiple 5GC Deployment Options

The definition of a 5G network slice includes all resources, including radio, mobile core, network data center/virtualization platform, and the transport network. As shown in Figure 7-7, orchestrating a slice successfully requires tight coordination between these different domains, although these domains are very different in their functions, as are the tools needed to control and manage them.

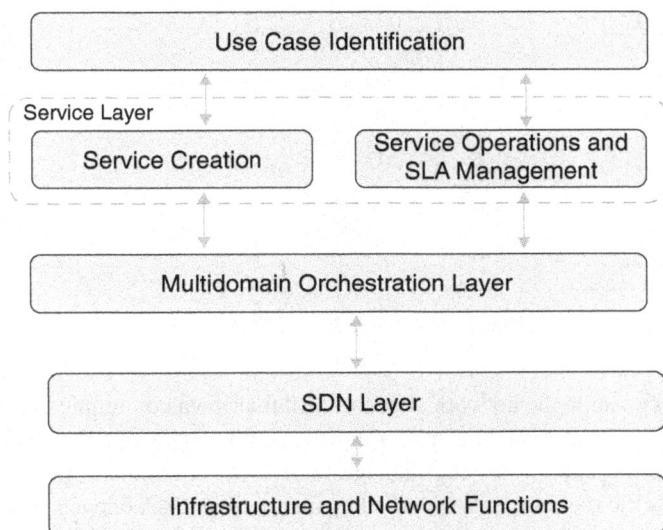

FIGURE 7-7   Depiction of Multiple Domain Coordinating for a Single Use Case

For the orchestration to be successfully deployed, it needs to follow a new approach to succeed by simplifying the control plane, reducing the protocol complexity, and becoming simpler to provision, secure, manage, and monitor. For a pragmatic deployment of a network slice, you need to understand the customer requirement from different industry verticals such as manufacturing, IoT, healthcare, automotive, and so on. Once the requirement is understood, you can then make the template for creating a service for the customer, which will include the type of 3GPP and non-3GPP NFs required and the type of configurations required in the infrastructure components such as the transport and RAN components.

Also, you can enable various commercial models to exploit the advantages provided by the network slice concept. Network slice as a service (NSaaS) is one commercial model you can offer to your customers. This method allows your customers to manage the network slice for themselves by accessing multitenant management of the network slice, where they can configure the required template, as illustrated in Figure 7-8.

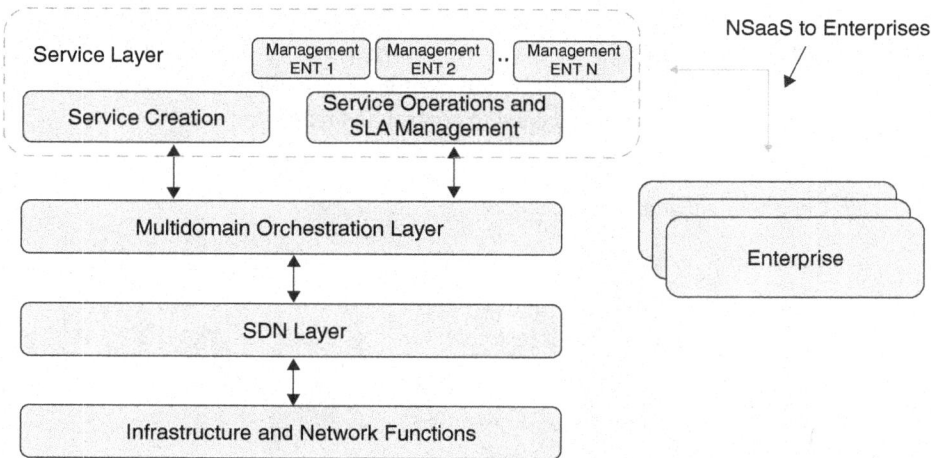

FIGURE 7-8 Multitenant Management of Network Slice for NSaaS End Customers

As Figure 7-8 illustrates, enterprises can consume the network slice to fulfill their own communication service needs. Management ENT 1–Management ENT N can be a multitenant management platform provided by the service provider so that the enterprises can manage their own requirement, or the service provider can manage the slice-specific requirements specified by means of an SLA between the service provider and the enterprise. Another model is where your customer uses the network slice as an offering to other customers.

These models are typically chosen by the Chief Revenue Officer (CRO) or the external product teams within your organization to monetize the investments in 5G. An example is parts of your network and features such as security, being offered as a service to an industry vertical or an enterprise. The level of NSaaS slice management would depend on the use case and the SLA agreed with your customer. The slice could be fully managed by you as a part of a managed services offering or could be managed by the customer themselves. For example, you offer slice 1 as a service to enterprise A. Enterprise A then resells it to industry vertical A and industry vertical B to fulfill their use cases. In this case, one of the scenarios could be you offering managed services to manage the slice for enterprise A, industry vertical A & B. Another scenario could be you providing access to industry vertical A & B to manage their own slices through a portal provided by enterprise A. Figure 7-9 illustrates an example of slice management when your network slice is used for reselling by your customers.

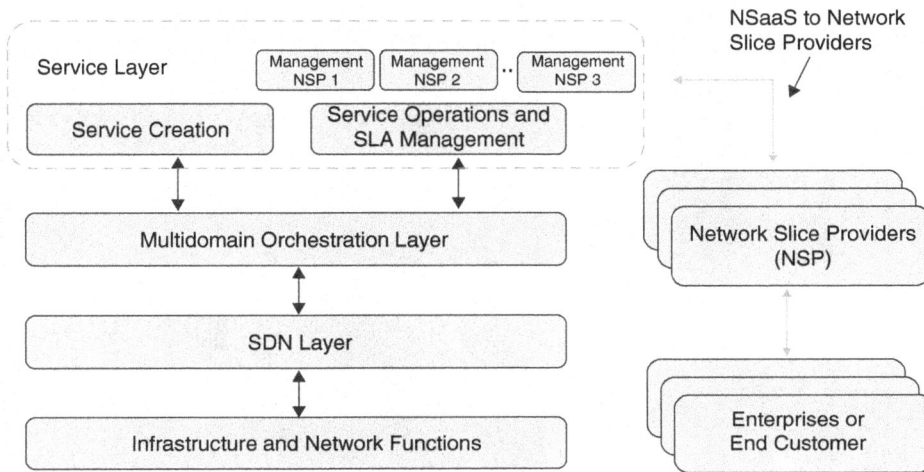

FIGURE 7-9 Multitenant Management of Network Slice for NSaaS Network Slice Resellers

As shown in Figure 7-9, the Network Slice Provider (NSP) can use the slice obtained from the service provider to deliver communication services to its end customers. NSPs can configure the slice requirements using the management platform provided, which is multitenant. Management NSP 1 in Figure 7-9 relates to the management layer, which NSP 1 can use to configure slices to provide communication services to enterprises or end customers. The NSaaS offering from the service provider could also include the management of NSP 1 being configured and maintained by the service provider using an SLA between the service provider and NSPs. Note that although NSP is a commonly used term among service providers, it is also called Communications Service Provider (CSP). Consumers of NSP / CSP are also called Communications Service Consumer (CSC).

Threat Surfaces in 5G Network Slice, SDN, and Orchestration Deployments

This section discusses the key threat surfaces related to the network slice and its dependencies, such as SDN, automation, and orchestration. Figure 7-10 illustrates the threat surfaces in a 5G network with network slice, SDN, orchestration, and automation.

Some of the key threat surfaces illustrated in Figure 7-10 are listed next:

- Orchestrators are primarily tasked with driving the scale and density of 5G workloads. This leads to risk factors such as workloads of differing sensitivity levels being deployed on the same host. An example would be one of the less sensitive workloads, such as an MEC application workload, with direct interface to the Internet being deployed alongside a critical workload, such as Unified Data Repository (UDR), which stores subscriber-related data, on the same host.

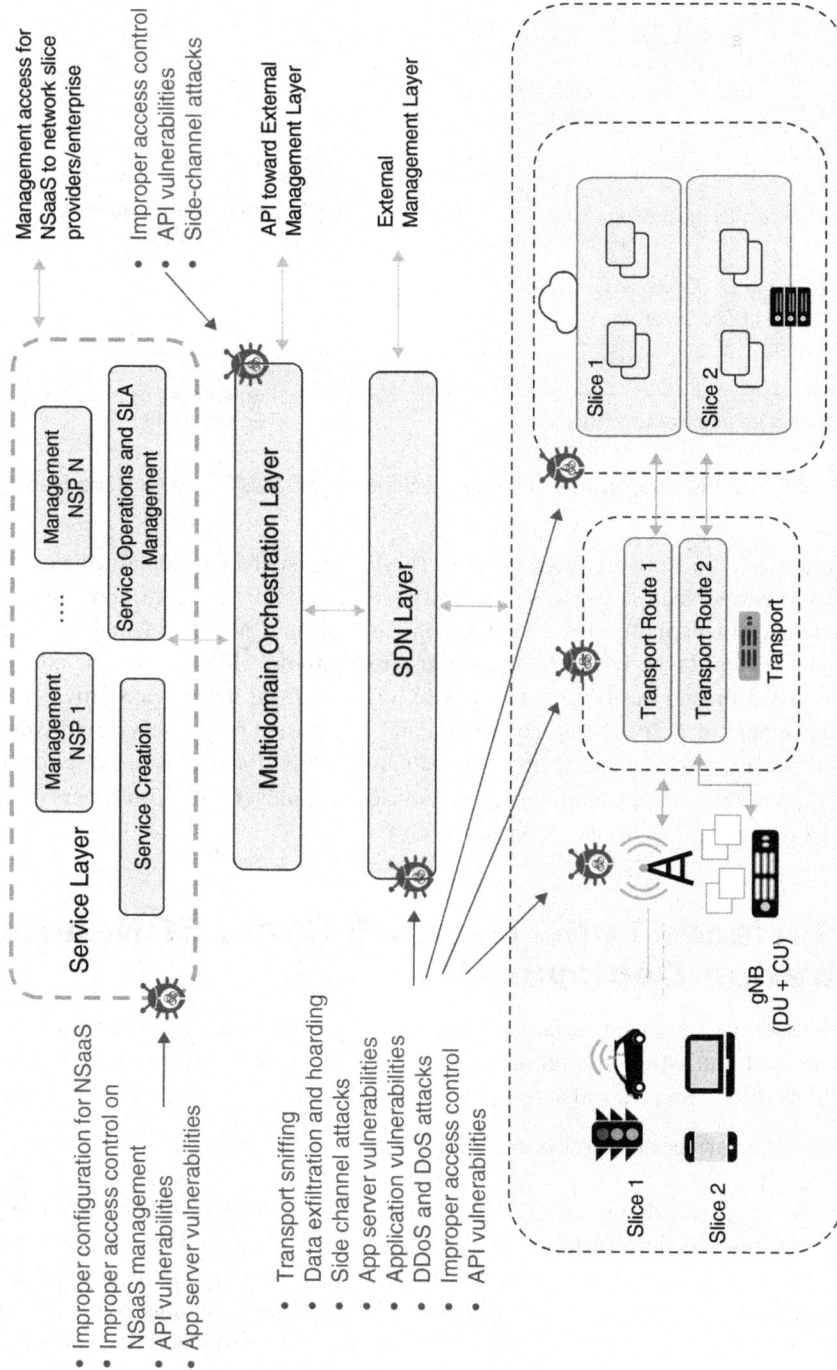

FIGURE 7-10   Multidomain Threat Vectors in Network Slice Deployments

- Orchestrators typically use API servers/web servers for API calls. In case of critical vulnerabilities in the web server, the entire orchestration components could be open to attacks.

- The SDN layer, automation, and orchestration layers being deployed to enable network slicing would require integration with the third-party application and other 5GC network functions using an API. The API, if not deployed following best practices, will bring in threat vectors related to insecure API security risks, such as user- and function-level authorization, excessive data exposure, Broken Object Level, and so on. This becomes more prominent in MEC deployments for 5G.

- Aside from traditional attacks against servers, DoS and DDoS attacks on the SDN/orchestration/ automation layers will impact the service creation leaving your network crippled and will also disrupt the services of the network slice providers and enterprises that are dependent on your network. The vulnerabilities that might cause this to happen would be through traditional attacks on hardware components of the infrastructure, application vulnerabilities, APIs that are not properly secured, and rogue nodes within the architecture.

- There is an emphasis on flexible software-based architecture enablers, such as software-defined networks (SDNs), to allow optimized slicing and routing for transport traffic; one of the key threats here is the manipulation of the routing table.

- Because all the traffic traverses the network using the SDN network, data exfiltration, data hoarding, and data sniffing are other threats related to the risks in the SDN components.

- The SDN controller itself could be a virtual instance built with the cloud-native approach of using an open source software stack that is advantageous to you because it provides flexibility in deployment types, such as VMs and containers, but it also brings in increased risk due to vulnerabilities originating from using unpatched open source software components.

- Unauthorized user access is another risk related to the SDN, orchestration, and automation components within the 5G network. This is applicable to all the layers, such as management, control, and data plane.

- NSaaS offers the option of the service management being handled by the enterprises or external third-party Network Slice Providers and opens your network to higher risk if not managed properly.

- The network slice could use a shared 5G infrastructure and network functions, or you could have dedicated 5GC NFs for separate slices, but it could share transport and 5G access nodes such as gNB, as shown in Figure 7-11. Improper isolation of network slice components in inter-slice and intra-slice deployments in these types of shared deployments will increase the risk of threats being migrated between the slices or within the slices.

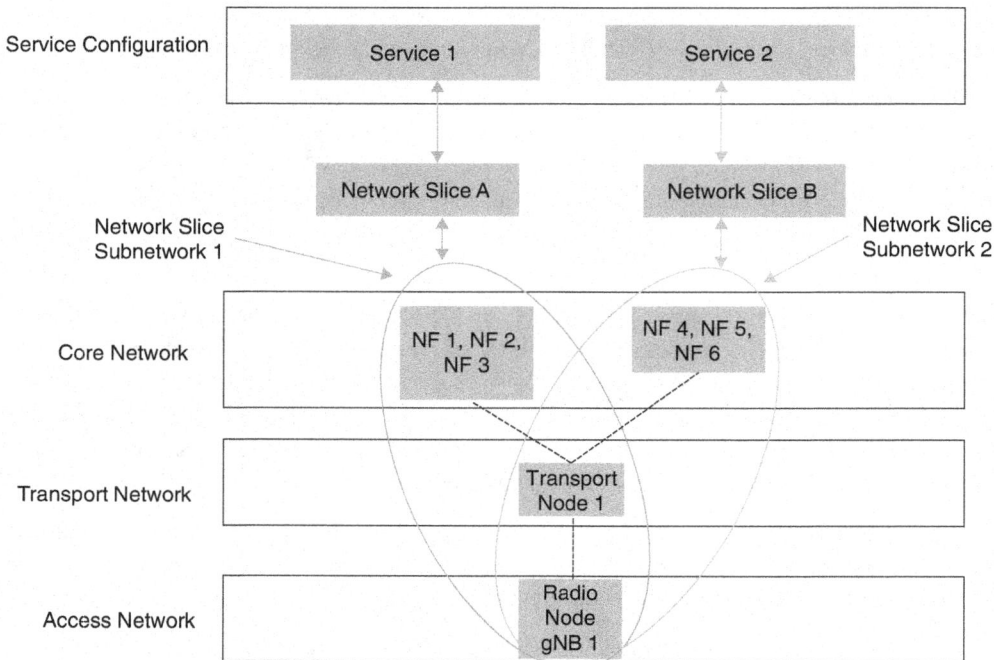

FIGURE 7-11   Network Slice Deployment Example with Shared Transport and Radio Access Node

As seen in Figure 7-11, two separate network slices are mapped to separate services. At the 5GC level, the same network functions can be allocated to dedicated NFs. The traffic between the access network (gNB) and the 5GC is catered to by a common transport network and, depending on the configuration, might also pass through a shared transport device such as an access and aggregation router as well as a shared access network component such as gNB.

Threats in the SDN Controller Layer

The SDN control plane handles all routing protocol control traffic. Protocols such as the Border Gateway Protocol (BGP) and the Open Shortest Path First (OSPF) Protocol send control packets between devices. These packets are destined to router addresses and are called control plane packets.

Any attacks on the control plane, such as TCP SYN flooding, Internet Control Message Protocol (ICMP) echo requests, and so on, will impact the performance of your SDN controller. Some of the key impacts are as follows:

- High CPU utilization or saturation of the route processor
- Memory and buffer resource exhaustion

- Impact on the services being catered to by the SDN, such as NSaaS, 5G slices catering to your customers (like IoT slices), and so on

- Crashing the SDN controller or destabilize routing and causing data disruption

- Dropping of legitimate packets due to CPU buffer and memory exhaustion

Figure 7-12 illustrates the threat surface due to an attacker exfiltrating data using the common SDN network used for different network slices. Depending on your network deployment, there might be multiple protocols being used by the SDN controller to communicate with the network elements and their Element Management System (EMS) or Network Management System (NMS), if any configuration changes are required. Although multiple protocols will be used, such as Locator/Identifier Separation Protocol (LISP), OpenFlow, and Simple Network Management Protocol (SNMP), the majority of the southbound communications from the SDN network toward the network elements would be based on different API architectures such as Representational State Transfer (REST), JavaScript Object Notation – Remote Procedure Call (JSON-RPC), Extensible Markup Language – Remote Procedure Call (XML-RPC), and Simple Object Access Protocol (SOAP). Although APIs have been deployed in enterprise and web servers for quite some time, they have not yet been widely deployed in the telecom part of the infrastructure, leading to insecure implementations of the APIs.

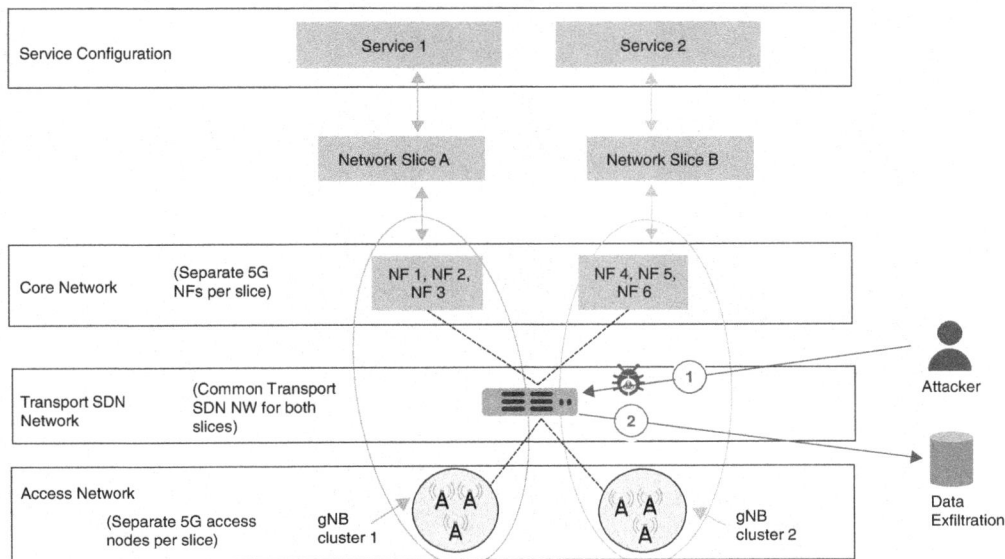

FIGURE 7-12 Sample Threat in Network Slice Deployment Example with Shared Transport and Radio Access Node

The attack surface shown in Figure 7-12 is explained for two sample attacks—one using an API and the other using topology poisoning—in the following list:

1. **API vulnerability on programmable transport device:** As shown in Figure 7-12, the attacker exploits the API interface in the programmable transport device for vulnerabilities such as broken authentication and broken access control. Many of the programmable transport devices support deployment of containers in the device hardware and allow communication using APIs. These third-party containers, which use the built-in general CPU, memory, and storage within the transport device, are susceptible to API vulnerabilities due to weak API implementation with little to no security controls. Some of the weak security mechanisms usually seen in API implementations allow the attacker to compromise passwords, keys, or session tokens, or to exploit other implementation flaws to assume other users' identities due to incorrect implementation of authentication and session management functions. Incorrect implementation of authentication and authorization could also allow attackers access to unauthorized functionality and/or data, such as accessing other users' accounts, viewing sensitive files, modifying other users' data, and changing access rights within the container deployed on the programmable transport device.

   One of the main foundations for the SDN architecture is the centralized SDN controller, which has visibility over the entire topology of the underlying network. One of the key attack surfaces on the SDN controller is the *topology poisoning attack*, as shown in Figure 7-13.

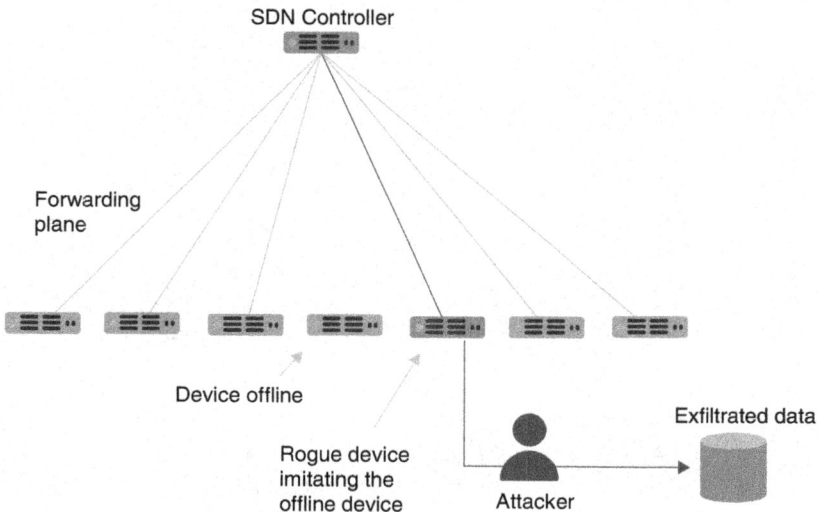

FIGURE 7-13   Depiction of the Topology Poisoning Attack

Topology poisoning attack vectors rely on the functionality of the SDN controller, which allows network devices such as switches to be unplugged and replaced with a new device or allows changing the location of the device. In this attack vector, the attacker probes the ports to check the state of the device and identify if any devices have been turned offline. Once the legitimate device is detected as being offline, the attacker uses the spoofed network identity of the legitimate device and shows it in a new location. The SDN controller sees this rogue device as a legitimate device and recomputes the flow table to forward the traffic to this rogue device. The rogue device will now act as the man in the middle (MitM). All the traffic flowing through this rogue device can be inspected by the attacker and can be used for malicious activities.

2. **Data exfiltration:** Both of the aforementioned attack surfaces—exploiting API vulnerabilities and the topology poisoning attack—will lead to data exfiltration or data hoarding, where sensitive information is exfiltrated out of the network. This data could be analyzed by the attacker to understand the identities of the network components and then launch an attack on the network using the most vulnerable components of the infrastructure.

DDoS Attacks on the SDN Control Plane

Distributed denial of service (DDoS) is one of the most popular attacks on the SDN infrastructure to disrupt the traffic and services of the targeted server or network component. Figure 7-14 illustrates DDoS attacks on the SDN control plane.

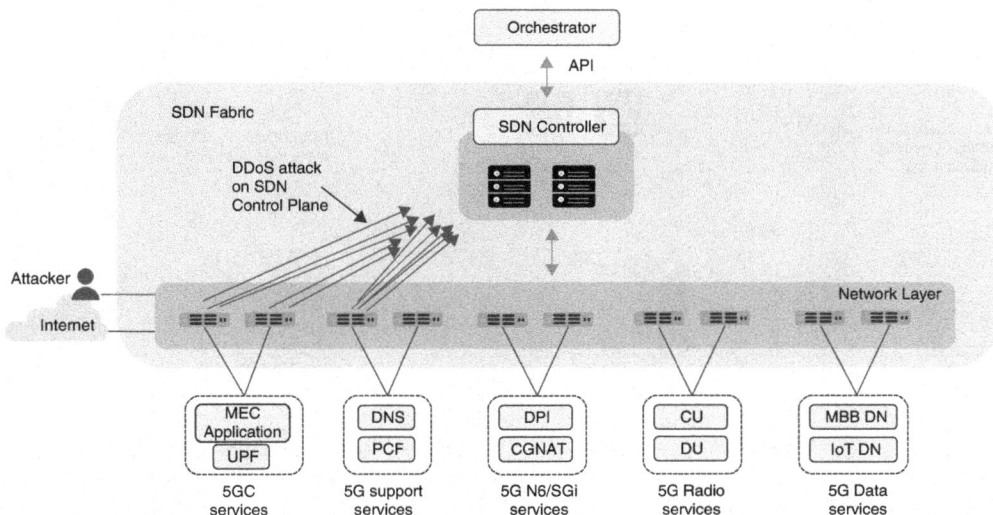

FIGURE 7-14   DDoS Attacks on the SDN Control Plane

Figure 7-14 illustrates an SDN deployment with the data plane providing Internet access to specific network functions and infrastructure components. In this illustration, the network layer allows Direct Internet Access (DIA) for 5G NFs like UPF, including dependent services like DNS. Due to the lack of security awareness with the programmers, the security implementation within the API calls is weak and has increased risk due to weak implementation of authentication and access controls.

The attackers exploit this weak API implementation to deploy malicious code that takes complete control of the switch. Apart from the traffic redirection, which is explained in the entry "API vulnerability on programmable transport device" in a list earlier in this section, the attacker could also launch a DDoS attack on the SDN controller. These DDoS attacks can saturate the CPU within the SDN controller with more requests than it can handle. This in turn cripples the SDN controller, leading to denial of service (DoS) for legitimate requests from other switches and transport devices within the infrastructure being catered to by the SDN controller.

Threats in the SDN Data Plane

The SDN data plane handles all the data traffic. The data plane for the SDN layer in 5G deployments enables data transfer to and from 5G NFs such as User Plane Function (UPF), Centralized Unit (CU), Distributed Unit (DU), and services supporting 5G NFs such as deep packet inspection (DPI) and the data network (DN) for each of the slices.

Figure 7-15 shows the threat surface due to DDoS attacks on the SDN data plane.

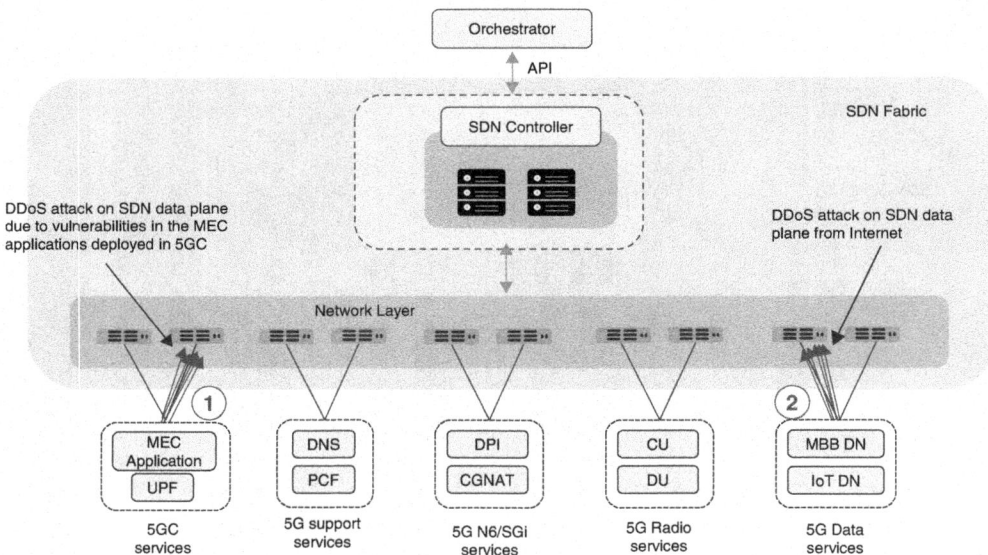

FIGURE 7-15   DDoS Attacks on the SDN Data Plane

Figure 7-15 illustrates two separate attacks on the data plane of the SDN controller, which caters for the data forwarding for the underlying 5G NFs and infrastructure components. Descriptions of the two separate attacks follow:

■ **DDoS attacks originating from internal network:** The SDN data plane within the transport SDN network allows Direct Internet Access (DIA) for User Plane Functions (UPFs) for fulfilling ultra-reliable low-latency communication (URLLC) use cases. Figure 7-16 illustrates the attack on the SDN data plane by devices in the internal network.

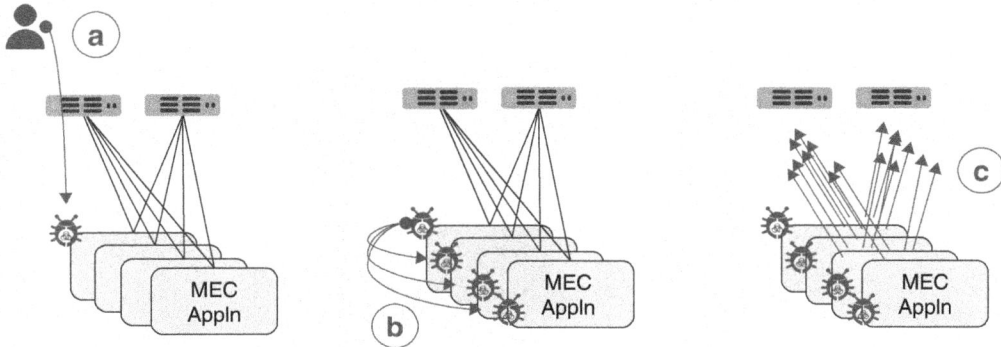

FIGURE 7-16 DDoS Attacks Due to Malicious Code Injection in MEC Applications

The steps in the attack illustrated in Figure 7-16 are as follows:

a. The attacker exploits the vulnerabilities in the deployed 5GC NFs (non-3GPP or 3GPP). The attacker can exploit the vulnerabilities in the API implementation or can exploit the vulnerabilities in the open source software stack being used to build 5G NFs.

b. The malware deployed by attackers are of multiple variants. Some malware methods could be fileless. In fileless malware-based attacks, the malware is loaded into memory and then executed. This method could also open a backdoor to the attackers or command and control (C2C) servers, which can now control all the NFs/MEC applications.

c. The C2C servers can now launch a DDoS attack toward the SDN data plane that can saturate the CPU or exhaust the device's resources such as memory and storage. As the DDoS attack is initiated from the NFs deployed within the infrastructure, it can be considered a DDoS attack originating from your internal network.

- **DDoS attacks originating from external network:** The SDN data plane within the transport SDN network enables the connectivity of various 5G NFs with different data networks (DNs) based on the slices, such as IoT and so on. One of the key risks from the networks serving the IoT devices are the DDoS attacks from botnets that take control of the IoT devices due to unpatchable vulnerable firmware, nonexistent or inadequate passwords, or unsecure interfaces on configuration management (CM), performance management (PM), or fault management (FM) communications. These DDoS attacks could be directed at the SDN data plane that provides connectivity between the DN and the NFs.

Threats in Orchestration Layer

Figure 7-17 illustrates the criticality of risks related to the orchestration layer. You will see these risks explained in many different use cases throughout the chapter.

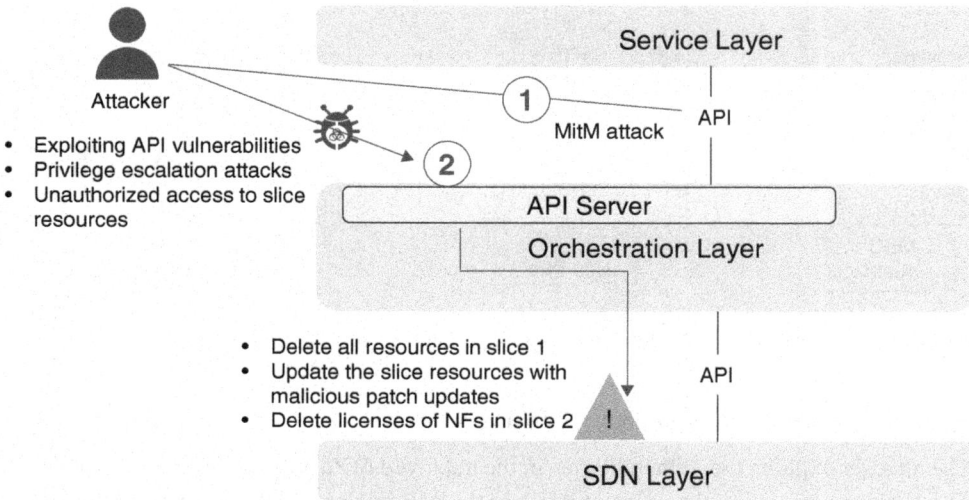

FIGURE 7-17   Threat Surfaces in Orchestration Layer

Figure 7-17 illustrates some of the key risk vectors related to the orchestration layer in 5G network slicing deployments, as explained in the following list:

1. The attacker could perform a man in the middle (MitM) attack in the API calls in the north-bound interfaces (NBIs) implemented from the orchestration layer toward the service layer, where the attacker deliberately indicates that the application in the service layer only supports a weaker encryption protocol. The communication packets sent using the API are then encrypted by a weaker protocol, and the attacker can now easily decrypt the encrypted packets and exfiltrate the information for further analysis and then carry out attacks against the critical infrastructure.

2. Another risk shown in Figure 7-17 is the attacker gaining unauthorized access to the orchestration layer due to weak API security or by exploiting the weak security controls implemented in the API server within the orchestration layer. Some of the key risks related to this attack surface are unauthorized access gained by exploiting the vulnerabilities in the web server being used for API calls and the weak implementation of the API itself, as explained in the following list:

   ■ Once the attacker gains unauthorized access to the orchestration layer, a privilege escalation attack could be performed whereby the attacker has admin rights and can perform actions like deleting the network functions related to a critical slice.

- Another risk vector could be the attacker choosing to force a patch update of the network functions in a specific slice. The patch update can be forced with a malicious patch that inserts malicious code into the network function, and the attacker could then start exfiltrating sensitive information from the network to an external server.

- The attacker could also perform actions such as deploying license exhaustion codes in the infrastructure, wherein the entire license file is erased in perpetuity, which means that each time any software is deployed on a specific host, the malicious code deletes the license files. These kinds of attacks are hard to detect, and the user of the host might find it difficult to understand that malicious code is performing such actions from the orchestration layer.

Insufficient Slice-Level Isolation

When you deploy slices to cater for the 5G use cases, there might be many devices with different capabilities within the individual slices. Some of the devices within the slice served by 3GPP access networks will perform primary authentication between the 5G device and the 5GC for PLMN access, authorization, and authentication. Other devices that are served by non-3GPP access networks will be able to access your 5GC, in which case a new 5G function called the non-3GPP interworking function with IPsec capabilities needs to be deployed. This unified authentication framework in 5G enables access network–agnostic authentication for the devices being served by your 5G network but does not entail slice-level authentication.

Figure 7-18 shows the threat surface due to insufficient isolation at the network slice level.

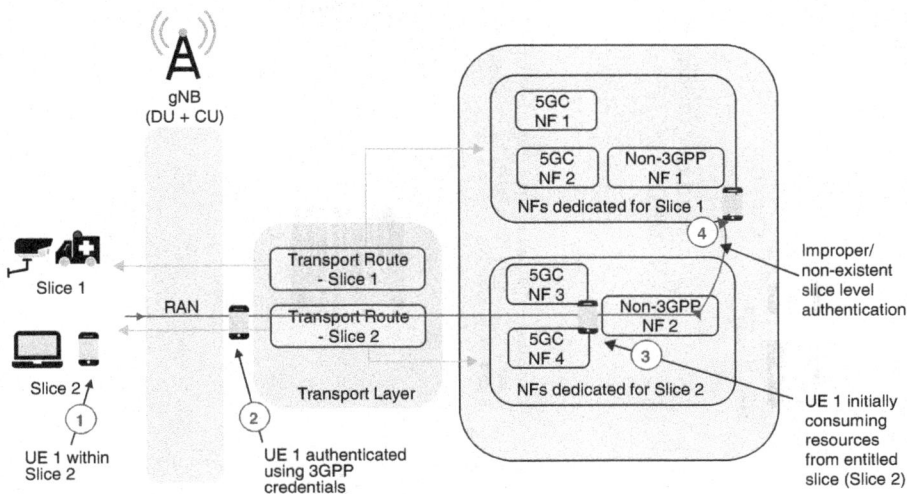

FIGURE 7-18   Threat Surfaces Due to Insufficient Isolation

To provide better clarity on the attack vector in actual deployment, Figure 7-19 illustrates the physical diagram of the functional diagram illustrated in Figure 7-18.

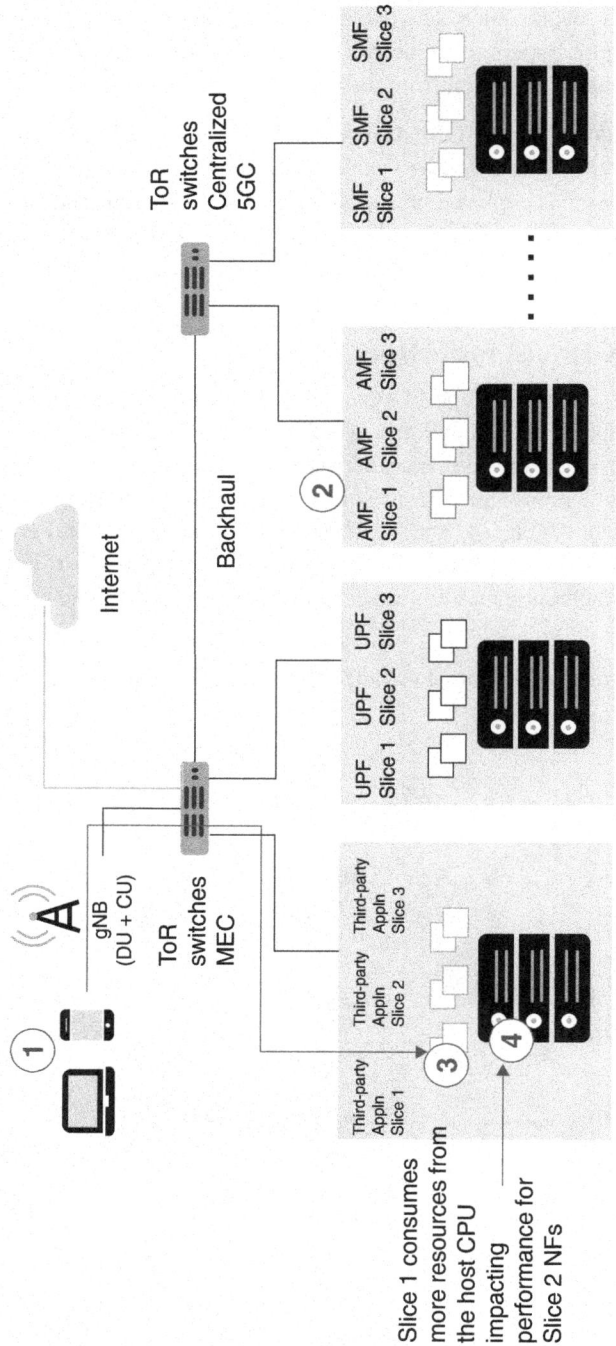

FIGURE 7-19 Impact of Insufficient Isolation

As shown in Figure 7-19, in real scenarios, similar functions are grouped together in the same NFVi. Due to inadequate isolation between slices, unauthorized devices or user equipment (UE) will be able to access other slices that those UEs are not entitled to access. This will lead to consumption of resources of the network slice planned to cater for another use case, thereby causing denial of service (DoS) to the devices that had been originally entitled to access the impacted network slice. Figure 7-19 is further explained in following list:

1. User Equipment 1 (UE 1) is a device within network slice 2, which is authenticated using the 5G unified authentication framework defined by 3GPP, as shown in Figure 7-20.

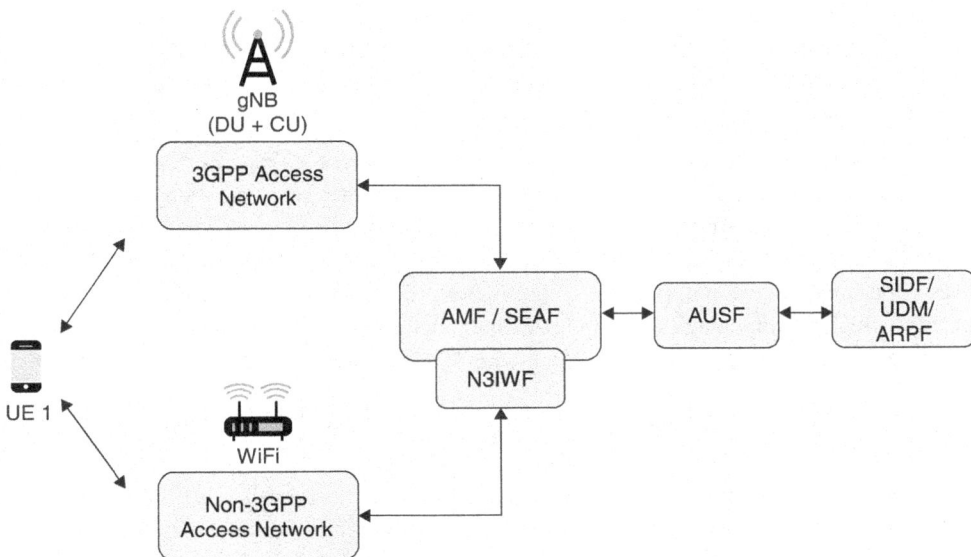

FIGURE 7-20   3GPP and Non-3GPP Network Authentication for UEs

The network components shown in Figure 7-20 and the network authentication options are explained in the following list:

- **UE:** User Equipment

- **3GPP Access Network:** Access network with cellular technologies specified by 3GPP (for example, 4G and 5G)

- **Non-3GPP Access Network:** Access network with wireless technologies not specified by 3GPP (for example, Wi-Fi)

- **AMF:** Access and Mobility Function

- **SEAF:** Security Anchor Function

- **N3IWF:** Non-3GPP Interworking Function, which is responsible for interworking between untrusted non-3GPP networks and the 5G Core

- **AUSF:** Authentication Server Function, which performs authentication of UE in the home network

- **SIDF:** Subscriber Identity De-concealing Function, which is responsible for decrypting Subscriber Concealed Identity (SUCI) to reveal the Subscriber Permanent Identifier (SUPI)

- **UDM:** Unified Data Management, which performs user verification handling

- **ARPF:** Authentication Credential Repository and Processing Function, which is responsible for generating 5G home environment authentication vectors (5GHE AV) based on the subscriber's shared secret key.

2. As shown in Figure 7-20, when UE 1 is served by the 3GPP access network, the UE is authenticated using a protocol such as Extensible Authentication Protocol – Transport Layer Security (EAP-TLS) by encapsulating it in Non-Access Stratum (NAS) messages between UE and AUSF for verifying the pre-shared key (PSK) or verifying each other's certificate, with SEAF functioning as a transparent EAP authenticator. When UE 1 is being served by the non-3GPP access network, then N3IWF will provide secure interworking between the UE and the 5GC. UE 1 performs registration using the IKEv2 SA establishment procedures; once registered, UE 1 will then carry on with the NAS signaling with 5GC network.

3. The device UE 1 would consume the User Plane Function (UPF) resources from the allocated slice, which is preconfigured with the required QoS.

4. The network deployed does not use any slice-specific isolation mechanisms that allow devices from either slice to consume the services of other slices, as shown in Figure 7-19. The services could be edge applications and any function such as UPF. Due to improper slice-level isolation and the nonexistence of slice-level authentication, UE 1, which was originally dimensioned for slice 2, now consumes the resources allocated to slice 1. This unauthorized access of slice 1 by UE 1 will cause a DoS for legitimate user equipment trying to access slice 1 due to resource exhaustion.

In deployments where slice-level authentication and isolation are not provided between critical infrastructure slices and consumer IoT slices, the impact on the network could be severe, leading to revenue-impacting outages and service impacts.

Threats in NSaaS Deployments

5G network slicing allows allocating dedicated or shared network resources to fulfil requirements of a particular application or use case. You might also offer customized services to the service consumers based on the network slice as a service (NSaaS) model, as described in 3GPP TS 28.530. The customized services offered are characterized by the network slice's properties, such as radio access technology

(RAT), bandwidth, latency, reliability, guaranteed/non-guaranteed QoS, and so on. However, security-related properties are not addressed by TS 28.530.

You could use the NSaaS model to monetize your network slices by providing the aforementioned customized services to either network slice providers (NSPs) or to end enterprises directly. Service providers could also provide managed services such as configuration and maintenance of slices for NSPs or enterprise customers.

Figure 7-21 shows the threat surface due to improper security controls for NSaaS services.

While providing the option for the NSPs and enterprises to manage their own slices, service providers will usually provide a multitenant management platform that can be accessed by the users (NSPs and consumers of NSPs). The configuration changes can include modification of QoS configurations, RAT selection, and other customizations for the slices. Due to improper access controls being implemented by the management platform, it introduces increased risk of users from one of the enterprise slices performing an unauthorized modification in another slice. As depicted in Figure 7-21, the user or an attacker from an enterprise slice first performs a basic access control and accesses the enterprise slice. The attacker then goes ahead to access the management service layer of a critical infrastructure slice such as the defense network slice and makes changes to the slice, such as deleting the slice or changing the QoS, which might lead to service degradation or cause a DoS for the users and devices of the critical infrastructure slice.

APIs are used for the exchange of information between external entities such as enterprises, network slice providers, third-party application developers, and the service providers. APIs are very efficient mechanisms if implemented in the right way following the development and deployment best practices; however, a majority of the API implementations are done improperly and increase the risk to your infrastructure.

Figure 7-22 shows the threat surface due to improper API implementation with external applications and web servers.

As shown in Figure 7-22, API calls will be used by external users, devices, and servers to communicate with the different entities of your network exposed to the Internet. Here are some examples:

1. Service layer management functions can be accessed by an authorized external entity to perform configuration changes as required by the slice provider or slice owner. The key threat vectors in the service layer are data leakage and access to the management of the unauthorized slice due to insufficient authentication and authorization within the API implementation.

2. If the service layer is provided by an external management vendor, then northbound API integration of the external management vendor and the orchestration layer needs to be implemented, as shown in Figure 7-23.

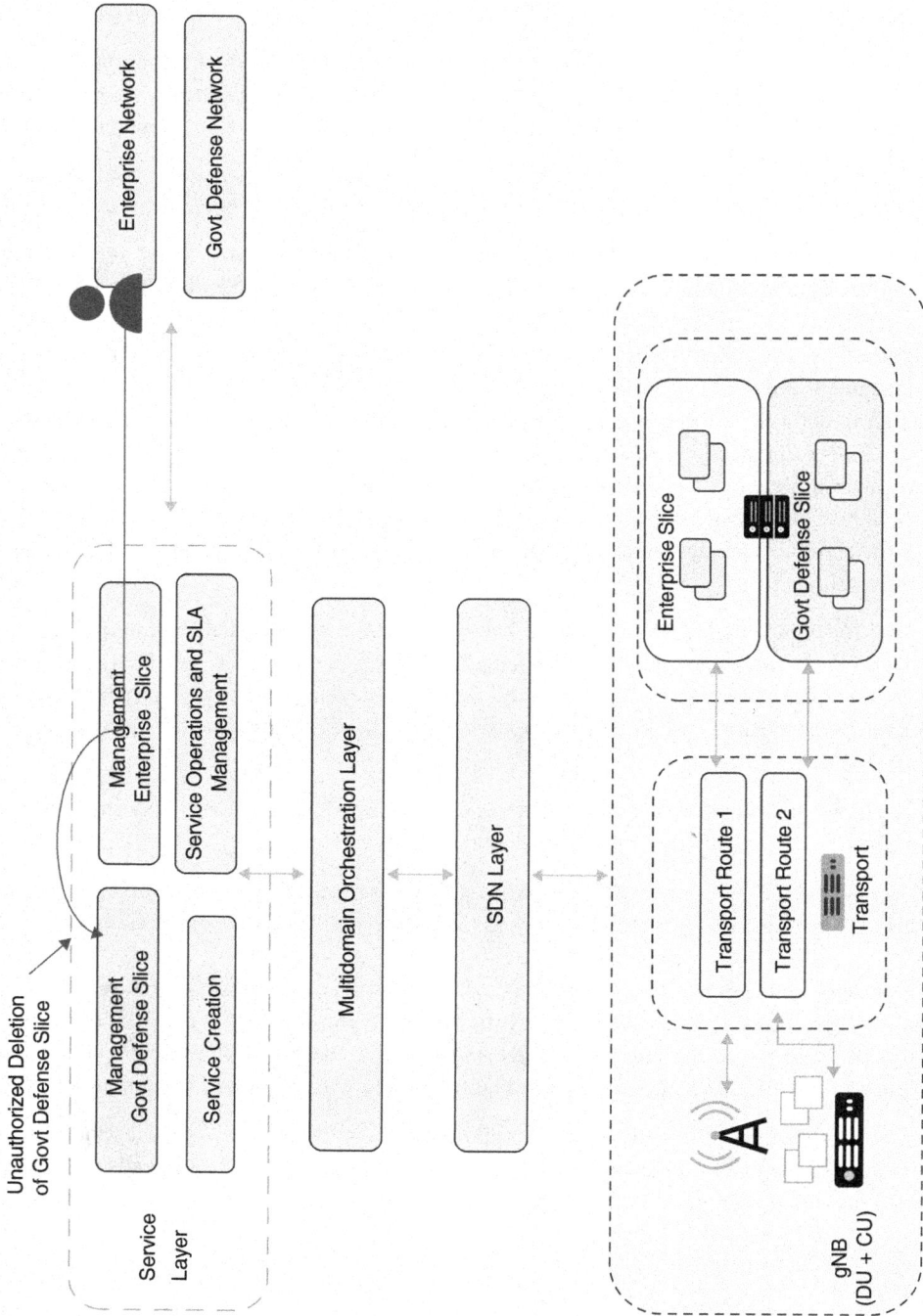

FIGURE 7-21   Unauthorized Slice Access in NSaaS Deployments

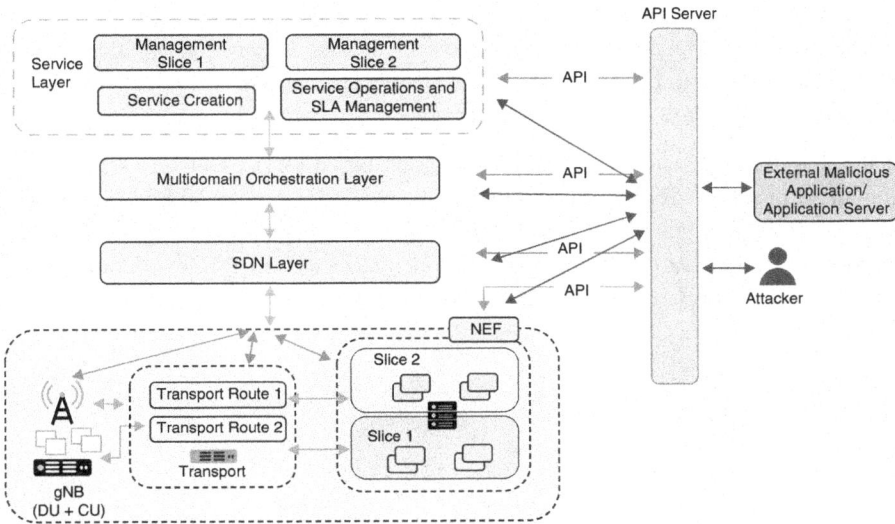

FIGURE 7-22 Exploiting API Vulnerabilities in NSaaS Deployments

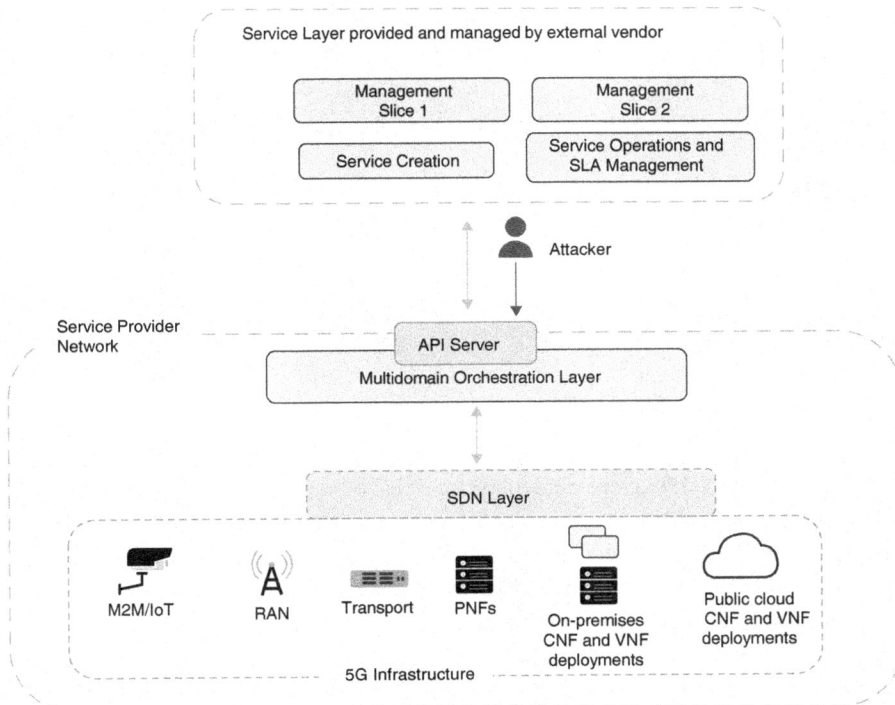

FIGURE 7-23 Exploiting API Vulnerabilities in Multidomain Orchestrator

Improper implementation of the API in this layer could lead to unauthorized modifications in the orchestration components. An attacker could exploit the broken authentication vulnerability within the API implementation to perform unauthorized actions such as deleting NFs, thus causing service interruptions, DoS, or network-level outages.

3. The SDN and automation layer can be integrated with cloud-based services such as security functions using an API interface. Northbound APIs can be used to integrate the SDN controllers to cloud-based automation stacks external to the network, such as Puppet, Ansible, Chef, and so on. Improper security controls during API implementation at the SDN layer could lead to successful DDoS attacks. Figure 7-24 shows the threat surface due to improper implementation of internal API calls.

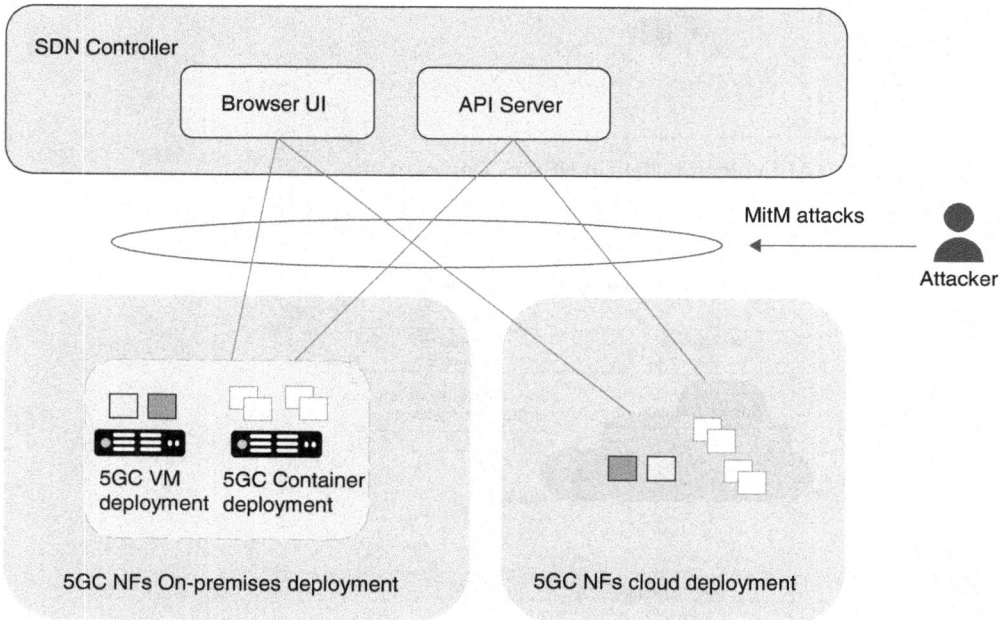

FIGURE 7-24 Exploiting API Vulnerabilities in SDN

As shown in Figure 7-24, the SDN layer you deploy for 5G should cater for the on-premises network functions and dependent infrastructure components and the 5G 3GPP and non-3GPP components deployed on the cloud. Users should be able to access the SDN controller using the user interface (UI) or by utilizing API calls. Many APIs lack encrypted communications between the API client and API server in SDN deployments. Attackers exploit such vulnerabilities through man-in-the-middle (MitM) attacks. Attackers can also intercept unencrypted or poorly protected API transactions between the API client and API server in 5G SDN deployment to steal sensitive information between the 5GC NFs and SDN or to alter data being transferred between the 5GC NFs and the respective SDN layers.

4. Network Exposure Function (NEF) is used to expose the 5GC NFs to external Application Functions (AFs) using an API. This implementation helps the third-party application ecosystem to evolve and innovate various applications that can be built for 5G use cases. Improper implementation of the API could lead to attackers exploiting the API vulnerabilities, such as passing on more information in the error codes as well as improper access and authentication controls in APIs. These could provide more information about the underlying 5GC NFs and could lead to data exfiltration and data hoarding by the attackers.

Mitigation of Threats

Figure 7-25 illustrates the key components and features of the end-to-end (E2E) secured architecture based on true zero-trust principles.

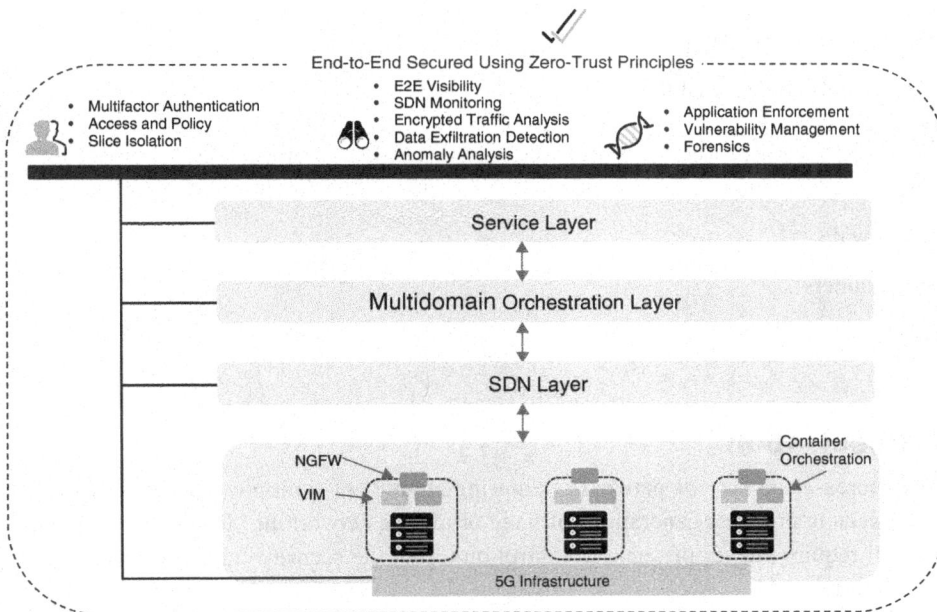

FIGURE 7-25   Key Security Controls for E2E Secure Network Slice Deployments

The key components of the end-to-end secured architecture for SDN, orchestration, and automation are explained further in the sections that follow.

Trusted Components

SDN, orchestration, and automation should be secured via a defense-in-depth approach and should have undergone several security verification cycles by following the secure development lifecycle

(SDL) process. This is the strong foundation of any zero-trust principle. The SDL process will ensure an increase in product resiliency and trustworthiness. The combination of tools and processes introduced during the development lifecycle promotes defense-in-depth, provides a holistic approach to product resiliency, and establishes a culture of security awareness. This secure product development and deployment practice should include inherent design and development practices, testing the implementation, and creating a set of recommendations for deploying with maximum security. The software development process should be ISO 27001 certified.

Your chosen SDN, orchestration, and automation vendors should harden the software and hardware with attention to the following:

- Minimum attack surface (no unnecessary open ports)
- On-demand password reset
- Randomly generated passwords
- Transparent updates of security patches
- SELinux (host level) always enabled
- SSL support
- TLS support
- Host-based firewalls
- Immutable containers
- Non-root containers

Securing Orchestration

5G will massively increase the size of networks, requiring support of multiple domains, and bring levels of change not seen in previous generations of fixed or mobile networking. To secure the orchestration layer, you will require a layer of security control on top of the security capabilities offered by the orchestration solution itself.

Your chosen multidomain orchestration vendor should provide you with the following security capabilities:

- Comprehensive lifecycle service automation to enable you to design and deliver high-quality services faster and more easily across different domains, such as transport, network data center, and mobile core, all of which could be components of a 5G network slice.
- The ability to provision assurance systems, deploy virtual probes, and so on to ensure that service monitoring is in place.

- The ability to embrace and collect massive amounts of data from different data sources and protocols, including leveraging of streaming telemetry for real-time visibility of the network infrastructure.

- The ability to discover the network status for each service and render it logically as well as geographically for the operations team.

- Logs such as developer logs, audit trail logs, web interface logs, device communication logs, and XPath logs for any forensics or auditing purposes.

- If segment routing is used, the orchestration solution should provide views for segment routing overlay, segment routing policies, and the underlay network, as well as deliver aids and reports to you, which are key for avoiding issues or for expedited remediation when needed.

- For secure CI/CD, the orchestration solution you choose should provide the following functionalities:

  - All of your source code must be stored in version control.

  - Automatic builds must be triggered on each commit.

  - Builds must be stored in a special artifactory.

  - An automated deployment from build to production should be provided.

Apart from the preceding security capabilities supported by the orchestration solution, you will require the key security controls illustrated in Figure 7-26 and explained in the list that follows to protect your orchestration layer:

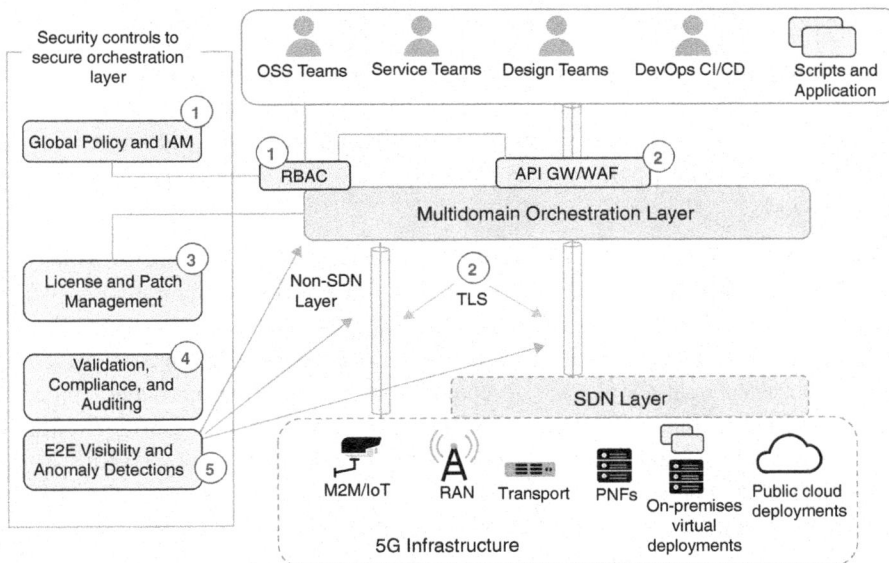

FIGURE 7-26   Key Security Controls for Secure Orchestration

1. **Granular access control:** The orchestration system's role-based access control (RBAC) lists providing different access privileges to different users should be integrated with your external IAM solution to provide granular access control based on your global policy. This will ensure that you have a consistent policy applied across your entire network and allow you visibility on the users accessing the orchestration solution.

2. **Securing interfaces:** The orchestration layer will consist of northbound APIs and software interfaces from programmatic or Remote Procedure Call–based protocols (such as NETCONF/ RESTCONF) to language bindings such as Erlang, Java, Python, and C, to human-to-machine interfaces, such as a web-user interfaces (UIs), and a set of command-line interfaces (CLIs) that allow straightforward integration into existing business systems and operational tool chains such DevOps Continuous Integration and Deployment (CI/CD) pipeline. To mitigate any MitM threats, these APIs are recommended to be encrypted using TLS—preferably TLS1.3, as it provides better security over TLS 1.2. An API GW or web application firewall (WAF) should be used to mitigate unauthorized use of APIs. If there are any untrusted network interactions on the northbound interface (NBI), such as with external third-party developers, then it should be secured using an API GW or WAF.

3. **License and patch management:** The multidomain orchestration solution you choose should have web interface, CLI, database, and API implementations that are version aware, so the correct license for the specific version of the 5G component under consideration for orchestration and automation can be checked for any vulnerabilities and patches applied before configuration changes or fresh deployment. This is very critical for CI/CD processes as well. This would require integration of the orchestration layer to your license and patch management solution. You should leverage advanced machine learning and artificial intelligence techniques to identify any unpatched software in the slice components and get the relevant patches for necessary updates.

4. **Validation, compliance, and auditing:** The orchestration solution you choose should have built-in support for compliance reporting that will check that all devices and services are configured as expected. It should also show details for any discrepancies, such as a misconfigured VPN on an interface. The report also includes details about all changes that have been performed in the network.

The orchestration platform should also be able to provision services like VPNs, ACLs, BGP peers, and so on. It should perform network audits to detect if any device configuration has changed with respect to the desired service configuration. The difference can be analyzed by the validation, compliance, and audition solution, and the service can be redeployed if needed. The orchestration platform should not just fire off commands to the network but rather confirm that all changes in the transaction are deployed correctly at the device level by getting a confirmation from the validation, compliance, and audition solution. If at any point of the series, a device or a CNF attribute cannot be changed, the entire transaction should be automatically rolled back. This ensures that there is always a consistent state in your network. The orchestration should follow an ETSI-compliant implementation of NFV orchestration and enable easy integration by offering a flexible interface into which a third-party VNFM or container orchestration can be included.

5. **End-to-end (E2E) visibility and anomaly detection:** The E2E visibility and anomaly detection engine should be integrated with the orchestration layer and provide you with E2E visibility, including the transactions in the encrypted layer. This will ensure that you have visibility of any malicious traffic on the encrypted layer. Some solutions available today provide you this level of visibility without having to decrypt the encrypted layer. This is done by collecting extra bytes from the packet header and understanding the behavior of the client/server handshakes and the selection of the crypto suites during the handshakes. Once such malicious packets are found, the hosts can be automatically put into an isolated quarantine segment of the network to prevent lateral movement of malware. Such implementations in your network would secure the orchestration layer and help you detect and mitigate any malicious behavior.

Securing the Software-Defined Network (SDN)

One of the key challenges you face today are the multiple layers of SDN being planned on for 5G deployments. You might already have a software-only network overlay approach based on host virtualization that offers limited visibility, performance, and scale as well as requires separate management of underlay and overlay network devices and security policies. This requires you to have multiple deployments of SDNs, making the management and operation of the SDN layer a hindrance to the deployment of 5G instead of simplifying it. The traditional security policy model for SDN deployments for 4G is based on a static network topology (bound to network connections, VLANs, network interfaces, IP addressing, and so on) and manual service chaining. This model, if followed for 5G deployments, will require policy configuration across multiple security devices (web security, firewalls, IPSs, and IDSs), slows application deployment, and is hard to scale for 5G because applications in 5G are dynamically created, moved, and decommissioned.

You should instead take a deployment approach that addresses the security needs of the 5G CNF, VNF, and hardware-based NFs by using an application-centric, unified, and automated approach to security policies in the on-premises data center and cloud infrastructure that is decoupled from the underlying network topology. The chosen SDN deployment method should also support 5GC NF mobility, offer real-time compliance lifecycle management of the CNFs, and reduce the risk of security breaches by having a robust API security approach. You should follow the concept of application policy, which enables a new open security policy framework that expresses policies using the language of the application rather than network. Policies are defined based on a language that is natural to application owners and not in terms of classical networking constructs like VLANs and IP and MAC addresses. This group policy approach decouples security policy and segmentation from the underlying network topology. An example of such an open policy framework is Open Policy Agent (OPA), which is part of the Cloud-Native Computing Foundation (CNCF) project. OPA is a vendor- and domain-agnostic general-purpose policy engine that helps unify policy enforcement across a wide range of technologies. It gives you the capability to offload and decouple policy decision-making from policy enforcement so that the people responsible for policy can read, write, analyze, version, distribute, and in general manage policy separate from the service itself, without sacrificing availability or performance. It provides a high-level declarative language that lets you specify policy as code and simple APIs to offload policy decision-making from your software. You can include OPA as a library or a daemon to enforce policies in microservices, Kubernetes, CI/CD pipelines, API gateways, and more. Using such an open policy framework, the SDN vendor you choose will be able to apply segmentation on the application, network, and user levels. The SDN vendor should also support all variants of 5G deployments—monoliths, VMs, containers, and cloud-deployed 5G CNFs and applications—along with different 5G services such as 5GC services, 5G support services, 5G N6 services, 5G radio services, and 5G data networks, as shown in Figure 7-27.

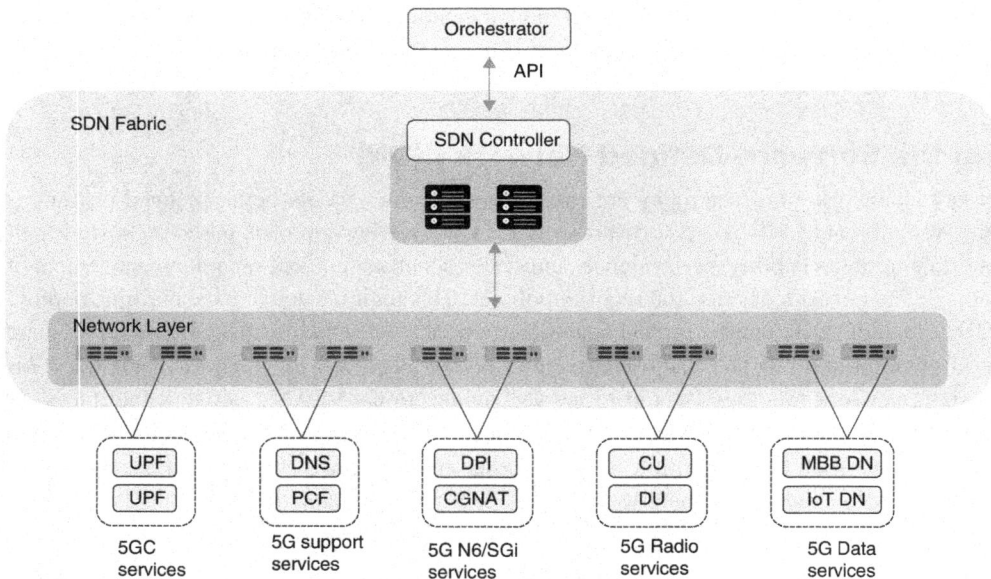

FIGURE 7-27   Multiple 5G Services Chained with the SDN Network Layer

Securing the Control Plane

Protecting the control plane is important for the SDN solution, as any attack at this layer can cripple the whole SDN network, as discussed previously in the "Real Scenario Case Study: Threats in the 5G Network Slice, SDN, and Orchestration Deployments and Their Mitigation" section.

Here are some of the common attack vectors for network devices:

- Denial-of-service (DoS) attacks where excessive traffic is directed at the device interfaces

- Exploiting API vulnerabilities to create disruption, such as slowing down the network

- Rerouting packets to malicious servers and creating malicious VNF/CNF for data exfiltration and data hoarding

To mitigate these risks, it's important to have a built-in control plane protection layer. This can be achieved by having multiple layers of built-in SDN attributes, allowing for the specification of parameters for each protocol that can reach the control processor to be rate-limited using a policy engine within the SDN control plane, called Control Plane Policing (CoPP), as shown in Figure 7-28. The policy is applied to all traffic destined to any of the IP addresses of the router or Layer 3 switch. This policy should be deployed on a sub-interface level, thereby providing you with granular protection. CoPP protects the control plane, which ensures network stability, reachability, and packet delivery.

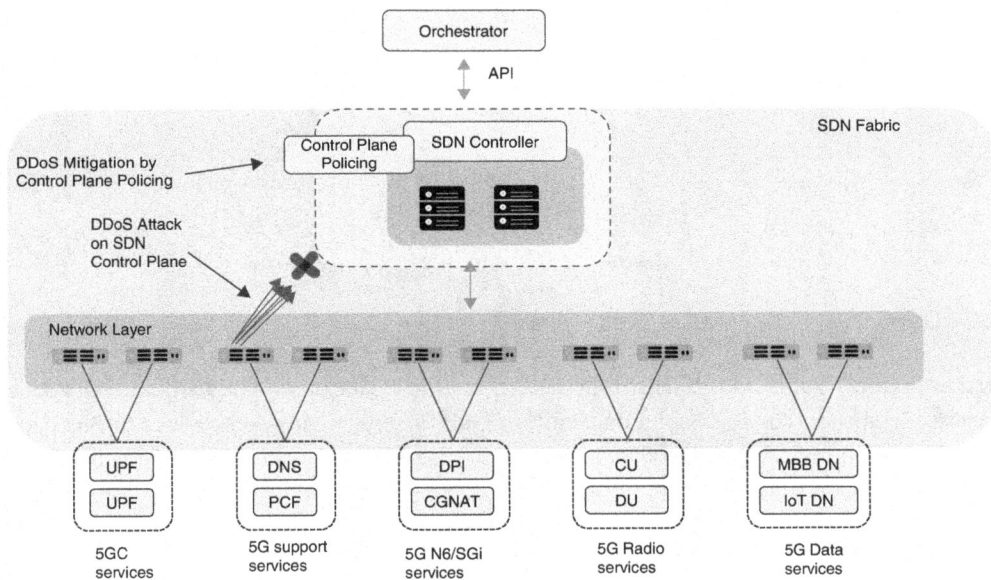

FIGURE 7-28   Using CoPP to Secure the Control Plane

As shown in Figure 7-28, CoPP functionalities would protect the SDN control plane layer from DDoS attacks using features such as rate limiting.

Securing the Data Plane

The data plane of the SDN layer is susceptible to attacks such as volumetric DDoS attacks. Protection of the SDN data plane should be provided by the Data Plane Policing (DPP) built in to the SDN solution provided by your chosen SDN vendor. DPP can be used to manage bandwidth consumption on the SDN data plane and can be applied to egress traffic, ingress traffic, or both. DPP monitors the data rates for a particular interface. When the data rate exceeds user-configured values, the SDN fabric can either drop the packets, as shown in Figure 7-29, or mark QoS fields in them.

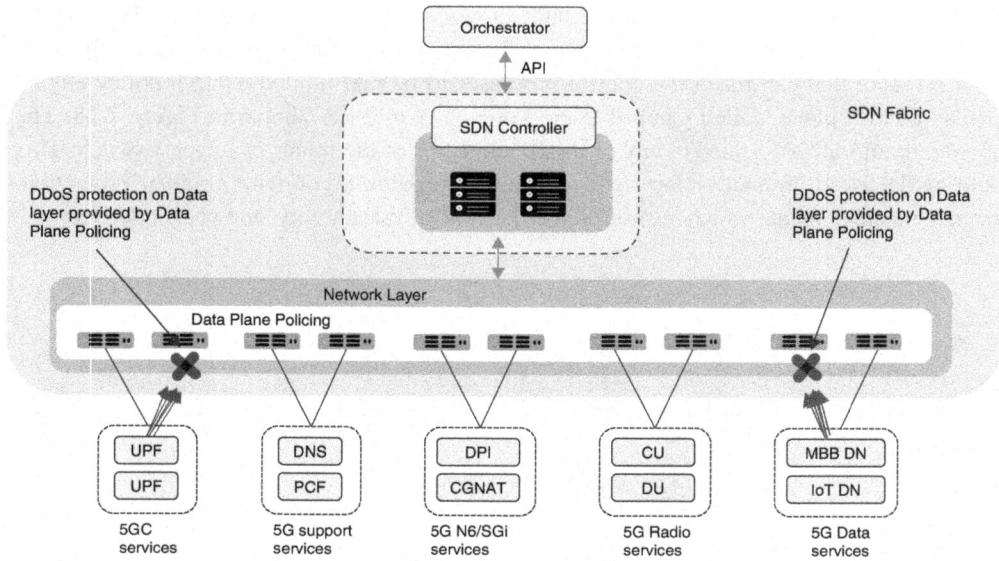

FIGURE 7-29   Using DPP to Secure the Data Plane

As illustrated in Figure 7-29, DDoS attacks from the UPF and IoT data network toward the SDN data plane are guarded against by the SDN solution's built-in DPP functionality, which checks for the data rate in the ingress and egress of the SDN data plane and drops packets once the threshold is reached/increased or a specific QoS is applied to it.

Securing the Management Plane

Securing the management plane should be done using RBAC integration with an MFA mechanism, thereby providing robust access control based on zero-trust principles. This method should be followed for all the components of the 5G deployment, including the network slice management, SDN, automation, and orchestration components.

The selected SDN and orchestration solutions for 5G deployments should have integration with identity providers (IdPs) to enable a use case such as single sign-on (SSO) authentication. Granular access control should be provided with privileges managed per resource using RBAC. The roles implemented should have privileges mapped to them, as explained in the following list:

- **Account administrator:** Full control and management capabilities for the accounts and devices under management

- **Read-only:** Read-only visibility into resources under management

- **Network device technician:** Administrative device actions

- **Network device administrator:** Administrative device actions including creation and deletion of devices

- **Server administrator:** Server lifecycle and policy-based management

- **User access administrator:** User, group, and IdP configuration

Figure 7-30 shows the out-of-band (OOB) control plane, which separates management data from IT production and application data.

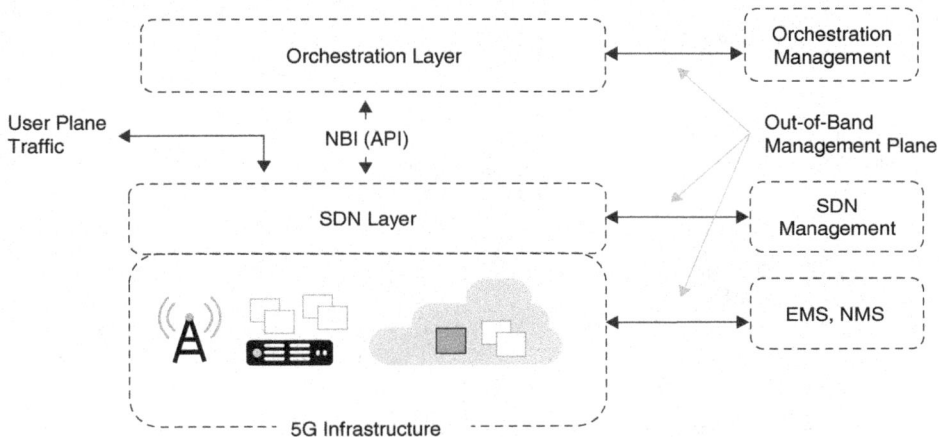

FIGURE 7-30 Using OOB Management Plane to Secure Management

As illustrated in Figure 7-30, the use of an out-of-band architecture means that users are not affected if devices are unable to communicate with the SDN controller due to Internet or other service disruptions. Users can still access local management and production networks, and all policies and settings continue to be enforced. In addition, local user authentication is unaffected, and local configuration tools remain available.

You should follow strict policies on data collection. For all data collected, the following additional security practices should be implemented:

- One customer's data should be kept separate from other customers' data through virtual data segregation. Data requests by SDN management services should return data specific to the customer account only, and per-customer encryption keys should be used for access.

- Long-term persistent data should be encrypted at rest.

- You should also have robust audit logging on the database to understand "who" did "what" and "when."

- Granular access control policies should be implemented and third-party access to data should not be permitted.

Mitigating Data Exfiltration

In 5G deployments, the SDN layer will cater for multiple modes of NF, such as on-premises. Figure 7-31 illustrates one of the methods to mitigate data exfiltration at the SDN layer, and the list that follows explains the data exfiltration mitigation technique using anomaly detection.

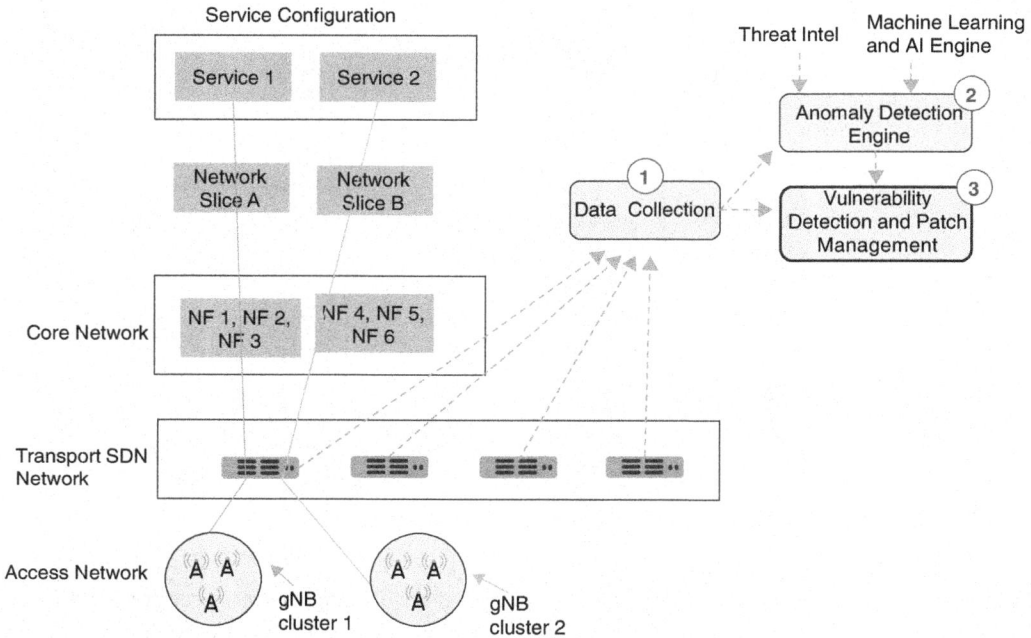

FIGURE 7-31   End-to-End Visibility and Anomaly Detection Solution

1. **Data collection:** Data collection should leverage telemetry from devices deployed in the 5G SDN layer such as NetFlow, Internet Protocol Flow Information Export (IPFIX), and flow data from the existing infrastructure such as routers, switches, firewalls, endpoints, and other network infrastructure devices. The data collector should also receive and collect telemetry from proxy data sources, which can be analyzed by the cloud-based, multilayered machine learning engine for deep visibility into both network and web traffic.

2. **Anomaly detection:** The anomaly detection engine (ADE) for the SDN layer closely monitors the activity of every device and network function on the SDN network. The ADE then creates a baseline of normal behavior across the entire network. In addition, the chosen solution for anomaly detection should include integration with threat intel coupled with the machine learning engine, which has an understanding of known bad behavior. It should apply heuristics that look at various types of traffic behavior, such as scanning, beaconing host, suspected data exfiltration, suspect data hoarding, and so on. These security events should feed into high-level

logical alarm categories. The system should then be able to determine what kind of attack might be in play and also tie it to a specific network device within the SDN network layer. These measures would help you identify any attacks or data loss such as any data being exfiltrated out of your infrastructure. Once the risk has been identified, you can take corrective measures to mitigate the risk or let the ADE solution create automated mitigation steps for you. For example, if you identify that the peering switch or router is causing the data exfiltration due to backdoor configurations done by a malicious user or due to malicious code during upgrade, the impacted switch or router can be moved to a segmented network with restricted service. Such segmented networks are also called a quarantine segment or quarantine service network. Moving the impacted devices to quarantine layer will restrict the network access of the devices, thereby mitigating any major impact to the network.

3. **Vulnerability detection and patch management:** The data collected can also be used to identify any vulnerabilities within the devices in the SDN layer by co-relating the version of the used software with the CVE IDs of the device vendor. Any device found with vulnerabilities should be indicated, and the exact location of the impacted device should be marked in the topology or in any identifiable format for the security team to act on it. This will help you to establish controls and processes to help identify vulnerabilities within the 5G SDN infrastructure and SDN-dependent components that could be exploited by attackers to gain unauthorized access, disrupt business operations, or steal or leak sensitive data. The patch management system should cater for patching any vulnerabilities identified by the vulnerability management system, thereby providing appropriate protection against threats that could adversely affect the security of the 5G SDN layer or data entrusted on the 5G services. Effective implementation of these controls will create a consistently configured environment that is secure against known vulnerabilities in operating system and application software.

Securing Network Slices

Figure 7-32 illustrates the multilayered security controls implemented to secure the network slices and prevent devices from accessing unauthorized slices.

As shown in Figure 7-32, the key techniques to mitigate the risks of unauthorized slice access are as follows:

1. Identity and Access Management (IAM) security control at the per-slice level

2. Segmentation and isolation between slices

3. Data collection and anomaly detection at the slice level

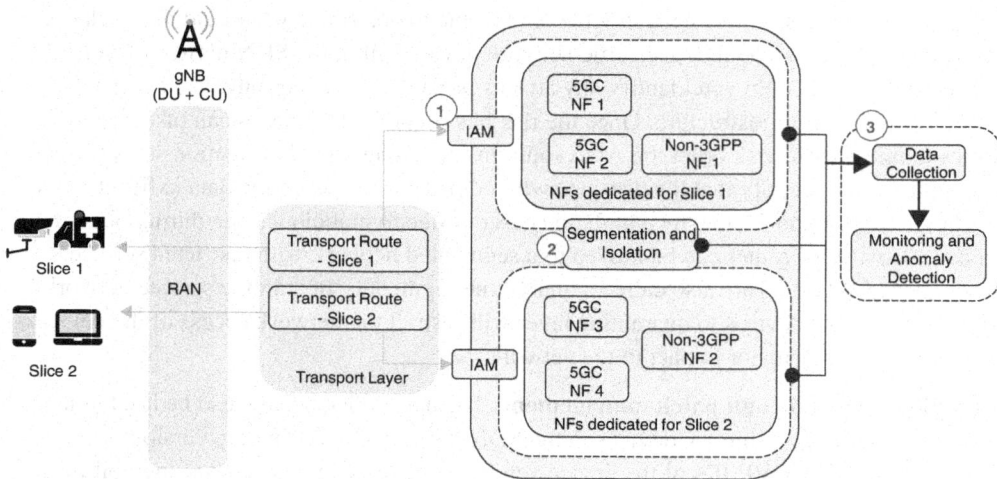

FIGURE 7-32   Securing the Network Slice

Identity and Access Control

To prevent unauthorized slice access after the primary authentication, there should be a separate slice-level authentication. This slice-level authentication should be performed by an authentication mechanism separate from the one used for the primary authentication. Figure 7-33 illustrates a slice-level authentication mechanism.

Figure 7-33 shows one of the methods of ensuring slice-level authentication explained in one of the solutions provided by 3GPP TR 33.813. It is discussed in detail next:

a. The UE is authenticated by primary authentication and based on the EAP framework (IETF RFC 3748), where SEAF/AMF takes the role of the authenticator. During the course of authentication, the UE also informs the AMF about its subscription ID and the list of Network Slice Selection Assistance Information (NSSAI) it wants access to. Using the subscription ID, AMF obtains the subscription information for the UE from the UDM to check whether the UE requires further slice-level authentication. During the course of the registration accept message, AMF then sends the UE the list of allowed NSSAI, including the S-NSSAI, which requires slice authentication.

b. This step will be required based on the subscription information for whether slice authentication is required. If slice-level authentication is not required, then step b is skipped. If slice-level authentication is required based on the list provided by AMF to the UE, then AMF directs the UE to the IAM/AAA server. EA-based authentication is performed between the UE and the specific IAM server or AAA server. In addition, there will be instances in network slice deployments where the IAM server is deployed externally by a consumer of the network slice. Figure 7-34 shows one of the methods of how the slice-level authentication could be accomplished in these deployments.

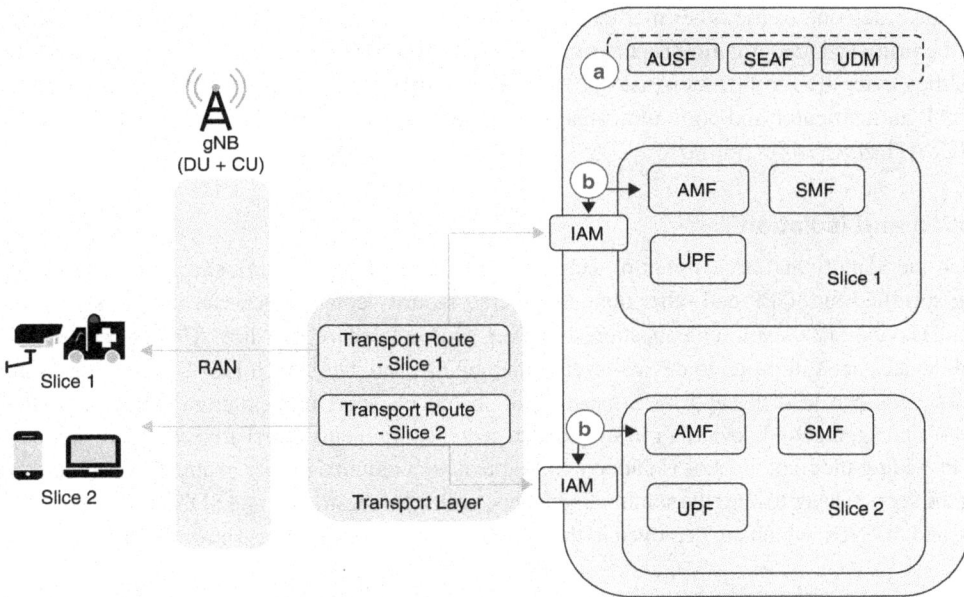

FIGURE 7-33   Multilayered Access Control for Network Slice Deployments

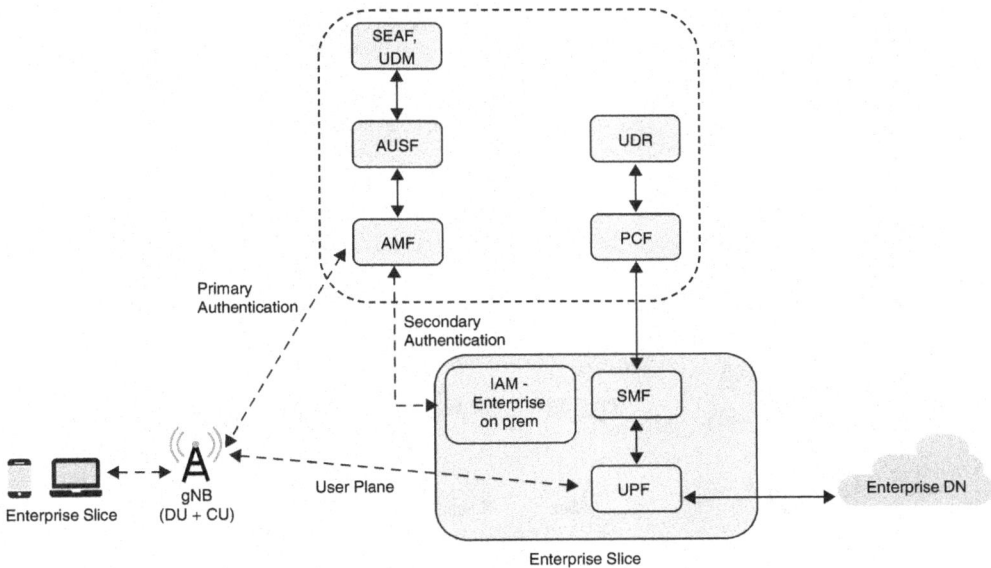

FIGURE 7-34   Network Slice Access Control for IAM Deployed at Enterprise Premises

Figure 7-34 illustrates one of the other methods to achieve slice-specific authentication of the device. In this case, the primary authentication is performed by the EAP framework, and the secondary authentication uses the device ID for the enterprise slice-specific authentication. Here, the AMF performs the role of the EAP authenticator and communicates with the enterprise on-premises-based IAM solution using RADIUS/HTTPS.

Segmentation and Isolation

As you saw in the identity and access control techniques for securing the network slice, there are mechanisms being specified by 3GPP and other optional layered security controls you can use for slice-level authentication. Having slice-level authentication, however, will not secure the slice. This is because the authenticated devices are still prone to device-level vulnerabilities and can launch DDoS attacks on to the slice resources. This can lead to capacity exhaustion in shared transport infrastructure components like transport access and aggregation devices or in shared data center components like shared compute, memory, and storage. Providing the right level of isolation and segmentation requires proper planning and should be applied from the service layer toward the southbound layers, such as orchestration and SDN, and then the 5G infrastructure and 5G NFs, which are deployed in the cloud and on the premises, as shown in Figure 7-35.

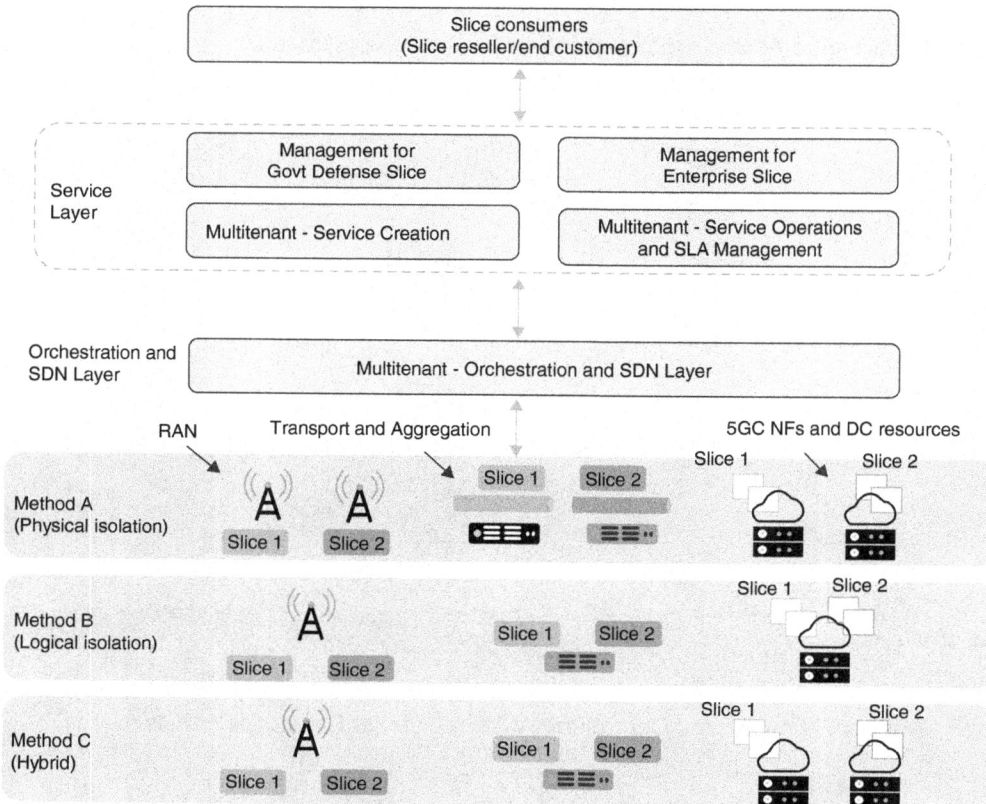

FIGURE 7-35   Methods for Isolation in Network Slice Deployments

Figure 7-35 shows the isolation at the service management layer and different methods of isolation at the 5G infrastructure and NFs layer. The service management isolation can be provided by using a multitenant platform and granular identity and access control mechanism. The methods of isolation for the 5G infrastructure and NFs are explained as follows:

- **Service layer isolation:** The isolation in this layer is provided by the separation of management by using multitenant solutions where service-level requirements can be defined by the consumer of the slice or you can configure the service-level requirements for the consumer of the slice based on the SLA between you and the slice consumer. If the slice is being utilized by a slice reseller, then the SLA will be between you and the slice reseller. The service layer management will enable you to create end-to-end slice and QoS requirements for the required slice. The multitenant management solution should also provide RBAC, where only people with privilege access can modify the end-to-end slice requirements.

- **Orchestration and SDN layer:** Multidomain orchestration and SDN ensure the required configurations of NFs and infrastructure components within a slice. They also cater for slice-level isolation and segmentation via different methods of deployment, such as physically isolated, logically isolated, or a mix of both, and are explained in detail next:

  - **Method A:** This method shows physically separate slices, such as having separate radio access technologies or separate gNBs deployed for specific slices, separate transport access aggregation devices and fiber for each slice, and separate DC infra, such as separate host hardware and host software for specific slices.

  - **Method B:** This method shows logically separate slices, such as using the same radio access technology (RAT) or the same RAT gNBs deployed for all slices, where different slice messages are broadcast with time separation in the air interface, the same transport devices and fiber are used for all slices and logically separated using methods like tagging the slices separately, and using the same host with separation of host resources using VM and security zones, application-based policy control, and microsegmentation.

  - **Method C:** This method shows a hybrid approach, where slices are separated logically, such as using the same radio access technology (RAT) or the same RAT gNBs deployed for all slices, where different slice messages are broadcasted with time separation in the air interface, the same transport devices and fiber are used for all slices and logically separated using methods like tagging the slices separately, and separate DC infra is used, such as separate host hardware and host software for specific slices.

You can have more iterations of the segmentation and isolation methods, but the hybrid options are the most likely used ones. Most of the deployments we are seeing now use a single NF type per host (for example, UPF deployed in one host, with SMF deployed in a separate host).

Data Collection and Anomaly Detection

Once you deploy the slice-level isolation and segmentation, it is important to provide the slice-level visibility and monitoring to detect any anomalies in the slice deployment. This can be achieved by collecting telemetry from network functions and network devices and creating a baseline of normal behavior for each slice. Any variation from this normal behavior can be considered an anomaly and is alerted for further investigation or automated corrective actions planned, as illustrated in Figure 7-36.

Figure 7-36 shows you a method of slice-level visibility, and the steps involved are explained in detail in the following list:

a. **Data collection:** This step consists of collecting telemetry from various parts of the infrastructure. The data collected includes telemetry such as NetFlow, IPFIX (Internet Protocol Flow Information Export), and other types of flow data from 5G infrastructure such as routers, switches, firewalls, endpoints, proxy data sources, and other network infrastructure devices. The 5G infrastructure component will require integration with the legacy network. To prevent any blind spots, it will be necessary to provide visibility into the devices or segments of the switching and routing infrastructure that can't generate NetFlow natively. This will require solution components in the monitoring layer, such as mirroring port or network tap, and generate telemetry based on the observed traffic. Apart from the network layer visibility, it is important to provide Layer 7 application visibility by gathering application information. This includes data features like RTT (round trip time), SRT (server response time), and retransmissions. The solution you choose for monitoring the network slice layer should support even the largest of network demands. It should perform well in extremely high-speed environments and can protect every part of the network, regardless of size.

b. **Anomaly detection and patch management:** This step consists of performing analysis on the collected data to perform corrective actions, enhance operational efficiency, and reduce costs by identifying and isolating the root cause of an issue or incident within seconds. The solution you choose for the anomaly detection should provide you with the following capabilities:

- The solution should be able to detect anomalies within the slice in public cloud, on-premises, and hybrid deployment modes.

- The solution should be able to monitor all hardware and software components of the slice.

- The solution should be able to consume external threat intel or integrate with a robust threat intelligence feed.

- The solution should be using machine learning and a statistical modeling of networks, creating a baseline of normal activity, identifying anomalous traffic, and pinpointing command-and-control communications and data exfiltration within the slice.

- This layer should have the patch management capability within the anomaly detection system or have a separate patch management solution with the capability of distributing and applying updates to the required network functions.

- The patch management should be automated selectively to ensure that critical software is patched only in the maintenance window.

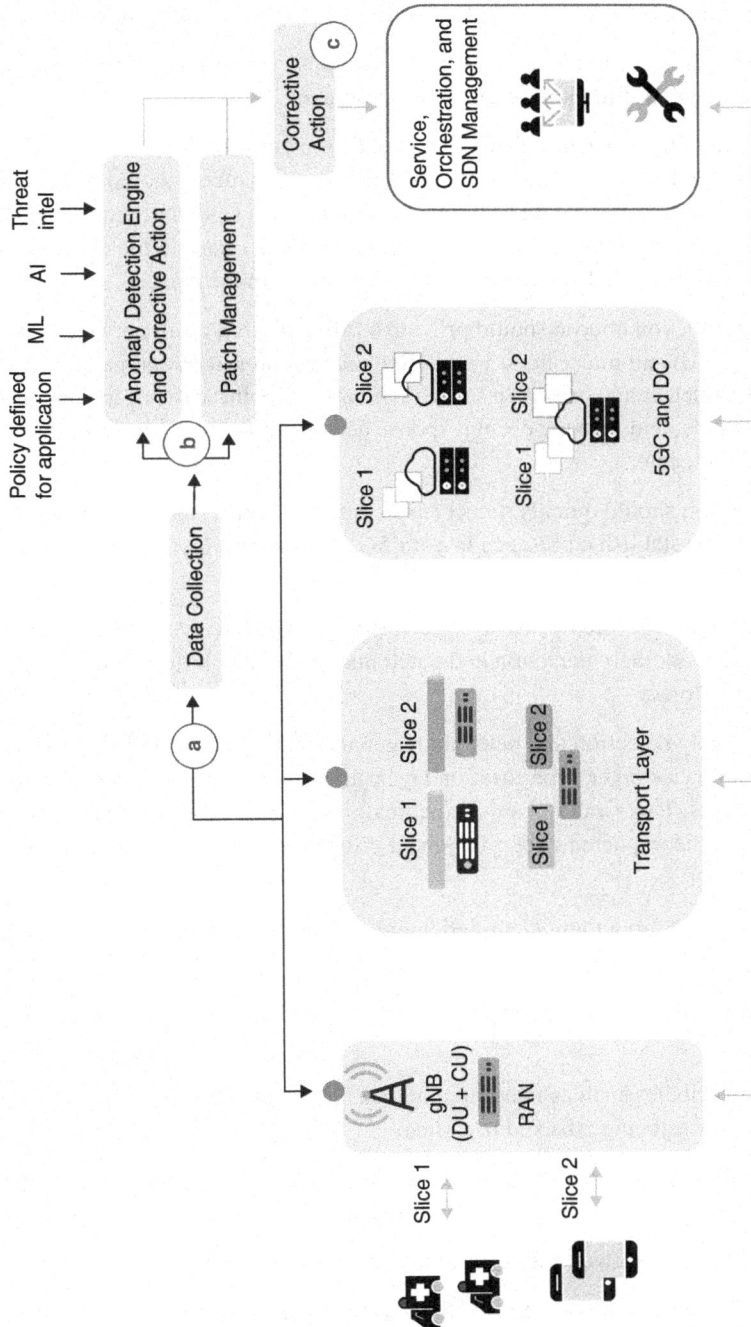

FIGURE 7-36   End-to-End Visibility in Network Slice Deployments

- The anomaly detection and monitoring layer should provide graphical views of the current state of the 5G control plane traffic of multiple slices within the network.

- Administrators should be able to construct maps of their network based on any criteria, such as location, function, or virtual environment.

- Other capabilities, such as creating a connection between two groups of hosts, where you can quickly analyze the traffic traveling between them, should be provided by the solution. An example could be visibility on increased signaling requests coming in from specific clusters in the IoT slice might indicate some issues in the cluster or a DDoS attack, with some IoT devices in the cluster being compromised.

- The solution you choose should provide a full audit trail of all network transactions within a slice and inter-slices. For more effective forensic investigations, it should be able to rapidly detect and prioritize security threats, pinpoint network misuse and suboptimal performance, and manage event response across tall network slices—all from a single, centralized center.

- The solution should quickly detect suspicious web traffic and/or NetFlow and respond to attempts to establish a presence in your 5G environment and to attacks that are already under way.

c. **Corrective action:** This step consists of performing corrective actions based on the analysis performed by the anomaly detection and patch management solutions. The key corrective actions are as follows:

- The anomaly detection and patch management solutions should be capable of integrating with the service layer, orchestration layer, SDN layer, and also directly with the slice components. This would give you the flexibility of allowing the corrective actions to be applied to the impacted slice component directly or indirectly, based on your preferred method.

- Send out alerts about any identified anomalous traffic within the slices to your security team.

- Quarantine any of the impacted slice components by restricting them to a minimum service slice.

- Modify attributes such as maximum allowed requests to prevent the CPU exhaustion within devices being attacked in a slice.

- Update the software with patches for devices with known vulnerabilities in multiple slices.

- Block any unauthorized traffic escaping the slice network.

These security controls will help you enhance the slice-level security.

Securing NSaaS Deployments

Network slicing will enable 5G networks to provide tailored services to enterprises that adhere to an SLA established with the 5G network. These tailored services could include network capabilities such as latency, throughout, RAT, bandwidth, latency, reliability, guaranteed/nonguaranteed data rates, security services, and so on. One of the deployment methods is to provide the management capability of configuring the slices to the enterprise itself. This requires multilayered security control to mitigate any threats related to unauthorized slice management access, as illustrated in Figure 7-37.

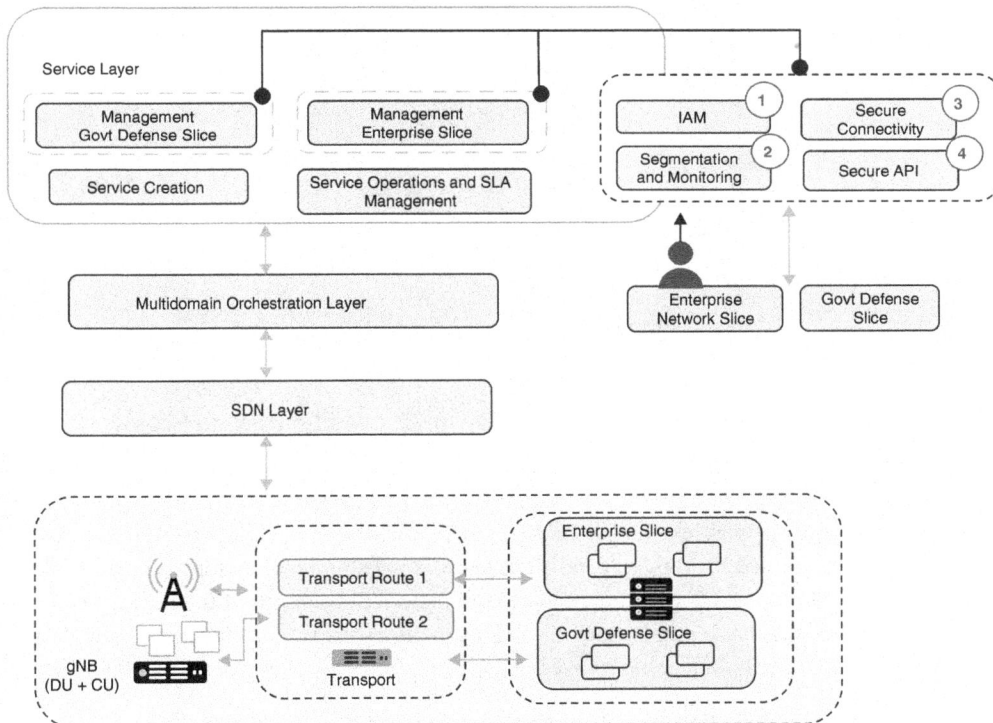

FIGURE 7-37   Method to Mitigate Unauthorized Access to Slice Management Layer

This section will take you through the key security controls to mitigate unauthorized slice management access. The key security controls are described in the sections that follow.

1. Granular Identity and Access Management

It is recommended to use the zero-trust principle-based granular access control to provide the right level of access control to the admin of the slice provider and the slice consumer.

Figure 7-38 illustrates the key security controls required for ensuring granular Identity and Access Management and is explained in detail in the list that follows.

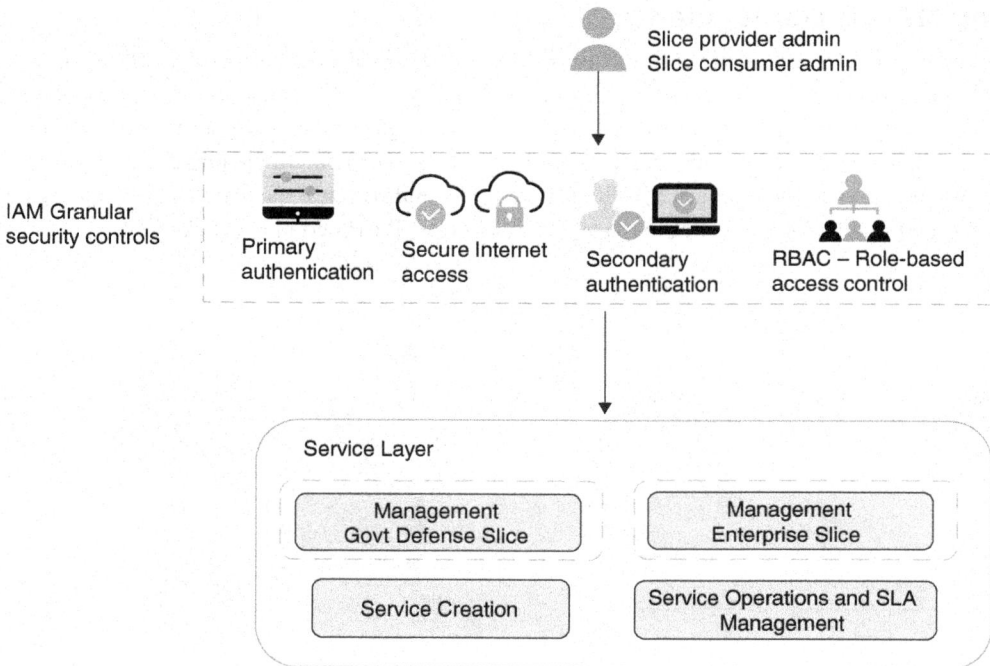

FIGURE 7-38   Multilayered Access Controls to Mitigate Unauthorized Access to Slice Management Layer

- Based on the zero-trust principles, the access to the network slice service layer should be dynamic and be determined by various conditions such as the software version and type of the device, network location, time/day of the request, and historical behavior of the requester, such as previously observed behavior and credentials. This includes the use of multifactor authentication to access the NSaaS management layer.

- Slice-specific authentication and authorization policies should be defined such that each is mapped to the allocated slices and can operate only in authorized slices. This function can be performed by using an identity management system that is responsible for creating, storing, and managing user accounts and identity records.

- As shown in Figure 7-38, RBAC is also implemented to ensure that different users from the same slice can have varying levels of access, as per their roles. For example, the slice admin can modify the attributes of the slice, whereas a non-admin user will have read-only access to the management system.

■ Based on the zero-trust principles, the access to the network slice service layer should be dynamic and determined by various conditions, such as the software version and type of the device, network location, time/day of the request, and historical behavior of the requester, such as previously observed behavior and credentials.

2. Segmentation and Monitoring of the Management Layer

Segmentation of the NSaaS layer should also include the NSaaS management access policy, which is microsegmented per NSaaS slice management. The idea behind using microsegmentation is to divide the data center into small zones that can then be effectively managed. This will allow granular security control of the users accessing the NSaaS management plane. Segmentation is also critical at the SDN layer for the NSaaS services.

The out-of-band control plane in the SDN platform you choose must separate management data from application data. Management data, such as configuration and monitoring information and statistics, must be directed from the 5G devices within all slices to the SDN controller. Application data from all the network functions should be sent directly to its destination on data networks configured for the respective slices.

The use of an out-of-band architecture provides resiliency for the users. It means that users are not affected if devices are unable to communicate with the SDN controller due to intentional or unintentional service disruptions. Users can still access local management and production networks, and all policies and settings continue to be enforced. In addition, local user authentication is unaffected, and local configuration tools for the 5G infrastructure remain available.

Figure 7-39 illustrates the microsegmentation layer for policy mapping and enforcement for the users accessing the NSaaS management layer, as further described in the list that follows:

a. **Global Policy Engine:** The Global Policy Engine should be able to take policy inputs from multiple sources, including manual and automated components. This will improve your policy engine for the right decision, such as granting, denying, or revoking access to the management of the NSaaS slice. The global policy engine should also be able to log the action taken by it for each of the requests.

b. **Policy per NSaaS slice management:** This layer is responsible for dynamically granting access to individual requests toward the NSaaS slice management layer per slice. When the management is deployed as a virtual machine or a container, the policy should be applied at the application layer so that the policy sticks to the virtual instance whether deployed on the cloud or on the premises.

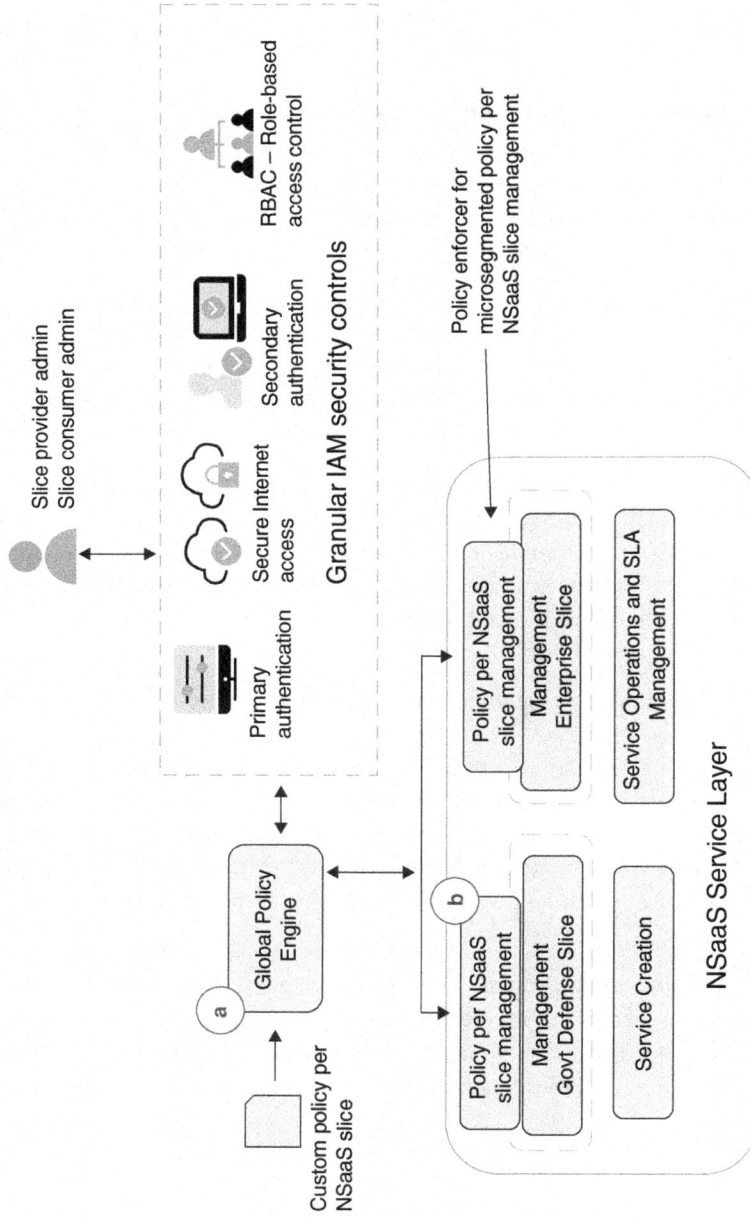

FIGURE 7-39 Microsegmentation Layer for Policy Mapping and Enforcement at the Application Layer

3. Connection Security

This section will take you through the different methods to secure the API communication for NSaaS and network slice components. All APIs in the 5G service, SDN, and orchestration layers should be secured using TLS, as shown in Figure 7-40.

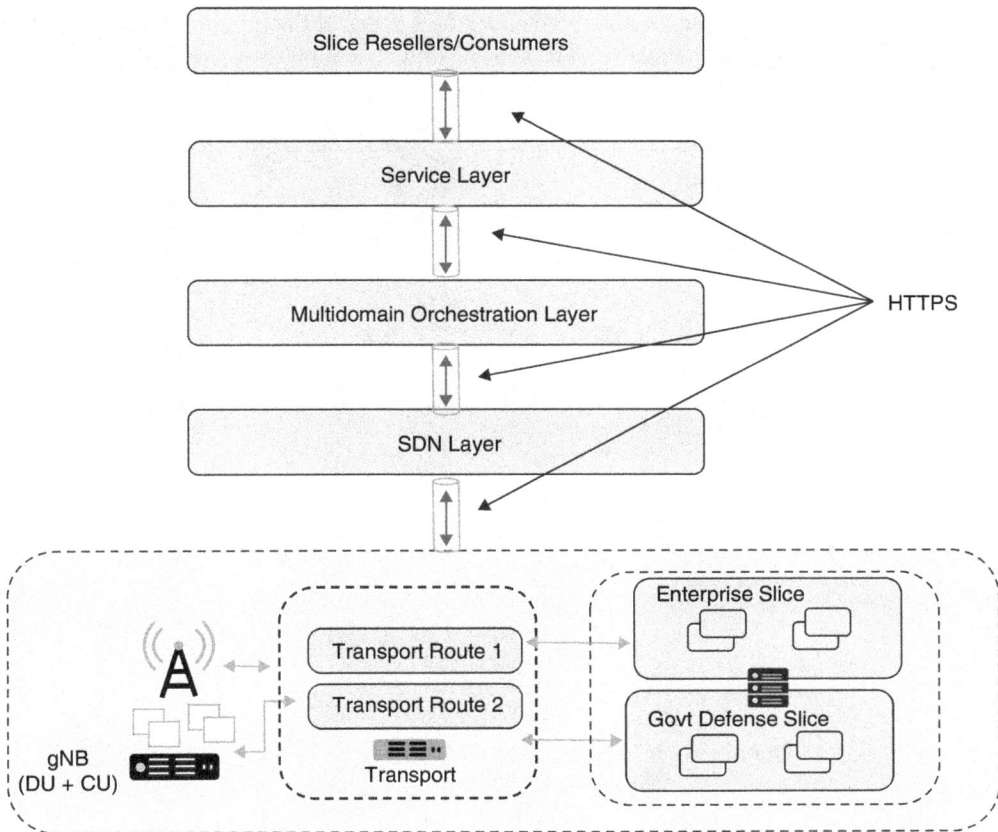

FIGURE 7-40   Secure API Connections

As shown in Figure 7-40, all management communications and traffic flow between the different layers should be secured. Because APIs would be used mostly in the exchange of data, encrypting the traffic using Transport Layer Security (TLS) is recommended.

Figure 7-41 illustrates encryption of the API communication between the SDN controller and the on-premises/cloud components using HTTPS.

As shown in the figure, all data exchanged between devices within the slices and the SDN controller should use industry-standard encryption and security protocols. Connected devices should use TLS with restricted ciphers and HTTPS on the standard HTTPS port 443. All data sent to the SDN controller should be encrypted using the Advanced Encryption Standard (AES) with a 256-bit, randomly generated key that is distributed with a public-key mechanism to prevent man-in-the-middle attacks.

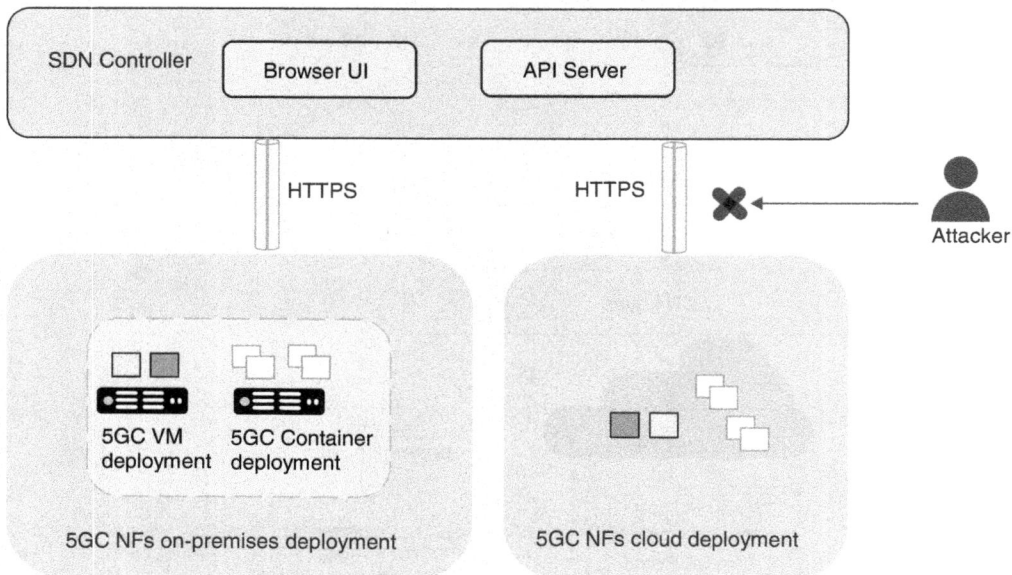

FIGURE 7-41 Secured API Connections Mitigating MitM Attacks

In addition, every device connecting to the portal should be authenticated with a cryptographic token so that only legitimate devices can be managed, thus closing a potential Trojan horse attack vector.

All connections should be initiated from the device. Thus, firewalls can block all incoming connection requests. Only HTTPS port 443 needs to be enabled for outbound connections. As a result, firewalls do not need any other special configuration to enable SDN controller connectivity. All network devices and network functions can be configured to use HTTPS proxy servers to add an additional layer of security through indirection.

4. Secure API

A majority of the communications in the 5G NSaaS infrastructure is based on APIs. This includes the API-based communications between the 5G network functions in the SBA architecture defined for 5G, northbound interfaces (NBIs) from the programmable transport devices toward the SDN controller, NBIs from the SDN controller toward the orchestration layer, and NBIs from the orchestration layer toward the NSaaS service layer. The external integrations, such as from the slice consumer and the slice reseller, are also based on the API calls. This makes APIs one of the critical components in NSaaS deployments. API connection security using encrypted API communication was covered in the section "3. Connection Security," earlier in this chapter. Figure 7-42 illustrates other API security methods to safeguard NSaaS deployments from attacks.

The key security controls for NSaaS deployments are as follows:

Step a.   API GW/WAF: Any API calls originating from an external network or from an untrusted zone should be inspected by the API gateway (API GW) or the web application firewall (WAF). Here are some of the key capabilities required at this layer:

■ **Authentication:** Improper or broken authentication within API communication is widespread due to the design and implementation of most identity and access controls in API calls. Attackers can detect broken authentication using manual means and exploit it using automated tools with password lists and dictionary attacks. To secure your network from these kinds of attacks, the following steps should be considered:

  ■ You should ensure registration, credential recovery, and API servers are hardened and updated with latest patches.

  ■ You should apply rate limits and thresholds on API requests, such as repeated requests for login, to mitigate DDoS attacks. You should log all failures and alert administrators when brute force and other attacks are detected. Use a server-side session manager that generates a new random session ID with high entropy after login. Session IDs should not be in the URL; instead, they should be securely stored and invalidated after logout, idle, and absolute timeouts.

  ■ You should use JSON Web Tokens (JWTs) to secure the REST API. JWTs are URL-safe JSON-based security tokens that contain a set of claims that can be signed and/or encrypted. JWTs are being widely used and deployed as a simple security token format in numerous protocols and applications, both in the area of digital identity and in other application areas. The most common use case for JWTs is to use them as access tokens and ID tokens in OAuth and OpenID Connect flows.

  ■ You should deploy MFA to prevent unauthorized users and devices access to slices. Do not ship or deploy with any default credentials, particularly for admin users.

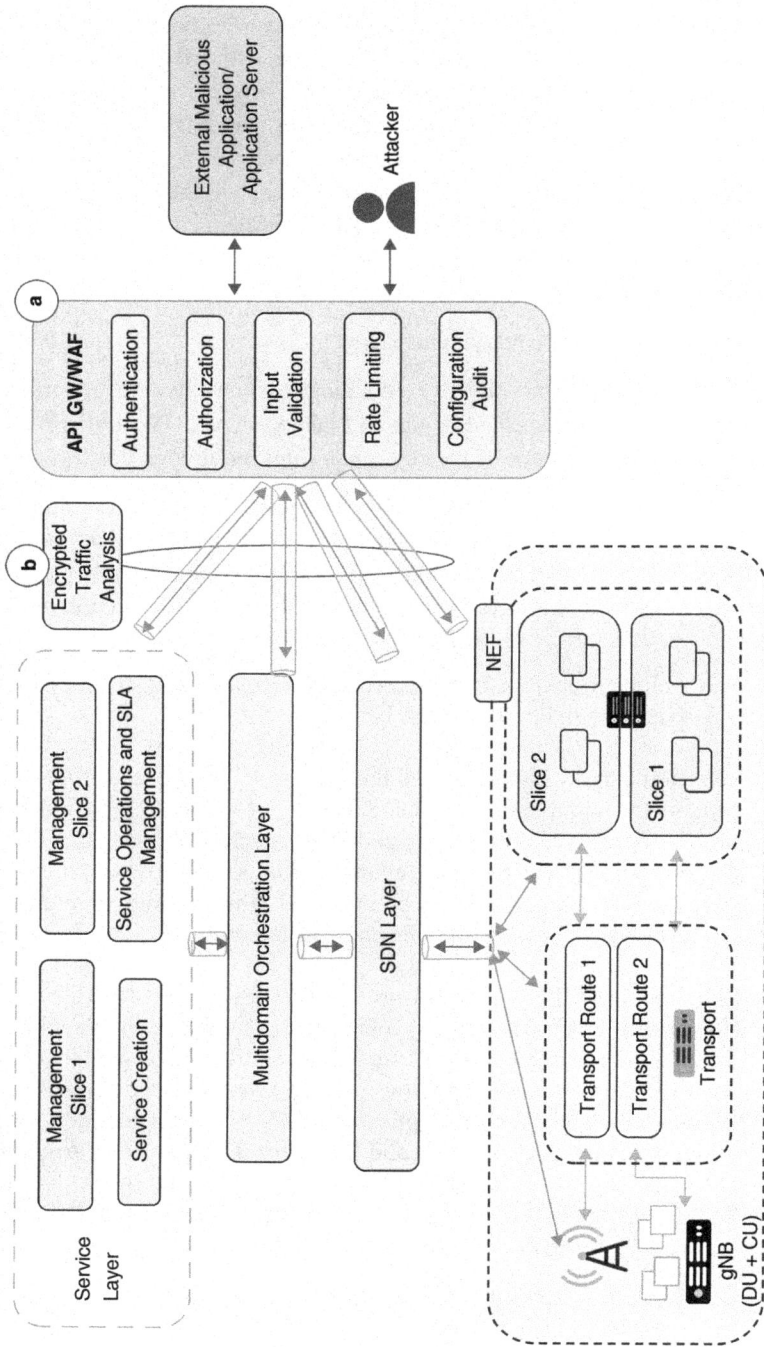

FIGURE 7-42   API Security Methods to Secure the NSaaS Deployments

- **Authorization:** APIs are also prone to weak or broken authorization due to lack of proper functional testing by the application developers. This might result in the MEC applications used by the NSaaS services being exploited, leading to unauthorized access to slice resources by illegitimate users.

  - To mitigate such unauthorized access, you should conduct regular audits and implement authorization checks for users. The global policy engine should also be audited to check user policies and hierarchy.

  - The authorization should be implemented on both the client side and the server side. Just having client-slide authorization, such as relying on client IDs sent from client, is not recommended.

  - Each time you have a request from a client or a user, the authorization policy for accessing the database should be checked.

- **Input validation:** Input validation is another area that should be carefully planned. 5G introduces openness to your network by allowing third-party software and application integrations with your 5G network. For NSaaS deployments, this means having the 5G slice integration with the required third-party applications or having the third-party applications deployed within the slice for fulfilling the use case for the network slice reseller or network slice consumer. Input validation prevents malformed data from entering your 5G network. In such deployment scenarios, it is difficult to detect a malicious user who is trying to attack software applications.

  - To mitigate such attacks, you should have a security control to check and validate all input entered into a system, such as validating the user input for any unexpected attribute values. For example, many organizations use automatic binding of HTTP request attributes to server-side objects. This can allow an attacker to update server-side objects that were not meant to be modified. The attacker can possibly modify their access control level or circumvent the intended business logic of the application with this feature.

  - To prevent such attacks, you should avoid automatically binding inputs directly and put a rule in place that allows only specific fields or attributes that are allowed to be auto-bound.

- **Rate limiting:** APIs do not impose any limitations on the size or number of resources that can be requested by the client/user. Depending on the type of API attack, it could not only degrade API server performance, leading to denial of service (DoS), but also could leave the door open to authentication flaws such as brute force. You might think that the brute-force attacks could be mitigated by locking accounts after a limited set of attempts, but the attacker would use this exact method against you.

  - For example, consider a brute-force attack in the NSaaS service management layer, where the slice configuration parameters are set. An attacker might perform a brute-force attack on all the accounts being used for the NSaaS slice resellers or slice consumers and intentionally lock all of them out.

- It is recommended to design your security architecture in such a way that if the rate-limiting solution fails or is unreachable, your service should fail open and try to serve all requests for critical services required by the NSaaS slices. The clients should then go through the multilayered access control mechanisms before being allowed to access the system, including the input and output validations. Failing closed leads to a complete outage, versus failing open, which leads to a degraded condition.

- **Configuration audit:** Having a proper validation process will ensure that the data being sent to the API is restricted and that only the required data is being sent. This will reduce the risk in sending too much information over the API for the attacker to gather more information and then plan an attack on your infrastructure. To enhance the security of the API, the following configuration audits can be performed:

- Define and enforce all API response payload schemas, including error responses, to prevent more information being passed during error exception, which might allow attackers to have more information about the type of software and web servers being used for the NSaaS service layer.

- A repeatable hardening process leading to fast and easy deployment of a properly locked-down environment for any impacted NSaaS slice.

- Continuous review and update configurations across the entire API stack for slice resellers and slice consumers, such as orchestration files, API components, and cloud services.

- An automated audit process to continuously assess the effectiveness of the configuration and settings in all environments such as on-premises and public cloud.

Step b.   Analyzing encrypted traffic: Encrypting API traffic in 5G networks is recommended to prevent MitM attacks in the NSaaS service layer, orchestration, SDN, and 5G service-based architecture deployments. But encrypted traffic also creates an additional challenge for security teams. They now have to address a massive influx of traffic that they cannot look inside without decryption technology. The hackers have quickly learned to use data encryption to their benefit to conceal delivery, command-and-control activity, as well as data exfiltration. Thanks to encryption, they can now break into the network and stay undetected for months. This is critical for NSaaS deployments, as there might be a slice for critical infrastructure deployments that needs to be protected from such attacks. To address the challenges of data encryption, it is important to identify any malicious packets in the encrypted traffic. The most widely used method for this purpose is to use SSL decryption engines to decrypt the encrypted packets and then run anti-malware or HTTP inspections. However, this requires multiple encrypted traffic to be decrypted, thereby breaking the security posture of your network.

Another way of addressing this challenge is to use solutions available today that identify malware within an encrypted layer without having to decrypt it. This is achieved by solutions that collect the packet flow from multiple network devices and then look into the header of the packets to identify necessary information such as cipher suites being used, cipher suites being chosen by the server, and so on. This information is then analyzed by ML and AI engines along with threat intelligence feeds to determine any malware within the encrypted packets. The key functionalities that should be supported by the encrypted traffic analytics solution are as follows:

- The chosen solution should provide insight into threats in encrypted traffic and contextual threat intelligence with real-time analysis correlated with user and device information.

- There should not be any need to decrypt the traffic. Analysis should be performed without the need of decrypting and then re-encrypting the packets.

- The solution should be able to provide cryptographic assessment, such as compliance with cryptographic protocols and visibility into and knowledge of what is being encrypted and what is not being encrypted on your 5G network.

- There should be quick identification and isolation of infected devices and users (faster time to response).

- The solution should be able to integrate with your existing security information and event management (SIEM) solutions.

Real Scenario Case Study: Threats in the 5G Network Slice, SDN, and Orchestration Deployments and Their Mitigation

The previous sections described the threats and mitigation techniques for specific areas of concern while deploying a 5G network slice that is enabled by the SDN and orchestration layers. This section focuses on a multidomain attack and discusses methods to deploy security controls at different parts of the infrastructure to mitigate attacks in a real-life environment. Although the 5G network slice was not widely deployed at the time of writing this book, this section is based on real-life attacks in similar environments or proof of concepts and lab tests done by various customers.

To explain the attack scenarios, Figure 7-43 shows an example of the V2X deployment as a reference for this section.

The 3GPP and non-3GPP network functions are as follows:

- **AMF (Access and Mobility Function):** Responsible for handling all connections and mobility-related tasks for the UE

- **AUSF (Authentication Server Function):** Manages UE authentication

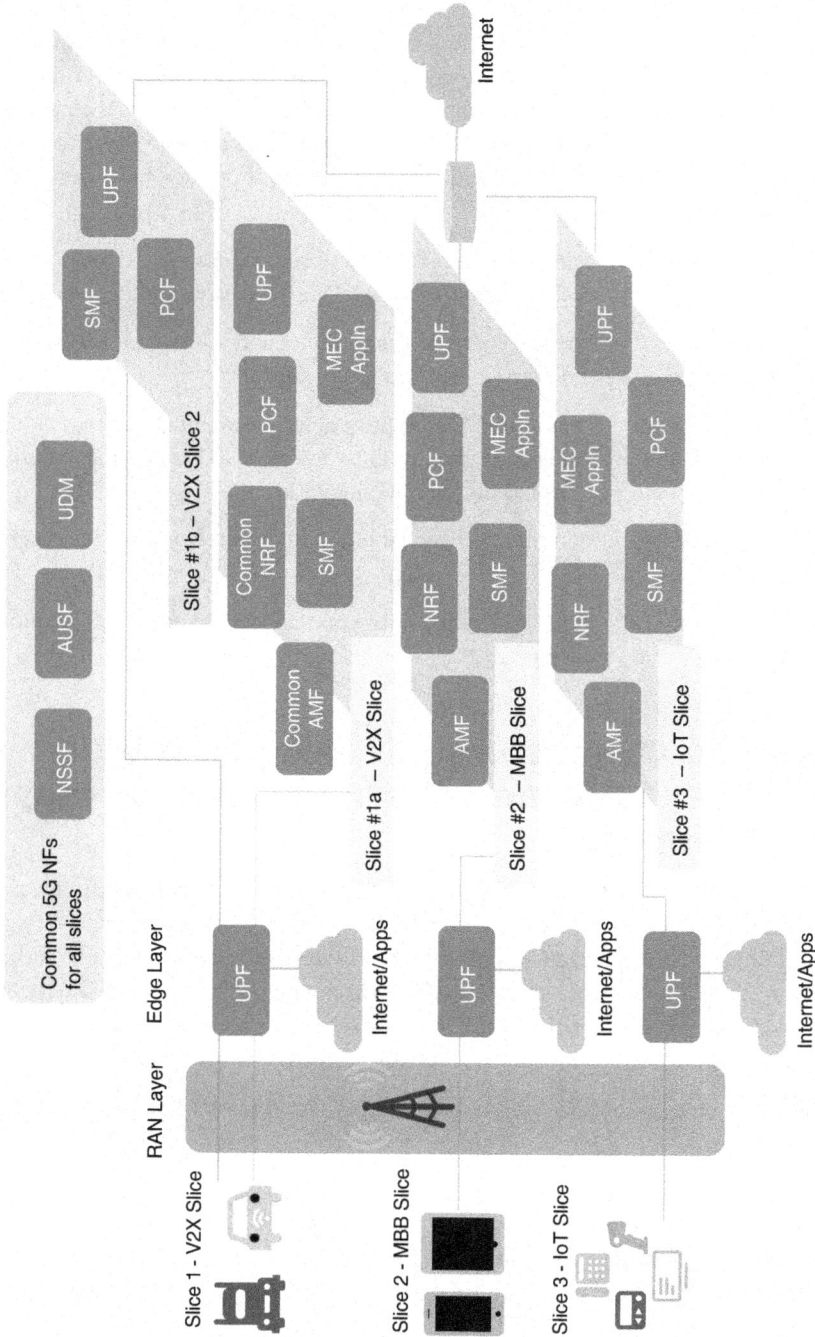

FIGURE 7-43   Example of the V2X Deployment

- **NRF (Network Repository Function):** Centralized repository for all the 5G network functions

- **NSSF (Network Slicing Selection Function):** Helps in selecting the network slice available for the service requested by the user in the 5G environment

- **SMF (Session Management Function):** Responsible for managing session context with the User Plane Function (UPF) and creating, updating, and removing Protocol Data Unit (PDU) sessions

- **PCF (Policy Control Function):** Provides policy rules to control plane functions to enforce them

- **MEC Application (Multi-access Edge Compute Application):** Application created by third-party developers deployed on the edge network components

- **UDM (Unified Data Management):** Manages data for access authorization, user registration, and data network profiles

- **UPF (User Plane Function):** Responsible for packet routing and forwarding, packet inspection, QoS handling, and external PDU session for interconnecting data networks (DNs) in the 5G architecture

Figure 7-43 shows the logical design of the 5G deployment for the multiple network slices. All the slices share common NSSF, AUSF, and UDM 3GPP network functions. Each slice has the following dedicated components:

- **Slice 1:** Slice 1a and 1b in Figure 7-43 cater for vehicle-to-everything (V2X) use cases and have shared AMF and NRF network functions. This slice has a dedicated SMF, PCF, UPF, and 3GPP-based network functions. It also has a dedicated third-party MEC application.

- **Slice 2:** Caters for mobile broadband (MBB) use cases. It has dedicated AMF, SMF, NRF, PCF, UPF, and 3GPP-based network functions. It also has a dedicated third-party MEC application.

- **Slice 3:** Caters for Internet of Things (IoT) use cases. It has dedicated AMF, SMF, NRF, PCF, UPF, and 3GPP-based network functions. It also has a dedicated third-party MEC application.

To understand the threats in real-life scenarios, let's look at how the slices are deployed (or are planned to be deployed) in actual 5G networks by service providers. Figure 7-44 shows the DC design of the 5G deployment for the multiple network slices in a real-life scenario.

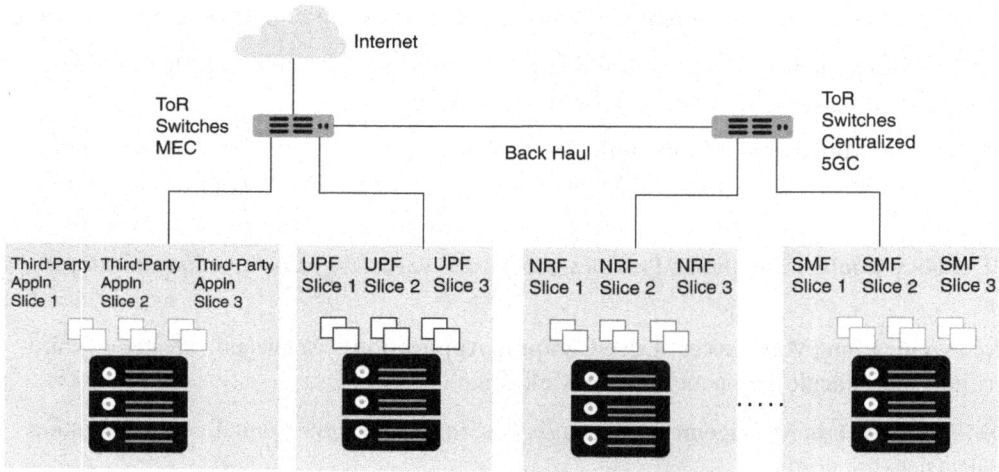

FIGURE 7-44   Example of the Actual Network NF Deployments per Slice

Figure 7-44 shows an example of on-premises network slice configuration for a 5G network. The same logic can be used when the 5G components are deployed in the public cloud, but the majority of the proof of concepts or deployments for network slices are initially on-premises.

To allow scalability and flexibility with the different orchestration solutions, the NFs are clubbed together in an x86 server. For example, Figure 7-44 shows you third-party applications together in one x86 server, and UPFs for all slices deployed together in another x86 server. This applies for all the other 3GPP NFs. Depending on the use case, all the 3GPP and non-3GPP NFs can be deployed in the same DC or distributed in multiple DCs. If the DCs are distributed, a back haul would be required between the edge DC and the centralized DC. If the DCs are not distributed, then the top of the rack (ToR) switches would be interconnected to allow user plane and control plane traffic flows. In real life, there will be multiple x86 servers handling a single NF; for example, for testing all three slices in a cluster, you will require anywhere from three to four x86 servers only for UPF NFs, depending on the traffic volume and throughput for the test cluster. Real-life deployments to cater for your entire subscriber base are a whole different scenario with multiple servers being dedicated to each NF.

Threats: Case Study

Let's look at the threat surfaces in these real-life scenario deployments. Figure 7-45 illustrates one of the threat vectors.

In this scenario, an attacker is exploiting the weak API implementation. One of the key attack vectors is the privilege escalation attack. Privilege escalation is a type of attack used to obtain unauthorized access to systems by exploiting a bug, design flaws in API implementation, and a misconfiguration in an application or operating system. To make the scenario a bit clearer, Figure 7-46 shows the actual physical deployment.

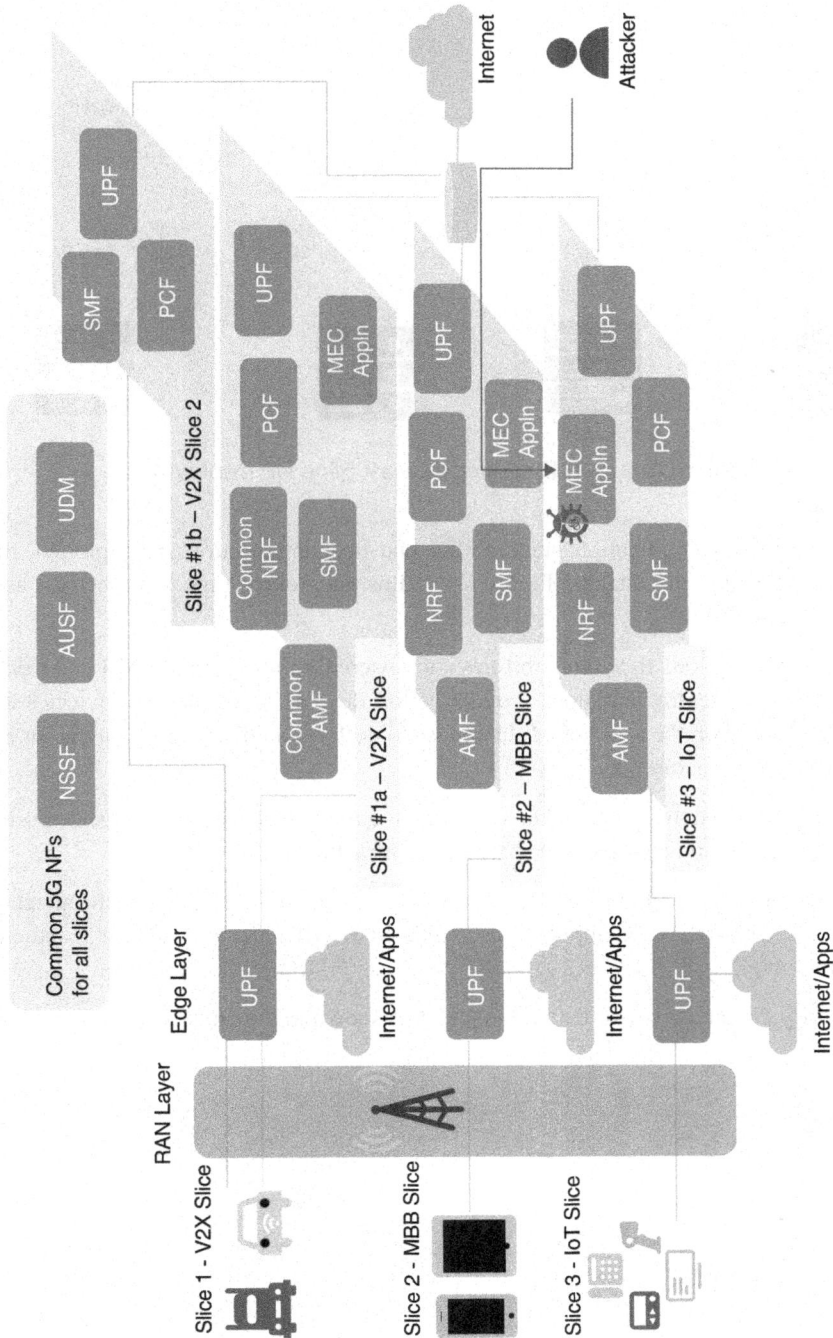

FIGURE 7-45   Attacker Exploiting the Weak API Implementation

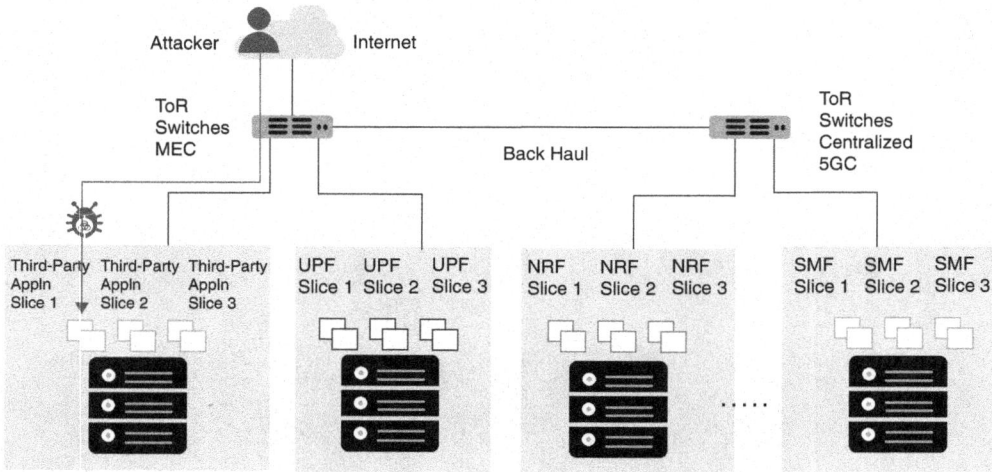

FIGURE 7-46   Illustration of Attack in Real-life Network Slice NF Deployments

Once the IoT slice MEC application vulnerability and misconfiguration are exploited, the attacker proceeds to perform multiple attacks. Figure 7-47 illustrates one of the risk vectors, as explained further in the list that follows:

Step 1. The attacker exploits the vulnerabilities and misconfigurations in the API and web server deployment to deploy malicious code in the IoT devices. Once the IoT devices are under the attacker's control, the attacker could choose to perform a DDoS attack on the service provider infrastructure.

Figure 7-48 illustrates the attacker spreading the malicious code in other slices using the privilege escalation attack in the shared host infrastructure.

Step 2. Once the attacker exploits the MEC application for the IoT slice, one of the attack vectors is to infiltrate other MEC applications dedicated to MBB and V2X slices, as indicated in Figure 7-48.

To clarify the attack vector better, the attack is shown in the physical design in Figure 7-49.

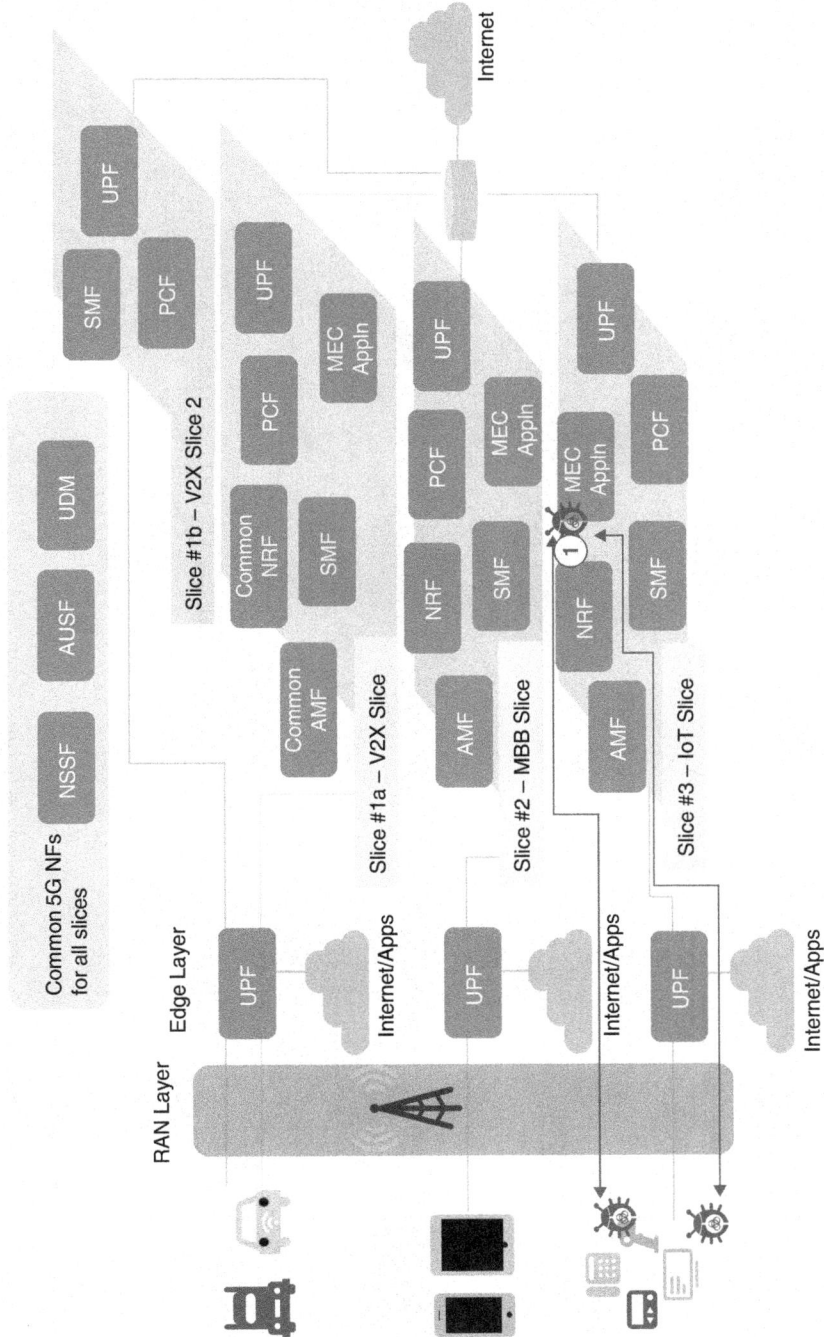

FIGURE 7-47 Illustration of IOT Application Infecting the IoT Devices with Malware

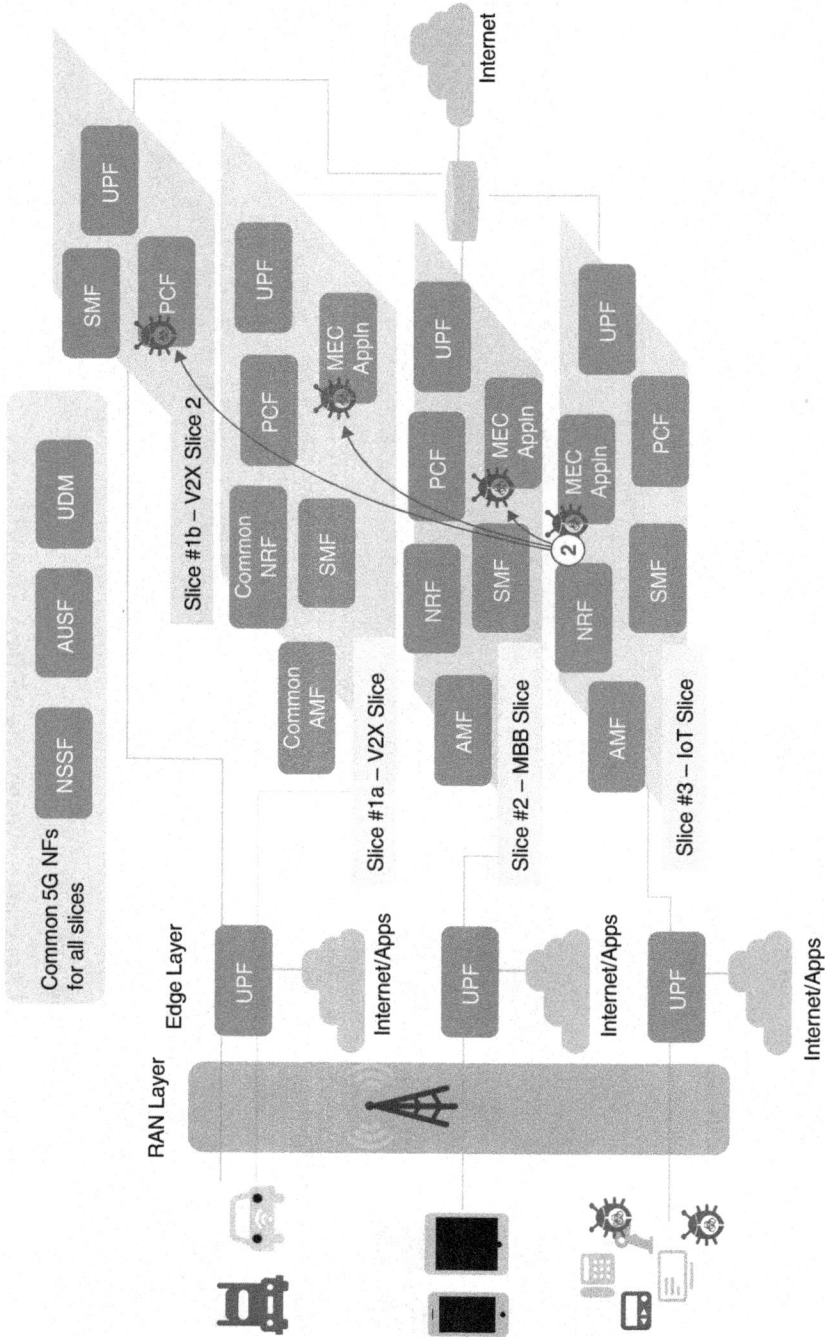

FIGURE 7-48   Illustration of Malware Spreading to Other Slices

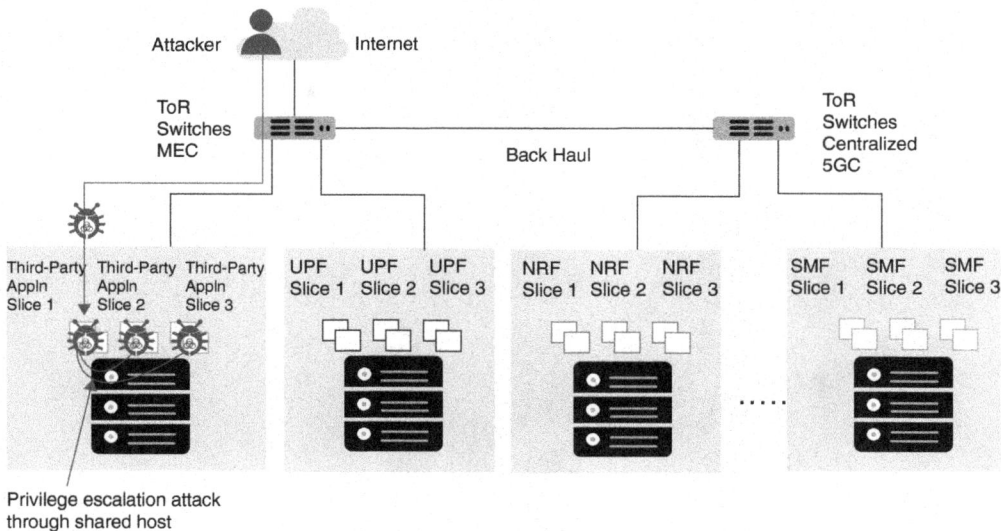

FIGURE 7-49   Illustration of the Actual Process of Malware Spreading to Other Slices

Figure 7-50 illustrates the steps the attacker performs to disrupt services to the critical V2X slice.

Step 3.   Once the attacker takes control of the MEC application dedicated for critical slices, such as V2X, it could lead to disastrous consequences. In the example shown in Figure 7-50, the attacker gains access to the V2X MEC application using the vulnerabilities in the host software and hardware.

Step 4.   As indicated in Figure 7-50, the attacker could perform a denial of service by rejecting all messages from the V2X application in the vehicles to the MEC application.

Figure 7-51 illustrates another threat vector due to inadequate security controls in network slice deployments.

Step 5.   Once the MEC applications for all the slices are under the attacker's control, the attacker could send unauthorized commands such as bulk exporting the parameters from the V2X application and exporting the information from the database connected to all the MEC applications to a C2C server.

Step 6.   The attacker could then analyze the collected data in the C2C server to understand your network and its vulnerabilities better to perform attacks on all the vulnerable software and hardware components in the slice network and applications of the devices being catered for by the slices.

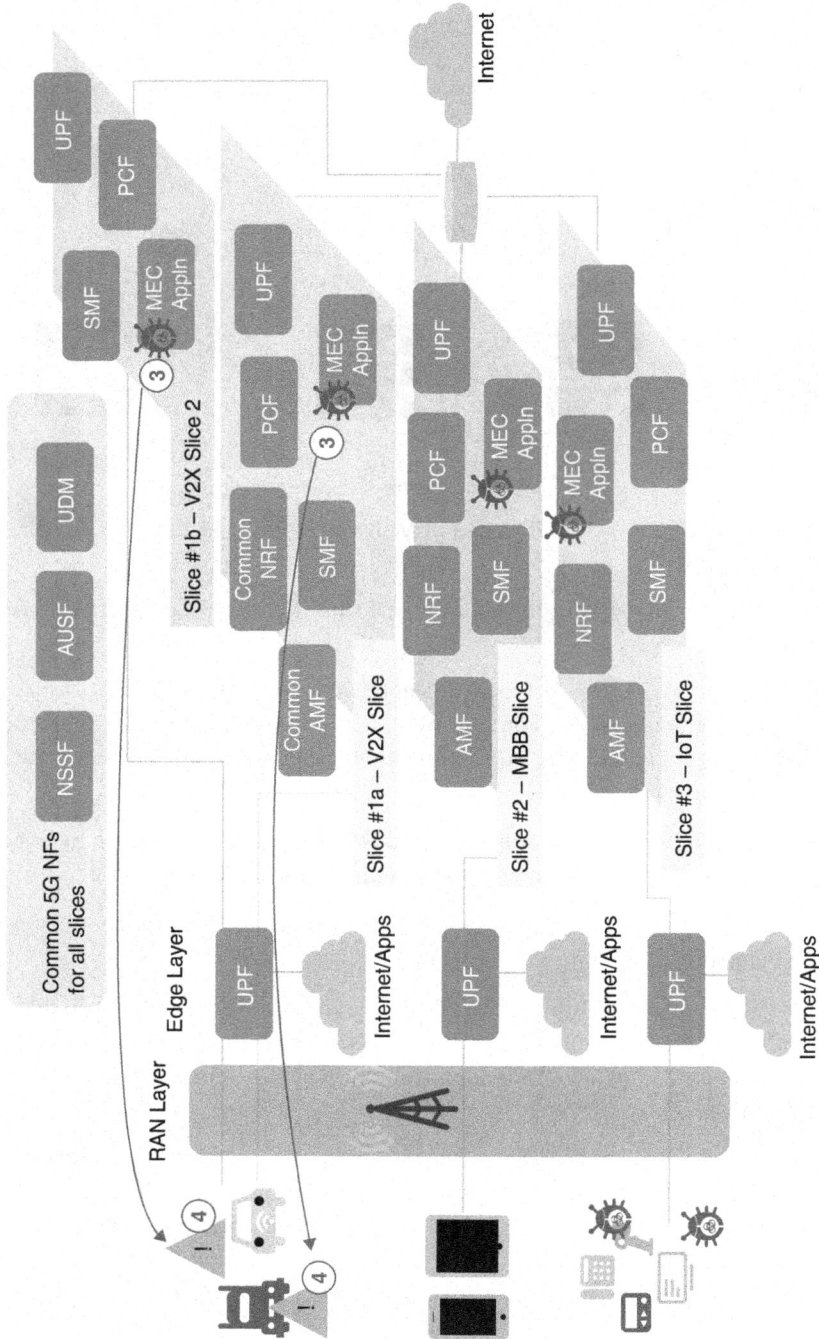

FIGURE 7-50   Attacker Aiming for the Critical V2X Slice

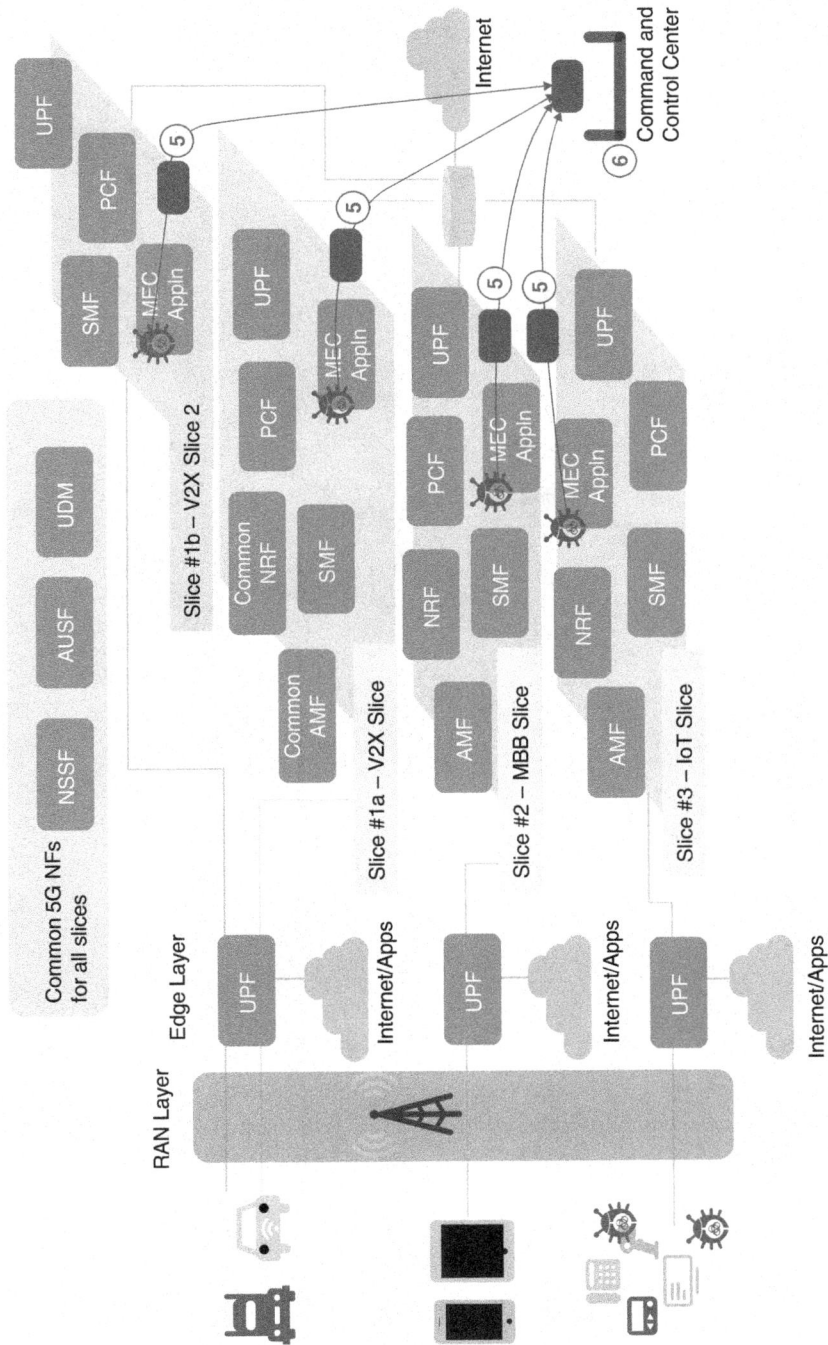

FIGURE 7-51   Data Exfiltration and Data Hoarding Due to Insufficient Security Controls

Mitigations: Case Study

This section will take you through the mitigation steps to secure the network slice deployments enabled by SDN and orchestration components. This section will be of particular interest to readers trying to understand the exact scope of the threats and what specific solutions could be used as a tactical method to mitigate the threat vectors discussed under the "Threats: Case Study" section earlier in this chapter.

Figure 7-52 shows the multilayered security controls.

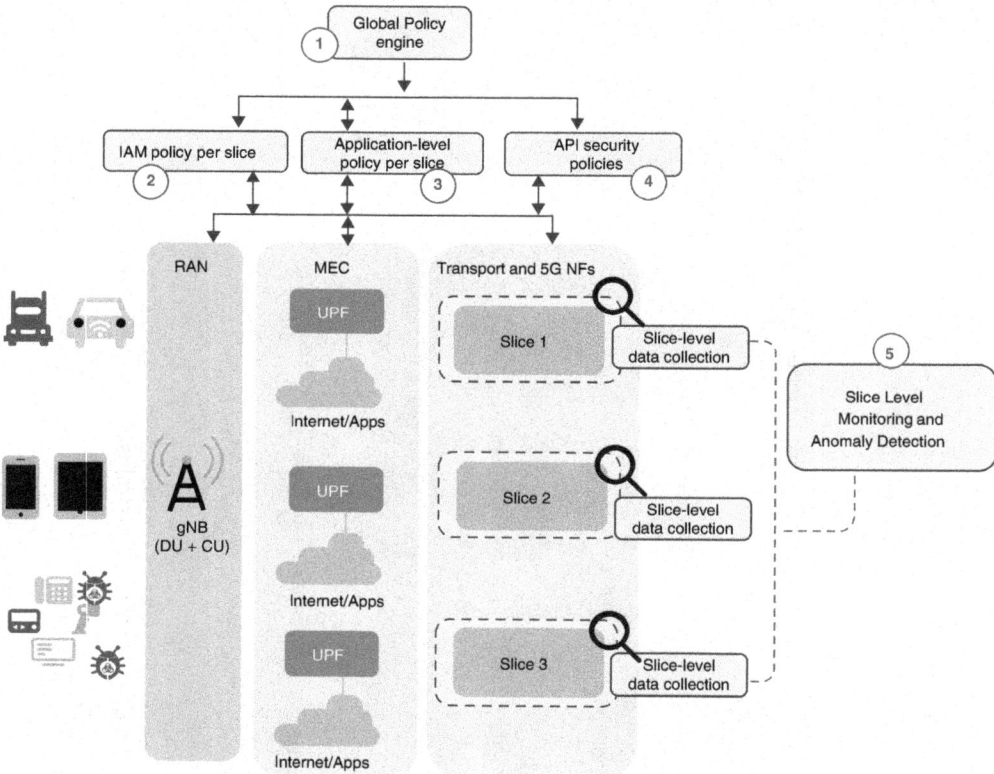

FIGURE 7-52   Key Security Controls to Secure the Network Slice in the Deployment Example

Figure 7-52 illustrates the five key security controls to mitigate the threats discussed in the "Threats: Case Study" section. The five key security controls are as follows:

1. Global Policy Engine

2. IAM policy per slice

3. Application-level policy per slice

4. API security policies

5. Slice-level monitoring and anomaly detection

To clarify the threat mitigation better, security controls are shown in the physical design in Figure 7-53 and described in the list that follows:

1. **Global policy:** The global policy engine can be considered the brain of the policy distribution and enforcement layers. You can apply custom rules to the global policy engine, which then is responsible for the final decision on granting access to users and devices. The global policy engine is then paired with the Identity and Access Management solution to ensure that the policies are applied to all the users and devices accessing the slice components, including the host on which the slice network functions are deployed. This topic is well covered in the section "Segmentation and Monitoring of the Management Layer."

2. **Identity and Access Management (IAM):** IAM ensures that repeated authentication and authorization based on zero-trust principles are applied to all the users and devices accessing the host software and hardware. It would generate any session-specific authentication and authentication token or credential used by a client to access an enterprise resource. This layer is also responsible for enabling, monitoring, and eventually terminating connections between users, devices, and the network devices and slice-level NFs. The IAM layer should be able to be integrated with local and external data sources such as industry-compliance systems, threat intelligence feeds, public key infrastructure (PKI), ID management systems such as LDAP servers, and security information and event management (SIEM) solutions.

3. **Application-level policy enforcement:** This layer ensures that application-level policy enforcement is applied to individual NFs deployed on the hosts. With the evolution of 5G deployments, the data centers will grow larger and much more complex, with hundreds or thousands of interdependent 3GPP and non-3GPP NFs. This will lead to a continual rise in complexity due to an increase in east-west traffic, application onboarding, virtualization, containerization, security threats, and cloud migrations for 5G-specific network functions. Having application-level security policies would help in enabling application-level access user and device policies instead of users and devices allowed access based on a pool of IP addresses. The chosen solution should also provide the following functionalities:

   ■ The solution should be able to gain complete visibility into slice-level NF and application components and their dependencies.

   ■ The solution should automatically generate microsegmentation policy based on application behavior.

   ■ The chosen application-level policy enforcement should provide a mechanism to integrate with your global policy engine.

   ■ The solution should enforce the application-level policy across all 5G multicloud workloads consistently. This will help to minimize any lateral movement of the attacker.

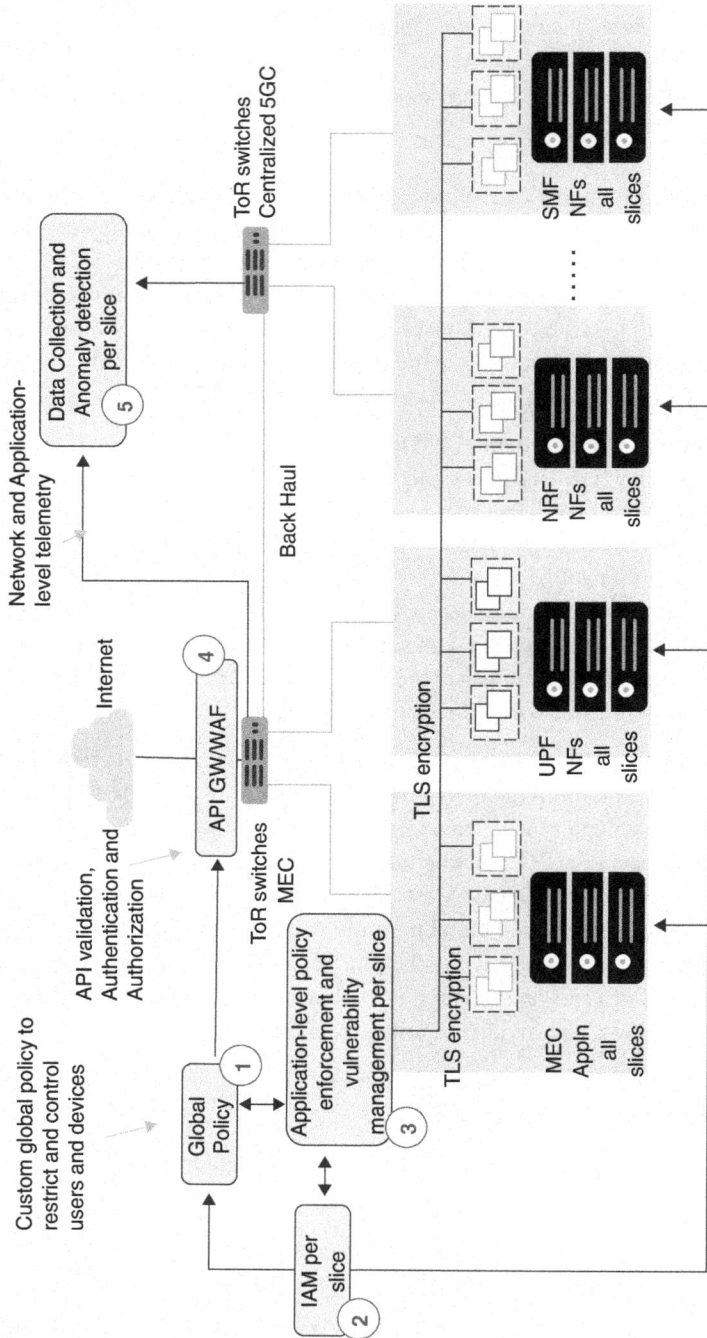

FIGURE 7-53   Key Security Controls Mapped to Actual Design of the Network

- The solution should be able to identify software vulnerabilities and exposures to reduce the attack surface.

- The solution should provide process behavior baselining and identify deviations for faster detection of any indicators of compromise (IoCs).

4. **API gateway:** The API gateway or web application firewall (WAF) layer is used to secure the API calls within the network slices. This includes the API calls from the NSaaS slice reseller and slice consumers toward the service providers. The API gateway will enforce access and authorization validation using the custom policies applied by the global policy engine on all API calls within the 5G network slice components using standard authentication, token generation, and multifactor authentication. Some of the key capabilities are as follows:

    - The solution should allow custom rules to be deployed or should be able to be integrated with an external global policy engine.

    - The solution should include multiple security models with ML and AI integrations to provide enhanced mitigation against zero-day attacks.

    - The solution should be able to terminate TLS sessions, such as HTTPS sessions.

    - The solution should be able to be implemented at any location (on-premises as well as public cloud).

    - The solution should provide visibility on the inbound and outbound web traffic.

    - The solution should provide multiple options to configure the rules (for example, using CLI, UI, and API).

   The key capabilities are further discussed in the "4. Secure API" section within this chapter.

5. **Monitoring and anomaly detection:** To ensure end-to-end monitoring and any anomaly in all network slices, a robust monitoring and anomaly detection solution should be deployed. Such solutions should provide end-to-end visibility at the slice level. You should be able to create rules per slice level, such as slice 1 should not interact with slice 2 and so on. Any variations from such rules should create an alert using which further investigations can be carried out. User-based per-slice rules should also be applied and any breaches identified. For example, User 1 usually interacts with Slice 1. An instance gets noticed when User 1 tries accessing the management of Slice 2. This is an anomaly in behavior, and an alert is sent to the security team and can then be investigated further.

Summary

As discussed in this chapter, network slicing will enable you to create an independent end-to-end logical network that runs on a shared physical infrastructure, capable of providing a negotiated service quality. The "Threats: Case Study" section took you through the different threats in the network slice deployment using SDN and orchestration, the mitigations of which were explained in the "Mitigations: Case Study" section. The section "Real Scenario Case Study: Threats in the 5G Network Slice, SDN, and Orchestration Deployments and Their Mitigation" took you through the multidomain threat examples and their mitigation.

Figure 7-54 shows you the multilayered security controls to secure the multiple layers for network slice deployments, including SDN, orchestration, and NSaaS layers.

Figure 7-54 shows the key security controls required to secure the various modes of network slice deployment. It has to be noted that you can share the security controls for other parts of the infrastructure if your 5G network deployment has other parts of the 5G infrastructure, such as 5G RAN, 5GC, and 5G MEC. The security control layers are summarized as follows:

- **Built-in hardening (hardware and software layer):** Network slice–related use cases are heavily dependent on the SDN and the orchestration layers. The vendors you choose for this layer should have a robust secure lifecycle process for developing the software, creating and applying the software update patches, and so on. SDN and orchestration layers will require automation to perform provisioning and dynamic scaling of the 5G 3GPP and non-3GPP network functions. Built-in hardening and security features of the vendor are important and should provide secure provisioning servers and secure communication with the provisioning servers. The SDN and orchestration solution being used should also encrypt data and separate management and production network traffic for additional isolation.

  The list provided in the "Hardening Hardware and Software" section in Chapter 5, "Securing MEC Deployments in 5G," is a good place to start the checks.

- **Enhanced access control layer:** Access to the network slice layers, such as the service layer, orchestration, SDN, and 5G network functions should be granular, role based, and should be applied at the slice level. Once your 5G network is mature, you will start deploying network slice as a service (NSaaS) business models. To provide mitigation of risk vectors such as unauthorized access in NSaaS deployments, it is recommended to have robust and granular access controls. Access for all the slice users should be restricted, and access should be provided only for their specific slice components. Any deviation of the user's behavior should have an alerting process involved to indicate the change in behavior.

- **Application protection and policy enforcement:** The solution used for application protection and policy enforcement should scan all images before they run and should enforce policy checking to ensure they are allowed to be executed in your environment. Such scans should also prevent the deployment of untrusted and vulnerable images and have the capability to block containers violating the runtime model based on the configured runtime rules. The policy enforcement layer should ensure you can optimize your multivendor firewall policy by updating it based on the application behavior, and it should remove access for third-party MEC applications that are decommissioned. The application protection and policy enforcement layer should also provide vulnerability assessment and any forensics capabilities to investigate any abnormal user or application behavior.

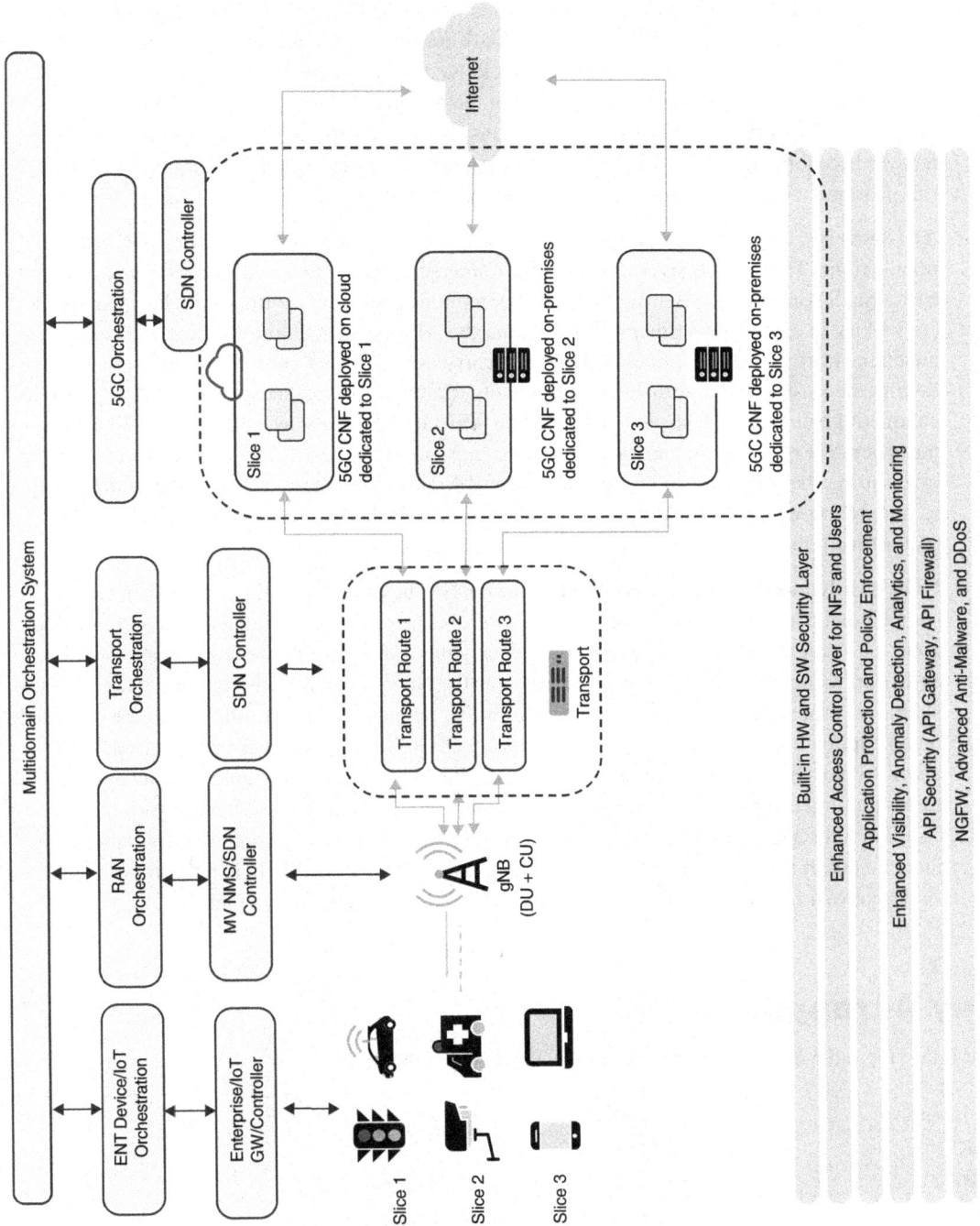

FIGURE 7-54 End-to-End Security Controls to Secure Network Slice Deployments

- **Enhanced visibility, anomaly detection, analytics, and monitoring:** The key to mitigating sophisticated malware attacks is to detect any anomaly within the MEC by using behavioral analysis, which could potentially detect the malware scanning, connection, and propagation throughout the network. Another type of attack that could be detected by enhanced visibility and monitoring is an attack taking place using the encrypted traffic, such as data being exfiltrated to a C2C server. The analytics part of this solution must also use the threat intelligence feed and combine this with the analysis of the gathered information using ML and AI algorithms to proactively detect any change in network behavior.

- **API security:** Orchestration and SDN layers use API for northbound interface (NBI) communications. The service layer uses an API to interact with the external NSaaS consumers and non-NSaaS consumers. Securing these API communications is very important because attackers will try to exploit the vulnerabilities within the API implementation and the web server to attack the critical network slices. You should carry out audits for your API communication, both in the trusted and nontrusted environments. Having a strong, secure API strategy, such as conducting penetration tests and audits on the APIs and applications using APIs, will ensure that your API is secure and any security holes/gaps are understood. The gaps can then be filled by ensuring the right feature or capability is enabled within the relevant security control, such as integration with the global policy engine to ensure granular access control for each API request from the NSaaS consumers.

- **NGFW, advanced anti-malware, and anti-DDoS:** The implementation of network slices should consider multiple layers of security controls due to the high interaction with untrusted zones and would require encryption of the data being transferred between the enterprise sites and your 5G network slice components, such as the Network Exposure Function (NEF), which should be protected by application-aware firewalls. Segmentation of traffic should be provided by an application-aware next-generation firewall (NGFW) that can filter traffic based on the application rather than just port and IP address. The NGFW should be able to utilize global threat intelligence feeds, consume telemetry data from applications and the network, and co-relate files and create context-aware knowledge to detect malware and identify patterns to indicate possibilities of zero-day attacks. DDoS protection should include rate limiting and should have a robust ML and AI engine to detect and mitigate attacks faster.

Key Acronyms

The following table expands the key acronyms used in this chapter.

Acronym	Expansion
5GC	5G Core
AMF	Access and Mobility Management Function
API	Application programming interface
C2C	Command and control

Acronym	Expansion
CA	Certificate authority
CAPIF	Common API Framework
CD	Continuous Delivery
CI	Continuous Integration
CNF	Cloud-Native Function
CoPP	Control Plane Policing
CP	Control plane
CU	Centralized Unit
CUPS	Control Plane User Plane Separation
DCT	Docker Content Trust
DDoS	Distributed denial of service
DPP	Data Plane Policing
DU	Distributed Unit
EAP	Extensible Authentication Protocol
ESP	Encapsulation Security Payload
FH	Front haul
FIPS	Federal Information Processing Standard
HSM	Hardware security module
HTTPS	Hypertext Transfer Protocol Secure
IAM	Identity and Access Management
IPsec	IP security
ISAKMP	Internet Security Association and Key Management Protocol
M2M	Machine-to-machine
MAC	Medium Access Control
MEC	Multi-access edge compute
MitM	Man in the middle
mMTC	Massive machine-type communication
NEF	Network Exposure Function
NR	New Radio
NRF	Network Resource Function
NSA	Non-standalone
NSaaS	Network slice as a service
NSSF	Network Slice Selection Function
RBAC	Role-based access control
SDN	Software-defined network
SLA	Service level agreement
TLS	Transport Layer Security

References

3GPP TS 28.530 V17.0.0

3GPP TR 33.813 V16.0.0

3GPP TS 24.502 V17.1.0

3GPP TS 23.501 V16.9

3GPP TS 23.502 V16.7.1

NIST SP 800-207

OWASP.org

https://www.cisco.com/c/en_uk/solutions/data-center-virtualization/application-centric-infrastructure/

https://www.cisco.com/c/dam/en/us/products/collateral/cloud-systems-management/network-services-orchestrator/nso-bridge-automation.pdf

https://www.openpolicyagent.org

Chapter 8

Securing Massive IoT Deployments in 5G

After reading this chapter, you should have a better understanding of the following topics:

- Threats in massive IoT use case deployments
- Securing massive IoT networks
- Real scenario case study examples of massive IoT threat surfaces and threat mitigation techniques

This chapter will take you through the threat surfaces in 5G massive IoT deployments and mechanisms to mitigate the threats.

This chapter will be of particular interest to the following teams from enterprise, industry verticals, Non-Public Networks (NPN), 5G service providers deploying 5G mIoT, and cybersecurity vendors planning product developments and new functionalities to secure 5G mIoT use cases.

- Mobile infrastructure strategy teams of service provider deploying mIoT in 5G

- Security strategy teams within service provider and enterprise verticals planning on deploying 5G mIoT

- Transmission and the packet core team within service providers and private 5G enterprises planning to deploy 5G mIoT

- Cloud computing and data center teams involved with 5G strategy and deployment

- Security architects and design teams looking at securing the public and non-public mobile infrastructure

- Solution and security architects deploying 5G mIoT on enterprises and industry verticals

- Enterprise solution and security architects using IoT services from mIoT service provider

- Government departments deploying 5G mIoT

- Cybersecurity vendor teams looking to secure mIoT deployments for their customers

- Product managers of cybersecurity vendors trying to identify use cases for new products or features to protect 5G mIoT deployments

5G represents a disruptive shift from just traditional consumer smartphones to advanced enterprise services, including ultra-reliable low-latency communication (URLLC)–based machine-to-machine (M2M) use cases. 5G is expected to be widely adopted in enterprise, industrial, and IoT use cases, enabling greater workforce mobility, automation, and countless new applications. Incorporation of 5G into these environments requires a deeper level of integration between end-user networks and 5G service interfaces, exposing both enterprise owners (in particular, operators of critical information infrastructure) and 5G service providers to new risks. Before we get into the risks and mitigation of risks, we will first need to look into the types of IoT use cases.

5G also sees a departure from the reliance on a single approach to authenticating all users onto the network-based SIM cards. The Third-Generation Partnership Project (3GPP) has addressed such shortcomings, with 5G now integrating the Extensible Authentication Protocol (EAP) framework, first adopted by Wi-Fi into WPA-Enterprise back in 2002, into its architecture. The 5G standard now provides examples of how to use EAP-TLS certificate-based authentication in 5G as well as other EAP methods that support mutual authentication. The list that follows outlines some of the key reasons why IoT threats are quite critical in 5G based on the excerpts taken from the Cisco Annual Internet Report (2018-2023):

- The number of devices connected to IP networks will be more than three times the global population by 2023. There will be 3.6 networked devices per capita by 2023, up from 2.4 networked devices per capita in 2018. There will be 29.3 billion networked devices by 2023, up from 18.4 billion in 2018.

- Globally, devices and connections are growing faster (10 percent compound annual growth rate [CAGR]) than both the population (1.0 percent CAGR) and the Internet users (6 percent CAGR). This trend is accelerating the increase in the average number of devices and connections per household and per capita. Each year, various new devices in different form factors with increased capabilities and intelligence are introduced and adopted in the market. A growing number of M2M applications, such as smart meters, video surveillance, healthcare monitoring, transportation, and package or asset tracking, are significant contributors to the growth of devices and connections. By 2023, M2M connections will constitute 50 percent of the total devices and connections.

- M2M connections will be the fastest-growing device and connections category, growing nearly 2.4-fold during the forecast period (19 percent CAGR) to 14.7 billion connections by 2023.

With this type of growth in the number of devices and spurts in new use cases such as M2M, an attack that successfully disrupts the network, or that steals or undermines the integrity of confidential data, could have a far greater economic and societal impact than previous generations.

IoT devices and applications have been around for quite some time and are not a new concept for 5G. There are networks today using LTE or NB-IoT technologies enabling IoT use cases. 5G offers flexibility in IoT deployment. The use cases aimed at 5G IoT are devices having different bandwidth requirements. Some require high bandwidth and transmit in burst, while some require low bandwidth and continuous connectivity. 5G offers this capability to support the massive number of devices with different bandwidth requirements. In addition, 5G also supports enterprise and industry use cases that have strict requirements on latency. This is one of the key reasons why the industry is looking at adopting 5G. The flexible mode of 5G deployment using network slicing and deployment of applications in the edge of the network can bring down the latency to 1ms or less, enabling ultra-reliable and low-latency use cases such as factory automation, enhanced vehicular technologies such as vehicle-to-everything (V2X), power and utility sector use cases such as smart energy grids, and other demanding use cases to become a reality.

There are different types of IoT use cases in 5G depending on the data consumption, energy consumption, and scale of deployment. When you take a step back and look at the use-case scenarios in 5G, we can split the IoT devices into smart devices and not-so-smart devices. Smart IoT devices are the devices that have some intelligence built into them and can make some decisions based on the input data. The not-so-smart IoT devices are the devices that just send the collected data and receive certain actions, such as stop data collection and a query to start data collection.

Use cases attributed to 5G such as smart cities would require the use of both types of devices, as shown in Figure 8-1, and have an artificial intelligence (AI), machine learning (ML), and an analytics layer to analyze the information from multiple devices and make a decision based on it. An example could be automated car parking in a busy area such as an airport parking lot, as shown in Figure 8-1.

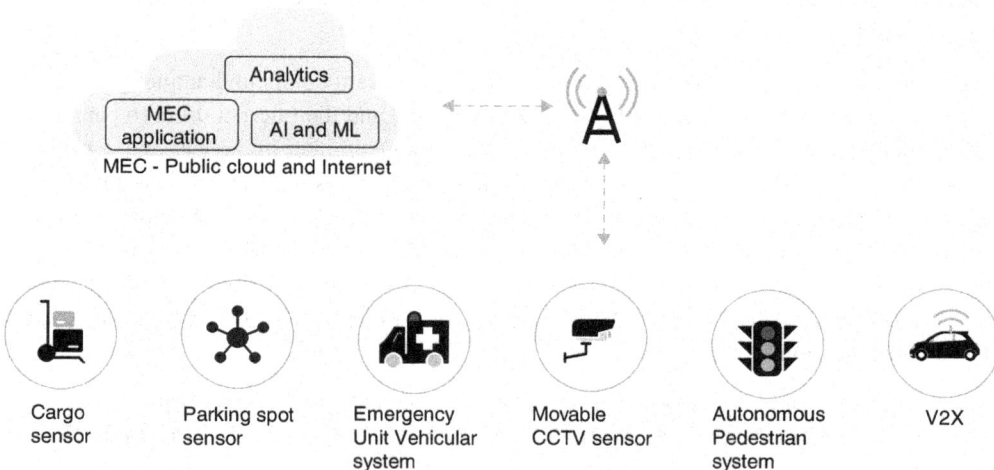

FIGURE 8-1   Different IoT Device Types to Enable a 5G Smart City

As shown in Figure 8-1, it would require different types of mIoT devices to enable the smart city use case. Table 8-1 lists the types of devices to fulfill the use case of finding a parking spot and the safest way to reach the parking spot.

TABLE 8-1   Different IoT Device Types

IoT Device	mIoT Device Type	Function
Cargo sensor	Not-so-smart device	Sends the geo-location metadata along with the speed
Parking spot sensor	Not-so-smart device	Indicates whether or not a vehicle is located in a parking spot
Emergency Unit Vehicular system	Not-so-smart device	Indicates whether an emergency vehicle is active in the location
Movable CCTV sensor	Not-so-smart device	Detects if there is movement near the parking spot
Autonomous pedestrian system	Smart device	Indicates any V2X application in the vicinity and broadcasts a message based on whether or not a pedestrian is crossing. Captures any speeding instances and sends data to the road safety officers. Indicates any collision and immediately broadcasts messages to the emergency health unit.
V2X	Smart device	Provides a road safety application such as intersection movement assist, provides emergency brakes, and also includes V2V (vehicle-to-vehicle) communications

As listed in Table 8-1, to fulfill this example of smart city–based parking, there is a need for both not-so-smart-devices and smart devices.

In this example, the cargo sensor, Emergency Unit Vehicular system, and autonomous pedestrian system are all part of the collision-prevention mechanism. The parking spot sensor and movable CCTV sensor are part of the parking detection mechanism. The V2X system is embedded within the vehicle for passing along the metadata to the MEC application.

All the data from the mIoT devices is then passed on to the AI and ML system and real-time (RT) analysis system. The AI, ML, and analytics system will then detect the free parking spot and the safest way to approach the parking spot and then help park the car or indicate the parking spot and the best way to reach it.

Massive IoT in 5G addresses the need to support billions of connections with a range of different services. IoT services range from device sensors requiring relatively low bandwidth to connected cars that require a similar service to a mobile handset. Network slicing provides a way for service providers to enable services to enterprises, giving them the flexibility to manage their own devices and services on the 5G network. mIoT, as the name suggests, is a category of use cases that is driven by scale.

Figure 8-2 illustrates an example of components that are part of the mIoT deployment.

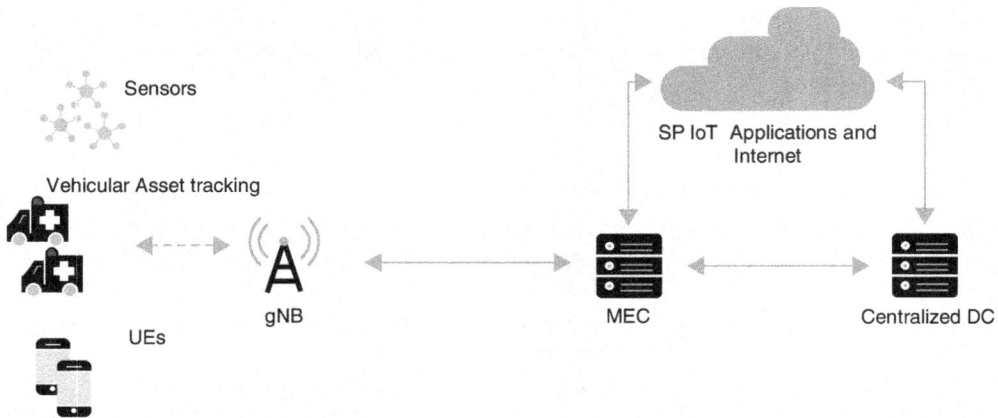

FIGURE 8-2   mIoT Deployment in 5G

Figure 8-2 shows an example of mIoT use-case deployment using 5G. The gNB serves geographically disparate devices such as sensors and vehicles that need to be tracked. mIoT would typically include devices that transmit and consume low data and are in the scale from hundreds to millions. Depending on the device type, it could be low-energy-consuming devices with limited access to power with a very light software stack for communications. There are device vendors in the market with 5G-capable chips with optimized power consumption.

This chapter will cover the 5G MIoT part. 5G IoT use cases based on smart devices (V2X, smart city, industrial IoT use cases, and so on) are covered in Chapter 9, "Securing 5G Use Cases.")

Massive IoT–Based Threats in 5G

Figure 8-3 shows the key threats for the device-based threats for the devices connecting to the service provider's 5G infrastructure. The devices in this case can be the 5G user equipment (UE), sensors, and IoT devices connecting to the 5G network provided by the service provider.

Figure 8-3 shows 5G multi-access edge compute (MEC), centralized 5GC (5G Core), public or private cloud-based SP applications, and the Internet access layer. Depending on the deployment plans of the service provider, the 5G User Plane Function (UPF) would be deployed in the MEC, along with any of the IoT applications that require caching. When the UPF network functions are deployed in the MEC, the N6 interface—the interface between the data network (DN) and the UPF—is also configured to allow UE and 5G devices to interconnect with the data network. Depending on the deployment scenario, the 5GC could host the 5G network functions that have low impact with higher latency, such as control plane functions, user plane functions for some IoT use cases, and the operations, administration, and maintenance (OAM) functions. Many service providers are also planning to have the configuration management (CM), fault management (FM), and performance management (PM) for the consumer IoT devices being catered to from the public/private cloud.

- Forced resource buffer overflows causing DoS
- Forced crash/shutdown due to malware injection casing DoS
- Interfaces from the application server to IoT Core components are based on the Restful API, which can be compromised leading to DoS attacks on 5G Core components

- Multiple new connection requests (artificial) causing control plane resources to exhaust or causing overloading on the 5G Core CP components leading to denial of service to legitimate devices

- DDoS attacks from IoT network

SP IoT Applications and Internet

MEC / Centralized DC

- Malicious code injected into the IOT applications during update causing IoT devices to cause DDoS attacks on 5G core
- Data exfiltration caused due to exploited code within IoT applications
- IoT device provisioning server (CM/PM/FM) exploits, leaving the entire IoT device ecosystem open to be exploited and taken under control

- Bot-based attack causing simultaneous response/request messages to/from IoT network
- MitM attacks in air interface due to weak encryption

- RFID/Bluetooth sniffing and eavesdropping on the IoT device causing messages to be intercepted, modified, and retransmitted with false information
- Spoofing another device on the network and collecting data (silently)
- Malicious code injection leading to same device to be seen on multiple locations with separate IP address
- Multiplying the number of nodes (artificially causing increased UL/DL signaling)

- Forced resource buffer overflows causing DoS
- Forced crash/shutdown due to malware injection causing DoS
- Malicious code injection on driver, which compromises HW causing DoS
- Compromised protocol on IoT device causing malicious code injection on primary device connected to IoT device
- Firmware OS hacking/code injection leading to compromised device

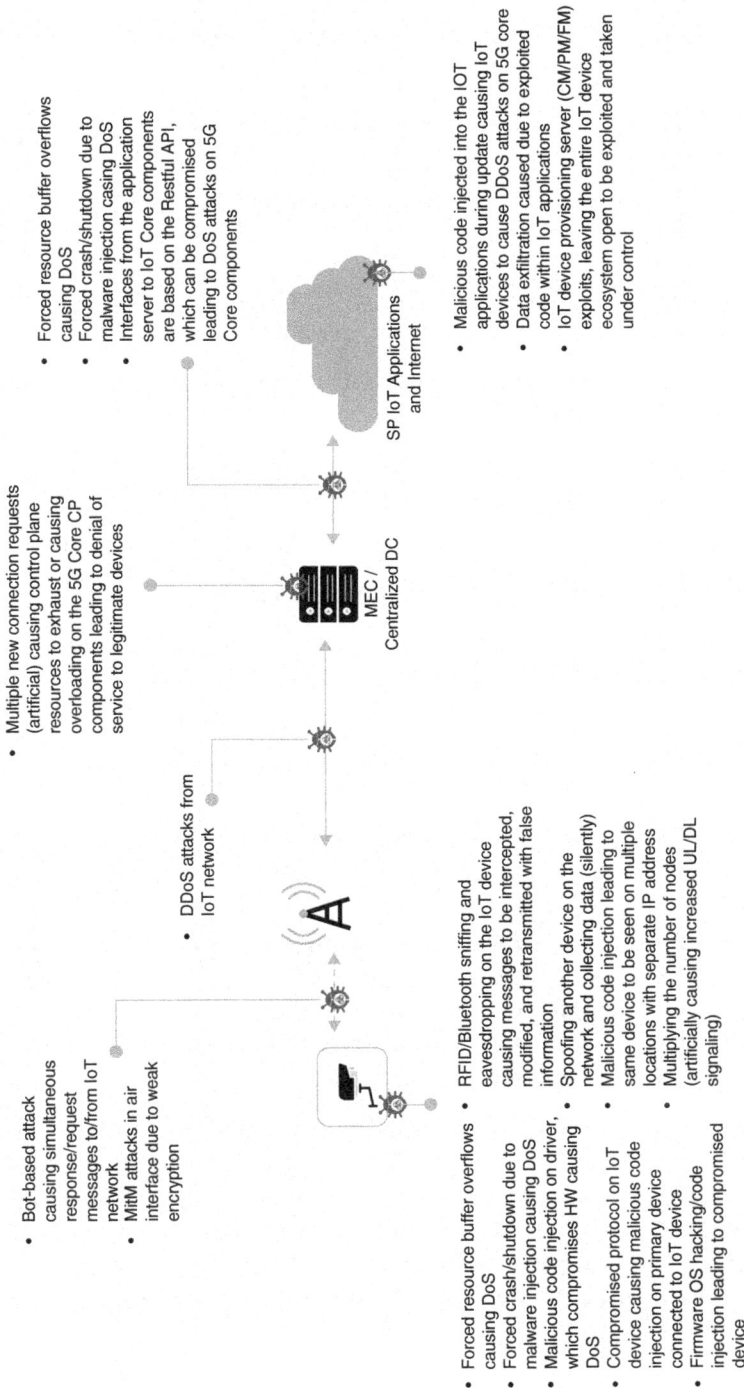

FIGURE 8-3 mIoT Threat Surface in 5G Deployments

The majority of the threat surfaces illustrated in Figure 8-3 are primarily due to the device vulnerabilities and the devices being compromised by the command and control (C&C) server.

Here are some of the key threats related to mIoT use cases within the 5G networks:

- C&C-based attacks

- Malicious code injection on the driver that compromises the hardware, causing a denial of service (DoS)

- Forced resource buffer overflows causing DoS

- Forced crash/shutdown due to malware injection, causing DoS

- Compromised protocol on an IoT device, causing malicious code injection on the primary device connected to the IoT device

- Firmware OS hacking/code injection, leading to a compromised device

- Radio-frequency identification (RFID)/Bluetooth sniffing and eavesdropping on the IoT device, causing messages to be intercepted, modified, and retransmitted with false information

- Spoofing another device on the network and exfiltrating data

- Malicious code injection leading to the same device being seen at multiple locations with separate IP addresses

- Multiplying the number of nodes (artificially), causing increased signaling in both UL/DL

Device Vulnerabilities Due to Weak Built-in Security

mIoT devices usually have very weak built-in security mechanisms due to lower price points of the devices to make them affordable to a large consumer base. The IoT deployment of any type, be it based on smart IoT devices or not-so-smart IoT devices, needs to be catered to by robust security controls to mitigate the vulnerabilities introduced by weak built-in security mainly due to the low cost and limitations due to the form factor. Non-mIoT use cases that are not geographically located would also need multilayered security controls to secure them from targeted attacks very specific to industry verticals, such as major automotive manufacturers or government utility verticals.

Spoofing, cloning, and eavesdropping on the 5G endpoints/IoT devices can be carried out by attackers impersonating an RFID or Bluetooth device and reading and recording the transmitted data from the 5G-enabled IoT device. This is primarily made possible due to weak access controls and poor authentication methods used by the IoT device. These kinds of attacks are more prevalent in verticals of IoT such as healthcare where the IoT devices use Bluetooth to transfer the patient's health statistics to a tablet where the vital stats of the patient can be checked/monitored by the healthcare workers.

Another type of attack mentioned in Figure 8-3 is where the devices are compromised. In this instance, all the data from the impacted devices is dropped or redirected instead of being transmitted to the intended receiver for further forwarding or analysis. The data from such devices can then be analyzed by the attacker for any valuable data points, such as the IP address of the receiver, which can then be targeted for DoS.

These kinds of attack methods can also be referred to as *sinkhole attacks* or a form of *routing attack*. This is because the method of attack used in such instances is to route the packets away from the main intended receiver. To prevent the detection of such attacks, the data can be mirrored to the malicious data collection server using a method very similar to port mirroring or Switch Port Analyzer (SPAN), which is used quite commonly in the network monitoring environment of the service provider networks. SPAN copies (or mirrors) traffic received or sent (or both) on source ports or source VLANs to a dedicated destination switch port for analysis. You can analyze network traffic passing through switch ports or VLANs by using SPAN or Remote SPAN (RSPAN) to send a copy of the traffic to another port on the switch or on another switch that has been connected to a network analyzer or other monitoring solution.

Management layer–based attacks are another key concern for device-based attacks within 5G. In these attacks, the attacker tries to take control of the key management layers, such as CM, FM, and PM, by exploiting the existing vulnerabilities of the IoT vendors' management platform or the open source components used in the vendors' IoT platform. Once the vulnerability has been successfully exploited, the attacker gains access and control over all endpoints catered for by the IoT vendor for the service provider. This can now be used for DoS and distributed denial-of-service (DDoS) attacks. One of the methods the attacker could also use here is to change the encryption type or level (from encrypted to null encryption), which makes the entire IoT network susceptible to man-in-the-middle (MitM) attacks.

The key threat surfaces and vulnerabilities are discussed in more detail in the sections that follow.

Supply Chain Vulnerability

Supply chain vulnerability is a well-known issue across different industry segments. The challenge of supply chain vulnerabilities becomes more prominent in 5G, as it enables attaching millions of low-cost IoT devices to the network. 5G also introduces critical infrastructure–based use cases and caters for use cases like smart cities, defense, and so on. These critical infrastructure 5G IoT use cases attract more nation-state attackers and thus are under higher levels of risk for cyberattacks. Supply chain is one of the weak links in security. If not secured properly, it opens the door wide for attacks, and the impacts of the attacks could be devastating, depending on the use case where the vulnerable IoT device was used. This section will take you through the vulnerabilities in the IoT supply chain related to manufacturing and distribution, as shown in Figure 8-4.

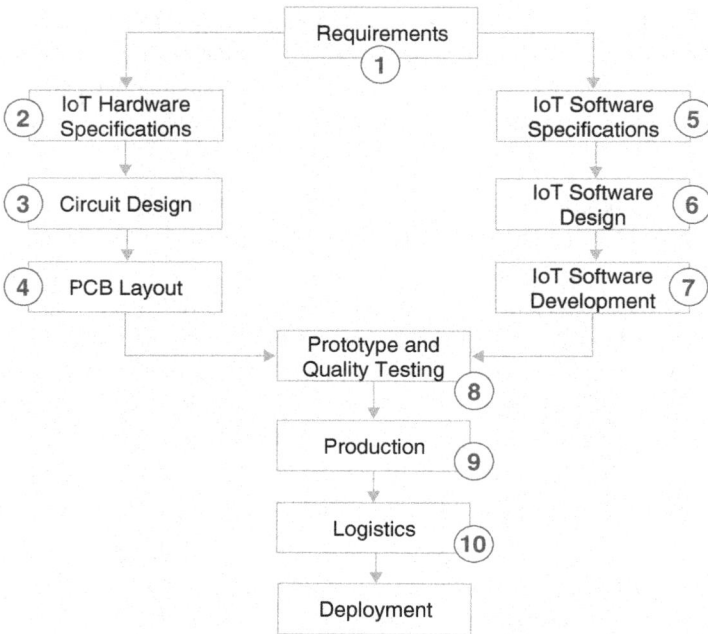

FIGURE 8-4   Vulnerabilities in Different Stages of the Supply Chain

Key vulnerabilities and threat vectors for the IoT supply chain related to manufacturing and distribution are explained in the list that follows:

1. The requirement stage is when you send the requirements for your IoT device to the vendor. This will include details like maximum energy consumption, dimension of the unit, maximum/ minimum temperature, pressure (depending on use case), software or platform requirements such as integration options using API, and so on. The threat vector here is the requirement that is actually passed on to the vendor product R&D and manufacturing team. An attacker might add a couple of details in the requirements not actually requested by you. These newly added details are aimed at creating the backdoor using hardware or software remodifications to the original design, which can then be exploited by the attacking entity once deployed.

2. The hardware specification team would normally take the requirements from the customer and map them to the required hardware, including deciding what sort of components should be used in manufacturing the device. Typical considerations are values to withstand humidity, temperature, power consumption, and so on. The threat vector here is that an attacker could choose certain components that will fail when a certain condition is met. For example, the malicious actor or the attacking entity could intentionally choose a substandard electronic component or a customized component that fails after a certain temperature or humidity level is reached.

3. Once the components are finalized, the design team would make a schematic of the design that will be used as a blueprint for the printed circuit board (PCB) manufacturing for the IoT device. This is a very important part of the manufacturing process, as all the further checks on quality and so on would be referred back to the schematic. The attacking entity or the malicious actor could alter the design to include an eavesdropping component to leak sensitive data to a predetermined destination such as a C&C server.

4. The PCB layout process and component soldering are the next steps after the circuit design process. Here, the key vulnerabilities and threat vectors are due to the attacker choosing counterfeit electronic components causing intermittent failures that are difficult to find and correct.

5. IoT software specifications are taken from the requirements list you have provided to the IoT vendor/manufacturer. A member of the IoT software specification team or an attacker working in the software specification team could be directed to modify the specification for the software. The software specification will also be used in the software quality process for validating the software and to ensure that the designed software meets the software specifications. Any modification done in the software specification process will be considered as the software blueprint for the device.

6. The software design team would follow the specifications set by the software specifications team and specify the architecture and software technology to be used. In this process, the vulnerabilities are mainly due to the lack of knowledge about security leading to weak software for the device.

7. The software development team programs the IoT device with the chosen software language. With attacks aimed at software vulnerabilities on the rise, it is imperative that the software team follows secure software design and avoids known vulnerabilities such as buffer overflows, which occur when there is more data in the buffer than it can handle, leading to software crash and thus creating a point for cyberattack. This can be intentionally implemented by an attacker within the software development team. Another threat vector is when a team member of the software development team is instructed by an attacker or an attacking entity to include malicious code within the program to allow a backdoor entry to the device or to the private network where the IoT device is deployed.

8. In the post-PCB layout and software development process, the IoT device manufacturer would validate whether the hardware prototype and software fulfill the requirements set by your (or your customer's) IoT device requirement. This is the last part of the process when a vulnerability can be identified and patched. If the quality team is compromised by an attacker, the specific vulnerability that is planned to be exploited by the attacker/attacking entity will be overlooked and will not be patched. This will leave the IoT device open for any attacks.

9. One of the key vulnerabilities in production is shadow production. Shadow production is where the real production numbers are hidden and used to flood the market with IoT devices with backdoors and vulnerabilities, making the devices open to attacks. Another threat vector is where the Joint Test Access Group (JTAG) ports are left unsecured. JTAG is an interface

that provides an option for debugging, reprogramming, and so on. In many gaming consoles, the JTAG ports are unsecured and open to user access. If you had the common interface cable for JTAG, you could plug it into your computer, use manufacturer default credentials, and play pirated games with some modifications on the attributes using the JTAG ports. The same unsecured JTAG port in an IoT device can allow an attacker to have unauthorized access and possibly have access to the private network where the IoT devices are deployed. The physical attacks, such as injecting malicious code into the IoT network, can be made possible by tampering with an IoT endpoint, gaining control over it, and then using that endpoint to gain access into the central IoT network. Attackers also exploit the JTAG interface used by manufacturers for debugging purposes. JTAG is an industry standard for on-chip instrumentation in electronic design automation (EDA). JTAG is also used to program field-programmable gate arrays (FPGAs). Most CPU vendors still use JTAG for debugging purposes. If JTAG ports are left unprotected, this interface can become a critical attack vector on the system.

10. Logistics is the other vulnerability in the supply chain that is prone to sabotage or modification of the IoT devices while in transit. Though this is not the most preferred attack vector for IoT devices in the supply chain, for critical infrastructure use cases, logistics needs to be carefully monitored. Your supply chain risk management (SCRM) should ensure that you have the right controls, such as choosing validated and security-cleared logistics vendors for shipping and transportation of IoT devices from production to deployment.

The attacks are primarily aimed at data exfiltration, tampering with the files within the IoT network, and gathering information. With the control garnered over the IoT network, the attacker could control the operations and the data flow between the IoT network and the 5G network components, such as a radio (gNB) or storage/configuration in the MEC layer of the 5G network. With the control over the IoT network, the attackers can damage the IoT devices and disrupt the IoT service, thereby causing DoS to service providers' IoT services. This is not a new threat vector for 5G technology specifically; it is prevalent in legacy technologies such as 2G, 3G, and 4G, but it's critical for 5G technology, as it is aimed at enabling IoT use cases such as mIoT that would impact different government and private sectors.

Command-and-Control Servers and Botnets

A command-and-control server (also referred to as a C&C, C2C or C2 server) is an endpoint/device that is compromised and controlled by an attacker. Devices on your network can be commandeered by a cybercriminal to become a command center or a botnet (a combination of the words "robot" and "network") with the intention of obtaining full network control. Establishing C&C communications via a Trojan horse is an important step for attackers to move laterally inside service provider networks, infecting machines and servers with the intent to exfiltrate data.

One famous example of botnet malware is Mirai, which causes its infected devices to scan the Internet for the IP address of IoT devices by using a table of common factory-default usernames and passwords. The Mirai malware then logs in to the IoT devices and infects them with the Mirai malware.

This method of malware infection can impact millions of IoT devices, turning them into botnets and launching DDoS attacks toward the service provider infrastructure.

Devices connected to 5G networks can implement botnet C&C protocols using a number of methods, such as Telnet, Internet Relay Chat (IRC), peer to peer (P2P), and domains, as described in the sections that follow.

Telnet

Telnet is an application-layer client/server protocol that provides an interactive and bidirectional text-oriented communication facility using a virtual connection terminal. Telnet originated in 1969 and was specified in RFC 15, with an extended version in RFC 855, and standardized as Internet Engineering Task Force (IETF) Internet Standard STD 8.

Telnet, by default, does not encrypt data sent over the connection, including passwords, which makes it susceptible to eavesdropping. Authentication is also not widely used in Telnet networks, thus leading to MitM attacks aimed at using the information collected later for malicious intent. Telnet botnets use a simple C2 botnet protocol in which bots connect to the main command server to host the botnet. Bots are added to the botnet by using a scanning script, and the scanning script is run on an external server and scans IP ranges for Telnet and SSH server default logins. Once a login is found, it is added to an infection list and infected with malicious code via SSH from the scanner server. When the SSH command is run, it infects the server. The infected server is now completely controlled by the control server and becomes part of the bot network. Once servers are infected, the bot controller can launch DDoS attacks of a high volume.

Internet Relay Chat (IRC)

IRC, created in 1988 and specified in the RFC 1459 standard, is an application layer protocol that works on the client/server networking model and facilitates communication in the form of text. IRC was designed for file sharing, group chats, and private chats, as illustrated in Figure 8-5.

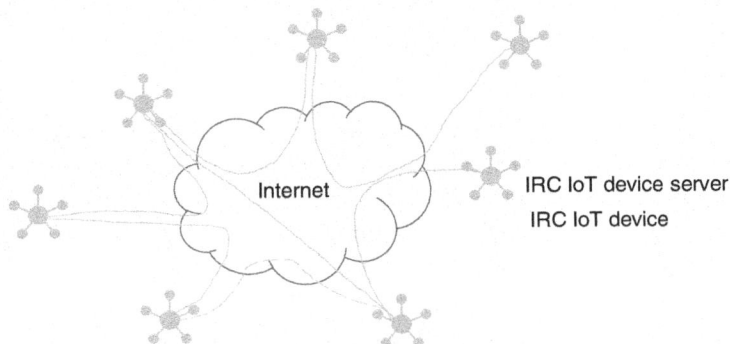

Internet

IRC IoT device server
IRC IoT device

FIGURE 8-5 Internet Relay Chat

Because of its simple and low-bandwidth methods of communication, IRC has been exploited to coordinate DDoS attacks that continuously switch chat rooms (also referred to as conversational channels or IRC channels) to avoid being neutralized. Figure 8-6 shows how the infected IRC devices capable of 5G can launch an attack on the 5G infrastructure.

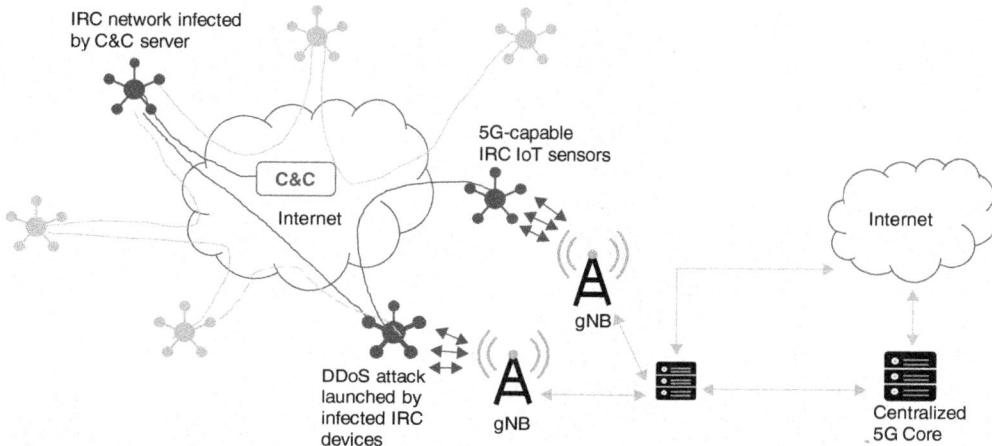

FIGURE 8-6   Infected IRC Devices Launching DDoS Attack on 5G Network

The IRC devices can be scanned by the IRC C&C servers looking for devices that have default usernames and passwords in its list. If the 5G-capable IRC devices have factory-default usernames and passwords, they can be infected and then scan the network for other vulnerable devices. This will cause a chain reaction and impact all the devices and endpoints in the sensor network. Once the attacker is in control of the sensor network, a DDoS attack can be launched using the devices to force signaling toward the 5G network, or the data from the sensor network can be exfiltrated. Some of the malware existing today can also be stored in the memory of the device, making it immune to mitigation via rebooting.

Peer to Peer (P2P)

P2P is a distributed application architecture that distributes tasks between peers. In a P2P network, the "peers" are endpoints (which can act as a file server and client) and are connected to each other via the Internet. The files can be shared directly between them without the need of a centralized file-sharing system, as illustrated in Figure 8-7. Hackers are now aiming IoT botnets based on the P2P network, as a majority of the IRC domains are being taken down and are no longer viable targets.

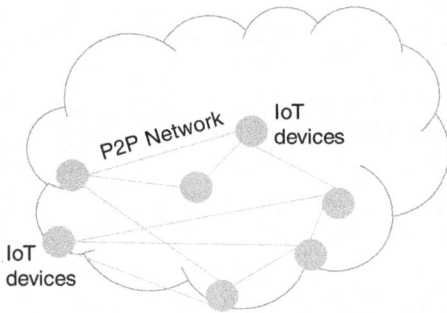

FIGURE 8-7   Peer-to-Peer Network

P2P technology allows lots of flexibility in consumer devices for easy access to the consumers, but it also makes the devices quite vulnerable. Figure 8-8 illustrates the threat surface in a 5G network using P2P. P2P technology, if not implemented with proper authentication, can allow any potential attacker to establish a P2P connection with the device and bypass security controls such as firewalls.

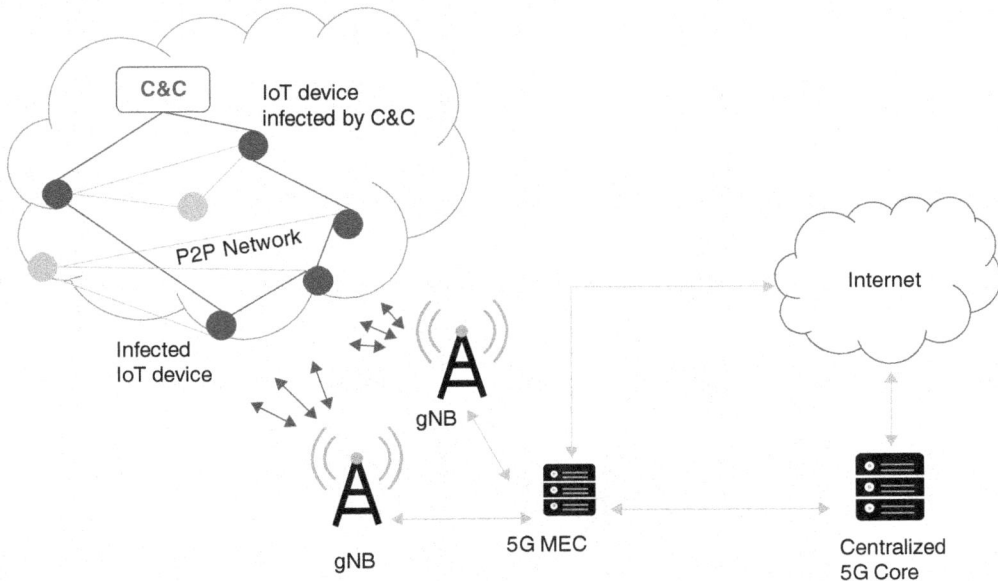

FIGURE 8-8   Infected P2P Devices Launching DDoS Attack on the 5G Network

Many of the IoT devices use the factory-default credentials (usernames and passwords), and the software used in these devices is quite old and mostly has no built-in security mechanisms such as authentication and encryption. The P2P communication used by many of the P2P devices is necessary for Plug and Play (PnP) features, as this offers a very simple way of deployment, without much configuration, and usually pulls a pre-filled template for Day 0 configurations from a configuration server usually deployed in the public cloud.

When such devices are deployed with 5G chipsets to enable communication with the 5G network, or if the devices communicate using non-3GPP mechanisms such as Wi-Fi and are then integrated into the 5G network, there is a high probability of attacks from such vulnerable devices. The Mirai malware has proved this attack method to be successful, as it used just the factory-default credentials to infect devices around the world.

DNS-Based Attacks

Attacks that leverage the Domain Name System (DNS) mechanism as part of their overall attack strategy, such as cache poisoning, are considered DNS attacks. The DNS-based attack is the most scalable method used by hackers in their attempts to deploy bots that are aimed at launching DDoS attacks or causing mass-scale data exfiltration. Although this method is scalable, it is easily identified due to the large bandwidth consumption by a specific domain name. DNS blocking techniques, where specific domains are blocked, are used to mitigate these attacks. A threat surface that IoT devices could use to circumvent the DNS blocking is the Fast Flux DNS technique, as illustrated in Figure 8-9.

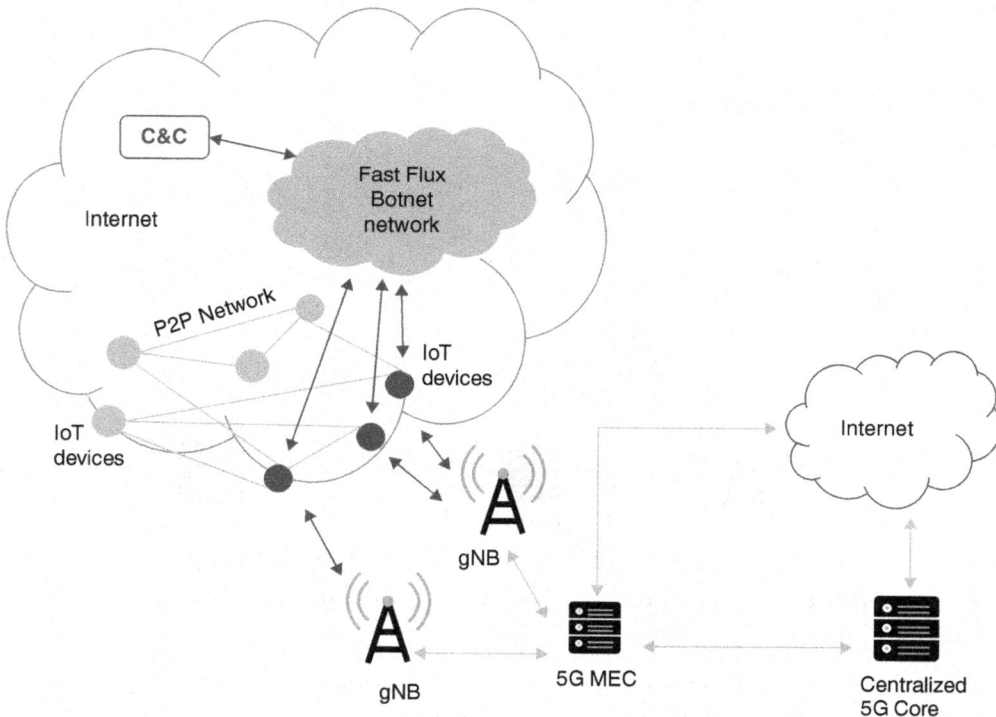

FIGURE 8-9   Devices Impacted by Fast Flux Botnet Network Launching a DDoS Attack on the 5G Network

To prevent the malicious control servers from being tracked down, the Fast-flux DNS technique is used by botnets. Fast-Flux DNS allows the fully qualified domain name (FQDN) to have multiple IP addresses associated with it that are very dynamic in nature. This is made possible by the Fast-flux DNS technique, as it uses a combination of load balancing, C&C, proxy redirection, and P2P networks. The malicious control servers can also hop between DNS domains by using domain-generation algorithms that generate new domains for the control servers.

Securing mIoT Deployments in 5G Networks

Mitigation of the device-based threats in 5G requires architectural strategy, rather than just deploying mitigation mechanisms for the threats from vulnerable devices.

The topic of mitigating the threats becomes important when there is an attack detected from the devices or when data exfiltration is detected in the service provider's public or private cloud infrastructure due to the vulnerabilities from the IoT device. A security framework that ensures that such projects are reviewed by the security team needs to be established and (more importantly) closely followed and reviewed (at least annually). The ETSI Technical Committee on Cybersecurity (TC CYBER) has also unveiled ETSI EN 303.645, which is a standard for cybersecurity in IoT that establishes a security baseline for IoT-connected consumer products and provides a basis for future IoT certification schemes.

Figure 8-10 illustrates the different layers of security required to secure the IoT environment in a 5G network.

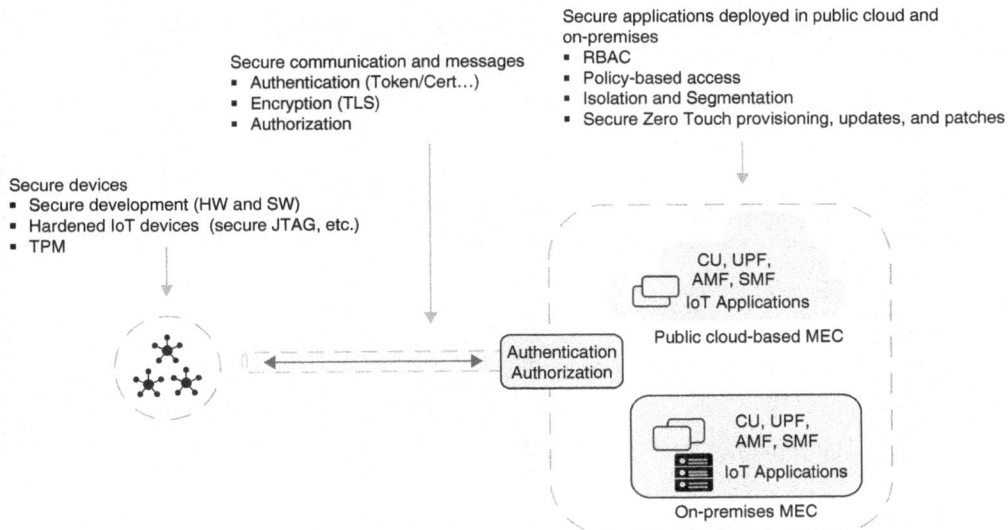

FIGURE 8-10   Multiple Layers of Security Required to Secure the IoT Device Communication with the 5G Network Functions

For the technical mitigation techniques, there are multiple layers of security capabilities and controls that should be put into place to ensure that the service provider's infrastructure is secure. A lot of the security controls depend on the deployment model chosen by the service provider. This section focuses on the mitigation techniques and also provides graphical illustrations for clarity. Due to the deployment models in 5G, such as non-public/private 5G deployment, enterprise use cases are also explained in this section.

Built-in Hardening of the Device

Service providers planning to deploy 5G devices and consumer IoT devices should ensure key built-in security capabilities are catered for, as shown in Figure 8-11 and detailed in the following list:

- The device provides anti-tampering functionality.

- The device should provide a hardware root of trust.

- The device should support secure boot capabilities as specified by the Unified Extensible Firmware Interface (UEFI).

- The hardware running is authentic, and processes are in place to protect network devices from being tainted and modified.

- The operating system is trusted and hasn't been modified.

- The management of the network device is properly authenticated, audited, and monitored.

- Any secrets stored within the network device can't be extracted or changed without proper authorization, thus providing secure data-at-rest functionalities.

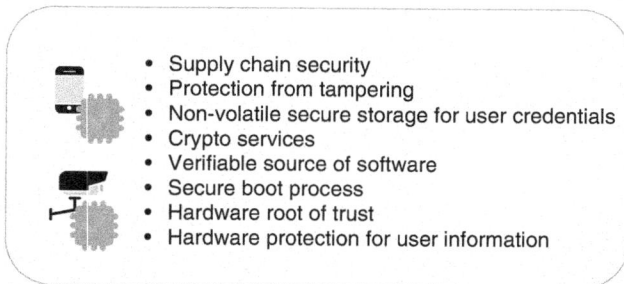

- Supply chain security
- Protection from tampering
- Non-volatile secure storage for user credentials
- Crypto services
- Verifiable source of software
- Secure boot process
- Hardware root of trust
- Hardware protection for user information

FIGURE 8-11   Built-in Hardening of the 5G Devices

Securing the supply chain is one of the most important topics in IoT security. This becomes more important for mIoT devices because they are manufactured with cost in mind. This leads to many IoT service providers choosing vendors and manufacturers with the lowest list price. Although this makes perfect business sense, it could lead to a disastrous consequence, with the service provider network being attacked, causing network outage, or worse, it could lead to sensitive customer information being leaked and sold to malicious entities. To secure the mIoT deployments, there should be multilayered security controls and a selection of manufacturers and vendors with secure supply chain processes.

The key methods of hardening IoT devices and securing supply chain for mIoT deployments, as illustrated in Figure 8-11, are discussed further in the sections that follow.

Securing Supply Chain

Securing supply chain risks and vulnerabilities is not a new topic and has been researched for multiple industry verticals such as pharmaceutical, defense, and so on. Specific departments are often dedicated to verifying vendor selection and checking to validate that the vendors conform to regulations such as ISO9001, ICH Q10, Q9, and so on. This requires the use of approved suppliers and a *defined supply chain*. Even in these verticals, however, the awareness of the subject has been historically poor and effort is lacking in the direction of securing the supply chain, unless certain leakages have been published in the media. These checks and validations are also very seldom done when IoT vendors and manufacturers are chosen by the service providers. Again, the key reason is cost; in this instance, it is cost savings by service providers to reduce the resources required to choose vendors and manufacturers. Several IoT vendors and manufacturers are working hard to improve their defense against supply chain attacks and vulnerabilities. This section will help you with some key supply chain methodologies that can then be verified against the IoT vendors and original equipment manufacturers (OEMs) you plan to use as part of mIoT deployments:

- The IoT device manufacturer/vendor should have a dedicated list of supplier bases for components through which the source of the components used for manufacturing can be identified.

- There should be strict background checks and verifications for the IoT device manufacturer/vendor employees and contractors working on the hardware and software design and development process.

- The IoT device manufacturer/vendor should have a dedicated team with risk management training.

- The IoT device manufacturer/vendor should have visibility of the entire process and be able to provide you with an example of what has been done for a similar IoT device without having to expose the name of another client. In other words, there should be an identifiable process to identify the engineers and architects working on a specific IoT device manufacturing process.

- The schematic of the IoT device following your requirements should be verified and audited by an independent electronic design auditing organization to ensure that there is no unnecessary component design or the use of unnecessary components.

- Components such as chipsets used in the IoT device should not be end of life (EOL) or near EOL for at least 5 years from the date of manufacturing. You will need to work with the manufacturers for the component selection to enhance the life span of the IoT device.

- The IoT device manufacturer/vendor should have a software vulnerability tracking and scanning program in place as part of the software testing process and should be able to provide you with a vulnerability report that can then be audited by an independent software auditing company. There are many instances when the software testing team sees the programs throwing up security exceptions, and the manufacturer will just ask the team to find a workaround instead of resolving the security exceptions. This will be difficult to detect. By having a report on the independent audit undertaken on the software by the manufacturer, you will have a closer view of exceptions that do not exist in the final software/firmware being deployed in the IoT device.

- The IoT device manufacturer should add a hardware security module (HSM), such as a secure element or Trusted Platform Module (TPM), on the device. Having an HSM within the IoT device will give you the ability to generate private keys and sign X.509 certificates.

Hardware Root of Trust

Having IoT devices that use signed images and hardware-anchored secure boot prevents malicious or compromised code from booting. Anchoring the first code in the boot sequence in hardware establishes a "chain of trust" and is the foundation of the secure boot process. Methods such as secure boot, which is one of the features of the UEFI specifications, is used to cryptographically verify the authenticity of the OS boot loader and the OS kernel as part of the boot process. Secure boot is commonly used to protect against BIOS rootkit attacks in server operating systems like Linux and Microsoft Windows Server.

A hardware root of trust is the most important component of a hardened device. Software alone can't prove its integrity; truly establishing trust can only be done in hardware, by using a hardware-anchored root of trust. To be effective, this root of trust must be based on an immutable hardware component that establishes a chain of trust at boot time. Each piece of code in the boot process measures and checks the signature of the next stage of the boot process before the software boots. Without a hardware root of trust, no amount of software signatures or secure software development can protect against a compromise of the underlying system.

Trusted Platform Module (TPM), specified by the Trusted Computing Group (TCG) and standardized by the International Organization for Standardization (ISO) and the IEC (International Electrotechnical Commission) in 2009 as ISO/IEC 11889, is a specific processor designed to secure the hardware by using integrated cryptographic keys. Each TPM has a unique and secret Endorsement Key (EK) assigned to it during production. This enables the primary scope of ensuring the integrity of the platform.

Apart from verification of the device integrity, TPM also helps in device identification, authentication, and tamper-resistant encryption. Today, the TPMs are mainly standalone crypto chipsets and can be implemented by various device manufacturers (including consumer IoT devices) into their devices. As cost is the primary factor while manufacturing the consumer IoT devices, specific devices that are supposed to cater for 5G can be implemented with TPM capabilities.

Just having TPM capabilities doesn't mean that the devices have to be blindly trusted. As threat surfaces are always evolving and becoming more sophisticated, service providers must ensure that the monitoring of the TPM and secure boot functionalities is also taken into consideration while deploying 5G devices/5G consumer devices in the network.

Some device manufacturers use Trust Anchor module (TAm), which tries to mitigate supply chain threat vectors by implementing highly secure storage for user credentials, passwords, and settings. Specifically, to mitigate supply chain attacks, some manufacturers also use the Secure Unique Device Identifier (SUDI), which is inserted during manufacturing and removes any requirement for manual intervention.

As discussed in this section, TAm provides a permanent and secure unique identification of the device that does not require manual intervention, thus providing mitigation against supply chain attacks if implemented properly, and it can be used by manufacturers in the mass-production of devices. Adoption of TAm will require a very close look at the secure manufacturing process of the vendor/manufacturer, which should follow a very stringent auditing process to ensure robust supply chain security. TPM, on the other hand, is more flexible because it doesn't necessarily require the vendor-specific unique identifier. In turn, it provides hardware protection for user certificates and integrity information.

The most crucial point for service providers would be to verify their existing device ecosystem that has been deployed (or is planned to be deployed) and understand the threat vectors. The less hardened the device is, the more layers of isolation it requires.

Identification, Authentication, Access, and Certificate Management

Device authentication is one of the key security controls that needs careful planning. This is because, depending on your network plans (that is, marketing team plans), the amount of mIoT devices might be in the hundreds of thousands or even in the millions in a couple of years. You might even need to cater for the IoT devices that are deployed in private 5G/5G NPN networks, depending on the SLA between you and the NPN customer. Looking at the growing ecosystem of partner models for service providers using NSaaS business models, you need to ensure that there are multiple authentication layers. An example could be reauthenticating the devices in a slice using an AAA method specific for the slice, even though these device have been authenticated using 3GPP 5G primary authentication mechanisms. This all sounds simple when it's written in a few words or shown in a figure with a few colorful boxes, but it really needs brainstorming sessions to understand the use cases and then to plan the device authentication layers. Also, you need to ensure that the device has end-to-end communication secured. If the device authentication layer is neglected or not thought through, you can spend millions in building your security architecture and still leave your network open to various attacks.

While discussing the topics of authentication with many service providers, one thing became very clear. There are many different types of IoT devices deployed already that support different kinds of protocols, many of which don't even support certificate-based authentication—forget having an IPsec stack or any form of software-based authentication or hardware-based authentication/trust and so on. So, for having a pragmatic view, it is important that we have a couple of different authentication and isolation mechanisms planned for the deployed IoT devices and the IoT devices being planned for 5G.

In 5G-capable devices, for identification, the methods of international mobile subscriber identity (IMSI) and network access identifier (NAI) are used. For authentication, 5G provides at least two variants of Authentication and Key Agreement (AKA), which are 5G AKA (evolution of 4G authentication method) and EAP-AKA' (based on the EAP). With slicing, a user device sends requests for a service to the 5G Core Access and Mobility Management Function (AMF), which is responsible for service allocation after the verification of Network Slice Selection Association Information (NSSAI). Unified data management (UDM) supports the Authentication Credential Repository and Processing Function (ARPF) and stores the subscription information and long-term security credentials used in authentication for AKA. The Network Slice Selection Function (NSSF) selects the set of network slice instances serving users and determines the NSSAI corresponding to applicable network slice instances.

For non-5G consumer IoT devices (primarily due to cost constraints and existing legacy deployments) being served by the service provider, multiple layers of protection need to be required, such as identification verification, authentication, access, and certification management at the user, device, and application layers. Figure 8-12 illustrates authentication, authorization, and encryption for IoT devices before accessing the IoT MEC applications.

Figure 8-12 illustrates the deployment of DNS- and IP-capable and non-capable IoT devices. The device capable of IP and DNS with 5G-capable chipsets can use the 5G spectrum for access. The user plane (for example, API calls) from the IoT devices are terminated in the 5G UPF. The API calls from the devices are then authenticated by the API gateway (API GW) or API firewall (API FW) deployed at the public-hosted or on-premises data center.

The unique identity (UID), which is configured in the IoT device during manufacturing, will also be used by the device identity/device authentication layer in the IoT initial configuration by the provisioning server to verify the authenticity of the IoT device. X.509 certificate–based authentication can also be used to authenticate the IoT devices, as shown in Figure 8-13.

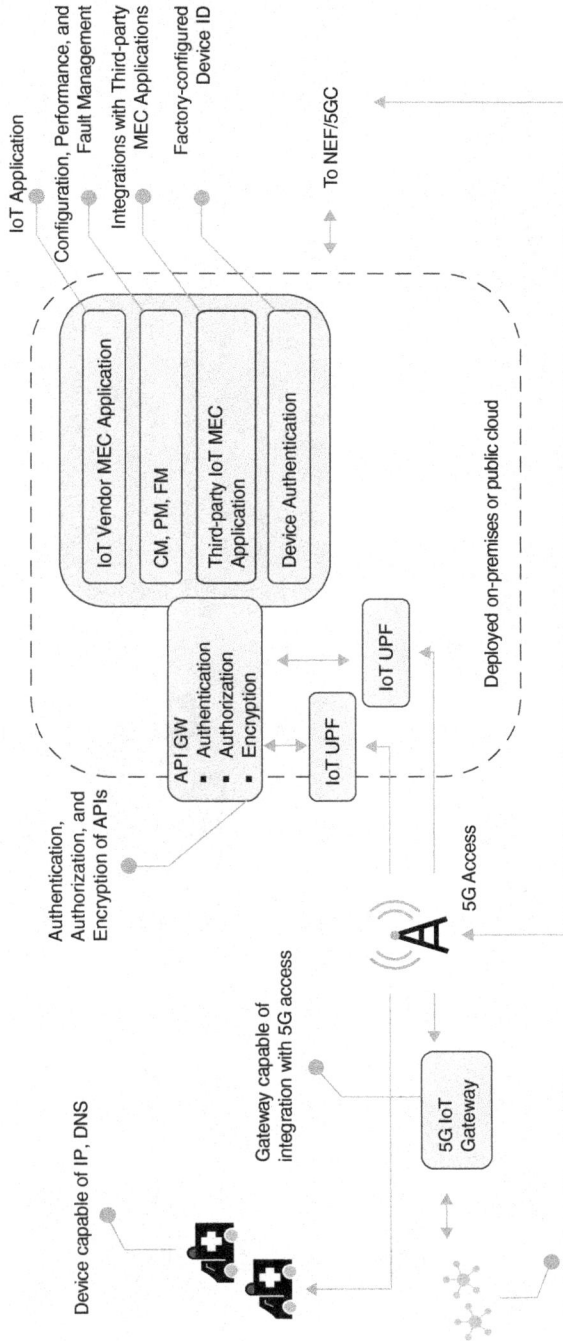

FIGURE 8-12 Authentication, Authorization, and Encryption Layers for IoT Devices Before Accessing IoT MEC Applications

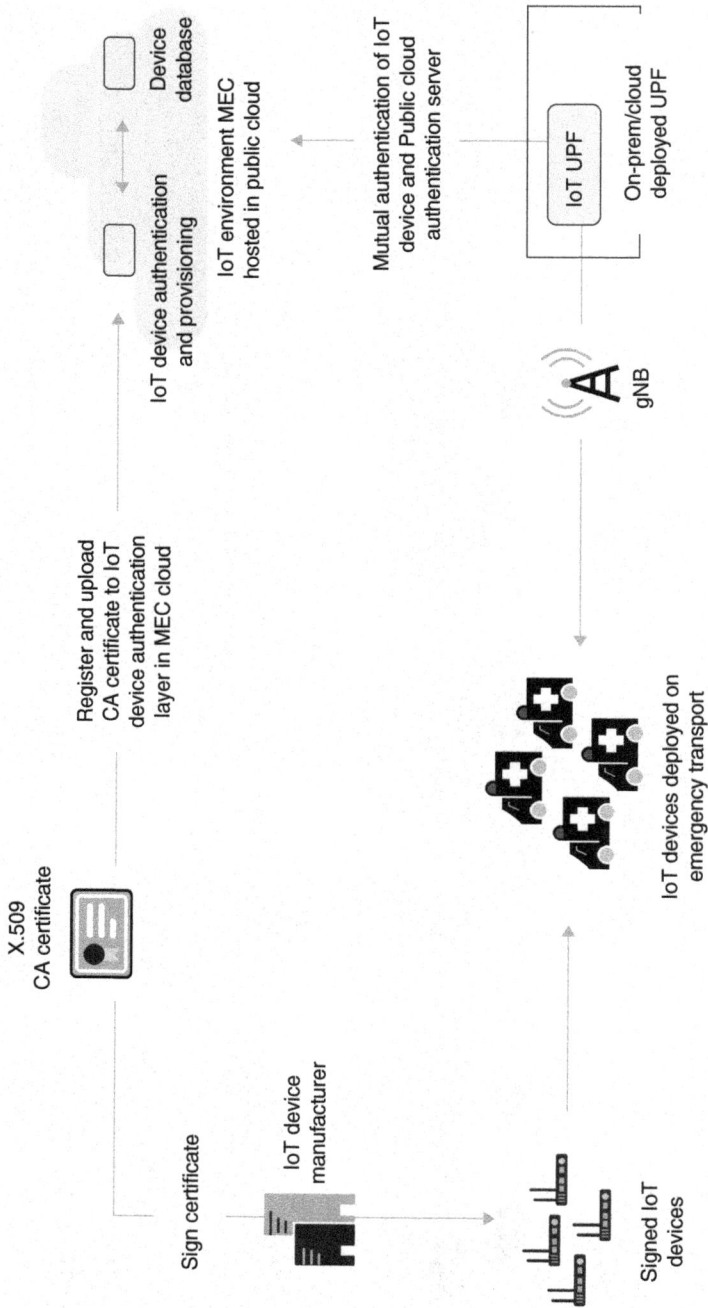

FIGURE 8-13   X.509 Certificate-Based Authentication for IoT Devices

The CA certificate is registered and uploaded in the cloud-based IoT device provisioning and authentication server. The signed certificate is used by the IoT device manufacturer to sign multiple IoT devices. These IoT devices are then deployed in the emergency transport vehicle. When turned on, the IoT device installed in the emergency transport vehicle automatically connects to the pre-assigned configuration and provisioning server. The IoT device will use the X.509 certificates to perform mutual authentication with the public cloud–deployed IoT provisioning and authentication server. This is one of the methods that can be used for scalable authentication of the IoT devices. As an added security layer, a TPM could be used by the IoT device manufacturer to store cryptographic keys for authentication purpose.

Configuration management (CM), performance management (PM), and fault management (FM) can be deployed in the cloud or on-premises data centers. Security controls related to APIs, discussed in Chapter 5, "Securing MEC Deployments in 5G," should be followed to ensure mitigation of API vulnerabilities for any API calls between the IoT devices and the management layer or third-party applications.

Due to the scale of mIoT deployments, the IoT authentication environment you choose should provide you with the option of identifying, provisioning, and onboarding a large number of devices based on the capabilities of the devices. Some device types have their unique X.509 certificate and private keys on them before being sold to the end customer, while others do not have the capability due to reasons such as the cost of implementing X.509 certificates and private key configurations during manufacturing.

If the manufacturing chain allows the device maker to provision unique credentials into the device at manufacturing time or in distribution, the IoT authentication environment can make use of them for identification and authentication purposes. If the device does have unique certificates, then the IoT authentication mechanism, such as the assignment of temporary certificates and temporary private keys, should be used to provide initial access to the provisioning server. Once the IoT device connects to the identity and authentication management using the temporary certificates and keys, it can be exchanged with a signed unique certificate and private key.

Another method of authenticating the IoT devices is by using the software development kit (SDK) provided by many IoT service providers, as shown in Figure 8-14, which illustrates IoT device integration with the public cloud using an installed SDK.

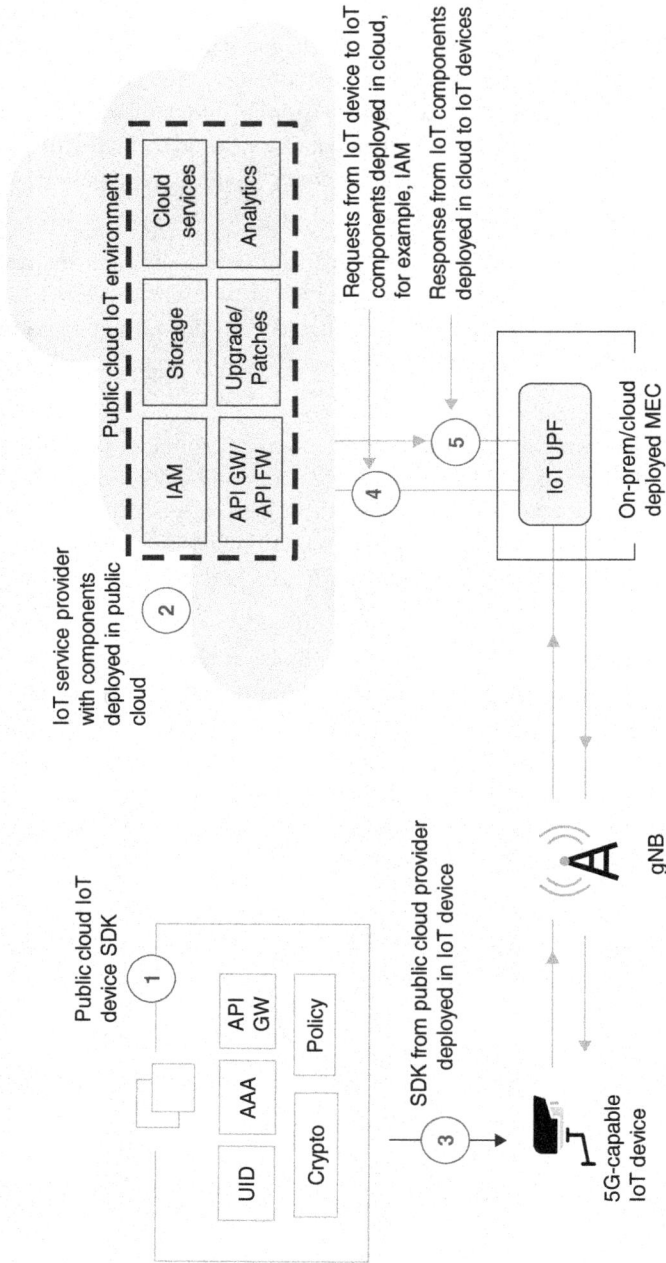

FIGURE 8-14   IoT Device Integration with Public Cloud Using SDK

Figure 8-14 illustrates the authentication of the IoT device using the installed SDK and is explained as follows:

1. The SDK will include open source libraries. The recommended practice for low-powered devices is to use an SDK that supports device connections that use Message Queuing Telemetry Transport (MQTT). The SDK will include a basic set of functionalities and policies to access the cloud-based IoT provider.

2. The key functionalities of the IoT service provider are deployed in the cloud. One of the key components is Identity and Access Management (IAM), which is used for authenticating the IoT devices. The API gateway (API GW) is used to protect the IoT applications from API vulnerabilities, such as providing rate-limiting functionalities and enhanced authentication and authorization functions.

3. Installing SDKs in the IoT devices will help you integrate IoT products to your choice of IoT providers deployed in public cloud.

4. The SDK deployed within the IoT device will initiate an HTTPS request toward the authentication, authorization, and accounting (AAA) component of the cloud-based IoT provider. The HTTPS request includes the X.509 certificate, which is verified by the AAA component to authenticate the IoT device.

5. Once the mutual authentication is performed, initial configuration can be downloaded to the IoT device. One of the other functions that can be performed is to attach a policy for the device, such as allowing the device to connect to the analytics engine, enabling you to enhance the services being offered to the IoT use cases.

In pragmatic deployment considerations, you also need to consider integration of hundreds of thousands or even millions of devices, which might require AAA to be deployed in the public cloud, as illustrated in Figure 8-15.

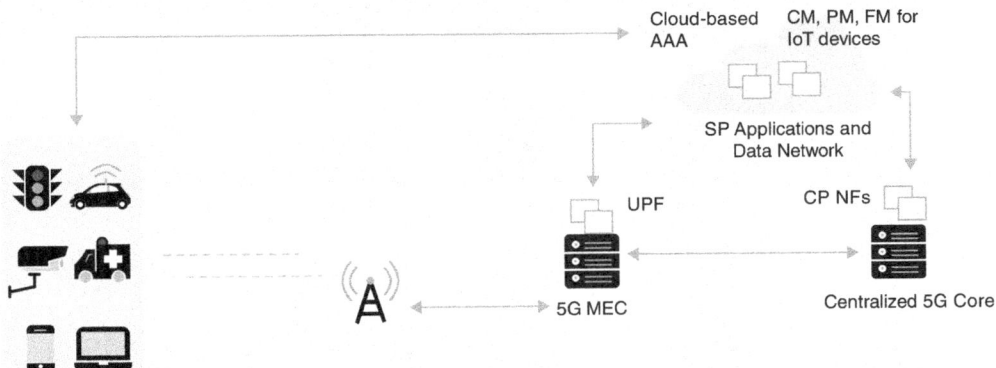

FIGURE 8-15   Cloud-Based Authentication for IoT Devices

One way to tackle the issue of identifying millions of devices is to build a strategy around having a unique ID (UID) assigned during the manufacturing process that can be used to identify and authenticate the device. Having a unique ID will also allow service providers to have proper lifecycle management, including tracking the software and hardware changes. Any infection or abnormal behavior can be easily tracked down to a specific device or group of devices.

Network Slice Isolation and Segmentation

Network slicing is one of the key evolutions of the network deployment brought in by 5G technology. Network slicing is the ability of the network to (automatically) configure and run multiple logical networks as virtually independent business operations on a common physical infrastructure. Network slicing is a fundamental architecture component of the 5G network, fulfilling the majority of the 5G use cases. Many operators are considering the offer of a network slice per enterprise, which is not that dissimilar to the per access point name (APN) offer for an enterprise in play today. As we consider the points where the enterprise then touches the 5G slice, a number of security aspects must be addressed— one of them being slice-level isolation, as illustrated in Figure 8-16.

Network slicing architecture, which allows the ability to run multiple logical networks as virtually independent business operations on a common physical infrastructure, also requires high isolation between the slices. Isolation within the components of the slice prevents the vulnerabilities from spreading to other components within the slice and between the slices in the case of any malicious attacks.

Intra-slice and inter-slice isolation should be implemented for both public and non-public networks (NPNs). The network slices should also allow a quarantine slice for identified malicious hosts, which provides isolation and restricts the spread of malware due to lateral movement.

Intra-slice can be provided by ensuring that the CNFs serving the slice are deployed on separate hosts. This ensures high availability for the slice.

Inter-slice isolation can be provided by deploying 5GC CNFs on separate hosts and then implementing network segmentation between slices. This mitigates malware propagation between slices of different sensitivity, such as a slice serving critical infrastructure (considered a highly sensitive slice) and a slice serving IoT devices (considered a less sensitive slice).

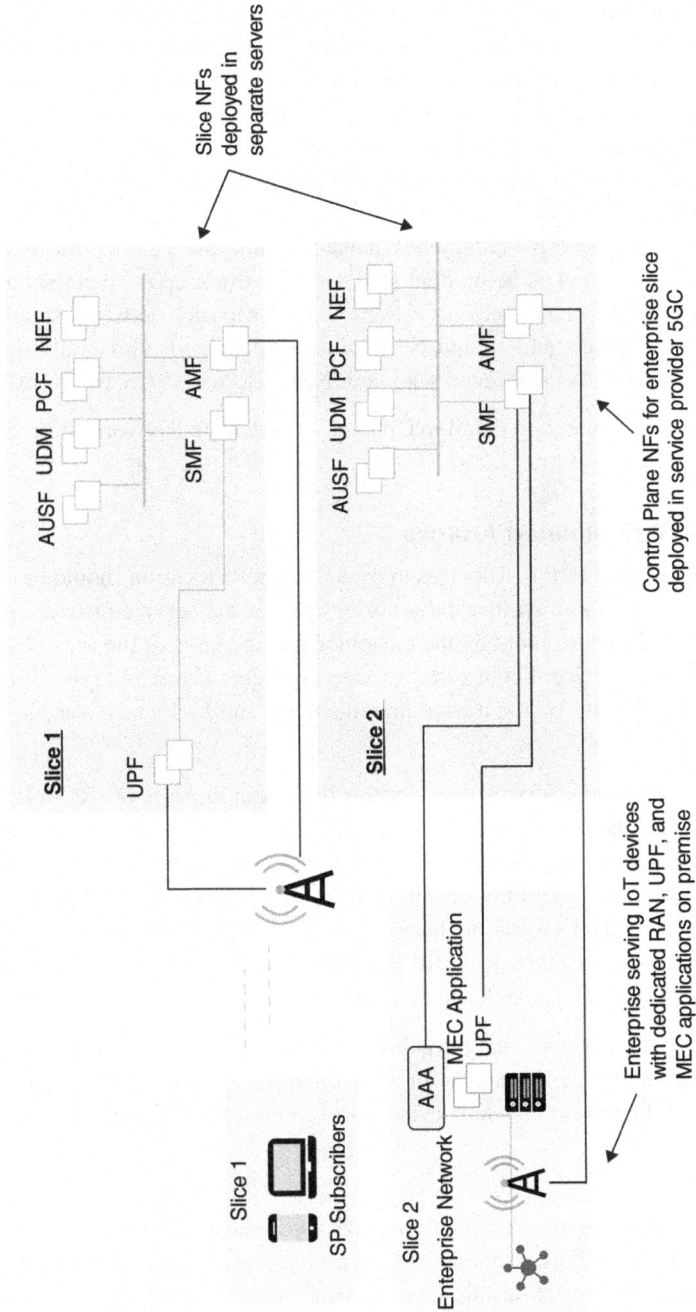

FIGURE 8-16   Slice Isolation and Segmentation for IoT Devices

Segmentation and isolation mechanisms used for the IoT deployment will vary depending on your deployment mode to cater for the mIoT use cases. If network slice mechanisms are used to provide access to the IoT device, you should ensure that the 3GPP 5G functions are isolated from other slices. This can be done by using separate x86 servers for deploying mIoT slice NFs. You should also architect your network such that web-facing applications are in a separate security zone and are not deployed in the same x86 server. This will ensure physical separation of the NFs and will reduce the probability of any side-channel attacks exploiting the vulnerability of the host OS and hardware (HW). If the mIoT devices are being deployed in the NPN network, then you should ensure that you have the mIoT network and the operational technology (OT) completely isolated from your IT network using a demilitarized zone (DMZ). In fact, if the mIoT is being deployed for critical use cases, there should be integrations with the IT network only if it is really necessary. Remote access to such networks should follow stringent identity and access mechanisms and should be continuously audited. This could be done by using a next-generation firewall (NGFW) integrated with your Network Access Control (NAC) and IAM layers.

Securing network slices is covered in detail in Chapter 7, "Securing Network Slice, SDN, and Orchestration in 5G."

Mitigating IRC and P2P-Related Attacks

In general, deploying IRC and P2P IoT devices in the subscriber's location should be avoided. But pragmatically speaking, it is well known that the security team of the service provider is rarely informed of IoT devices being sold to customers by the customer-facing teams of the service provider. To solve this issue, recommended practice dictates that service providers check the type of device, secure the development lifecycle followed by the device manufacturer, and look at the supply chain lifecycle of the device manufacturer.

If the existing devices within the service provider use IRC, then in cases of IRC-related botnet attacks, each bot client must know the IRC server, port, and channel. Anti-malware solutions available today can detect and shut down these servers and channels, effectively halting the botnet attack. If this happens, clients are still infected, but they typically lie dormant since they have no way of receiving instructions. A botnet can also consist of several servers or channels. If one of the servers or channels becomes disabled, the botnet simply switches to another. It is still possible to detect and disrupt additional botnet servers or channels by sniffing IRC traffic, which can be catered for by anti-malware and monitoring solutions.

If the existing devices within the service provider use P2P, for mitigating the P2P attacks that use the firewall pin-holing technique, then granular firewall configurations to block traffic on specific ports should be used. This would prevent infected devices from communicating with the malicious P2P servers.

Zero-Touch Security

Many of the consumer devices aimed at enabling IoT use cases use Zero Touch Provisioning (ZTP) to allow the PnP capabilities. This is done to allow easier deployment for the customer and provide a better user experience. Before choosing such devices from a manufacturer or vendor, the service provider should check whether the device manufacturer or vendor uses ZTS as a model for the ZTP process. Depending on the vendor, the method of ZTS is also called *secure zero touch* or *zero touch secure identity*, or other variants.

Implementing ZTS by the device vendor is quite critical, as it secures the device and authenticates and encrypts its communication with the cloud-hosted provisioning and configuration server or PnP servers and provides a secure lifecycle thereafter, including secure auto-deployment of patches, secure auto-installation of updates, and so on.

ZTS techniques used by the vendor should also ensure continuous authentication if any anomaly in behavior is detected or if reauthentication of the device occurs at certain intervals without interrupting the device functions. During assessment of the device vendor by the service provider, scalability of the solution should also be verified. Quite a few vendors in the market today use artificial intelligence (AI) to detect anomalies in behavior and can initiate the detection and response capabilities automatically depending on the behavior of the devices, including triggering the reauthentication of the devices and moving the devices with anomalous behavior to an isolated segment.

DNS Security for 5G IoT Devices

The Domain Name System (DNS) plays a very important role in the IoT ecosystem. The 5G devices enabling consumer IoT would primarily be using cloud-based provisioning servers for PnP, which is usually configured using an FQDN that will have the URL of the provisioning server configured or hard-coded. Using this configuration, the device will connect to the provisioning server, get authenticated (depending on the device vendor), and then connect to cloud services to transmit and receive data.

One of the key threats is DNS cache poisoning attacks, where a malicious or fraudulent IP address is logged in the local memory cache. The device configuration can also be modified for it to connect to a malicious server. This is because the devices trust the domain names to be secure. If an attacker changes the original domain name within the configuration template of the device or can change the hardcoded domain name to a malicious one, the device will try connecting to that domain name. The attacker can then insert a rogue update to the device, potentially taking full control of the device and targeting it against the service provider infrastructure, causing a DDoS attack or taking down the infra-structure, causing a DoS attack.

DNS, although scalable, does not include any inherent security mechanisms such as encryption, which makes it vulnerable to MitM attacks for interception and manipulation. Domain Name System Security Extensions (DNSSEC) and DNS over HTTPS (DoH) improve the security capability of DNS. DNSSEC is becoming more important for IoT devices due to the fact that it secures parts of the supply chain system as well. When an IoT device is manufactured, many of the device vendors use the cloud-based configuration for shipping and initial factory configuration. This is because many of the orders from service providers can be customed labeled so that when the customers receive their devices, they will be in the name of the service provider. This requires some changes at the manufacturing end, and many of these processes are automated in the industry these days. Secure DNS solutions can also be used by the service providers to enhance security for the IoT devices. This is further explained in detail in this section.

DNSSEC

DNSSEC is a set of extensions to DNS that provides a security chain of trust and protection from DNS vulnerabilities. DNSSEC provides DNS clients with cryptographic authentication of DNS data by using cryptographic keys to validate connections between the DNS client and a domain name.

Having DNSSEC as part of the device capability will ensure that the device is routed and connected to the authentic server.

Although DNSSEC adds integrity and trust to DNS, it does not provide confidentiality (DNSSEC responses are authenticated but not encrypted), which means that the DNSSEC responses can be intercepted. As the attacker can attempt to use DNSSEC mechanisms to consume a victim's resources, it does not provide complete mitigation against DoS attacks.

DoH

DNS over HTTPS (DoH) caters for DNS resolution using the HTTPS protocol. Using HTTPS, DoH provides better user privacy and prevents MitM-type attacks because it includes encryption between the DoH client and the DoH-based DNS resolver. DoH is published by the IETF as RFC 8484.

DoH works just like a normal DNS request, except that it uses Transmission Control Protocol (TCP) to transmit and receive queries. DoH takes the DNS query and sends it to a DoH-compatible DNS server (resolver) via an encrypted HTTPS connection on port 443, thereby preventing third-party observers from sniffing traffic and understanding what DNS queries users have run or what websites users are intending to access. Because the DoH (DNS) request is encrypted, it's even invisible to cybersecurity software that relies on passive DNS monitoring to block requests to known malicious domains.

If service providers plan to use DoH-based endpoints, there are certain mechanisms the security team can put into place to ensure that the devices use specific browsers. Browsers such as Chrome ensure that DoH will only be enabled when system DNS is observed to be a participating DNS provider. After DoH is enabled in Chrome, the browser will send DNS queries to the same DNS servers as before. If the target DNS server has a DoH-capable interface, then Chrome will encrypt DNS traffic and send it to the same DNS server's DoH interface.

Secure DNS

In many cases, consumer IoT devices today are not yet fully DNSSEC or DoH capable. One of the mitigation mechanisms from DNS cache poisonings and malicious DNS configurations is to use a cloud-based DNS security layer that ensures that the DNS request is not resolved to a malicious domain. There are vendors in the market today that integrate the secure DNS resolution along with the threat intelligence, anti-malware, and antivirus capabilities.

As illustrated in Figure 8-17, when the DNS security layer receives a DNS request from a 5G-capable IoT device, be it for the provisioning or PnP layer or for CM, PM or FM, it should use threat intelligence to determine if the request is safe, malicious, or risky—meaning the domain contains both malicious and legitimate content. Safe and malicious requests can be routed as usual or blocked, respectively. Risky requests can be forwarded to an inspection layer for deeper inspection. The secure DNS layer should also inspect the files attempted to be downloaded from the sites using antivirus (AV) engines and anti-malware protection, and based on the outcome of this inspection, the connection should be either allowed or blocked.

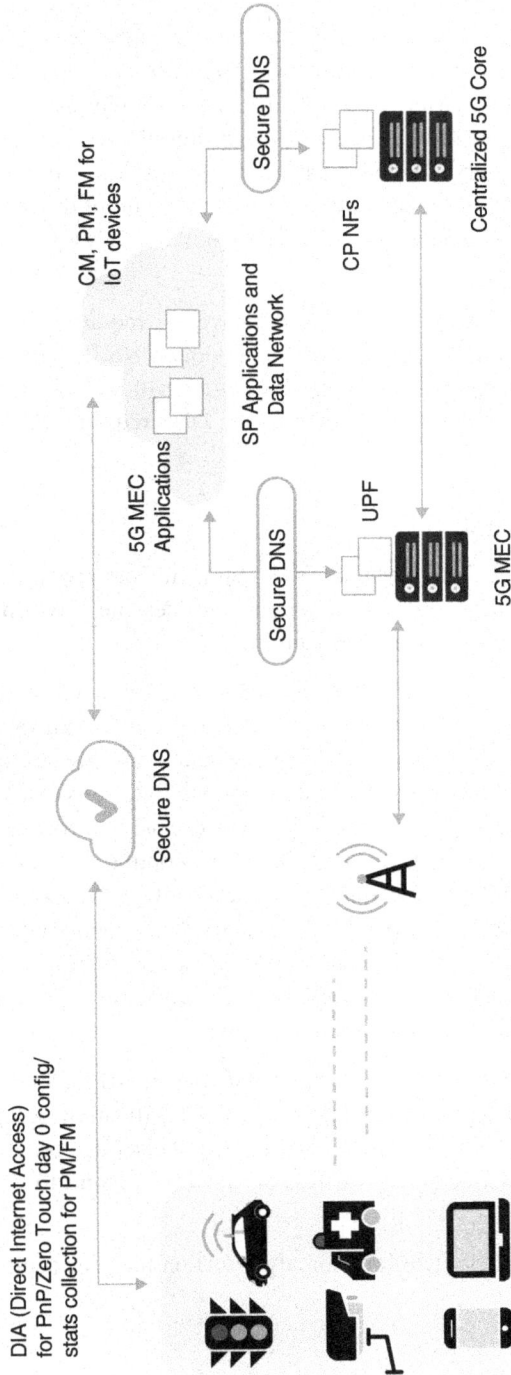

DIA (Direct Internet Access)
for PnP/Zero Touch day 0 config/
stats collection for PM/FM

Secure DNS

CM, PM, FM for
IoT devices

5G MEC
Applications

SP Applications and
Data Network

Secure DNS

UPF

5G MEC

Secure DNS

CP NFs

Centralized 5G Core

FIGURE 8-17   Securing DNS Layer Communication of IoT Devices

This is one of the most effective methods that will lead the security teams to remediate fewer instances of malware, and the threat is mitigated even before the devices are impacted or an attack is launched. Service providers selecting vendors or partners for secure DNS solutions should ensure that they have extremely good threat intelligence to ensure high efficacy. They should also ensure that the vendor providing such solutions has a robust machine learning algorithm that allows the solution to predict attacks. Many of the recursive DNS service providers resolve millions if not billions of Internet requests every day, and they have ML algorithms analyzing the massive amount of data to understand patterns and co-relate patterns by running statistical and machine learning models to identify attacks and thus uncover the attacker's infrastructure.

The secure DNS layer is also easy deployable and doesn't have any requirements on the device itself. It only requires the DNS IP address to be changed from a previous DNS IP address to the secure DNS provider's IP address. Any DNS request coming from the device will now be redirected to the secure DNS vendor's cloud network, which will then resolve all the DNS requests and block any request to the malicious domains.

Enhanced Visibility and Monitoring

One of the most important security capabilities that's required in any organization is enhanced end-to-end monitoring to understand the communication among the devices and between the devices and the network elements, including monitoring the encrypted traffic.

After discussing and deploying proof of value (PoV), which is a marketing term used by many vendors to make solution validation in service provider networks sound cooler, a number of service providers see very little value in aggregating and tapping the user plane data of the devices. In 5G, the user plane data from devices (eMBB slice-related devices) will be in the terabytes of volume. Having a solution for end-to-end user plane (UP) monitoring is not viable due to cost and technical reasons. Control plane, service plane, and OAM are the key layers that should be monitored at minimum. By validating this method in multiple service providers, it is quite clear that many of the anomalies can be detected by monitoring the control plane, service plane, and OAM layer. Once the monitoring for these layers is established, the service provider can pick and choose the UP-layer visibility for specific use cases. IoT devices (related to machine-to-machine use cases), as such, are not user plane intensive, so having granular visibility would not be a major hurdle in terms of cost.

Before investing in an end-to-end monitoring system for the consumer IoT, service providers should try to build a unique ID system, as explained in the section "Identification, Authentication, Access, and Certificate Management" in this chapter. This will also help the service providers in reducing the mean time to repair (MTTR), as the service provider can quickly respond to the unplanned device breakdown.

Figure 8-18 illustrates the monitoring system for anomaly detection for your deployments.

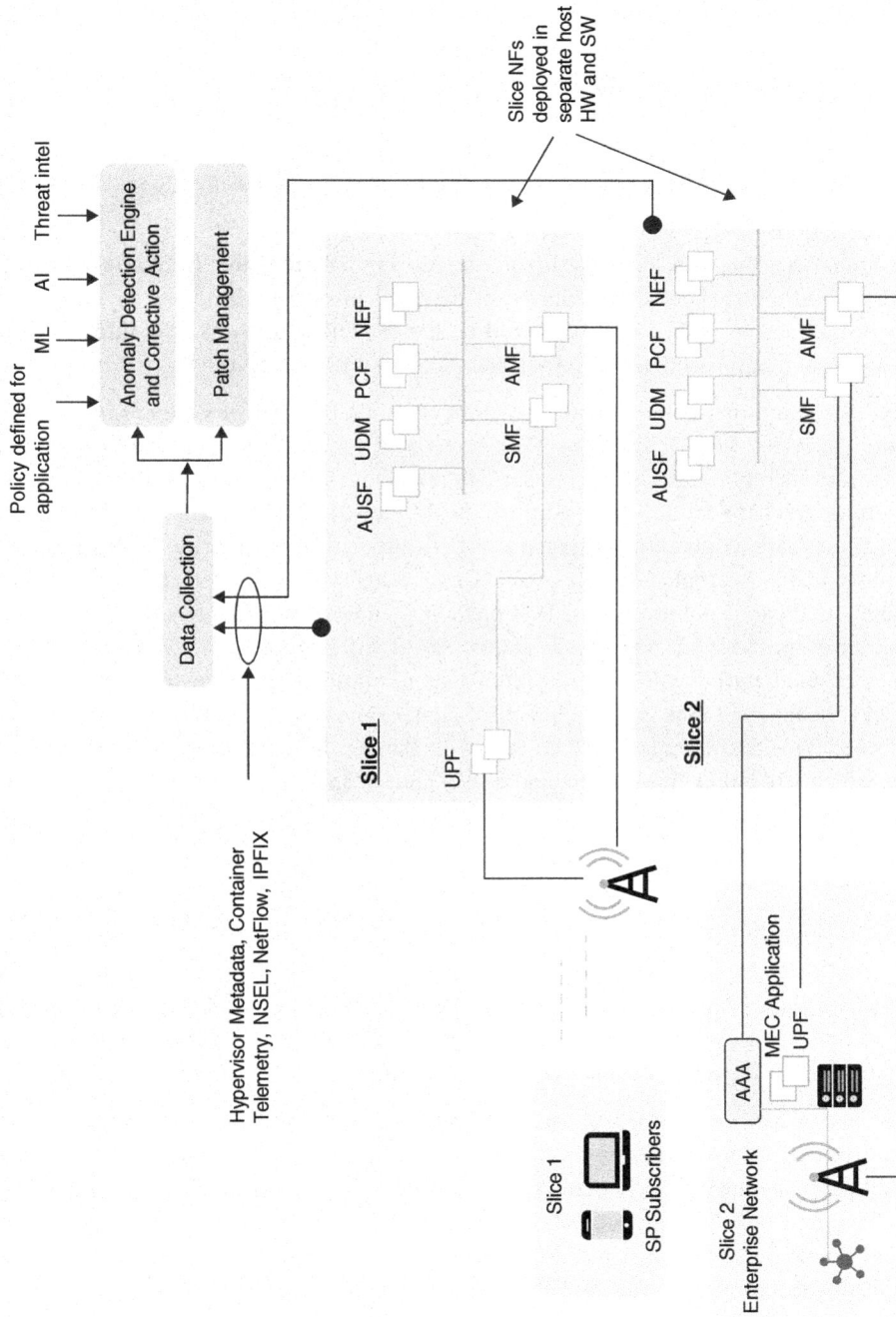

FIGURE 8-18   Enhanced Visibility of IoT Device and IoT Slice Layers

As shown in Figure 8-18, the monitoring solution should also cater for enterprise use cases, as 5G allows easier integration into the enterprise networks using methods such as multi-access and edge computing (MEC) and network slicing. Due to the flexibility in deploying the use cases, the monitoring solution should also follow flexibility and scalability. There are monitoring solutions available in the market today that allow for multivendor packet flow collection (without the need for physical probes) and then analyze the data collected after packet de-duplication and VXLAN striping. Having such monitoring solutions would also support other use cases, such as reusing the same solution for IT and telco DC infrastructure monitoring.

It is also recommended that you look at utilizing monitoring solutions that have integration with the products with capabilities such as responding to any detected anomalies within the device or the device network. The minimum possible response should be the capability to isolate the infected devices or push the devices into a segment that will have access to only critical services.

The visibility and monitoring layer, though very critical, might become very expensive for you if you don't plan it properly for the IoT use cases. One of the methods you could use here to optimize is to consider enhanced visibility and monitoring for control plane, service and management layer of the network functions, and network devices specific to the IoT network. If the IoT network and devices use API-based communications that are encrypted using Transport Layer Security (TLS), it is important to have visibility in the encrypted layer as well. Using a decryption engine and then analyzing the packets, though effective, is not always the best method, as multiple decryption points will reduce the effective security posture of your network. In such cases, it will be more effective to perform malware detection in encrypted traffic without decryption using solutions available today that analyze the encrypted packet header and look at the behavior of cipher suites and so on to determine any anomaly and malicious behavior. Some smart mIoT devices will also provide a basic telemetry with a couple of key counters, which will help you to understand if they have been tampered with. Such IoT devices can be blocked or reported to the IoT device user, depending on the SLA.

Access Control

Access control for 5G SIM or universal integrated circuit card–capable devices are catered for by the inherent 5G Identity and Access Management mechanisms. But many of the consumer IoT devices being deployed for quite some time will use non-3GPP technologies and legacy 3GPP mechanisms and connect to the 5GC using network elements like the non-3GPP Inter-Working Function (N3IWF), which is responsible for the interworking between the untrusted non-3GPP components and the 5GC.

There are various access control mechanisms used by service providers today, primarily role-based access control (RBAC), mandatory access control (MAC), access using security group tags (SGTs), attribute-based access control (ABAC), and so on. For the cloud-hosted IoT management functions such as CM, PM, and FM and provisioning servers catering for consumer IoT devices, a very strict RBAC schema should be applied as a minimum, which is then followed by using multifactor authentication (MFA) for the users and devices. There should be layers of access control for any remote configuration of the IoT subsystem (controller, server, device, and so on).

To ensure that only legitimate users with the right levels of access are accessing the management layer/operational technology (OT) of the IoT network, you should apply zero-trust principles and use mechanisms where you authenticate and re-authenticate the users at varying levels of time and network layers. For example, you should use mechanisms such as MFA, which is integrated into your existing Identity and Access Management (IAM) layer. This integration will ensure that any change in the user's role is mapped to RBAC. If the previous role of the user was admin with privilege access, once the person leaves the organization or changes role, the integration will ensure that the person does not have privileged access anymore. This layer, although foundational, is rarely designed properly due to multiple access control vendors and multiple MFA vendors being deployed at different departments of the service provider. In some cases, there are six to seven multiple IAM solutions deployed in the same domain of the service provider, thus unnecessarily complicating the access control and leading to improper configuration and blind spots.

Figure 8-19 illustrates the granular access control for IoT deployments by providing the secondary authentication mechanism for IoT devices using the enterprise AAA/IAM.

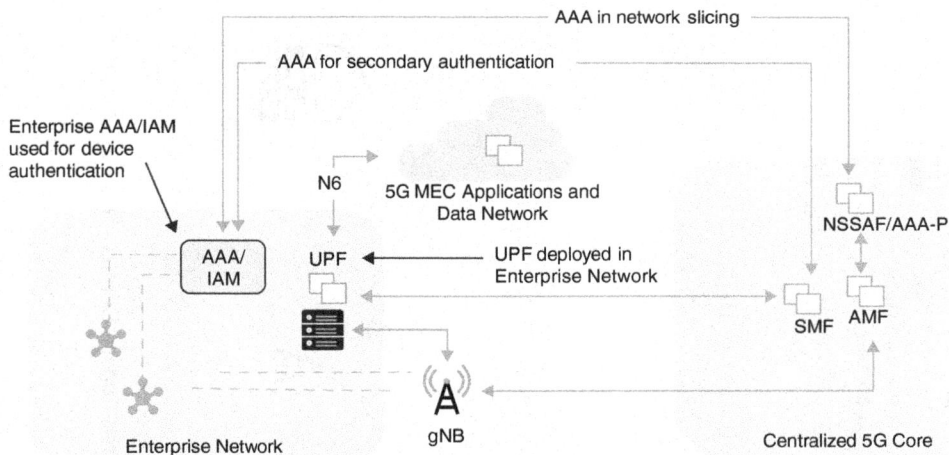

FIGURE 8-19   Granular Access Control for 5G IoT Network

As shown in Figure 8-20, the user will have to go through primary authentication, secondary authentication, secure Internet access, and a granular role-based access for accessing the device and the consumer IoT subsystem.

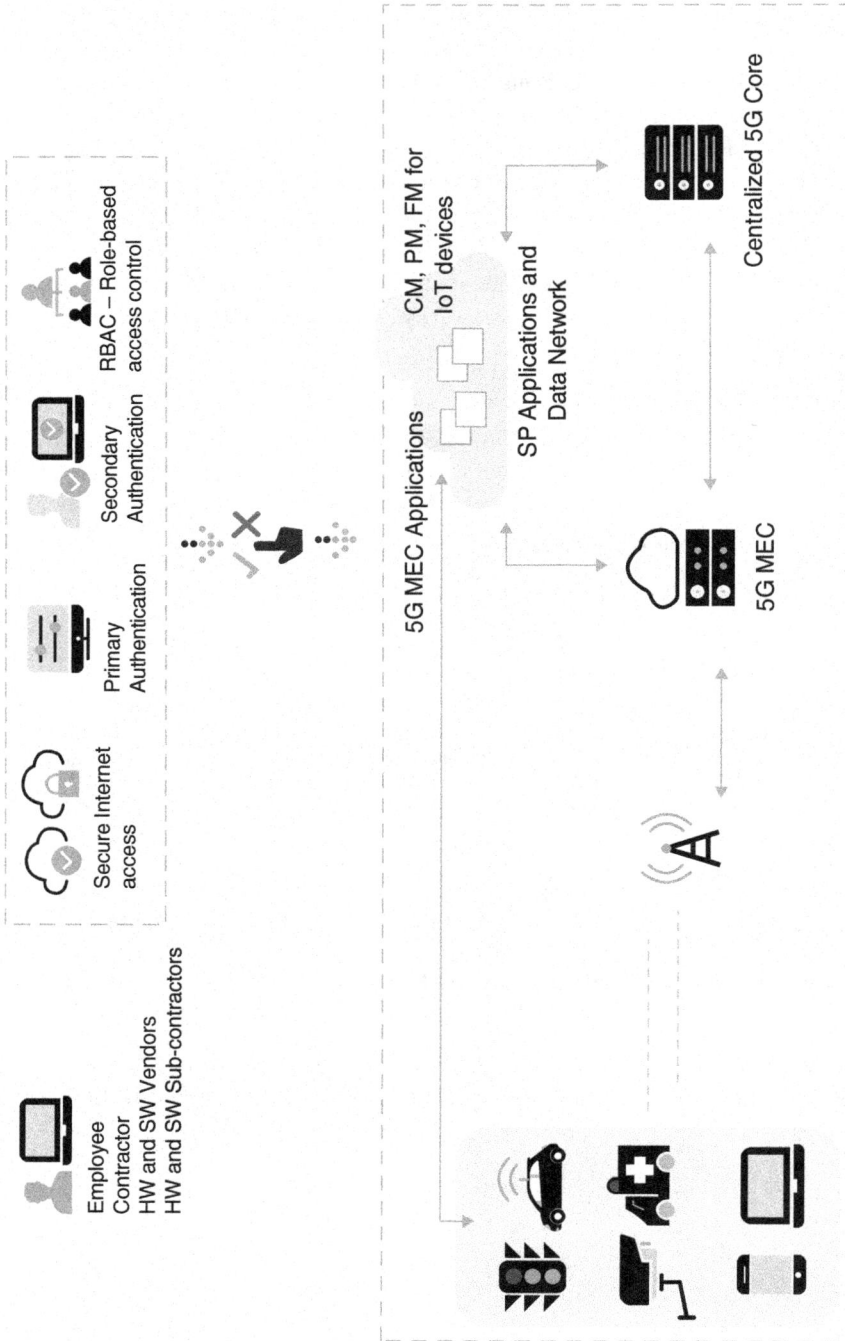

FIGURE 8-20 Granular and Multilayered Access Control for 5G IoT Network and IoT Devices

DDoS Protection

One of the key threat surfaces in 5G-capable IoT devices is DDoS, due to the sheer number of devices (both low bandwidth and high bandwidth) expected in the service provider's infrastructure in 5G.

Looking at the growing sophistication of DDoS malware, service providers need to implement the most sophisticated security solutions to protect against new threats and emerging attack types. Attackers are deploying multivector attack campaigns by increasing the number of attack vectors launched in parallel. To target an organization's blind spot, different attack vectors target different layers of the network and data center. Even if only one vector goes undetected, then the attack is successful, and the result is highly destructive. Attacks are becoming completely automated and more sophisticated, making it difficult to defend against them manually. New techniques like burst attacks and advanced persistent denial-of-service demand advanced detection and mitigation and underscore the need for automation. Figure 8-21 illustrates the location of the anti-DDoS solution to mitigate DDoS attacks.

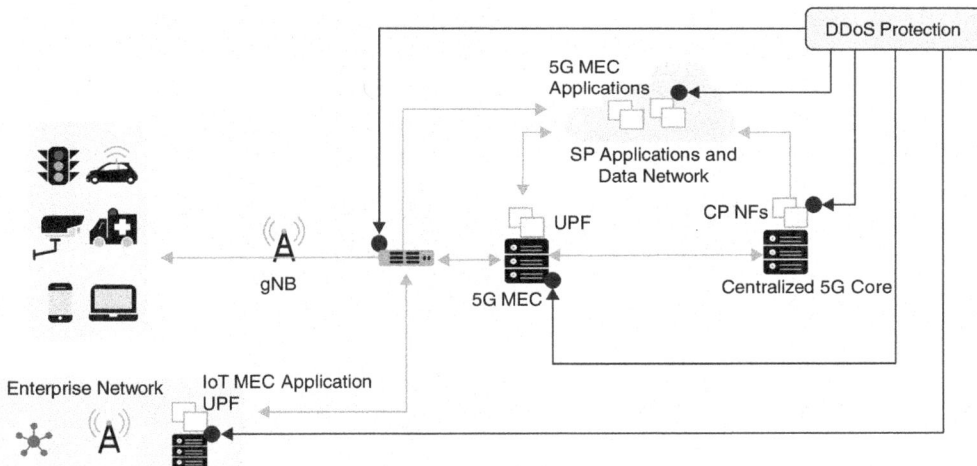

FIGURE 8-21   DDoS Protection for the 5G IoT Network

There are anti-DDoS mechanisms available in the market today that can detect the DDoS attacks by analyzing the packet flows and then taking the decision to drop the packets if they are identified as part of the DDoS attack. The DDoS protection solution would normally create baselines of a normal network, application, and user behavior and use these baselines to notice abnormal traffic and accurately detect attacks. When a new, previously unknown zero-day attack is detected, the solution creates a signature that uses the attack characteristics and starts blocking the attack. The mechanism described is true for any technology and for 5G. But in 5G the deployment of the anti-DDoS solution need not be deployed at the centralized 5GC layer but can instead be deployed at the MEC or distributed DC layer. This ensures that the DDoS packets are mitigated at the edge of the network and do not hog the back haul.

Using a next-generation firewall (NGFW), anti-malware, and DDoS protection functions is not new for 5G and has been part of the security ecosystem for quite some time. Although the functions have not changed, the type of deployment and the features required within these functions need enhancement to cater for the mIoT and other 5G use cases. One of the key changes is the container-based deployment of 5G functions, which means that a single IP address could have multiple NFs; this requires the application-level microsegmentation and access controls to be implemented. This should also cater for identifying vulnerabilities in the open source Cloud-Native Functions, which are the norm for 5G 3GPP NFs. Developers of non-3GPP-based 5G NFs aimed to be deployed in the MEC Layer are also building their applications using open source software.

Real Scenario Case Study: mIoT Threats and Their Mitigation

This section focuses on a real scenario case study of attacks from threat vectors in a 5G mIoT environment and discusses techniques for threat mitigation. Although 5G was not widely deployed at the time of writing this book, this section is based on real-life attacks in similar environments or proof of concepts and lab tests done by various customers. This section will take you step by step through malware infection of the device and then explain a mitigation technique.

Figure 8-22 shows the 5G IoT network used as the example for the attack scenarios discussed in this section.

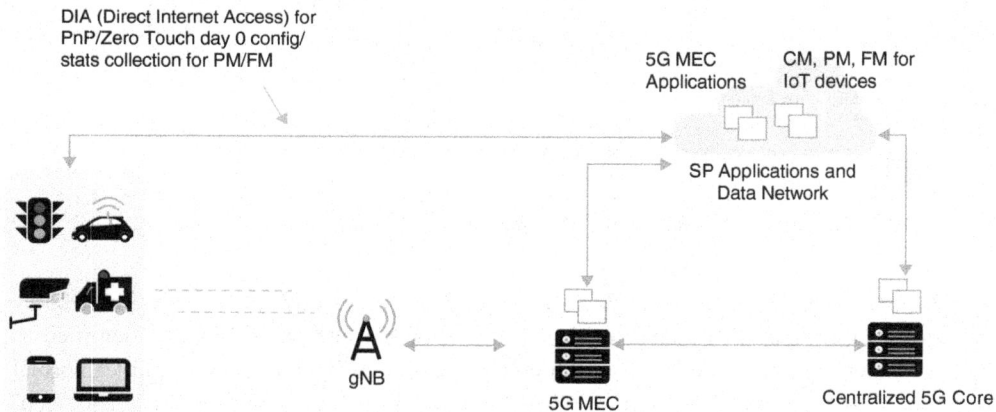

FIGURE 8-22   5G Network Serving IoT Devices

Figure 8-22 illustrates IoT devices that are connected to the IoT applications and management components, such as CM, PM, and FM, deployed in the cloud. The IoT devices are configured using a domain configuration that lets you specify a custom FQDN to connect to the public cloud IoT environment. The FQDN is configured using the CM deployed in the cloud. The IoT devices are factory-configured to resolve to a preconfigured server (IP address/URL). Once the IoT device resolves the URL to the PnP server, the initial custom configuration is then downloaded to the IoT device, which includes the details for configuration, performance, update servers, and patch servers.

Threats Example

In Figure 8-23, the consumer IoT devices connect to the Internet and discover the provisioning server based on the FQDN attributes provided by the CM of the IoT device. In step 1 of the threat, the CM is exploited by a vulnerability.

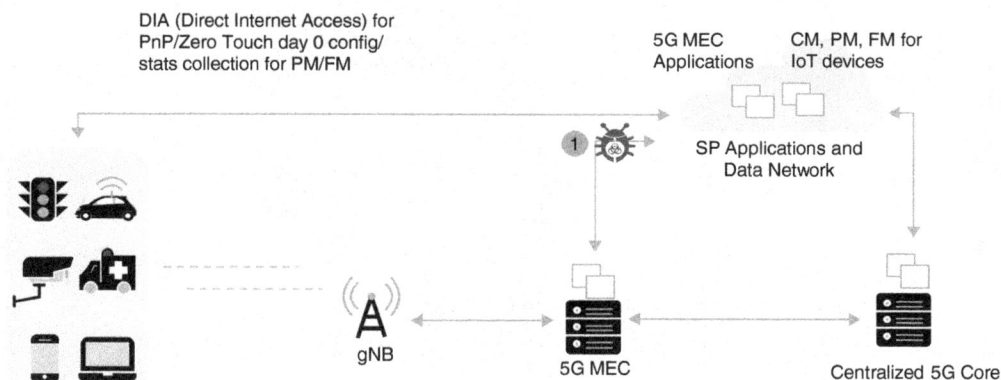

FIGURE 8-23 IoT Configuration Management Vulnerability Being Exploited

Figure 8-24 illustrates the IoT device FQDN being modified by the attacker.

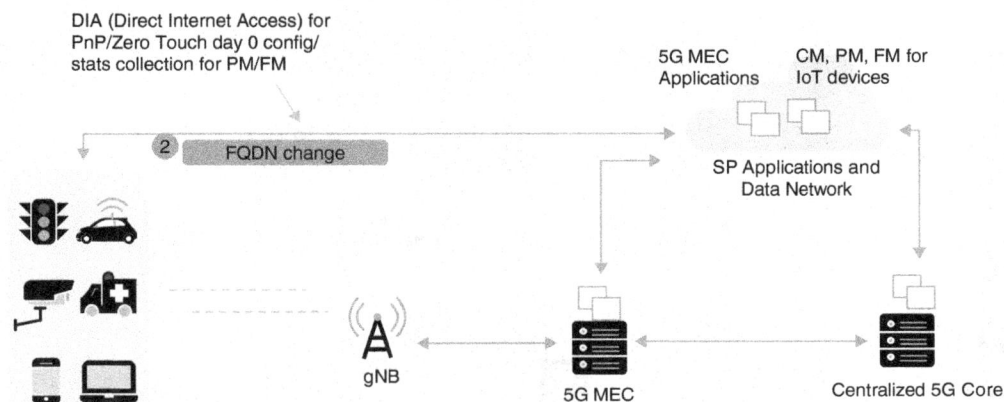

FIGURE 8-24 IoT Device FQDN Attribute Being Modified

In step 2 of the threat, the FQDN attribute is modified by the attacker exploiting the vulnerability in the public cloud–deployed configuration management virtual instance.

Figure 8-25 illustrates the redirection of the devices to the malicious command and control center.

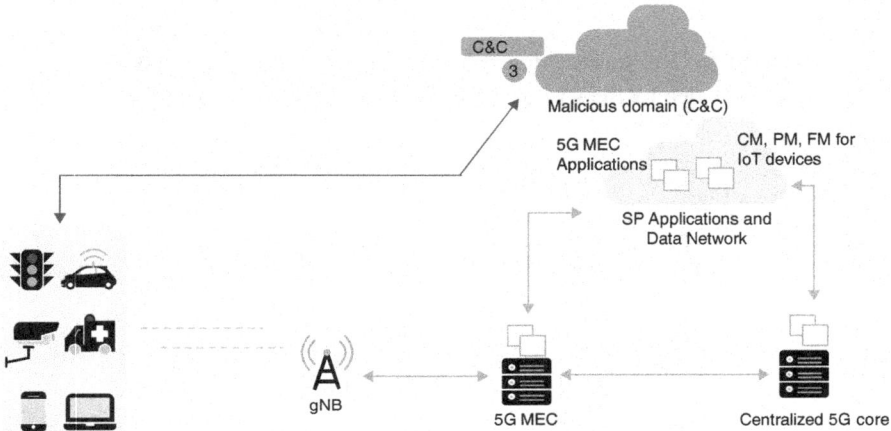

FIGURE 8-25   IoT Devices Being Redirected to the C&C Server

In step 3 of the threat, the IoT devices now reach out to the malicious domain instead of the CM/PM/FM domain of the device vendor. The malicious domain could also be C&C, which now will take control of all the devices.

Figure 8-26 illustrates an example of how the malicious C&C center takes over the IoT devices.

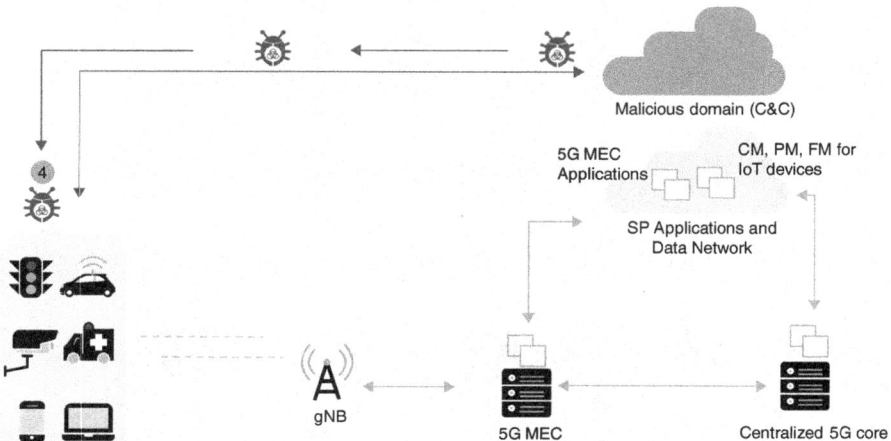

FIGURE 8-26   IoT Device Being Exploited by the C&C Server

The C&C node will now start pushing rogue updates with malicious code in it. The primary intention of the attacker is to understand the behavior of the system (for example, what's the busy hour of the network, what kind of data is being transmitted and received, and so on).

Figure 8-27 illustrates how IoT devices start a DDoS attack toward the 5G infrastructure.

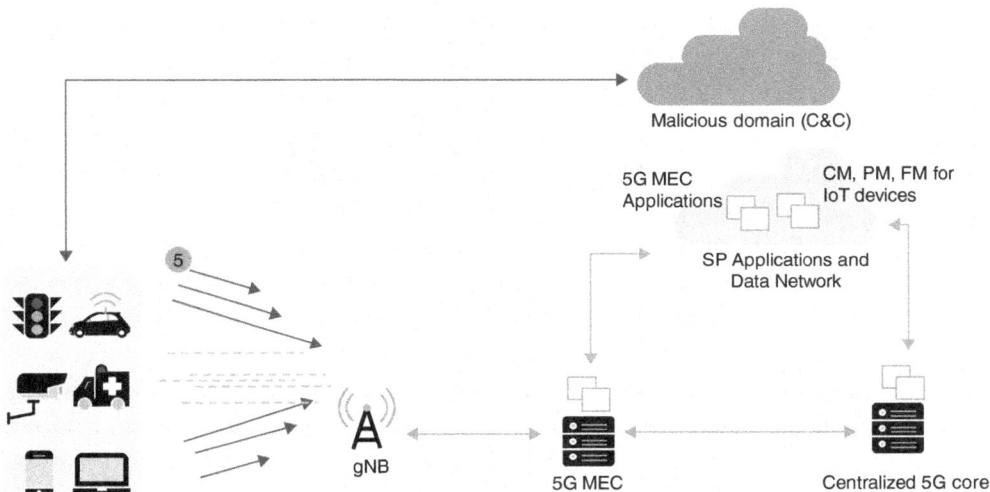

FIGURE 8-27   IoT Devices Causing DDoS Attack Toward the 5G Network

Once the attacker has enough of the devices under control, the devices can now be used to launch attacks on the service provider's network or any other network. Once the network infrastructure is compromised via DDoS, the attacker can launch signaling or paging attacks, which could overload the control plane of the service provider.

Mitigation Example

Some of the key methods to mitigate the device-based threats are the secure DNS solutions (DNSSEC, DoH, and Secure DNS) unless the device does not use DNS at all. A majority of service providers today are planning to deploy devices that use ZTP with the provisioning based on the cloud platform.

The mitigation technique illustrated in Figure 8-28 is that of the Secure DNS solution, which resolves the DNS query from the IoT device. If the FQDN is changed maliciously to a C&C domain, the secure DNS solution would block the domain and not allow the device to be connected to it.

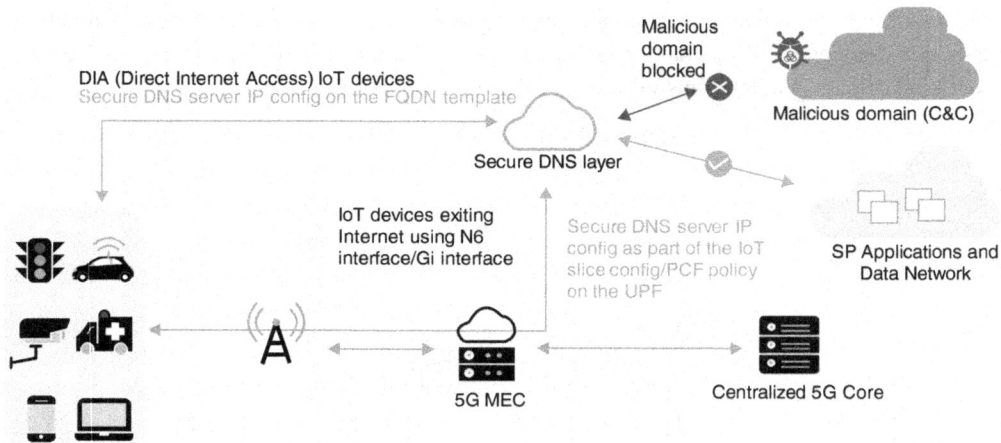

FIGURE 8-28   Securing DNS Layer to Mitigate the Attack Scenario

The advantage of the Secure DNS solution is that there is no dependency on the type of device or operating system. The only configuration the service provider has to perform is to use the Secure DNS solution IP address as the DNS IP address in the FQDN attribute template being pushed from the provisioning or configuration management toward the devices.

Summary

There needs to be multiple security layers to ensure mitigation of device-based threats. The threat surfaces and the security capabilities might vary by deployment. Depending on the sophistication of the attack and the network deployment method followed by the service provider, no single type of mitigation technique will be fully effective, as can be seen in the Figure 8-29.

When you're planning for consumer IoT deployments, which is very cost-sensitive, care should be taken to ensure multiple layers of isolation, device monitoring, and granular access control are applied. Figure 8-30 provides the summary of all the multiple security layers required to secure the network from device-based threats. Depending on the network infrastructure of the service provider that needs to be secured from the devices, and the type of devices catered for by the network, one or multiple security layers could be chosen from those highlighted in the figure.

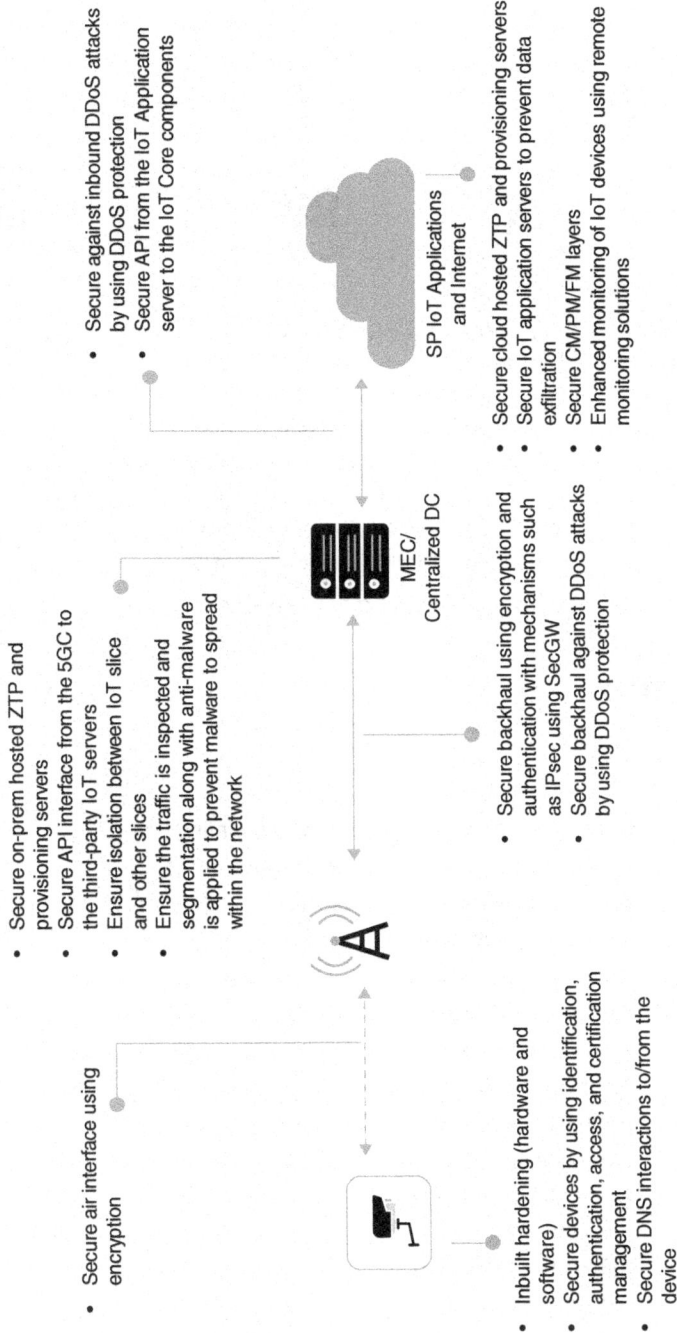

FIGURE 8-29   Multiple Layers of Security Required to Mitigate IoT Threats in the Given Scenario

- Secure air interface using encryption

- Secure on-prem hosted ZTP and provisioning servers
- Secure API interface from the 5GC to the third-party IoT servers
- Ensure isolation between IoT slice and other slices
- Ensure the traffic is inspected and segmentation along with anti-malware is applied to prevent malware to spread within the network

- Secure against inbound DDoS attacks by using DDoS protection
- Secure API from the IoT Application server to the IoT Core components

SP IoT Applications and Internet

- Inbuilt hardening (hardware and software)
- Secure devices by using identification, authentication, access, and certification management
- Secure DNS interactions to/from the device

- Secure backhaul using encryption and authentication with mechanisms such as IPsec using SecGW
- Secure backhaul against DDoS attacks by using DDoS protection

MEC/ Centralized DC

- Secure cloud hosted ZTP and provisioning servers
- Secure IoT application servers to prevent data exfiltration
- Secure CM/PM/FM layers
- Enhanced monitoring of IoT devices using remote monitoring solutions

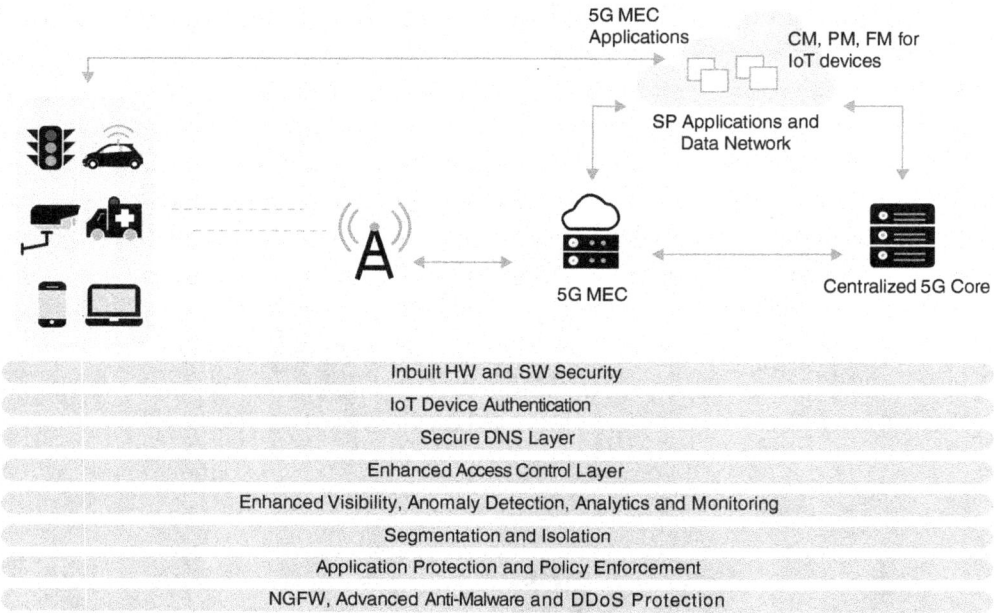

Inbuilt HW and SW Security

IoT Device Authentication

Secure DNS Layer

Enhanced Access Control Layer

Enhanced Visibility, Anomaly Detection, Analytics and Monitoring

Segmentation and Isolation

Application Protection and Policy Enforcement

NGFW, Advanced Anti-Malware and DDoS Protection

FIGURE 8-30   Multilayered Security for Securing IoT Devices

Key Acronyms

Acronym	Expansion
5GC	5G Core
AAA	Authentication, authorization, and accounting
AMF	Access and Mobility Management Function
AUSF	Authentication Server Function
C-V2X	Cellular – vehicle-to-everything
C&C	Command and control
C2C	Command and control
CNF	Cloud-Native Function
COBOT	Collaborative robots
CSC	Communication services consumer
CSP	Communication services provider
DDoS	Distributed denial of service
DMZ	Demilitarized zone
DNS	Domain Name System
DoS	Denial of service
DSRC	Dedicated short-range communications

Acronym	Expansion
eMBB	Mobile broadband
FQDN	Fully qualified domain name
IAM	Identity and Access Management
IIoT	Industrial Internet of Things
IoT	Internet of Things
M2M	Machine-to-machine
MEC	Multi-access edge compute
MFA	Multiple factor authentication
mIoT	Massive Internet of Things
MQTT	Message Queuing Telemetry Transport
NAC	Network Access Control
NF	Network Function
NPN	Non-public network
NSaaS	Network slice as a service
OBU	On-board unit
OT	Operational technology
PCF	Policy and Charging Function
PNI-NPN	Public Network Integrated NPN
RAT	Radio access technology
RBAC	Role-based access control
SBA	Service-based architecture
SBI	Service-based interface
SDN	Software-defined network
SMF	Session Management Function
SNPN	Standalone NPN
SSL	Secure Sockets Layer
TCG	Trusted Computing Group
TEE	Trusted Execution Environment
TLS	Transport Layer Security
TPM	Trusted Platform Module
UDM	Unified Data Management
UE	User equipment
UICC	Universal integrated circuit card
UPF	User Plane Function
V2X	Vehicle-to-everything
VNF	Virtual Network Function
ZTS	Zero-Touch Security

References

"Cisco Annual Internet Report (2018-2023) White Paper," https://www.cisco.com/c/en/us/solutions/collateral/executive-perspectives/annual-internet-report/white-paper-c11-741490.html

Pramod Nair, "Securing Your 5G Infrastructure to the Edge and Beyond," Cisco Live, Cisco Public, March 30, 2021

Pramod Nair, "Securing Your 5G Network End to End," Cisco Live APJC, Cisco Public, April 30, 2019

Pramod Nair et al., "Cisco 5G Security Innovation with Cisco," https://www.cisco.com/c/dam/en/us/products/collateral/security/5g-security-innov-wp.pdf

"Examining the Security of Wi-Fi 6 and 5G White Paper," https://www.cisco.com/c/en/us/solutions/collateral/executive-perspectives/white-paper-c11-743921.html

"What is Trust?" Cisco Public, https://www.cisco.com/c/en/us/solutions/service-provider/5g-what-is-trust.html

Robbie Grue, "DoH! What's all the fuss over DNS over HTTPS?", Cisco Public, https://umbrella.cisco.com/blog/doh-whats-all-the-fuss-about-dns-over-https

"Trusted Platform Module (TPM) Summary," https://trustedcomputinggroup.org/wp-content/uploads/Trusted-Platform-Module-Summary_04292008.pdf

3GPP TS 29.561 V17.1.0: Interworking Between 5G Network and External Data Networks

3GPP TS 28.807 V17.0: Study on Management of Non-Public Networks (NPNs)

Chapter | **9**

Securing 5G Use Cases

After reading this chapter, you should have a better understanding of the following topics:

- Key 5G use cases
- Key threat surfaces in different 5G use cases
- Key security controls for securing 5G use cases

This chapter will take you through the different 5G IoT use cases, their respective threat surfaces, and mechanisms to mitigate the threats.

This chapter will be of particular interest to the following members of teams from enterprise 5G consumers, 5G service providers, and cybersecurity vendors planning to provide security layers for 5G:

- Strategy teams of service provider and enterprise verticals looking at 5G use cases
- Security strategy teams within service provider and enterprise verticals deploying 5G
- Cloud computing and data center teams involved with 5G strategy and deployment
- Security architects and design teams looking at securing the mobile infrastructure
- Enterprise solution and security architects deploying standalone private/NPN 5G or utilizing a service provider's 5G slice network
- Government departments deploying 5G for smart city projects
- Cybersecurity vendor teams looking to secure 5G use cases for their customers
- Product managers of cybersecurity vendors

This chapter focuses on 5G use cases that can have a mixture of different deployment methods, such as standalone private 5G/non-public network (NPN), public 5G, and hybrid 5G deployments with integrations between the enterprise-deployed NPN and the service provider network. The hybrid 5G deployment is also called a public network infrastructure (PNI) NPN.

The industrial IoT (IIoT) use cases discussed in this chapter make use of enhanced mobile broadband (eMBB), machine-to-machine (M2M), and massive Internet of Things (mIoT) domains. Although Chapter 8, "Securing Massive IoT Deployments in 5G," is dedicated to mIoT/consumer IoT, some of the use cases in this chapter would cater for mIoT devices. Although the use cases mentioned in this chapter are yet to be fully realized, there are engagements that I am working on with service providers that made me realize that the key use cases require a broader set of security controls. The main reason being that any use case that will be deployed using 5G will need to cater for legacy and existing network infrastructures and devices in the network. This chapter is a sincere attempt to spread the message of the impending threat surfaces I've encountered during my discussions with customers and the mitigation mechanisms that need to be undertaken to enhance the security of the network.

Other use cases, such as securing smart cities, will include all the use cases discussed in this chapter. There are various industry frameworks, standardizations, and open source communities working toward achieving a secure smart city concept, and its realization is still quite a few years away. There are many more use cases, some selected for further study for 3GPP Release 18, which is expected to be released around December 2023.

With the aim of keeping the chapter as close to a real scenario as possible, this chapter includes the following topics, which have traction in the market today:

- Secure 5G smart factory and manufacturing

- Secure critical infrastructure (example considered for secure 5G energy utility deployments)

- Secure 5G-V2X

The example of a secure energy utility could be applied to other critical utilities with variations in the use case, such as securing water utilities. As discussed in Chapter 8, the IoT/IIoT devices being used in the 5G deployments to fulfill the use cases might be smart or not-so-smart devices. These devices will have varying levels of intelligence and security awareness and therefore require different levels of security controls.

Secure 5G Smart Factory and Manufacturing

5G technology is touted as the key enabler of Industry 4.0. This is due to the advantages that 5G provides, such as higher throughput, lower latency, flexibility in deploying the 5G components in the cloud or on the premises, openness in integration with applications made by third-party developers, and the flexibility in deploying the 5G infrastructure. With sensors that are deployed in factories, machines will benefit from this fast and reliable low-latency data transfer. Driverless transport systems within the manufacturing factory will benefit from the low-latency features of 5G. The 5G technology is still young and evolving. Companies are starting to market various products and services that will run on 5G. This transformation will be a multiyear journey, as production deployments require stable specifications, production-quality implementations, and supportive ecosystems to be in place before widespread availability.

One of the key enablers of improved efficiency in the manufacturing industry is the use of robotics. Automated manufacturing solutions using robotics improve efficiency and safety, automate repetitive tasks, reduce margins of error, and enable human workers to focus on more productive areas of the operation. Fully autonomous robots in manufacturing are commonly needed for high-volume, repetitive processes. Other manufacturing automation solutions include robots performing heavy tasks such as lifting, holding, and moving heavy pieces. The robots in the factory are in continuous production and movement, which requires near real-time implementation of commands.

Manufacturing factories also have sensors deployed all over for various reasons, such as collecting data from disparate devices and systems to enable different machines to talk to one another. This creates seamless connectivity throughout the factory for equipment monitoring, data collection from different systems to predict any equipment failure, and triggering protocols related to site and equipment maintenance.

Figure 9-1 shows one of the possible different methods of deploying 5G to cater for a manufacturing use case.

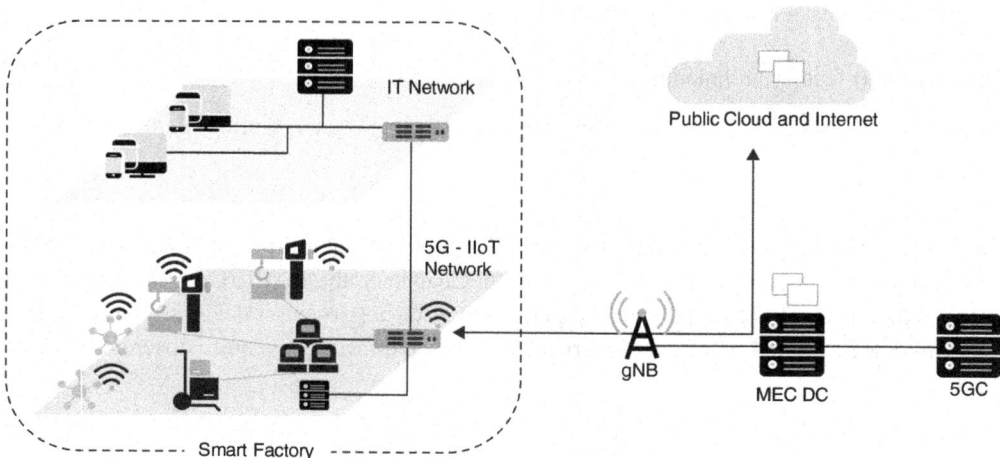

FIGURE 9-1 Example of a 5G Smart Factory Deployment

As shown in the Figure 9-1, the manufacturing entity, such as a smart factory, can be realized by using the service provider's 5G network coverage. The 5G coverage could be provided by Fixed Wireless Access (FWA), an open RAN deployment model where Distributed Unit (DU) and Centralized Unit (CU) or a micro/pico cell could be deployed at the factory premises. The service provider would then allocate a network slice and configure the required quality of service (QoS). The QoS would ensure that the required availability and other SLAs are met to provide the required service to the smart factory/manufacturing use case. This scenario could also be fulfilled by using the network slice as a service (NSaaS) method, where the manufacturing/smart factory entity is the communications service customer (CSC) and the service provider with the 5G network components is the communications service provider (CSP). The CSC will be provided access to the management of the service layer of the network slice provided by the CSP. A privileged user such as an admin from the CSC can then configure the required QoS for the 5G NFs within the network slice catering for the manufacturing/smart factory facility. Figure 9-2 illustrates the different IIoT devices and the IIoT network within the 5G smart factory considered in this example.

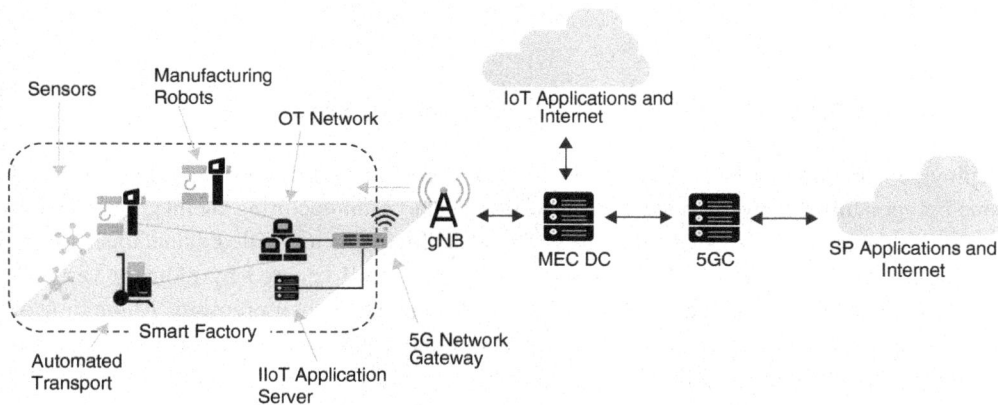

FIGURE 9-2   Components of the 5G Smart Factory

Figure 9-3 shows another option of 5G deployment by using private 5G or non-public network (NPN) to cater for 5G smart factory and manufacturing use cases.

In the deployment method shown in Figure 9-3, 5G radio nodes (gNB/CU/DU) and the key user plane components are deployed on the premises of the smart manufacturing facility, and the 5G control plane is deployed on the service provider's premises. These deployments are called public network integrated private 5G network/non-public network (PNI-NPN). This method of deployment is considered when ultra-reliable low-latency communications (URLLC) needs to be deployed. As URLLC is extremely latency sensitive, deploying the non-3GPP NFs, such as applications catering to the smart factory use cases, and 3GPP NFs, such as User Plane Functions (UPFs), closer to the URLLC devices would help you meet the stringent latency requirements. In such deployments, the enterprise/smart factory would retain complete control over the flow of 5G traffic, including the 5G air interface (Uu) communications between the 5G IIoT device/5G network gateway and the 5G access node (gNB) components. If all the components, including RAN, 5GC user plane, and control plane, are deployed and catered for by the enterprise, it is referred to as a standalone private 5G/NPN network.

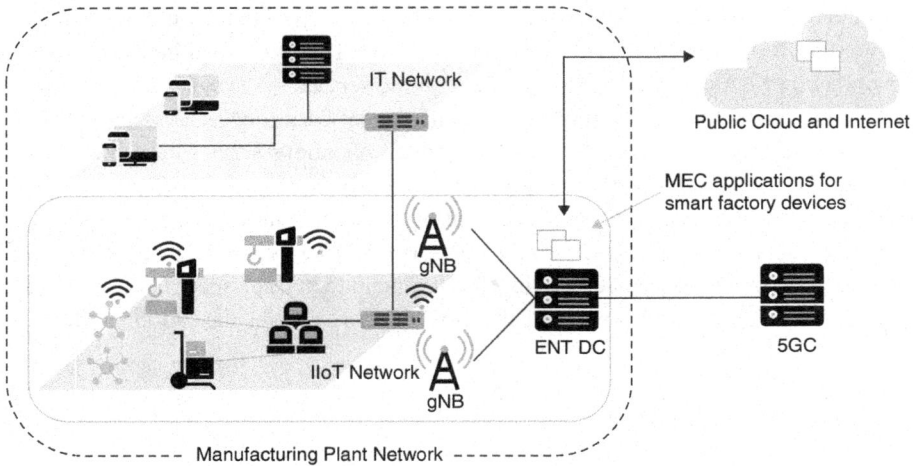

FIGURE 9-3   PNI Non-Public 5G/PNI Private 5G Deployment Example

Another deployment method, shown in Figure 9-4, has the entire 5G radio and 5G Core user and control plane components deployed on the premises of the smart manufacturing facility, and there is no dependency on the service provider's premises. These deployments are called standalone private 5G networks/non-public networks (SNPNs). These deployments will be used by industry verticals and enterprises that need the 5G use cases deployed and executed within their premises. At the time of writing this book, national security government organizations and smart factory industry verticals were undergoing initial standalone non-public 5G network deployments.

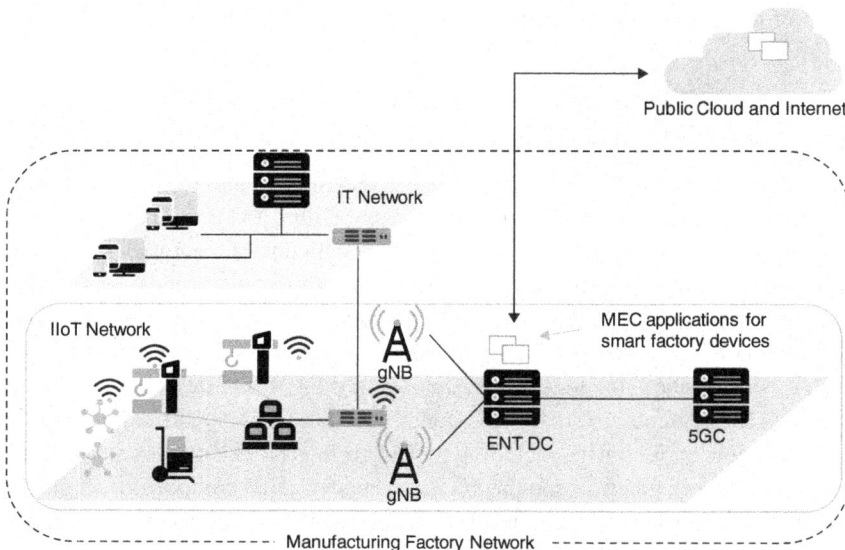

FIGURE 9-4   Standalone Private 5G/Non-Public (SNPN) 5G Deployment Example

Threats in 5G Smart Factory Deployments

5G introduces the technology advancement required to fulfill URLLC use cases in the smart factory and manufacturing sector, thereby improving productivity. 5G's enhanced deployment methods, such as multi-access edge computing (MEC), allow robots and collaborative robots (cobots) to be monitored and updated in near real time as the environment changes. Where, for example, customizations to specific products are needed, 5G IoT can be used to feed precise requirements for each product through to the robot or cobot to create personalized products and facilitate integration with third-party ecosystems using APIs. This openness in technology, however, also introduces an expanded threat surface, as illustrated in Figure 9-5 and described in the list that follows:

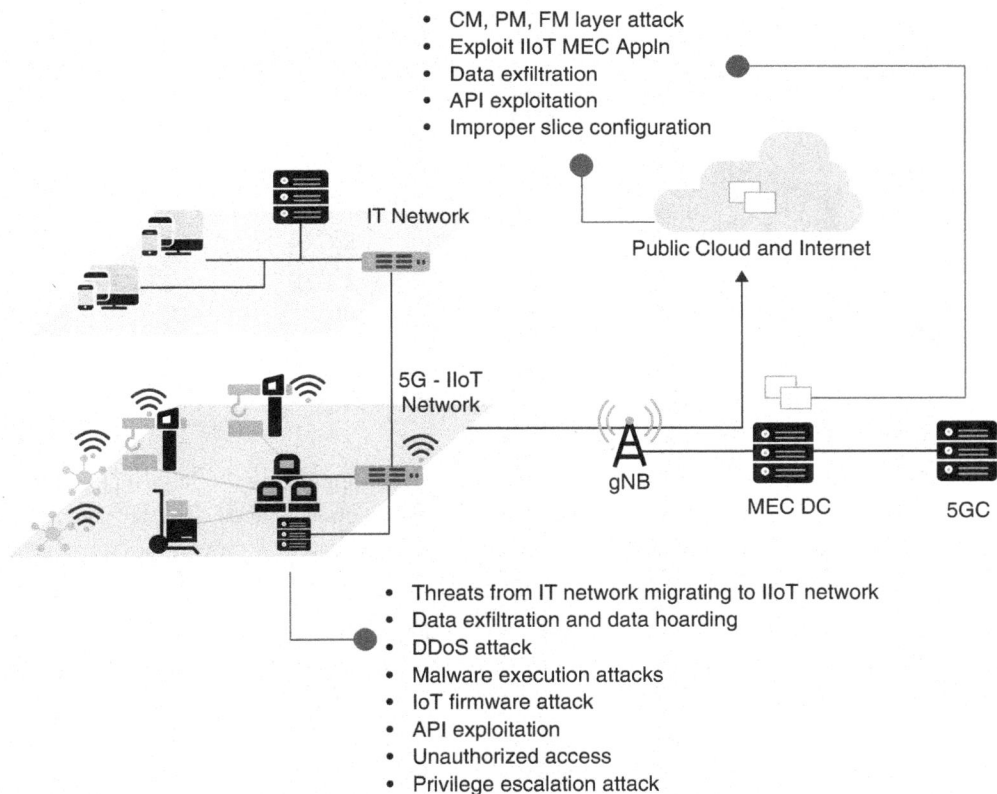

- CM, PM, FM layer attack
- Exploit IIoT MEC Appln
- Data exfiltration
- API exploitation
- Improper slice configuration

IT Network

Public Cloud and Internet

5G - IIoT Network

gNB

MEC DC

5GC

- Threats from IT network migrating to IIoT network
- Data exfiltration and data hoarding
- DDoS attack
- Malware execution attacks
- IoT firmware attack
- API exploitation
- Unauthorized access
- Privilege escalation attack

FIGURE 9-5   Key Threats 5G Smart Factory and Manufacturing Use Case

■ One of the key risks in today's manufacturing network that is also important in the 5G–smart factory/manufacturing use case is the migration of threats from the IT network to the IIoT network. The employees and guests in the IT enterprise network of the factory are susceptible to malware and other risks due to phishing and other such attacks on the users of the IT network, as shown in Figure 9-6. Although this is not a pure 5G threat surface, if this risk is not mitigated or the security control planned in the design, it will impact the 5G–IIoT network.

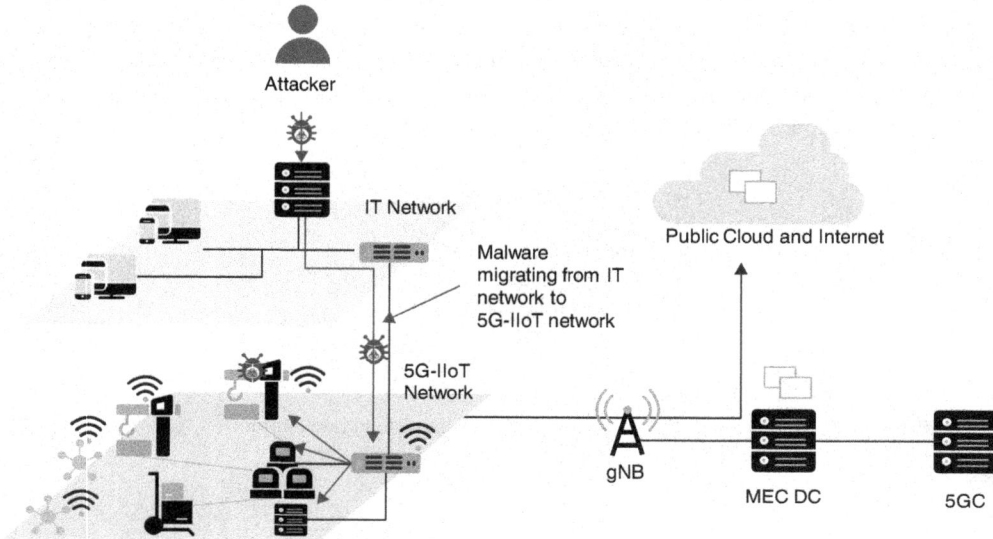

FIGURE 9-6   Example of a Threat Surface from an IT Infrastructure

■ The attacker uses the Internet-facing server in the IT network of the factory to launch an attack into the 5G–IIoT network due to improper isolation. Once the cobots and 5G–IIoT devices are under the control of the attacker, the attacker could launch a DDoS attack on your 5G infrastructure or conduct data exfiltration.

■ With regard to application-based threat surfaces, non-3GPP 5G NFs and 5G–IIoT devices are susceptible to patches or updates with embedded malicious code, as shown in Figure 9-7. This is basically done at a couple of levels. One of the primary insertion points of malicious code is the update server of the 5G–IIoT device and application vendor itself. The other key risk area for malicious insertion of code is the management network/operational technology (OT) network of the smart factory. Due to the implementation of weak authentication and authorization in the management layer of the 5G–IIoT network, the attacker could launch a privilege escalation attack on your network.

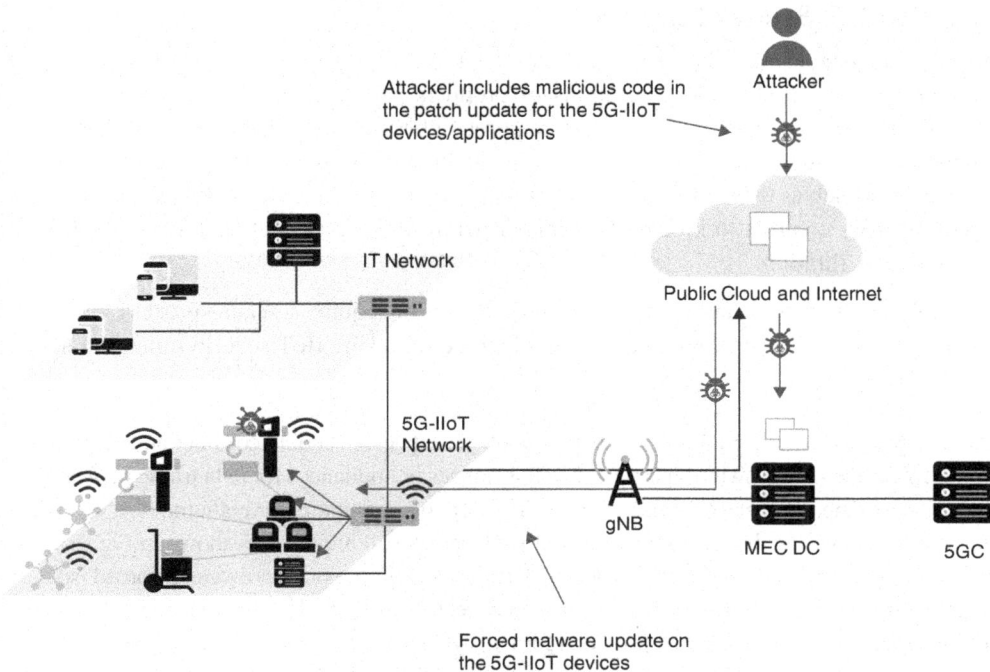

FIGURE 9-7   Embedded Malicious Code in Firmware and Application Updates

- API is another area that has an increased risk factor in 5G smart factory deployments. 5G brings in service-based architecture (SBA), which uses API-based communication both internally to the 3GPP 5G NFs and in interactions with non-3GPP NFs using the Network Exposure Function. API will also be the preferred mode of communication between the 5G–IIoT devices and the corresponding management layer. Although 3GPP has recommended the Common API Function (CAPIF), the security feature implementation of the API, such as granular authentication and authorization, is up to the developer. Weak API implementation, such as passing too much information in error messages, will lead to exposing the type and version of the web server to the attackers, which could then be exploited to launch a targeted attack on the API web server. Attackers have to gain access to only a few accounts, or just one admin account, to compromise the system, which could lead to the leakage of legally protected, highly proprietary, sensitive information in your smart factory network. Insecure communication links between the headquarters (HQ) and the multiple 5G smart factory sites are susceptible to man-in-the-middle (MitM) attacks.

Securing the 5G Smart Factory

5G technology will enable manufacturers deploying IoT technologies to migrate to Industry 4.0, optimize production, and build new generations of products and services. This deeper integration between 5G, IT, cloud, and industrial networks requires cybersecurity capabilities that can protect your production environment and assets as well as maintain operations. Service providers deploying technology enablers, such as MEC and network slicing, to implement smart factories or manufacturing should deploy security controls to protect their infrastructure from risks such as DDoS attacks, data hoarding, and data exfiltration.

Securing the 5G smart factory and manufacturing entities will require a secure process and multi-layered security controls. The secure process should include following IIoT security-related standards such as the following:

- **Segmentation and isolation:** 5G smart factory deployment design should consider an industrial demilitarized zone (DMZ) with separate firewall instances or appliances for data traffic between the enterprise/IT network and the wider factory IIoT network and jump servers/jump host server for managing and accessing the IIoT devices from the IT network, if required, as shown in Figure 9-8. The separate firewall/gateway for the IT and IIoT networks will provide hardware-enforced isolation, preventing arbitrary connections to act as attack vectors from the Internet toward the IIoT network. Separate protection should be provided for legacy devices. The network devices should have support for cybersecurity protocols such as access control lists (ACLs), NetFlow, TrustSec, IPsec, and MACsec. A network access controller (NAC) should be dedicated to creating and enforcing network security policies. The NAC solution you choose should provide software-based device-level segmentation to manage network access control at scale. Figure 9-8 illustrates the segmentation and isolation security controls for the 5G smart factory use cases.

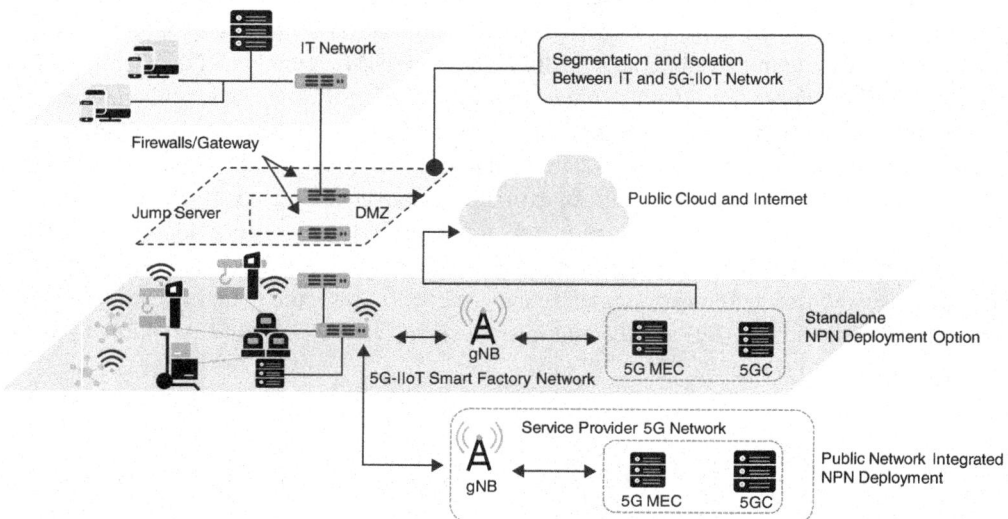

FIGURE 9-8   Segmentation and Isolation Between IT and 5G–IIoT Networks

■ **Enhanced visibility and anomaly detection:** Deploying security controls for your smart factory 5G–IIoT network can quickly become very complex, especially if your industrial network is dispersed across an entire country or many remote industrial sites. For your security deployment to be successful, the security solution you select must be able to scale easily and at a reasonable cost across your entire organization. To have an effective security strategy, recommended practice dictates selecting IIoT network devices with built-in security sensors that enable you to collect the information required to provide comprehensive visibility, analytics, and threat detection. Having a built-in sensor in your 5G–IIoT network automatically uncovers the smallest details of the production infrastructure, such as vendor references, firmware and hardware versions, serial numbers, and so on. The built-in sensor should also identify asset relationships and communication patterns as well as provide you with detailed analysis of inter-device communication and identify any anomaly in behavior, such as unexpected controller modifications or data exfiltration from the 5G–IIoT network to a new host or a new connection to an unexpected host. This will enable you to take action to ensure production continuity and maintain system integrity. The data from the built-in sensors should be consumable by an existing or newly deployed anomaly detection solution, as illustrated in Figure 9-9. The anomaly detection engine (ADE) should be customizable so that you can provide the policy defined for the IIoT devices. This custom policy must be used by the ADE to compare it with the real-time behavior of 5G–IIoT systems; this will ensure that any anomalies in behavior will be detected, thereby immediately improving the time taken to detect. The ADE should also use machine learning (ML) and artificial intelligence (AI) to enhance the efficacy of the threat detection. ML is trained by the security vendors by providing vast amounts of learning data with varying efficiency, depending on the vendor. Recommended practice dictates testing the efficiency of these ML algorithms by creating your own set of learning data and verifying if the ML algorithms actually detect the anomalies. End-to-end visibility on the enterprise IT network and the 5G–IIoT factory network can be provided by utilizing NetFlow telemetry data from the network, creating a baseline of the behavior of network devices, and then detecting any anomalies. Such a solution should also cater for encrypted packets without the need for decrypting the encrypted traffic. Figure 9-9 illustrates the enhanced visibility security controls for the 5G smart factory use cases.

■ **API security:** For multisite smart factory deployment, the alerts and security events should be sent to the centralized monitoring system in the HQ, which includes the configuration management (CM), fault management (FM), and performance management (PM) for the 5G–IIoT smart factory devices. The CM commands would be configured using the API interface. In NPN deployments, an API mode of communication would be used for any integration between the Network Exposure Function (NEF) and the Application Function (AF). This reasserts the importance of API security, including API connectivity security using Transport Layer Security (TLS), which enhances both privacy and data integrity, as shown in Figure 9-10. Many of the developers building applications do not enforce the right restrictions on what authenticated users are allowed to do. Attackers exploit these flaws to access unauthorized services and perform malicious actions, such as accessing other users' accounts, viewing sensitive files, modifying other users' data, changing access rights, and so on. To mitigate the risks due to improper authentication and authorization, there are a few key steps you could undertake. One of the key steps is to verify if your chosen vendor ships API web servers with default credentials for admin users. The other key method of mitigating weak API implementation is to ensure that the granular authentication can be implemented as shown in Figure 9-10, preventing automated brute-force attacks and credential stuffing attacks, which makes use of lists of known passwords. Figure 9-10 illustrates the API security controls for the 5G smart factory use cases.

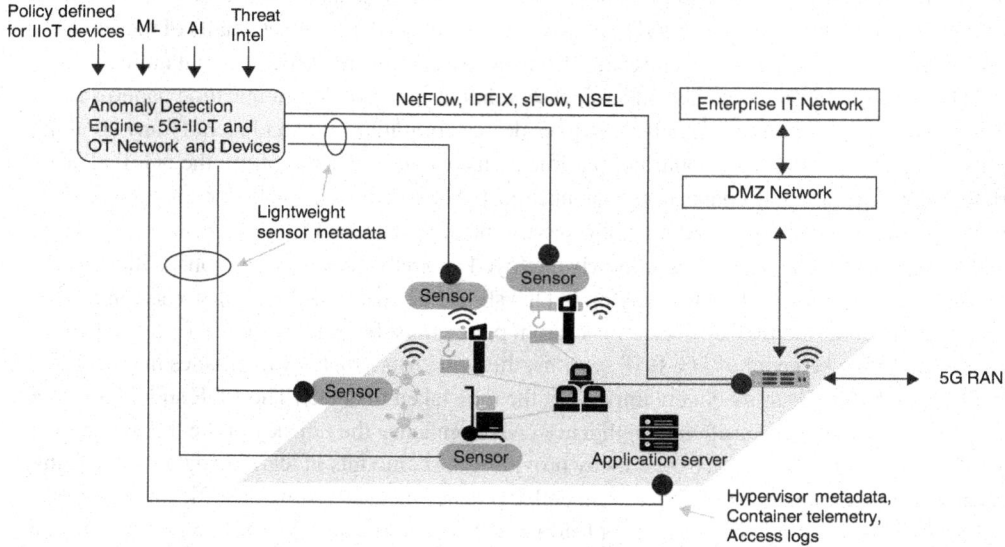

FIGURE 9-9   Anomaly Detection Engine for 5G–IIoT Network

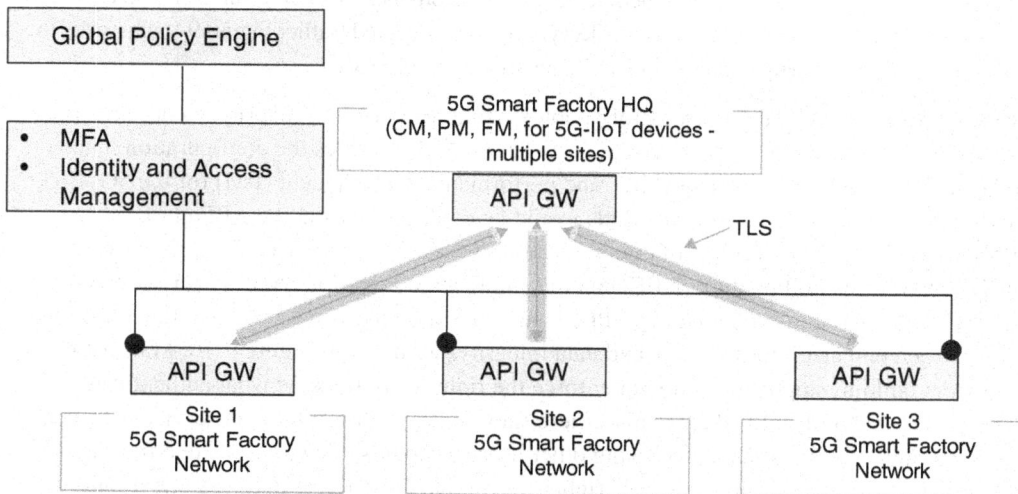

FIGURE 9-10   API-GW Integration with Identity and Access Management

As Figure 9-10 illustrates, multifactor authentication (MFA) and Identity and Access Management (IAM) should be integrated with a global policy engine that will use a custom policy provided by you as well as input from external sources as input to a trust algorithm to decide access to the resource.

Application-Level Security Controls

Depending on the 5G deployment opted for by the smart factory, the implementation of the application security controls will vary. If the PNI-NPN method is deployed, as shown in Figure 9-11, the deployment would be based on a specific slice for the smart factory use case. In the example shown in Figure 9-11, the user plane and some key 5GC NFs, like AMF, SMF, and PCF, will be deployed on the enterprise's premises with SMF integration with a local IAM solution for authentication of users and 5G–IIoT devices.

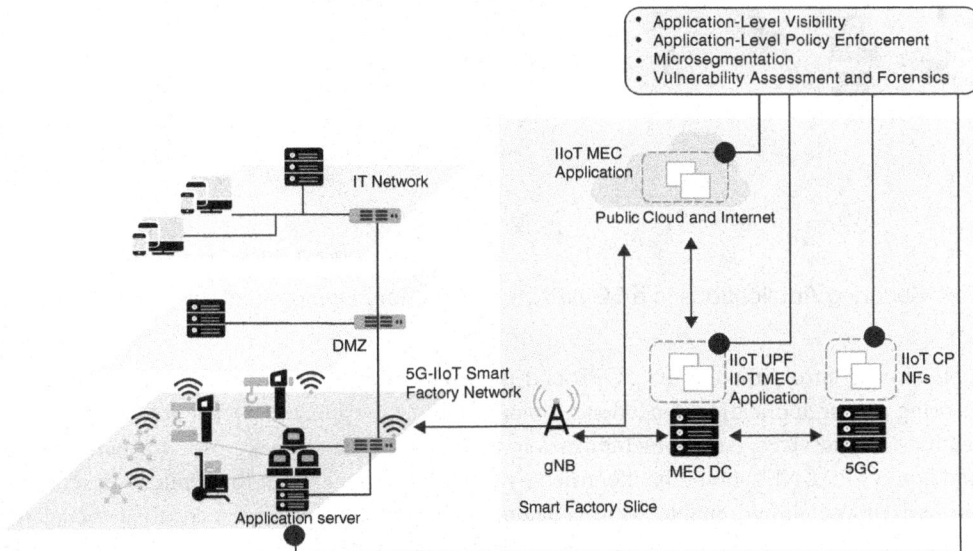

FIGURE 9-11   Securing Applications in the 5G and IIoT Network

If the enterprise-hosted NPN method is deployed as shown in Figure 9-12, it would be based on a 5G network fully owned and operated by the enterprise. In the example in Figure 9-12, the 5GC components are deployed on the enterprise's premises with integrations with a local IAM solution for authentication of users and 5G–IIoT devices.

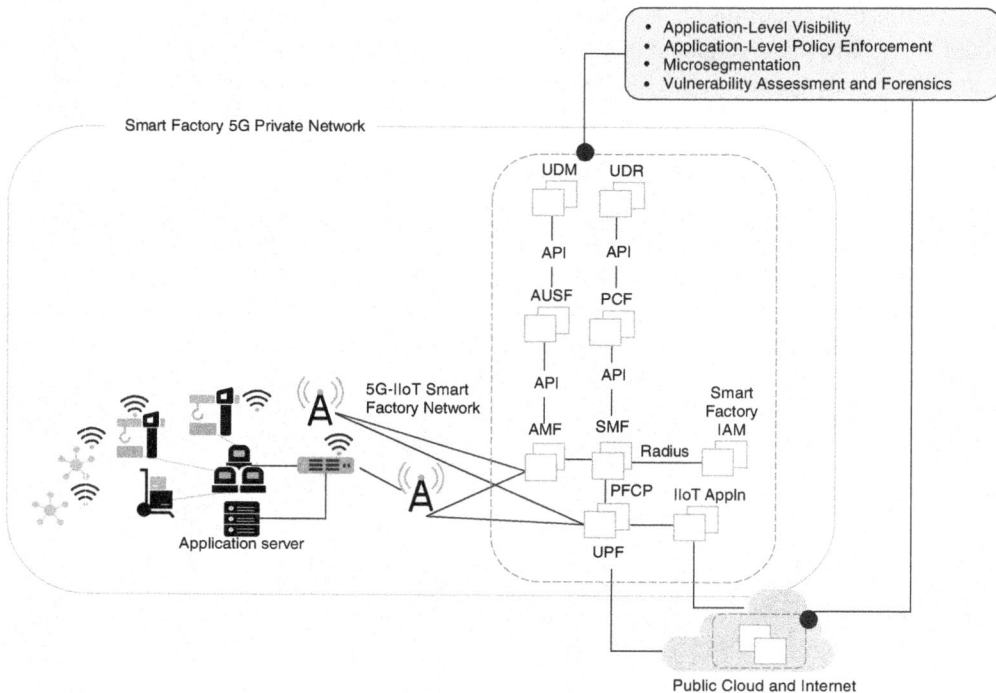

FIGURE 9-12   Securing Applications in a 5G NPN Smart Factory Deployment

The key application controls for the 5G 3GPP and non-3GPP Cloud-Native Functions (CNFs) are based on ensuring proper application dependency mapping, such as applications that are dependent on other applications and services. Any vulnerabilities in the open source software used for building the 5G 3GPP and non-3GPP CNFs should be identified by the solution you select for application security. The solution you choose should enable security teams to implement a secure, zero-trust model for workloads using microsegmentation. This layer of security control must automate the policy generation for microsegmentation using unsupervised machine learning and near-real-time application behavior analysis. It must also normalize these policies based on priority and hierarchy before enforcing the security policies.

Figure 9-13 shows the summary of the security controls for mitigating the threats for the manufacturing use case.

The aforementioned security controls, along with ongoing standardization efforts on private 5G/NPN 5G deployments, are key for establishing a reliable and secure smart factory and manufacturing solution. Smart factory 5G–IIoT devices such as 5G-enabled robotics and sensors would require integrations with non-3GPP applications at the edge of the network, enabling URLLC use cases for smart manufacturing. A multilayered security control, as shown in Figure 9-13, is necessary to reduce risk and provide a robust security layer.

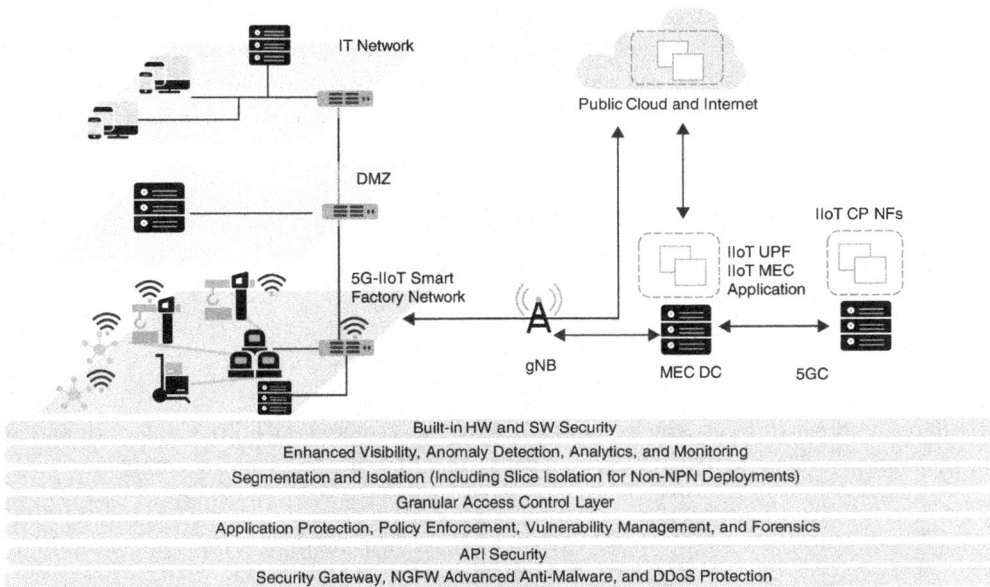

IT Network

Public Cloud and Internet

DMZ

IIoT CP NFs

IIoT UPF
IIoT MEC
Application

5G-IIoT Smart
Factory Network

A

gNB

MEC DC   5GC

Built-in HW and SW Security
Enhanced Visibility, Anomaly Detection, Analytics, and Monitoring
Segmentation and Isolation (Including Slice Isolation for Non-NPN Deployments)
Granular Access Control Layer
Application Protection, Policy Enforcement, Vulnerability Management, and Forensics
API Security
Security Gateway, NGFW Advanced Anti-Malware, and DDoS Protection

FIGURE 9-13   Summary of Security Controls to Secure 5G Smart Factory Deployments

Critical Infrastructure

5G brings advanced connectivity technologies to critical infrastructure providers such as the energy utility sector. 5G can play an essential role in enabling utilities to keep pace with rapid change. There are multiple ways the utility companies, such as energy providers, can make use of the advantages of 5G, such as low latency, network slicing features, and edge deployment modes. Service providers could also monetize the 5G network deployment by reselling parts of the network resources to utility providers. This section will take you through both views—from the energy utility company's perspective and the service provider's perspective.

5G Energy Utility

Energy utility companies are seeing demand from their customers to have near-real-time information on usage and pricing. They are also seeing internal demand for real-time identification of faults to reduce the network downtime or outage period. 5G technology with enablers like edge computing deployments for radio and packet core components will help in catering for the aforementioned new requirements from energy utility companies. They also have to cater for some areas of the energy utility company that might have legacy deployments of proprietary control systems, which need to be catered for until the time they are discontinued and taken out of operation, as they will be inefficient and expensive to operate and maintain. Grid operations have expanded over time, leading to a patchwork of legacy control and business environments. Utility operations have recently become a rich target for cyberattacks, given both their national visibility and the expanded attack surface across their increasingly connected assets. Connected assets also carry the risk of physical damage, utility downtime, and breaches of customer data and intellectual property. The energy grid in today's network is vulnerable to the following:

- Lack of compliance with evolving security standards

- Legacy nodes and infrastructure elements

- Blind spots in the network

- Harsh outdoor environments that affect transmission and delivery

Let's look at the new threat surfaces that are introduced by energy networks utilizing 5G technology. Figure 9-14 illustrates an example of a 5G use case deployment for an energy utility company.

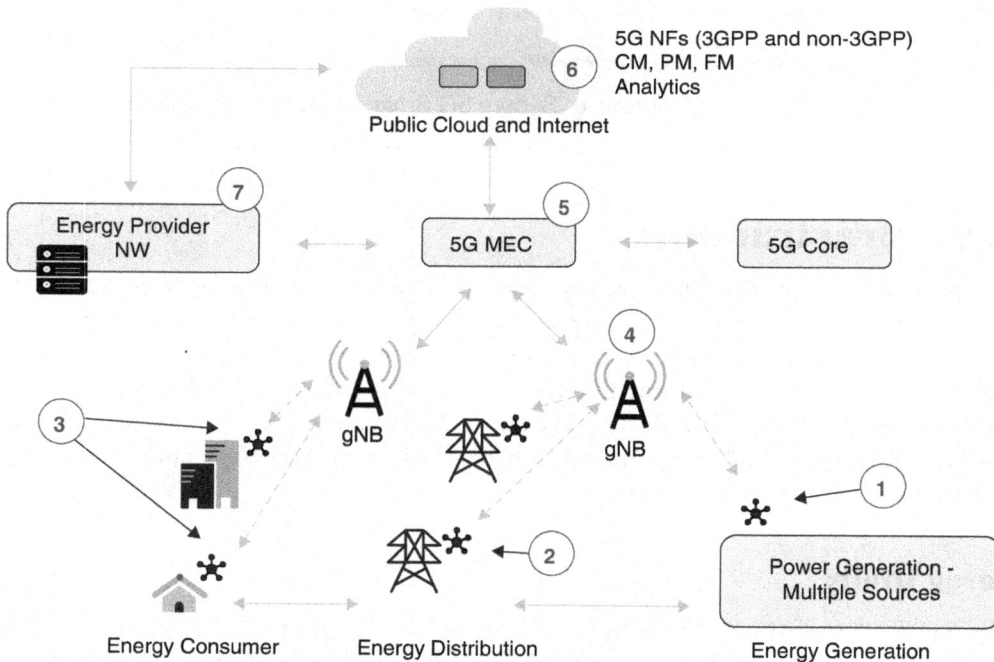

FIGURE 9-14   5G Use Case Example for the Energy Utility Sector

The key components of the energy utility use case shown in Figure 9-14 are explained in the list that follows:

1. **Smart sensors in energy generation plants:** These smart sensors are capable of listening and connecting to 5G radio nodes. These will be connected to on-premises equipment in the energy/power generation plants to monitor the critical infrastructure. Some of the use cases for these smart sensors in the location will be for monitoring the key statistics of critical infrastructure and sending them across to the centralized monitoring system.

2. **Smart sensors in the energy distribution centers:** The smart sensors in energy distribution locations are capable of monitoring and identifying critical outages in real time. This enables you to resolve outages much faster.

3. **Smart sensors in customer locations:** The smart sensors in customer locations will collect the real-time information of energy usage and send it to the centralized monitoring system for customer on-premises equipment. This allows you to provide more tailored solutions to your customers.

4. **5G gNBs:** 5G radio nodes (gNBs) in different form factors such as software radios, split gNBs such as Distributed Unit (DU) and Centralized Unit (CU), macro cells, micro cells, and pico cells could be used to provide radio coverage to the sensors deployed by the energy utility company. The radio nodes could be part of the private network, also called a non-public network, that is deployed and owned by the energy utility company.

5. **5G multi-access edge compute (MEC):** 5G MEC consists of 3GPP and non-3GPP network components required to cater for the use case. There are different ways to achieve these use cases. For example, the state-owned energy utility company could deploy its own private 5G network, which includes 5G radio nodes in different form factors, back haul, 5G packet core, and integration with third-party applications for analytics and fulfillment of specific use cases such as performance monitoring and resolution. MEC, as described in Chapter 5, "Securing MEC Deployments in 5G," can be deployed in public cloud, on-premises, or hybrid deployment mode. The NSaaS model could also be used if the energy utility company would like to reuse the service provider infrastructure. NSaaS is one of the key use cases being considered by service providers to monetize their 5G network deployments. The service provider, depending on its capabilities, could also offer to secure the IIoT network deployment, which might be deployed at the energy generation plants.

Figure 9-15 illustrates an example of the NSaaS model being used by the energy utility company.

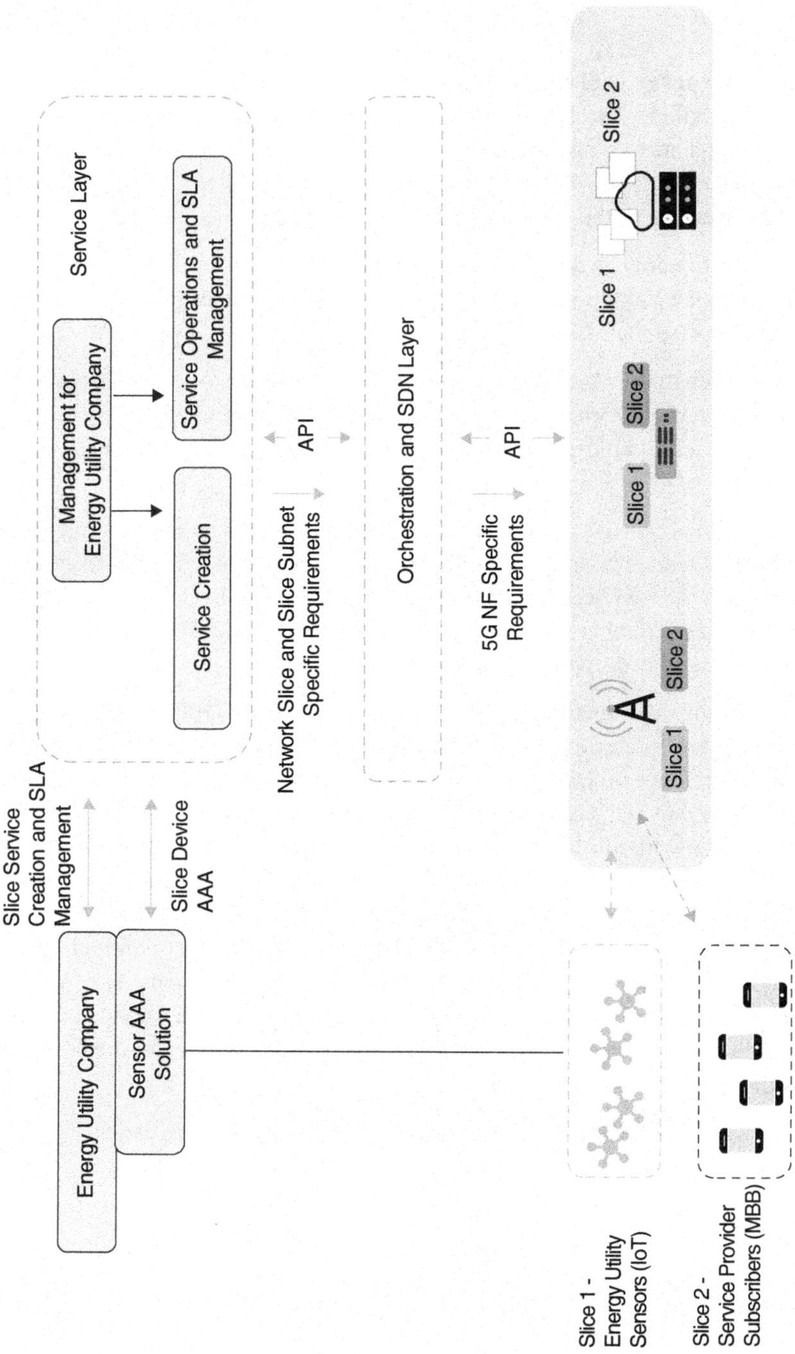

FIGURE 9-15   5G NSaaS Deployment Model for Energy Utility

Figure 9-16 illustrates the private network deployment model being used by the energy utility company.

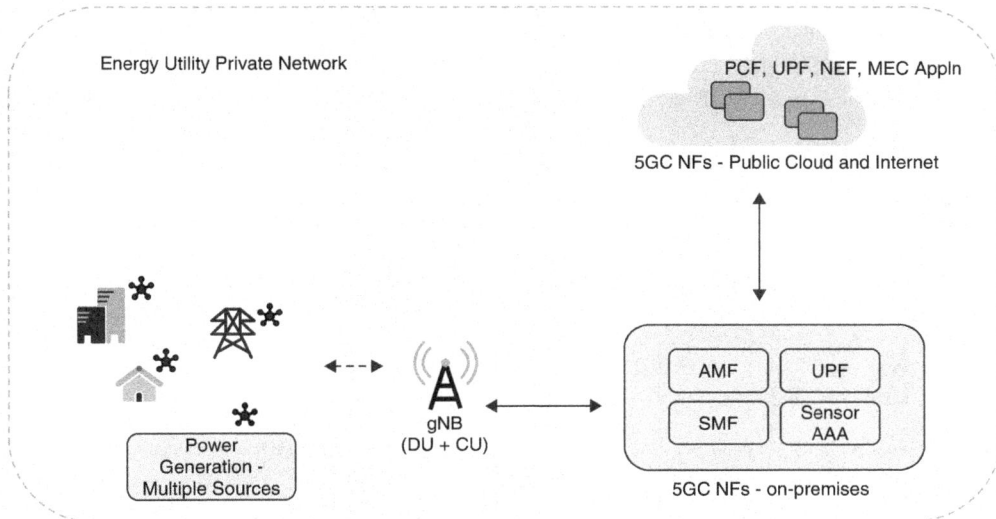

FIGURE 9-16 Example of a 5G NPN Deployment Model for an Energy Utility

6. **5G 3GPP and non-3GPP network functions deployed in the public cloud:** This allows the energy utility company to be flexible in integrating 5G NFs with third-party applications for real-time analysis.

7. When the NSaaS model is being followed, both the communication services provider (CSP) offering NSaaS and the communication services customer (CSC—in this case, the energy utility company) consuming NSaaS have knowledge of the existence of network slice instances. Depending on service offering, the CSP offering NSaaS might impose limits on the NSaaS management capabilities exposure to CSC (the energy utility company). The CSC can manage the network slice instance according to the limited level of NSaaS management capabilities exposed by and agreed upon by the CSP.

Threats in the 5G-Enabled Energy Utility

Critical infrastructures such as utilities have been the subject of sophisticated cyberattacks aimed at data exfiltration, mapping critical infrastructures and collecting detailed information about them to create detailed targeted attacks. Apart from the attack surfaces, legacy control systems also pose multiple risks. The legacy, often proprietary, control systems are no longer efficient to operate and are challenging to secure. If not adequately detected and contained, cyberthreats can go on for a long time and impact networks, monitoring systems, and consumer smart sensors. Targeted attacks, such as crippling the targeted country's entire electric grid, can cause havoc and complete chaos.

This section will discuss the threat surface in the 5G deployment for the energy sector use case. The threats will cover different deployment models, such as private 5G deployment by the energy utility company and the NSaaS model. This section will cover the threat surfaces for the energy utility company and the service provider. Figure 9-17 shows the key threat surfaces for the manufacturing use case.

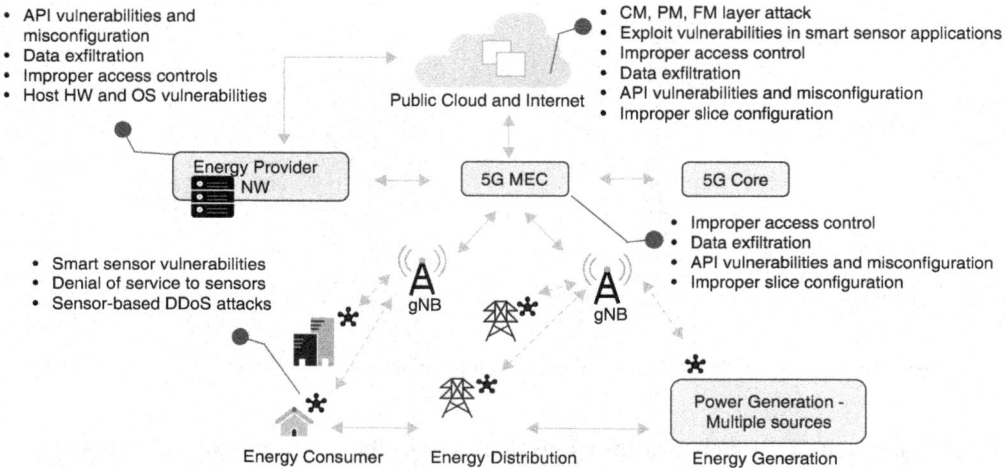

- API vulnerabilities and misconfiguration
- Data exfiltration
- Improper access controls
- Host HW and OS vulnerabilities

Public Cloud and Internet

- CM, PM, FM layer attack
- Exploit vulnerabilities in smart sensor applications
- Improper access control
- Data exfiltration
- API vulnerabilities and misconfiguration
- Improper slice configuration

Energy Provider NW   5G MEC   5G Core

- Smart sensor vulnerabilities
- Denial of service to sensors
- Sensor-based DDoS attacks

gNB   gNB

- Improper access control
- Data exfiltration
- API vulnerabilities and misconfiguration
- Improper slice configuration

Power Generation - Multiple sources

Energy Consumer   Energy Distribution   Energy Generation

FIGURE 9-17   Key Threat Surfaces for 5G Energy Utility Deployments

Inadequate Application-Level Security

All the 5G NFs are based on the cloud-native principles where the open source software stack will be used to build both 3GPP and non-3GPP Network Functions for 5G, which can then be deployed on the premises and in any public cloud offering.

Specifically for the energy utility use case, one of the key threats at the application level is the unauthorized access of smart sensor configuration management deployed in the public cloud. If any attacker has unauthorized access to the configuration management, it can lead to complete control of the smart sensors. The smart sensors can then be force-updated with a malicious code. The malicious code then can initiate distributed denial-of-service (DDoS) attacks toward the 5G infrastructure.

API Vulnerabilities

5G brings in service-based architecture (SBA), which uses API-based communication between the 5G NFs. If the NSaaS model is used to cater for the energy utility use case, then the energy utility company can manage the network slice instance via the management interface exposed by the service provider using an API. APIs might also be used by the smart sensors to communicate with the MEC application to detect and respond to any disruptions in service in the energy grid. Although APIs enable many of the interactions in the energy utility use case, they also bring in risk vectors due to improper implementation of APIs and security misconfigurations.

In NSaaS deployments, broken authentication of APIs can lead to unauthorized access to the slice management layer. This can lead to users from another slice accessing the management of the energy utility slice and deleting an entire slice subset. This can lead to loss of service or service outage.

Energy Provider Network Vulnerabilities

Improper isolation between the IT network of the energy utility company and the IIoT control network (through a smart sensor network, for example) can lead to attacks from the Internet or any malware from the IT network impacting the smart sensor network.

Host OS and hardware (HW) vulnerability is another threat vector. If the 5G energy utility use case deployment model consists of certain applications related to the smart sensors being deployed in the on-premises data center, it could be prone to side-channel attacks, leading to data exfiltration of sensitive data from the smart sensors.

Securing 5G-Enabled Energy Utility

The energy sector is identified as one of the most critical national infrastructures by many countries. This sector must maintain stringent reliability and availability of the power grid. In countries like the US, standards like North American Electric Reliability Corporation – Critical Infrastructure Protection (NERC-CIP) are designed to secure the assets required for operating the electric system.

While 5G does help in optimizing the total cost of ownership (TCO) by implementing innovative and effective use cases, it also brings in new risks. This section discusses the multilayered security controls to secure critical energy sector 5G use cases.

Securing 5G critical infrastructure use cases such as energy utility requires a holistic approach rather than just focusing on securing the 5G components. This is because the energy sector has been in operation for quite some time, and the 5G IoT/IIoT network will require integration with the legacy network. This integration is important because 5G use cases should complement the existing legacy framework to help you achieve cost optimization. Ripping out the legacy infrastructure is not always the right decision. It looks attractive in PowerPoint slides, but implementing a full overall "rip and replace" inherently has a lot of unforeseen challenges, leading to delay in the project or adding more complexity and cost. Let's discuss some of the key areas to secure and look at re-utilizing and enhancing the existing technology and infrastructure.

Looking at the energy sector, network transformations, transmission, and especially distribution involve investing in digital technology that is transforming the grid. IT and OT convergence is also transforming the utility market, requiring a blend of technology and expertise from both organizations to optimize the benefits of digitizing the utility. You should segment the operations environment, detect suspicious traffic flows and behavior, perform security policy audits and detect violations, and identify compromised devices. Across this security posture, automation is fundamental to facilitate the scale required in an OT environment. Further, utilities can reduce risks across all OT and IT traffic by creating visibility into how, when, where, and why users and devices connect to the network. Apart from multilayered security controls, you should work closely with your suppliers and ecosystem of industrial partners to deliver secure devices and deploy networks with best practices in mind.

Figure 9-18 shows the key security controls for mitigating the threats for the smart sensors within the energy utility use case.

The list that follows explains the key security controls illustrated in Figure 9-18 in more detail:

1. The Secure Connected Grid Router (S-CGR) enables secure connection to all the smart sensors deployed in various parts of the energy utility network. Depending on the capabilities of the sensor, the S-CGR would use TLS, IPsec, or any other mechanism to secure the communications with the smart sensor. The S-CGR should support Supervisory Control And Data Acquisition (SCADA) protocol (serial-to-IP) translation features to allow easy integration of legacy (non-IP) devices onto an IP network. It should also support integration with Advanced Smart Metering, Distribution Automation, Remote Workforce Automation, as well as integration of Distributed Energy Resources (DER) and secure wireless console access for remote access and resolution of critical faults.

   The S-CGR should also support integrated Global Positioning System (GPS), which will help with location mapping of the router. S-CGR should provide network segmentation and quality of service (QoS) features, allowing logical separation of application traffic with specific constraint policies applied on each traffic flow.

2. To allow authentication of the sensors and users, S-CGR should have integration with the user and device identity and management solutions, PKI infrastructure as well as allow Secure Zero-Touch Provisioning (S-ZTP). S-CGR should support mutual authentication and authorization of all nodes connected to the network, IEEE 802.1x-based authentication, role-based access control (RBAC) and certificate-based identity, and strong usernames and passwords. S-CGR should allow tamper-proof secure storage of router configuration and should use a hardware chip to store the router's X.509 certificate and other security credentials.

   The smart sensors being used in critical infrastructure deployments should allow S-ZTP Day 0 configuration options, include a tamper-resistant mechanical design, and generate security alerts if compromised.

3. The transmission distribution system includes fault location isolation and service restoration (FLISR), SCADA, and analytical layers to provide real-time analysis of any disturbances, enabling quicker and accurate identification of faults.

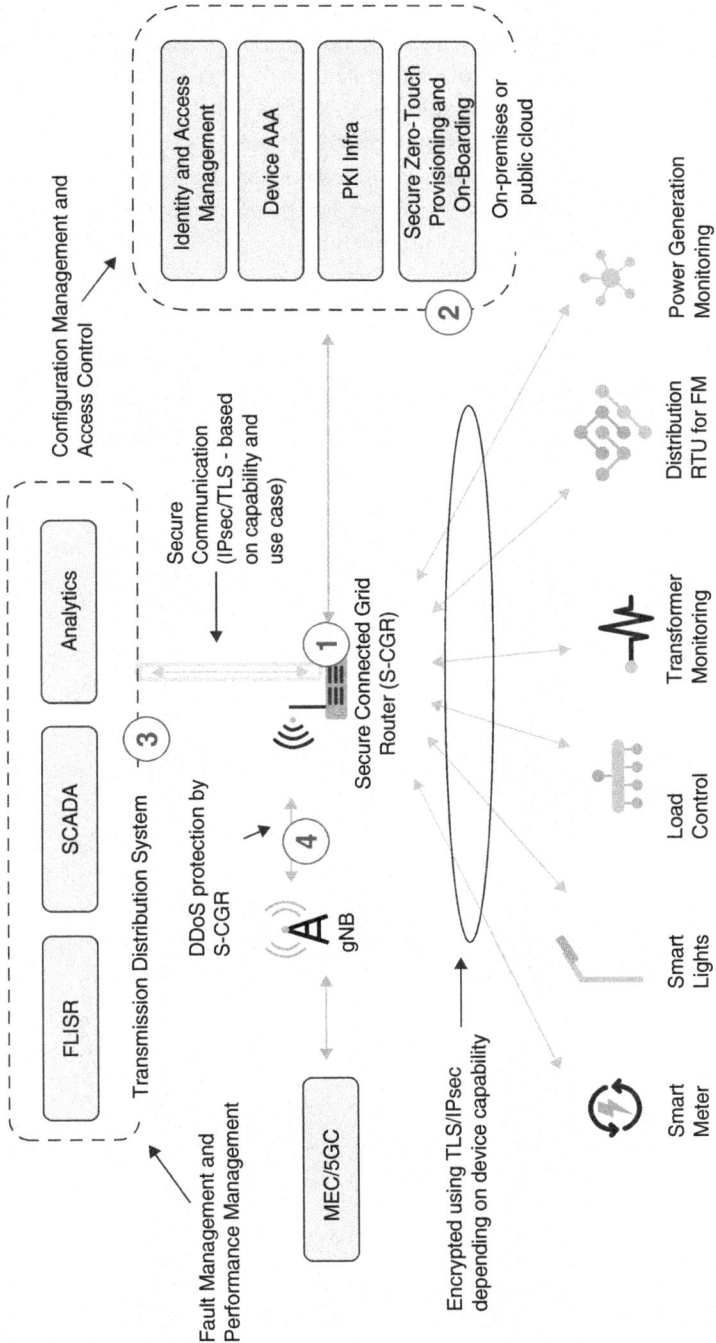

FIGURE 9-18   Key Security Controls for Energy Utility Being Catered by 5G

4. The analytical layer should include the closed loop anomaly detection, which can identify any anomalies in the network using telemetry from various components of the architecture, such as network telemetry from switching and routing devices, hypervisor metadata, container telemetry, and so on. The anomaly detection solution should be able to recommend appropriate action to mitigate any identified threat. For example, the analytical layer could identify the anomalies in the smart meter behavior, such as the smart meter device trying to reach out to an unexpected url. Such behavior could indicate that the device has been compromised. An appropriate action that the closed loop anomaly detection system could recommend is to isolate the device by allowing access to only a quarantine network. Once the device is inspected and seen to comply with the security policy, it can be introduced back to the non-quarantine part of the network.

5. The communication between the network devices and the transmission distribution system should be secured using secure mechanisms such as TLS or IPsec.

6. With regard to secure connectivity toward 5G, to mitigate the DDoS attacks against the 5G network, S-CGR will include rate-limiting mechanisms.

Figure 9-19 shows the summary of security controls for securing the critical energy sector 5G deployments.

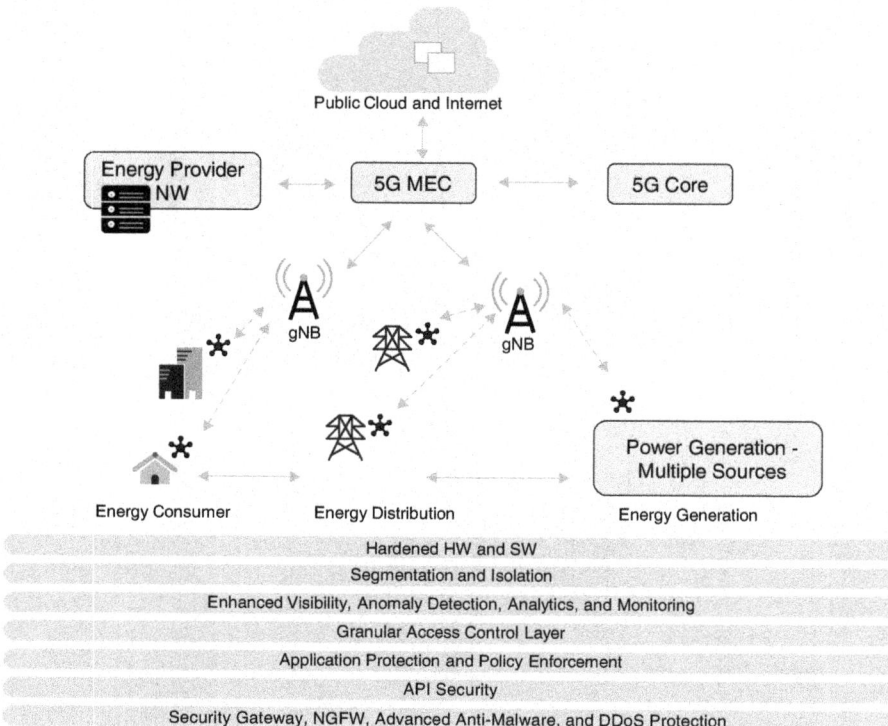

FIGURE 9-19 Summary of Security Controls for Securing Energy Utility Sector

5G Vehicle-to-Everything (5G-V2X)

Innovations in connected vehicles with the vision of autonomous transport systems require an infrastructure that can provide high throughput for in-car infotainment and ultra-reliable and low-latency services for assisted driving. In addition to existing IEEE 802.11p and dedicated short-range communications (DSRC) technology, 5G promises to deliver reliable, real-time communication at high speed on the distributed architecture needed to support transmission between vehicles, network, and transport infrastructure.

Vehicle-to-everything (V2X) communication enables data exchanges between vehicles, infrastructure, pedestrians, and applications running on the edge and cloud. The different types of V2X are as follows:

- **V2V:** Vehicle-to-vehicle
- **V2I:** Vehicle-to-infrastructure
- **V2P:** Vehicle-to-pedestrian
- **V2N:** Vehicle-to-network
- **V2S:** Vehicle-to-sensor

V2V, V2I, V2S, and V2P are short-range, direct device-to-device communications and can exist without 5G coverage. To enable these type of communications, 3GPP has specified the PC5 reference point/interface.

V2N is device-to-network communication utilizing 5G networks. To enable V2N communications, 3GPP has included attributes related to V2X in the specifications for UE, air interface, and 5GC NFs.

Figure 9-20 illustrates the non-roaming 5G system architecture for V2X communication over PC5 and Uu reference points as specified in 3GPP TS 23.287 V16.5.0.

In order to simplify interworking, interoperation, and interchange among different stakeholders, 3GPP has specified V2X Application Enabler (VAE). VAE will allow common functionalities to be utilized by different V2X applications, including the network situation and QoS monitoring, communication parameter (for example, PC5), and network resource management.

Apart from VAE, 3GPP has also specified the Service Enabler Architecture Layer (SEAL) in order to support vertical applications over the 3GPP system.

The interfaces in Figure 9-20 are as follows:

- **PC5:** Interface specified for V2X between the UEs. It includes the LTE-based PC5 and/or NR-based PC5. The PC5 reference point will use the Hypertext Transfer Protocol (HTTP), Message Queue Telemetry Transport (MQTT), or Advanced Message Queuing Protocol (AMQP).
- **Uu:** Interface between UE and NG-RAN (gNB).
- **V1:** Interface between the V2X applications in the UE and in the V2X application server.
- **V5:** Interface between the V2X applications in the UEs.

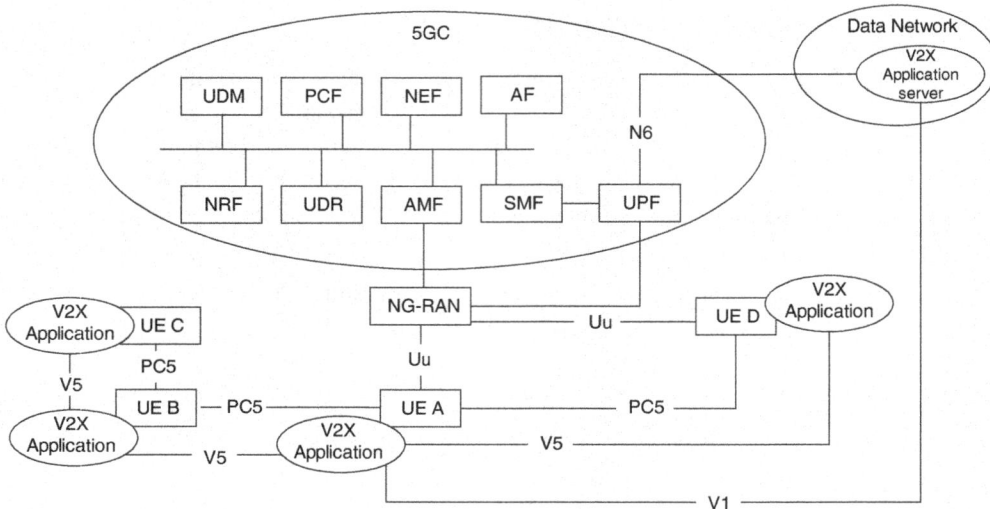

3GPP TS 23.287

FIGURE 9-20   Non-Roaming 5G System Architecture for V2X Communication

Apart from the preceding interfaces, the following interfaces and NFs will carry additional messages to enable V2X services:

- **N1** interface between UE and AMF will be used to convey the V2X policy and parameters (including service authorization) from AMF to UE and to convey the UE's V2X capability and PC5 capability for V2X information from UE to AMF.

- **N2** interface between NG-RAN and AMF will be used to convey the V2X policy and parameters (including service authorization) from AMF to NG-RAN.

- **NEF** services are used to enable communication between the 5GC 3GPP NFs and non-3GPP V2X application server NFs, as shown in Figure 9-21.

Figure 9-21 shows the high-level view of AF-based service parameter provisioning for V2X communications. The V2X application server may provide V2X service parameters to the PLMN via NEF. The NEF stores the V2X service parameters in the UDR. The 5GC reference points N37, N36, N30, and N15 shown in Figure 9-21 are all API-based service-based interfaces (SBIs). The V1 reference point will use HTTP or MQTT.

5G allows convergence of 3GPP and non-3GPP networks to 5G packet core (5GC), thereby allowing the coexistence of multiple technologies enabling V2X use cases. 5G enablers such as MEC, network function virtualization (NFV), network slicing, API-enabled SBA and open software, and stack-based Cloud-Native Functions (CNFs) will help you realize V2X use cases.

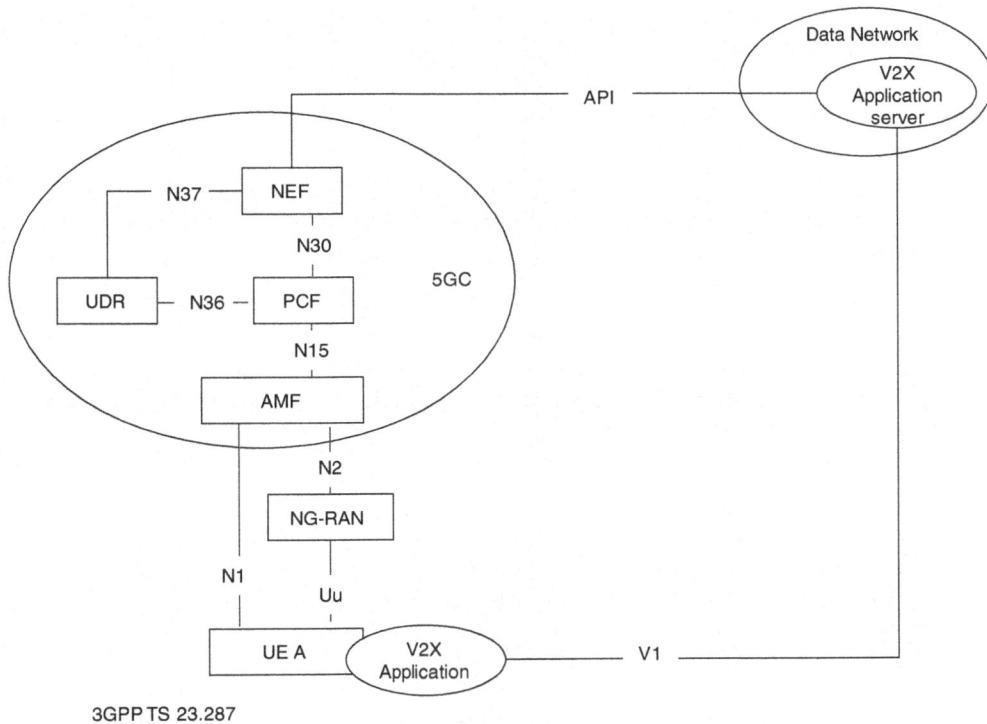

3GPP TS 23.287

FIGURE 9-21   High-Level View of AF-based Service Parameter Provisioning for V2X Communications

V2X will include use cases that impose varying latency requirements. For example, use cases such as real-time awareness would require URLLC with stringent high-availability requirements. V2X will also include use cases such as video streaming, as an in-car entertainment service, which requires high-bandwidth but lower criticality. 5G allows networks to move away from monolith mode of network deployment to a flexible virtual network deployment with resources being allocated based on the use case requirements using network slicing, as shown in Figure 9-22. 5G networks will use MEC to deploy the low-latency dependent non-3GPP applications and 3GPP NFs closer to the edge in the public cloud or on-premises regional/micro/mini data center. NFV, augmented by software-defined network (SDN) and orchestration, allows optimum distribution of NFs using Control Plane User Plane Separation (CUPS), optimum sharing of resources, and allows scalability based on actual capacity and performance requirements, thereby making the network more flexible and dynamic for different types of services.

Figure 9-22 depicts the slices configured:

- Slice 1 is the URLLC slice, which is latency sensitive and has low-bandwidth requirements.

- Slice 2 caters for in-car entertainment and has a requirement on high bandwidth and has less-stringent requirement on latency.

- Slice 3 caters for configuration, fault, and performance management, requires low bandwidth, and is not very sensitive to latency.

Applications that are latency sensitive, such as V2X MEC applications, are deployed in the edge of the network. They interact with the V2X applications in vehicles and direct them to the specific V2X application servers deployed in the public cloud in this example. Applications that are not latency dependent, such as applications for operations and maintenance, and 5G control plane NFs are deployed in the centralized 5G Core data center.

5G-V2X can be used in multiple ways to enhance road safety. Some of the key 5G-V2X use cases are as follows:

- **Real-time awareness:** Real-time awareness will enable vehicles to receive real-time information about speed recommendations based on traffic conditions and weather information, lane closures, traffic density, accident alerts, and other conditions that might necessitate adjustments to driving patterns. This 5G-V2X use case requires extremely low-latency communication, which is allowed by MEC deployment bringing the low-latency-dependent applications closer to the edge. Network slicing can be used here to reserve the required resources to allow real-time awareness, as this is a critical use case that requires high availability.

- **Intent-based driving:** Vehicles driving in proximity can share information with each other for activities such as lane changes, sudden stops, and so on to avoid collisions, improve traffic safety, and increase efficiency. Similarly, traffic information can be collected from roadside units, temporary roadwork blocking units, and pedestrians for vehicles to obtain a holistic view of the driving environment, making it increasingly autonomous.

- **Edge and cloud integration:** V2X applications on the UE, such as vehicles, will need to use network resources available at the edge for CDN caching content for improved quality of experience (QoE). V2X applications on the UE also require backend integration with the V2X application servers and cloud-hosted, backend systems to provide vehicle-to-network application services over the licensed cellular spectrum. Such applications require centralized control to operate the service subscription management necessary to enable vehicles to securely create a trust group with other vehicles in order to form a platooning convoy and have trusted data exchange. In addition, V2X applications, such as road safety, traffic management, and infotainment services running on the core, cloud, and third-party infrastructure, need secure management to ensure critical real-time information legitimacy. Trusted vehicles can subscribe to OEM services and third-party services for functionality updates, such as a new map for navigation, with a dynamic API interface allowed by the Network Exposure Functionality (NEF) on the core.

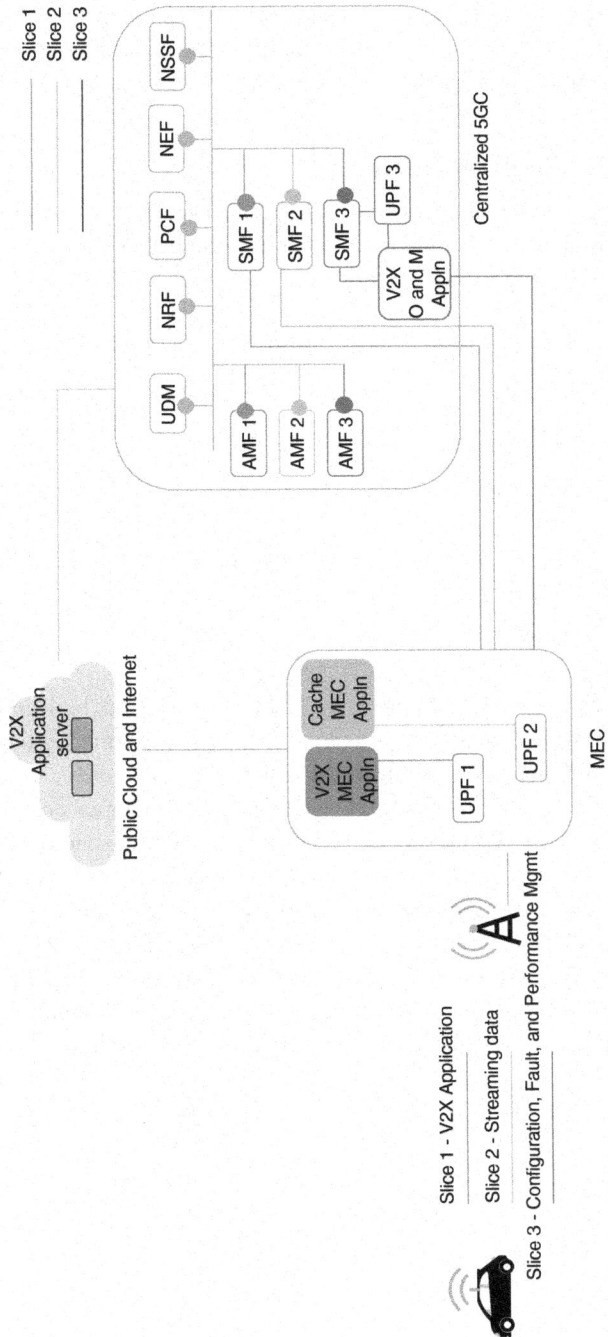

FIGURE 9-22　Example of 5G-V2X Deployment Utilizing Network Slicing

Figure 9-23 illustrates an example of V2X deployment, with the NSaaS model being utilized by the communication slice consumers (CSCs). The slices are being consumed by multiple entities, such as the car manufacturer, road policing unit, traffic management center, and media and entertainment for vehicles.

Threats in 5G-V2X Deployments

5G-V2X enables both safety and non-safety vehicular-related use cases. Many of the use cases, specifically safety related, require low latency and high reliability, and because they are classified as URLLC use cases, they will be susceptible to the following risks:

- Service availability and DoS to critical services such as road policing

- DDoS attacks from V2X devices

- Data exfiltration of sensitive user and device information

- Improper identification and access control of users and devices

This section will take you through some of the key threat surfaces in 5G-V2X deployment, illustrated in Figure 9-24.

The key threat surfaces for the 5G-V2X deployment shown in Figure 9-24 are as follows:

1. UE communication with 5G RAN is prone to threats such as weak built-in security of the on-board units (OBUs), which can be retrofitted to vehicles for aftermarket deployment or factory-fitted in cars with V2X capabilities. One of the key risk vectors here is the malicious code that could be used as a forced OBU firmware update. Once the malicious code is installed, it will take control of the OBU. The attacker can now force the OBU to send out sensitive information to the command and control (C2C) server, which can then be used for detailed targeted attacks. If the malicious firmware code is mass-updated on the vehicles using the vulnerable OBU, the attacker can initiate a DDoS attack toward the 5G network. V2X applications can be attacked using the same concept, as shown in Figure 9-25. The V2X applications installed on the UE are also prone to vulnerability in the programs made to build the V2X application. The vulnerabilities can be exploited by the attacker to exfiltrate sensitive information.

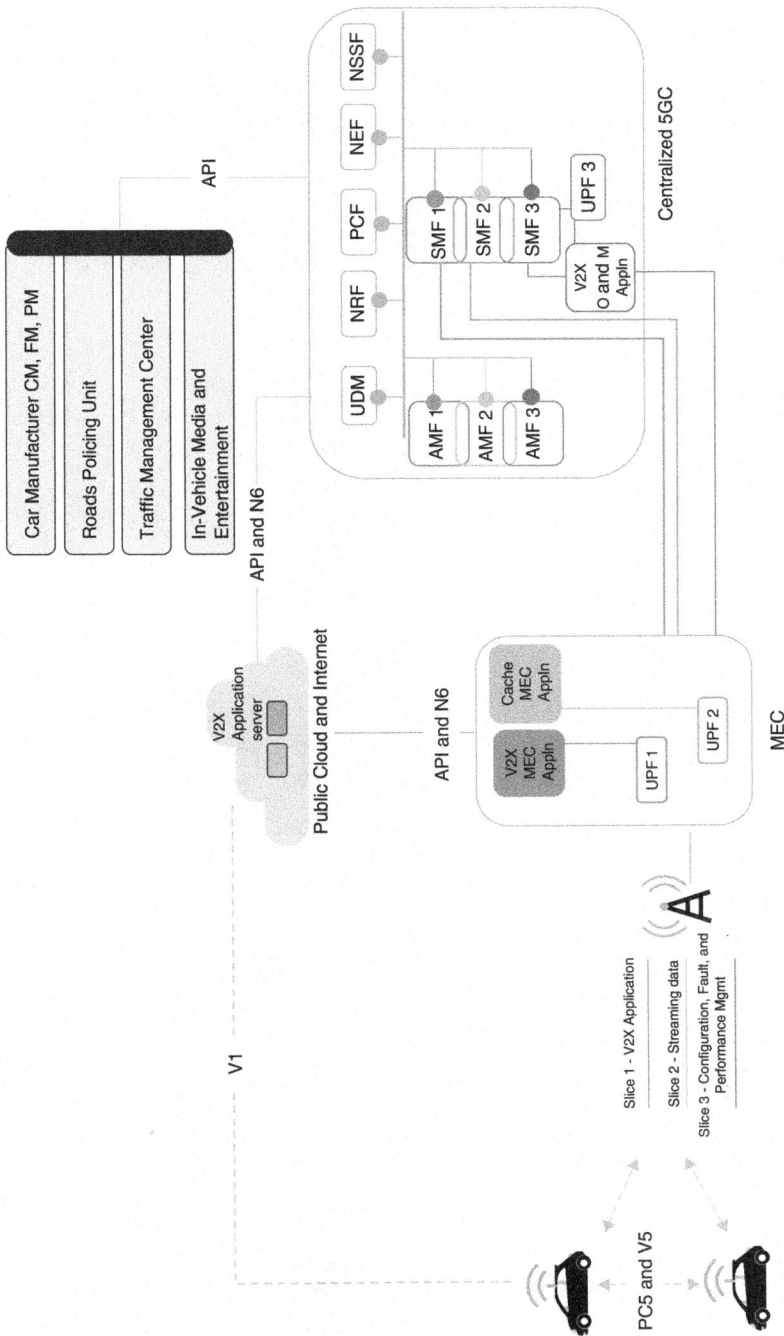

FIGURE 9-23   5G-V2X NSaaS Deployment

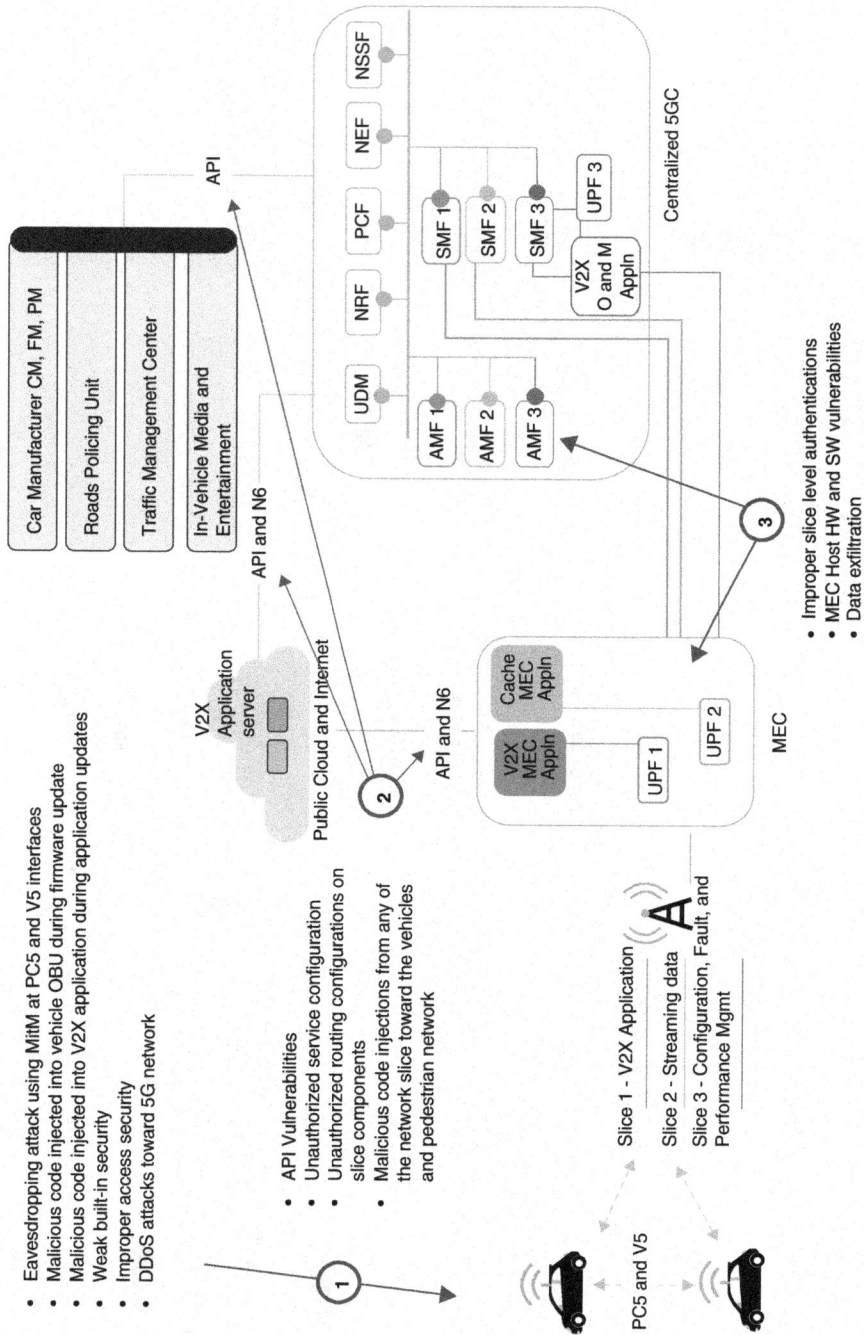

FIGURE 9-24   Illustration of Key Threats in 5G-V2X Deployment

Another risk vector at this layer is the possibility of the rogue UE providing incorrect information to deceive other UE, which might trigger unnecessary actions.

As shown in Figure 9-25, the attacker can exploit the vulnerabilities in the V2X application to insert malicious code as an update and push it to all the endpoints for V2X application update. Once the endpoints are updated with the malicious code, they can be controlled by the C2C server operated by the attacker. The attacker can then launch a DDoS attack toward the 5G infrastructure operated by the 5G-V2X provider/service provider.

2. Another area of risk is insecure API communication. For the V2X use cases to be realized, you will need to deploy the V2X application client and V2X application servers. The V2X application servers will be deployed in multiple edge compute sites, which can be located on the premises or in the public cloud, as illustrated in Figure 9-25. APIs are also used in the NSaaS deployment mode, where different slice consumers or CSCs such as car manufacturer, road policing unit, traffic management center, and in-vehicle media and entertainment will require management access to the service layer for configuring QoS and other service-level parameters. Admins from the CSC will use APIs to communicate with the service-level management layer. The 5GC NFs are exposed to third-party application functions, such as the V2X application layer, using the API calls. So, as you see here, any vulnerability in the API could have a negative impact on the V2X deployment. V2X deployments will reuse the 5G API-based communications in the MEC layer. This includes the API calls being used by the NEF to integrate with the external network functions like the V2X application server. V2X will also require integration with many entities providing services such as media and entertainment services and entities providing emergency services such as road policing. Car manufacturers will also require connectivity to the vehicles to provide firmware updates. All the aforementioned management and services provisioning can be performed by integrating the entities with a services provisioning layer in network slicing deployments. The access to the management for the services will be provided using APIs. These APIs are prone to attacks due to vulnerabilities within the APIs and security misconfigurations during API implementation. Improper implementation of APIs will also lead to unauthorized access to slices, from which the attacker might be able to exfiltrate the data and, in worse-case scenarios, delete the all the network slice–related functions, leading to DoS to all the impacted slice users and devices. APIs also do not impose any restrictions on the size or number of resources that can be requested by the client/user. This can lead to DoS and to risks such as brute-force attacks.

3. Another area of risk in 5GC deployments is data exfiltration. This is due to the threats arising from improper authentication in network slice deployments, vulnerabilities in the host hardware and software, where 5G 3GPP and non-3GPP NFs are deployed, and vulnerabilities in the software components used to build the NFs. Chapter 7, "Securing Network Slice, SDN, and Orchestration in 5G," covers threats due to networking slicing in more detail. To summarize the threats under this domain, the key risks arise from improper isolation of the NFs serving critical slices. As an example, if the NFs servicing critical slices (such as the road policing unit) are deployed in the same host hardware as the noncritical slices (such as in-vehicle media and entertainment), attackers accessing a noncritical slice could use the vulnerabilities in that slice to access the NFs in a critical slice and then exfiltrate sensitive data.

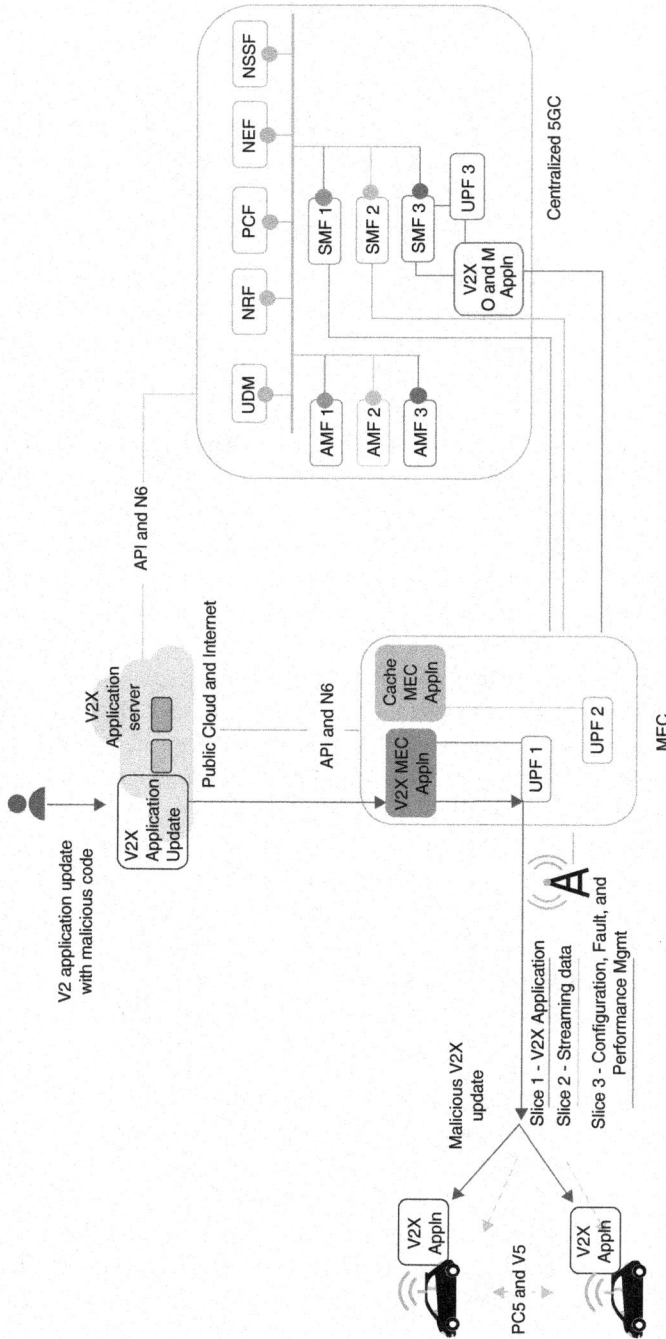

FIGURE 9-25   Example of V2X Application Being Updated by a Patch Embedded with Malicious Code

Securing 5G-V2X Deployments

5G-V2X use cases are both safety related and non-safety related. While safety-related use cases are primarily URLLC, non-safety-related use cases are eMBB oriented and provide value-added services to consumers such as in-vehicle media and streaming. While 3GPP caters for securing V2X over an NR-based PC5 reference point, as specified in TS 33.536, and V2X over a Uu reference point based on existing secure Uu mechanisms, as specified in TS 33.501, there are general security and privacy principles applicable outside of the 3GPP scope that need to be secured by multilayered security controls. This section covers these multilayered security controls, which are summarized as follows:

- **Enhanced visibility:** This layer should allow visibility for the control plane across the extended network, from the control plane on the V2X user equipment to the 5G packet core deployed on the premises and in the cloud. The enhanced visibility layer uses enterprise telemetry from the existing network infrastructure. It should provide advanced threat detection, accelerated threat response, and simplified network segmentation using multilayer machine learning and entity modeling. This layer should be able to provide you the capability to detect obsolete and insecure protocols. An example is the TLS protocols. Since their first iteration in 1995, SSL and TLS encryption protocols have had many versions. The most current version is TLS 1.3, which addresses vulnerabilities and supports stronger cipher suites; therefore, the older versions, such as TLS 1.1 and 1.2, should not be used. Network administrators need to be alerted when older versions are used on their networks. The solution chosen by you should create a baseline of the network and then allow advanced behavioral analytics, enabling you to know if any of the V2X-related slice NFs or V2X devices are behaving differently than usual. Figure 9-26 illustrates security controls for end-to-end monitoring, enhanced visibility, and patch management.

As Figure 9-26 illustrates, the end-to-end visibility layer should be able to provide you with end-to-end visibility, threat detection, and mitigation using the following listed functions.

1. Collect telemetry from different parts of your network, including cloud deployed NFs.

2. This should then be compared against the baseline of your network and the network security policy and be analyzed using ML and AI algorithms to identify any malicious traffic in the encrypted layer, specifically on external API calls from the 5G-V2X network to the cloud-deployed applications (specifically, for detecting any anomaly in the control plane).

3. The data collected should also be analyzed to identify any vulnerabilities.

4. Any vulnerabilities in the 5G 3GPP and non-3GPP V2X NFs should be patched. If the CNFs are container based, then the latest signed image should be downloaded from a trusted container registry to replace the existing container image with known vulnerabilities.

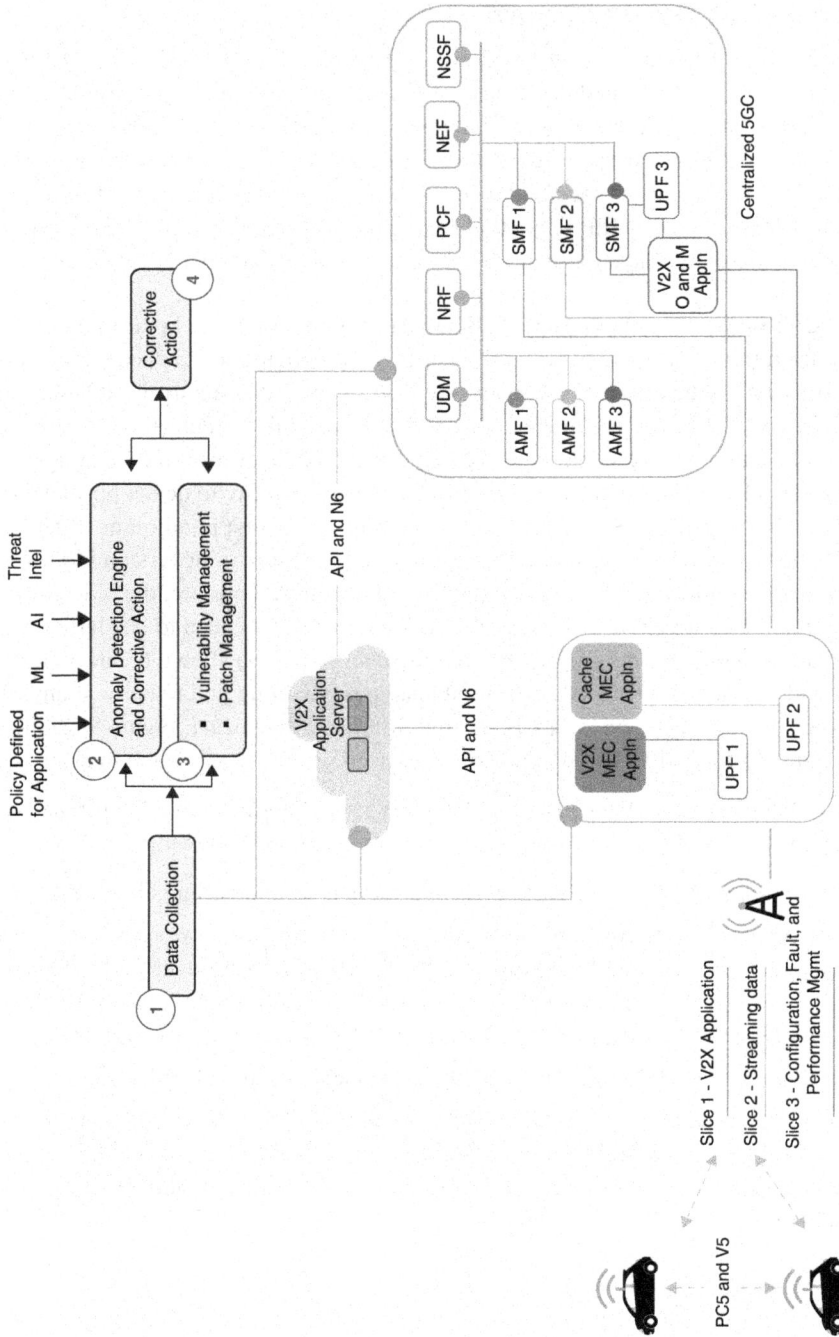

FIGURE 9-26   Detection of Anomalous Behavior in V2X Applications in Control Plane

■ **Application protection and policy enforcement:** A traditional perimeter-based security approach alone is ineffective in meeting the new 5G deployment modes to cater for V2X use cases.

- To address these security challenges effectively, a new approach is required to define, discover, and implement a consistent security policy at different layers. When dealing with modern distributed and dynamic applications for V2X being deployed at public cloud and on-premises MEC, you need insight into applications and their dependencies. You also require the capability to apply business context and automation to core security policy management processes, such as change management, risk and compliance assessment, and auditing.

- You will need a combination of application policy discovery and workload-based enforcement with a mixture of infrastructure-based segmentation and be able to apply security controls for applications running in any infrastructure and any cloud.

- You must also be able to dynamically enforce granular microsegmentation policies on the workloads while providing a consistent, real-time policy for enforcement across infrastructure-based elements such as firewalls and the network. This will allow you to realize seamless security policy across a diverse multicloud environment, with the demands of any application, from static, legacy operating systems to dynamic container-based microservices.

- To generate an accurate microsegmentation policy, your chosen solution should perform application dependency mapping on the entire network to discover the relationships between different application tiers and associated services. In addition, the platform should be able to simulate the policy application pre-enforcement to ensure ongoing application availability. The normalized microsegmentation policy should then be enforced through the application workload itself for a consistent approach to workload microsegmentation across any environment, including virtualized, bare-metal, and container workloads running in any public cloud or any on-premises data center. Once the microsegmentation policy is enforced, the solution should continue to monitor for compliance deviations, ensuring the segmentation policy is up to date as the application behavior changes.

- You should also have comprehensive scanning capabilities, including build, ship, and run CI/CD pipeline scanning and vulnerability scanning during runtime. The solution should also review overall and individual risk scores for vulnerability exploits during runtime and be able to continually scan your container deployments.

■ **API security:** API security requires focus on multiple areas. Some of the key API vulnerabilities include the implementation of the API itself, where verbose error messages contain sensitive information such as the type and version of the web server being used. This requires proper validation of the error messages. As shown in Figure 9-27, the APIs should be implemented with a robust API authentication mechanism, including standard authentication, token generation, and multifactor authentication. Best practices, such as denying all access by default, should be implemented as part of the access control mechanism. DDoS attacks from the V2X UE toward the V2X application servers can be mitigated using mechanisms such as rate limiting, payload size limits at the API layer, and ensuring that you configure limits on container resources at the configuration layer. Figure 9-27 illustrates security controls to secure API and user access.

Figure 9-27 shows you the different locations within the 5G network to deploy the API gateway/API firewall to secure the API communication.

- **Enhanced access control:** V2X deployment will use multiple vendors and contractors configuring the 3GPP and non-3GPP NFs. To prevent any malicious actors trying to access the system, apart from using role-based access control (RBAC) for the V2X NFs, a multilayered access control method using multifactor authentication and mapping of workloads to users is required to provide you with the right level of access control. The access control layer should also allow you to gain device visibility and establish trust with endpoint health and management status. Having an enhanced access control layer with adaptive and role-based access controls using the global policy, as shown in Figure 9-27, enables you to enforce access policies for every app.

- **Network security and anti-DDoS protection:** For ensuring secured connections within the network, one of the key steps is to implement TLS encryption on all API communication between the V2X public cloud and on-premises MEC. Any traffic flowing through the untrusted public network should be secured using IPsec provided by a security gateway. Anti-DDoS mechanisms can be provided by multiple components. Anti-DDoS mechanisms could also be provided by built-in features in API GW/API firewalls using rate limiting and so on, which will cater for mitigations against brute-force attacks and botnets. Figure 9-28 shows the end-to-end multilayered security control to secure your 5G-V2X deployments.

The aforementioned security controls, along with ongoing standardization efforts, are key for establishing a reliable and secure 5G-V2X solution. Solutions such as V2X would need integrations with many external non-3GPP functions, thereby requiring a robust multilayered security architecture, as illustrated in Figure 9-28. You might have realized by now that there are multiple standardization works, some still ongoing, such as 3GPP to ensure that 5G can really be used to realize V2X use cases. Enhancements to 5G-V2X capabilities such as the PC5 reference point show promise in the direction of real-world implementation on V2X. However, it seems there is still quite a long way to go to before mass deployments of V2X will be realized, as the business models or monetization models should also be dusted out before actual deployments take place.

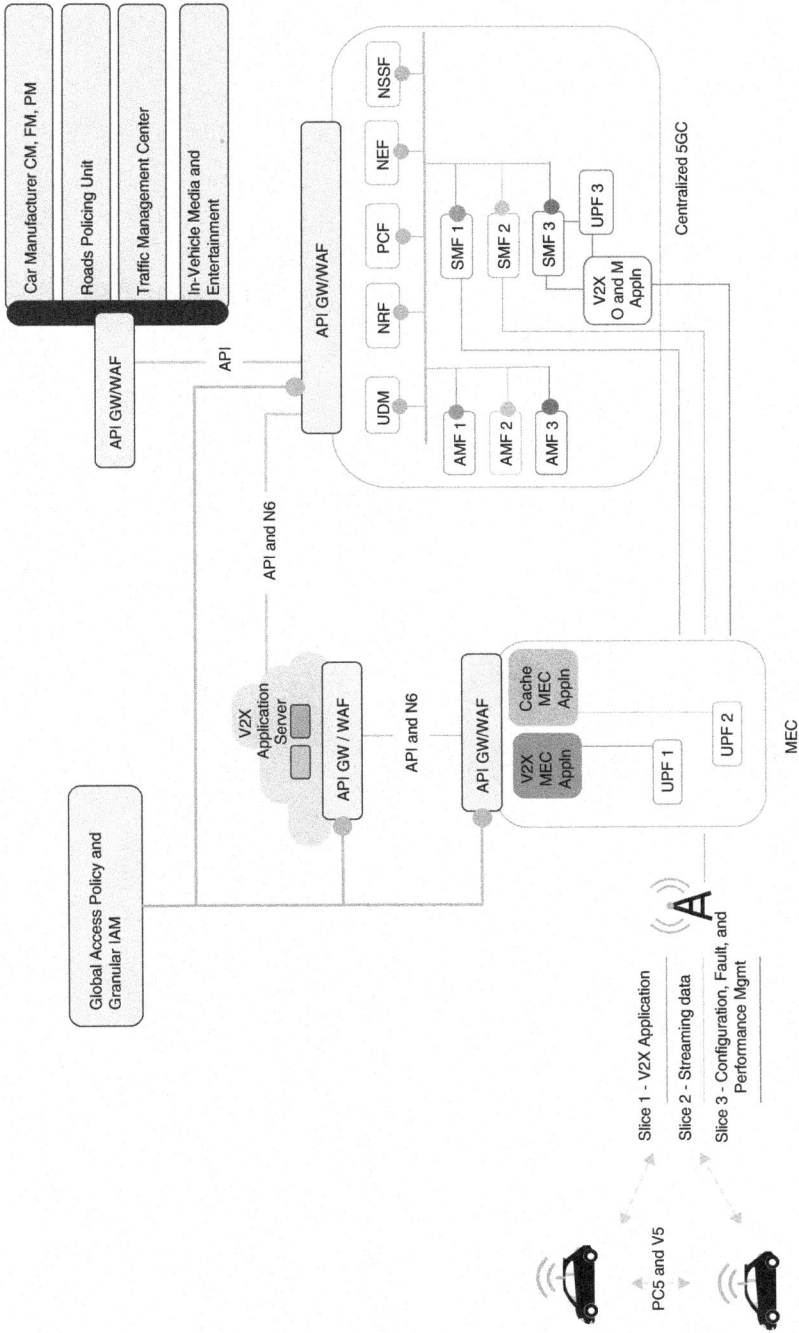

FIGURE 9-27   Securing API and Providing Granular Access Control

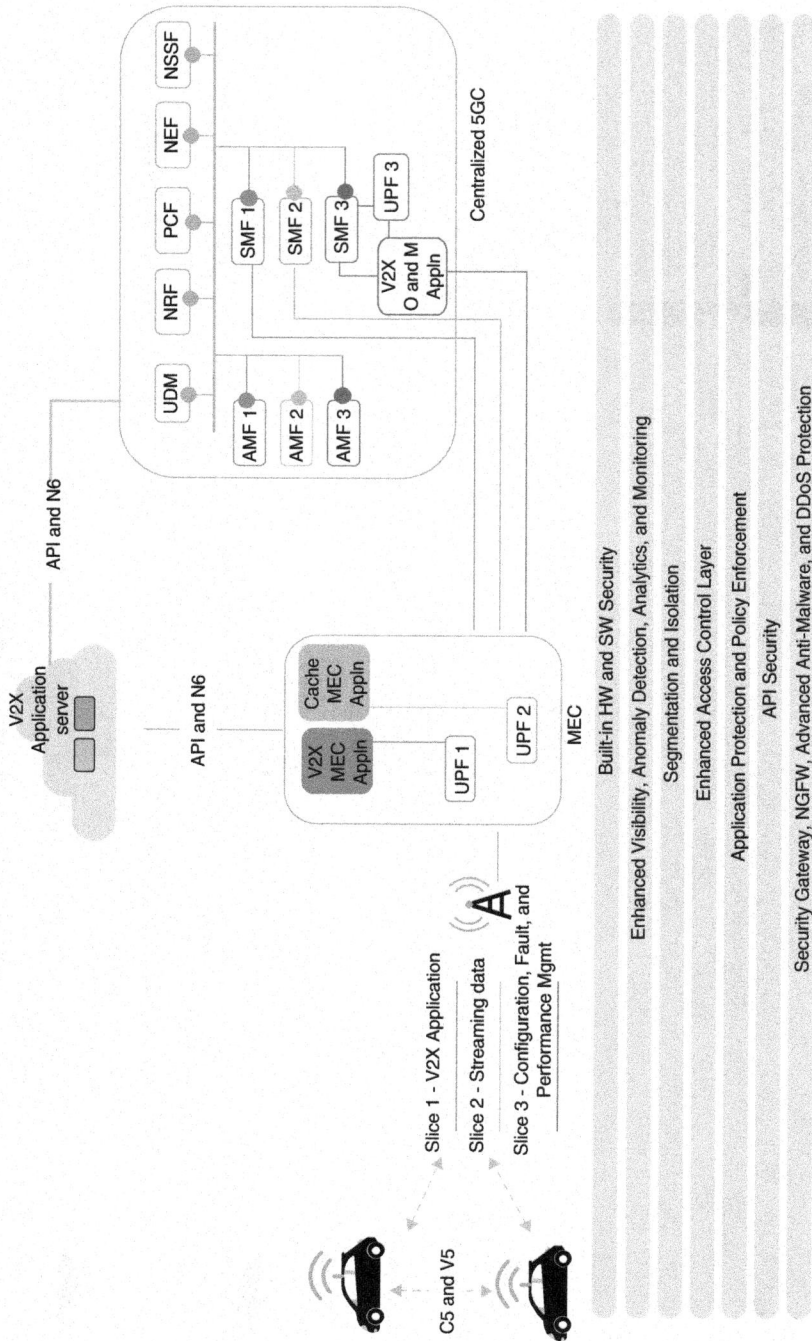

FIGURE 9-28   End-to-End Multilayered Security Control to Secure Your 5G-V2X Deployments

Standards and Associations

Adherence to the following standards will help you establish, operate, monitor, review, maintain, and continually improve the security posture of your 5G use cases. Note that the following list is not exhaustive and contains only a selected number of standards and associations that are of key relevance for secure 5G NPN deployments:

- **ISO/IEC-JTC1-SC27:** The scope of ISO/IEC-JTC1-SC27 is the development of standards for the protection of information and ICT. This includes generic methods, techniques, and guidelines to address both security and privacy aspects, such as the following:

  - A security requirements capture methodology

  - Management of information and ICT security; in particular, information security management systems, security processes, and security controls and services

  - Cryptographic and other security mechanisms, including but not limited to mechanisms for protecting the accountability, availability, integrity, and confidentiality of information

  - Security management support documentation, including terminology, guidelines, and procedures for the registration of security components

  - Security aspects of identity management, biometrics, and privacy

  - Conformance assessment, accreditation, and auditing requirements in the area of information security management systems

  - Security evaluation criteria and methodology

- **ISA/IEC 62443:** The International Society of Automation (ISA)/International Electrotechnical Commission (IEC) 62443 series of standards, developed by the ISA99 committee and adopted by the IEC, provides a flexible framework to address and mitigate current and future security vulnerabilities in industrial automation and control systems (IACSs). The committee draws on the input and knowledge of IACS security experts from across the globe to develop consensus standards that are applicable to all industry sectors and critical infrastructure. A new standard in the series, ISA-62443-4-2, "Security for Industrial Automation and Control Systems: Technical Security Requirements for IACS Components," provides the cybersecurity technical requirements for components that make up an IACS—specifically the embedded devices, network components, host components, and software applications. The standard, which is based on the IACS system security requirements of ISA/IEC 624433-3, "System Security Requirements and Security Levels," specifies security capabilities that enable a component to mitigate threats for a given security level without the assistance of compensating countermeasures.

- **ISO27001:** ISO 27001 sets out the specification for information security management system (ISMS). Using them enables organizations of any kind to manage the security of assets, such as financial information, intellectual property, employee details, or information entrusted by third parties.

- **FedRAMP:** FedRAMP stands for the Federal Risk and Authorization Management Program. It standardizes security assessment and authorization for cloud products and services used by U.S. federal agencies. The goal is to make sure federal data is consistently protected at a high level in the cloud.

- **Secure Controls Framework (SCF):** This comprehensive catalog of controls is designed to enable companies to design, build, and maintain secure processes, systems, and applications. The SCF addresses both cybersecurity and privacy so that these principles are designed to be "baked in" at the strategic, operational, and tactical levels

- **NIST 800-53:** NIST 800-53 is a regulatory standard that provides a catalog of security and privacy controls for information systems and organizations to protect organizational operations and assets, individuals, other organizations, and the nation from a diverse set of threats and risks, including hostile attacks, human errors, natural disasters, structural failures, foreign intelligence entities, and privacy risks. The controls are flexible and customizable and are implemented as part of an organization-wide process to manage risk. The controls address diverse requirements derived from mission and business needs, laws, executive orders, directives, regulations, policies, standards, and guidelines. Finally, the consolidated control catalog addresses security and privacy from a functionality perspective (that is, the strength of functions and mechanisms provided by the controls) and from an assurance perspective (that is, the measure of confidence in the security or privacy capability provided by the controls). Addressing functionality and assurance helps to ensure that information technology products and the systems that rely on those products are sufficiently trustworthy.

- **NIST 800-82:** NIST 800-82 is a regulatory standard that provides guidance on how to secure industrial control systems (ICS), including supervisory control and data acquisition (SCADA) systems, distributed control systems (DCS), and other control system configurations such as programmable logic controllers (PLCs), while addressing their unique performance, reliability, and safety requirements. It also provides an overview of ICS and typical system topologies, identifies typical threats and vulnerabilities to these systems, and provides recommended security countermeasures to mitigate the associated risks.

Due to the nascent stage of 5G NPN network implementations, other standard and partnership organizations, such as oneM2M, are looking at working closely together to focus on security and deployment aspects of IIoT for 5G SNPN and PNI-NPN deployments.

oneM2M is an association of eight of the world's leading ICT standard development organizations: the Association of Radio Industries and Business (ARIB) from Japan, the Telecommunication Technology Committee (TTC) from Japan, the Alliance for Telecommunications Industry Solutions (ATIS) from the USA, the Telecommunication Industry Association (TIA) from the USA, the China Communications Standards Association (CCSA), the European Telecommunications Standards Institute (ETSI), the Telecommunications Standards Development Society, India (TSDSI), and the Telecommunications Technology Association (TTA) from Korea.

Summary

This chapter discussed the mechanisms for securing the key use cases being planned to be enabled by 5G. The use cases discussed in this chapter will require 5G-capable IoT/IIoT devices or IoT/IIoT devices capable of being integrated with a 5G network gateway. *Supply chain vulnerability* is one of the key threats in the 5G use cases, specifically for the *critical infrastructure use case*s, as discussed in Chapter 8, "Securing Massive IoT Deployments in 5G," in the section "Supply Chain Vulnerability." The key methods to mitigate supply chain vulnerability are discussed in the same chapter, under the section "Securing Supply Chain." The use cases definition is still in the nascent stages, and trials are being conducted with very low footprints.

The key challenge most of the early adopters are having is the right design of also including the legacy infrastructure. If it was a greenfield deployment, it would be comparatively less complex because you need not consider the vulnerabilities in the older-generation equipment and workarounds, as implemented in network designs throughout ages of network implementations, migrations, and expansions. The security controls discussed in each of the use cases and adhering to the aforementioned standards comprise a good starting point in your design for securing the 5G use cases.

Key Acronyms

The following table expands the key acronyms used in this chapter.

Acronym	Expansion
5GC	5G Core
AMF	Access and Mobility Management Function
ARIB	Association of Radio Industries and Business
ATIS	Alliance for Telecommunications Industry Solutions
AUSF	Authentication Server Function
C-V2X	Cellular – Vehicle-to-Everything
CCSA	China Communications Standards Association
CNF	Cloud-Native Function
Cobots	Collaborative robots
CSC	Communication services consumer
CSP	Communication services provider
DDoS	Distributed denial of service
DoS	Denial of service
DSRC	Dedicated short-range communications
eMBB	Mobile broadband
ETSI	European Telecommunications Standards Institute
FLISR	Fault location, isolation, and service restoration

Acronym	Expansion
IAM	Identity and Access Management
IIoT	Industrial Internet of Things
IoT	Internet of Things
M2M	Machine-to-machine
MEC	Multi-access edge compute
MFA	Multifactor authentication
mIoT	Massive Internet of Things
NF	Network Function
NPN	Non-public network
NSaaS	Network slice as a service
OBU	On-board unit
OT	Operational technology
PCF	Policy and Charging Function
PNI-NPN	Public Network Integrated Non-Public Networks
RAT	Radio access technology
RBAC	Role-based access control
RSU	Road side unit
S-CGR	Secure connected grid routers
SBA	Service-based architecture
SBI	Service-based interface
SCADA	Supervisory control and data acquisition
SDN	Software-defined network
SMF	Session Management Function
SNPN	Standalone non-public network
SSL	Secure Sockets Layer
TIA	Telecommunication Industry Association
TLS	Transport Layer Security
TSDSI	Telecommunications Standards Development Society, India
TTA	Telecommunications Technology Association
TTC	Telecommunication Technology Committee
UDM	Unified Data Management
UE	User equipment
UPF	User Plane Function
V2I	Vehicle-to-infrastructure
V2N	Vehicle-to-network
V2P	Vehicle-to-pedestrian
V2S	Vehicle-to-sensor

Acronym	Expansion
V2V	Vehicle-to-vehicle
V2X	Vehicle-to-everything
VNF	Virtual Network Function
SCF	Secure Controls Framework
NIST	National Institute of Standards and Technology
IACS	Industrial Automation and Control System
IEC	International Electrotechnical Commission
ISA	International Society of Automation

References

3GPP TS 33.536 V16.2.0, "Security Aspects of 3GPP Support for Advanced Vehicle-to-Everything (V2X) Services"

3GPP TS 23.287 V16.5, "Architecture Enhancements for 5G System (5GS) to Support Vehicle-to-Everything (V2X) Services"

3GPP TS 23.434 V17.1, "Service Enabler Architecture Layer for Verticals (SEAL)"

https://www.cisco.com/c/dam/en_us/solutions/iot/demystifying-5g-industrial-iot.pdf

5G Automotive Association, "C-V2X Use Cases and Service Level Requirements Volume II," https://5gaa.org

5GAA TR A-200094, https://5gaa.org

https://www.cisco.com/c/en/us/td/docs/solutions/Verticals/Solution_Briefs/IA_Networking_Solution_Brief.html

Pramod Nair, " Securing Your 5G Infrastructure to the Edge and Beyond," Cisco Live, Cisco Public, March 30, 2021

Pramod Nair, "Securing Your 5G Network End to End," Cisco Live APJC, Cisco Public, April 30, 2019

https://www.isa.org/

https://www.iso.org/

https://csrc.nist.gov/

Chapter | 10

Building Pragmatic End-to-End 5G Security Architecture

After reading this chapter, you should have a better understanding of the following topics:

- Building a pragmatic security architecture for 5G networks
- Understanding key risks in the 5G ecosystem
- Key security controls and threat mitigations in the 5G ecosystem
- Applying zero trust principles to secure 5G deployments

The previous chapters discussed the key security controls required per domain for different deployment types such as public, hybrid, and on-premises. We also saw the controls used to secure the 5G architectures, such as securing use cases based on standalone private 5G/Standalone 5G non-public networks S-NPN, Public network integrated non-public network (PNI-NPN), service provider 5G networks, and network slice as a service (NSaaS). Understanding threats and mitigation techniques by each domain is very important if you want an enhanced and effective security posture. This chapter will take you through the considerations for building a pragmatic 5G security architecture.

This chapter will be of particular interest to the following individuals and teams from enterprise/5G service providers deploying 5GC as well as cybersecurity vendors planning product developments for 5G security use cases:

- Government departments looking at security impacts of 5G deployment and looking at implementing security measures

- Management consultants advising governments and service providers on 5G security strategy

- CSO and CTO teams from service providers looking at securing 5G deployments

- Enterprise teams deploying NPN/private 5G deployments

- Security architects looking at securing 5G deployments

- Enterprise solution architects and enterprise security architects working with enterprises integrated with service provider 5G networks

- Cybersecurity vendor security architects looking to secure 5GC deployments for their customers

- Cybersecurity vendor product managers looking for use cases or features to enhance security products to cater for secure 5G deployments

5G is not just a technology for providing better throughput to subscribers; it is the beginning of a technology revolution that will be the critical part of your country's infrastructure, allowing real-world use cases for government organizations, industries, enterprises, and citizens. 5G technology introduces a lot of flexibility in the way you can deploy the network as well as robust capabilities in creating and offering services to subscribers, which is unmatched by any of the previous technologies. Due to the many advantages that 5G provides, it will form the backbone of your country's economy and services provided by it. Once 5G is widely deployed, the networked government and private enterprises will be critically reliant on innovative new applications.

5G is the first real cellular technology aimed at open integration with various third-party applications and systems to allow the realization of the use case. It is the first real technology where the network functions can be peeled away and deployed at the edge or at the enterprise premises to allow true low-latency use cases. 5G network slicing will play a critical role for use cases such as the smart factory, automotive, and the like. This also enables service providers to offer their network to the enterprise for better connectivity and to fulfill their use cases.

As you see due to the importance of 5G in the ecosystem, any attack on the 5G infrastructure could cripple a country's economy. What we really have to understand is that 5G is just the beginning. Upcoming technologies such as 6G, which is already in the works, will reuse the openness brought in by the 5G specifications and build on top of that to provide closer integration with other standards and frameworks to have a seamless functionality across multiple use cases. An example could be vehicle-to-everything (V2X), where practical realization of the use case would require integrations at least among vehicle-to-pedestrian (V2P), vehicle-to-infrastructure (V2I), vehicle-to-sensor (V2S), and vehicle-to-network (V2N).

This chapter will take you through the key components and capabilities required by the security architecture to secure 5G deployments. The key topics covered in this chapter, such as the security architecture for open RAN (radio access network) and distributed RAN deployments, will be applicable to standalone private 5G/5G NPN networks as well as service provider networks.

Foundations of 5G Security

Building a security architecture for 5G will require you to have a completely different approach as compared to securing previous generations of cellular technologies due to the aforementioned applications of the 5G technology. Securing 5G deployments will require a holistic view of applying security. You will need to look at the 5G use cases your organization plans to implement and then build a security architecture which includes built-in security features and layers of security controls, ensuring that each of the use cases can be secured. This will require security architects to understand the use case first, which means involving the security teams early in the process and allowing the security principles to be integrated in the foundation of your 5G deployment.

Having an end-to-end 5G security architecture includes protecting 5G network infrastructure, IT and OT network, and consumers of 5G services in any public or non-public deployment, as illustrated in the Figure 10-1. These different network areas or domains are usually separated by security zones and are catered by different teams within the organization. For example, a service provider will have separate teams catering for the 5G network infrastructure, such as radio planning and optimization, transport, back haul, and mobile packet core. A separate team is responsible for the IT and OT network infrastructure, usually called the network operations team. The managed services provided by the service provider are usually catered by a value added services team or a team under the Chief Revenue Officer (CRO).

In legacy cellular networks, such as 4G, the 4G network infrastructure teams will not be much involved in the IT and OT network. Similarly, the IT teams within the organization will not be much involved in the cellular network operations.

However, for 5G, depending on the deployment mode (such as private 5G/standalone NPN 5G deployment), the teams responsible for IT and OT networks need closer interaction with the teams responsible for 5G infrastructure.

Figure 10-1 illustrates the key network domains to secure for public and non-public 5G deployments.

A typical service provider will have to cater for securing the three key domains shown in Figure 10-1. For standalone NPN and PNI-NPN 5G deployments for industry verticals, apart from securing 5G network components, providing isolation between the IT and OT network components is important to ensure that attacks are not migrated from the IT part of the infrastructure to the OT and IIoT network of the industry vertical. If deploying a NSaaS offering, you should ensure proper isolation between the slices and granular access control for users to prevent unauthorized access to customer slices.

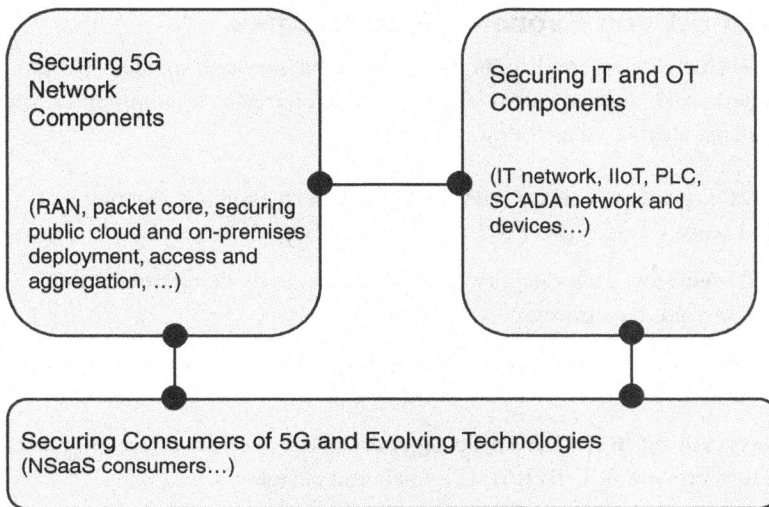

FIGURE 10-1   Key Network Domains to Secure within Public and Non-Public 5G Deployments

Securing 5G and Evolving Network Deployments

Securing 5G and evolving networks refers to securing the end-to-end components of the 5G infrastructure, which includes the RAN; access and aggregation, consisting of front-haul, mid-haul, and back-haul transport; multi-access edge (MEC) deployments in the public cloud and on-premises data centers; and Packet Core Network functions. It also includes the Control Plane, User Plane, and Management Plane between the 5G network functions.

Securing IT and OT

In a typical network environment of any enterprise or service provider, you will be mainly dealing with the IT part of security. OT security is used primarily to protect and control critical infrastructures such as power stations, transportation networks, and smart city deployments. 5G use cases related to industrial Internet of Things (IIoT) requires integration between the IT and OT networks. Attempts to exfiltrate data, steal intellectual property, and disrupt operations are steadily increasing as more cyber-attacks target critical infrastructure and industrial assets. Threat actors use weak security within IT networks to access and attack OT networks and exploit weak security implementation in OT networks to attack IT networks. These kinds of attacks are prevalent in critical infrastructure deployments such as oil and gas. To secure such critical infrastructure and industry vertical–related 5G deployments, you need to implement robust security controls on both the IT and OT networks.

Securing Consumers of 5G and Evolving Technologies

Service providers offering 5G services are referred to as communication service providers (CSPs). CSPs can also offer NSaaS to enterprise/industrial customers, which are referred to as communication service customers (CSCs). CSCs are categorized as follows:

- **Business-to-consumer (B2C) services:** This category includes data (messaging, Internet and web surfing, and so on) and voice services (5G voice) to end customers.

- **Business-to-business (B2B) services:** This category includes connectivity being provided to small, medium, and large enterprise businesses.

- **Business-to-home (B2H):** This category includes home broadband, parental controls, and VPN services.

- **Business-to-business-to-everything (B2B2X):** This category includes services offered to other service providers, who in turn provide B2C/B2B/B2H to their end customers.

Key Tenets of 5G Security Architecture

Figure 10-2 illustrates the key tenets of 5G security architecture.

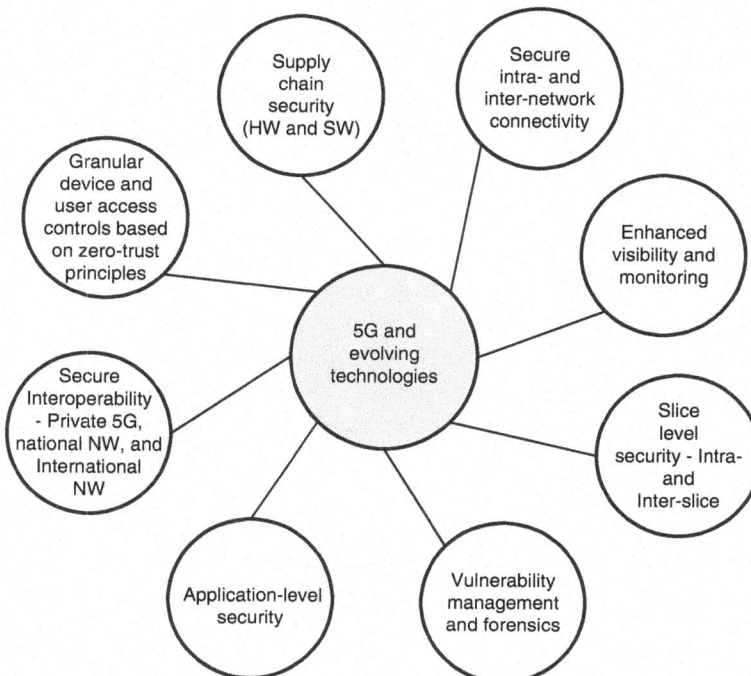

FIGURE 10-2   Key Tenets of 5G Security Architecture

Whereas each of the chapters in Part II, "Securing 5G Architectures, Deployment Modes, and Use Cases," went into detail on the key security tenets for the respective domains, as shown in Figure 10-2, the sections that follow look at these tenets from the perspective of an end-to-end security architecture for securing your 5G network along with its use cases.

Supply Chain Security

The significant risks of supply chain threats arise from nation-state actors who aggressively target service provider and public sector networks to exfiltrate sensitive information, modify certain parameters to compromise critical infrastructure, destroy information, or cause an outage in order to cripple a country or the private/public sector. Until the 2010s and early 2020s, supply chain attacks were related mainly to hardware and required a certain amount of sabotage, including coercing the hardware engineers and designers in printed circuit board (PCB) design and manufacturing teams to change the PCB schematic or modify it to include an extra track to include a component for sniffing purposes. Once the design team was infiltrated, it became difficult to identify any anomalies in the design because the quality team had the modified schematic. This can still be carried out and is actually difficult to detect because any given hardware related to telecom infrastructure has a complex PCB design. This is also difficult to detect during functional testing, as the components can remain quiet unless triggered remotely once deployed.

This risk has been present for ages in one form or another. So what makes it important now?

If you take a step back you will see that many of the communication companies providing voice and data services using legacy technologies had their critical network functions and devices air-gapped, meaning they were isolated and didn't have any interactions with the Internet. Of course, we had the SS7, GPRS Tunneling Protocol (GTP), and Diameter-based threats, but those fall under network and protocol security, which we can cater to with specialized firewalls.

Beginning with 5G, the entire ecosystem is based on openness. Various models such as NSaaS allow service providers to logically peel away parts of their network and offer them to industries, government sectors, and enterprises. Other models such as Cloud RAN/Open RAN/Virtual RAN, private 5G/NPN, and PNI-NPN can be used by any of the industry verticals to create their own 5G network using their own spectrum or a spectrum leased from service providers who own parts of the spectrum.

As the nation's entire ecosystem is run by 5G and other evolving architectures, robust security processes should be planned by the national cybersecurity agencies to prevent disruption to the country's economy. The security agencies should be looking at the virtualization scenario as a priority. Virtualization, although not new, has expansive use in telecom via the use of Cloud-Native Functions (CNFs), which have emerged only during the 5G evolution. The entire 5G network functions will now be based on CNFs, which use open-source software to create 5G network functions, which can then be deployed as a virtual machine (VM) or a container on any hypervisor on the premises or any multistack public cloud infrastructure. As an example, any government or industry vertical can now deploy its own 5G network with all the 5G network functions on the public cloud.

Here are some of the security processes that can be followed by the service provider, government entity, or any industry vertical deploying 5G:

- The cybersecurity agency of the country should lay down the key security controls required for the deployment of 5G, other than the security controls stated as mandatory by 3GPP specifications.

- The cybersecurity agency of the country should have an incident response (IR) team specifically for the national critical infrastructures and key industry verticals. Of course the industry verticals and critical infrastructure providers will have their own security operations center (SOC) and IR, but an overlay team is important to ensure the right skills are in place during a drastic nation-state attack.

- Test centers should be shared for different countries, reducing bottlenecks for service providers in deploying 5G networks. Instead of having separate testing centers of equipment for each and every country in Europe, there should be three to four test centers led by a joint effort for specific testing to certify the network functions and devices. This would improve the rate at which the NFs and telecom service equipment could be tested.

- The cybersecurity agency of the country should provide services to audit the 5G network for nation-state-specific risk vectors.

- Most software depends on third-party components (libraries, executables, or source code), but there is very little visibility into this software supply chain. It is common for software to contain numerous third-party components that have not been sufficiently identified or recorded. Methods such as software bill of materials (SBOM), which is a formal record containing the details and supply chain relationships of various components used in building software, will help you identify the "ingredients" of software installed on any system or device. This can save hundreds of hours in the risk analysis, vulnerability management, and remediation processes. The SBOM initiative is a National Telecommunications and Information Administration (NTIA) multistakeholder process on software component transparency.

- 5G deployments should have a solid network and application monitoring layer with tight integration capabilities with critical telecom service equipment and applications. This allows early detection of any attacks on the critical infrastructure.

Securing User and Device Access Using Zero-Trust Principles

Traditional networks are based on defined perimeters, where most of the mobile packet core functions are centralized. The evolving mobile packet core architectures such as 4G Control Plane and User Plane Separation (CUPS) and 5GC using MEC create software-defined perimeters. The 5G core is based on a cloud-native architecture that allows 5G core network functions to be deployed as a microservice in a private data center of the service provider or in a public cloud such as Amazon, Azure, Google Cloud, and so on. Multiple vendors will supply Network Function Virtualization Infrastructure (NFVI) components and Virtualized Network Functions (VNFs). In addition, many contractors and sub-contractors will require access to the network for support, configuration, and deployment purposes. Many of the interfaces in the 5G core rely on web-based communications, including the Representational State Transfer–based API, which doesn't have any predefined security methods, so developers need to define their own. The previously mentioned threat vectors and threat actors in 5G drive a requirement of having a robust security foundation while deploying 5G networks.

Zero trust for 5G consists of multiple layers of security, establishing trust in user identity, enhanced end-to-end visibility, and the trustworthiness of the user device accessing the data control network (DCN). Enforcement using risk-based and adaptive access policies while enabling secure connections to devices and applications can then be undertaken. Figure 10-3 illustrates the main models within zero-trust security for 5G deployments.

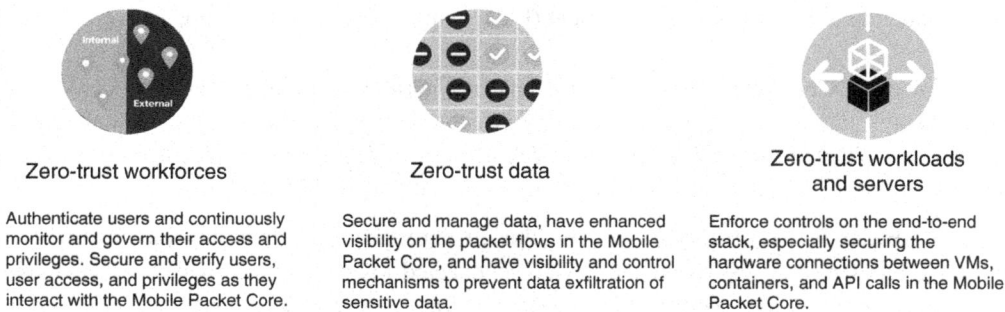

Zero-trust workforces

Authenticate users and continuously monitor and govern their access and privileges. Secure and verify users, user access, and privileges as they interact with the Mobile Packet Core.

Zero-trust data

Secure and manage data, have enhanced visibility on the packet flows in the Mobile Packet Core, and have visibility and control mechanisms to prevent data exfiltration of sensitive data.

Zero-trust workloads and servers

Enforce controls on the end-to-end stack, especially securing the hardware connections between VMs, containers, and API calls in the Mobile Packet Core.

FIGURE 10-3 Main Zero-Trust Security Models for 5G Deployments

Risk-based and adaptive access policies provide a control mechanism and the security policies for the users and devices. Some controls are taken proactively, while others are applied after an attack takes place. There are two types of attacks:

- Zero-day attacks are threats that are previously unknown.

- Day-one attacks are threats that have been communicated by the vendor but have not necessarily been patched in the production environment.

Typically, deviations in known-good behavior of the carrier cloud and the applications that request service and state from it are identified by the security controller, and then some action is taken to mitigate the attack or to gain additional visibility. An action is then taken to properly identify the miscreant and mitigate the risk. Day-one or n-day attacks are attacks where publicly acknowledged vulnerabilities are leveraged before the vendor releases the security patch or before the customer has applied the security patch to mitigate the vulnerability. These attacks can be reduced by ensuring that the latest patch is applied to minimize the damage caused by the vulnerability exposure.

One of the critical parts of your 5G network is the data control network (DCN), which provides management access to the packet core elements and is predominantly used by network management systems (NMSs) as well as third-party vendors for troubleshooting access and for internal command-line access for configuration and troubleshooting. Typically, the DCN provides the following:

- VPN remote access for third-party vendor troubleshooting and for internal remote support

- "Gray" management Virtual Routing and Forwarding (VRF) for transport across MPLS cores

- Routed and switched network infrastructure for DCN traffic (for example, logging, configuration, and performance management)

- A proxy or "jump host" to break external communication and provide proxied access

- Logging and reporting of user access and commands issued by each user

- Out-of-band access to lights-out management (LOM) and console ports (terminal server access)

Vendor access into the DCN to provide support assistance is a normal requirement; however, credential reuse (sharing a single password for multiple people within the vendor) and the potential for staff churn are both possible risks that allow unauthorized access to devices. This can be mitigated through the use of a VPN concentrator in combination with multifactor authentication (MFA) technology.

Once a specific vendor has authenticated and is allowed onto the DCN, it is not uncommon that they would have unfettered access to all the devices connected to the network, which includes the management access layer of other vendors. In a zero-trust methodology, we would want to limit vendor A to accessing only the elements they need to support, while also limiting the protocols allowed on that interface, as shown in Figure 10-4.

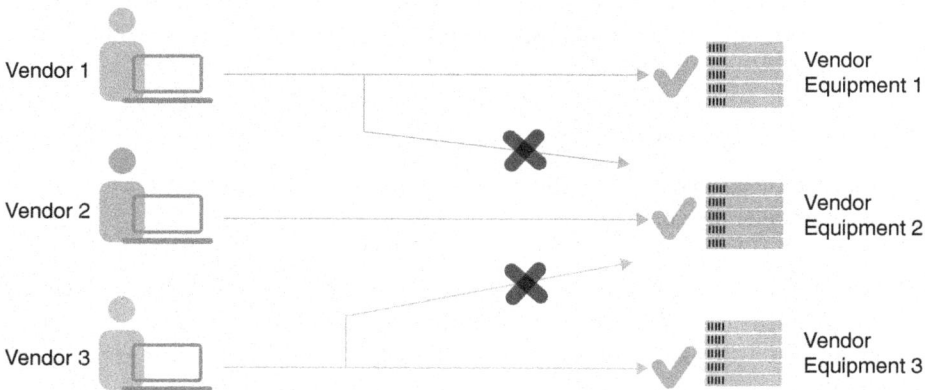

FIGURE 10-4 Applying Zero-Trust Principles for Vendor Specific Access

Visibility plays a key part in understanding security and can be enabled on the DCN through extensive logging of activity via RADIUS and TACACS+, with a focus on ensuring that log data is able to be stored for a long-enough period for forensic investigation purposes. Understanding network traffic profiles and potential data exfiltration can be achieved through an anomaly detection engine, which can automatically baseline normal traffic and identify data being exfiltrated. Figure 10-5 shows an example of multiple remote users accessing the service provider's network resources.

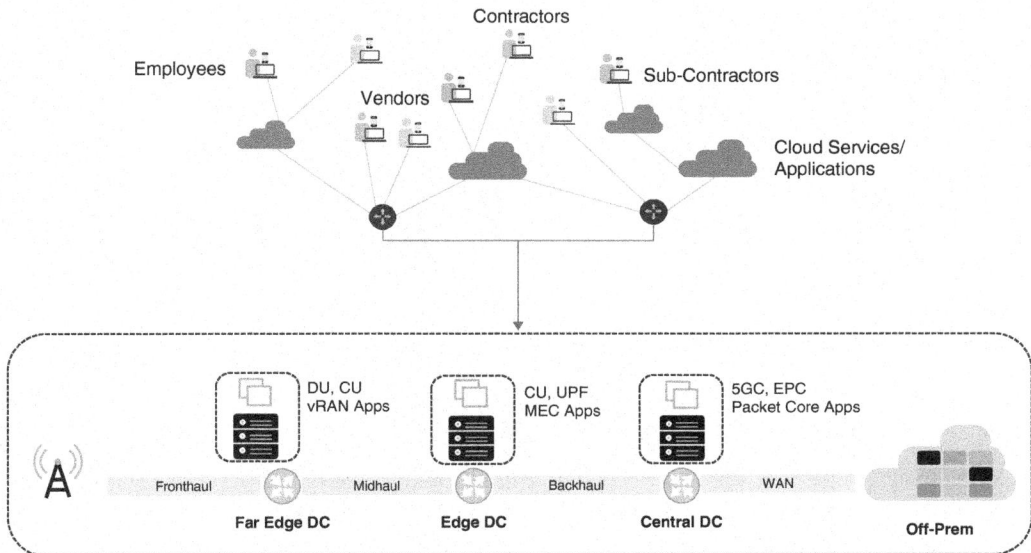

FIGURE 10-5 Applying Zero-Trust Principles for 5G Deployments

As shown in Figure 10-5, the network has multiple personnel (employees, vendors, contractors, and sub-contractors) accessing the infrastructure, which includes multivendor NFVI, multivendor VNFs, and containers for various provisioning, operational, and maintenance purposes. To prevent malicious actions intentionally and unintentionally in 5G networks, there are three steps through which the operator could deploy a zero-trust security architecture in 5G networks, which are listed next and illustrated in Figure 10-6:

Step 1. Implement a VPN layer.

Step 2. Implement multifactor authentication (MFA) and segmentation.

Step 3. Implement an enhanced visibility and threat mitigation layer.

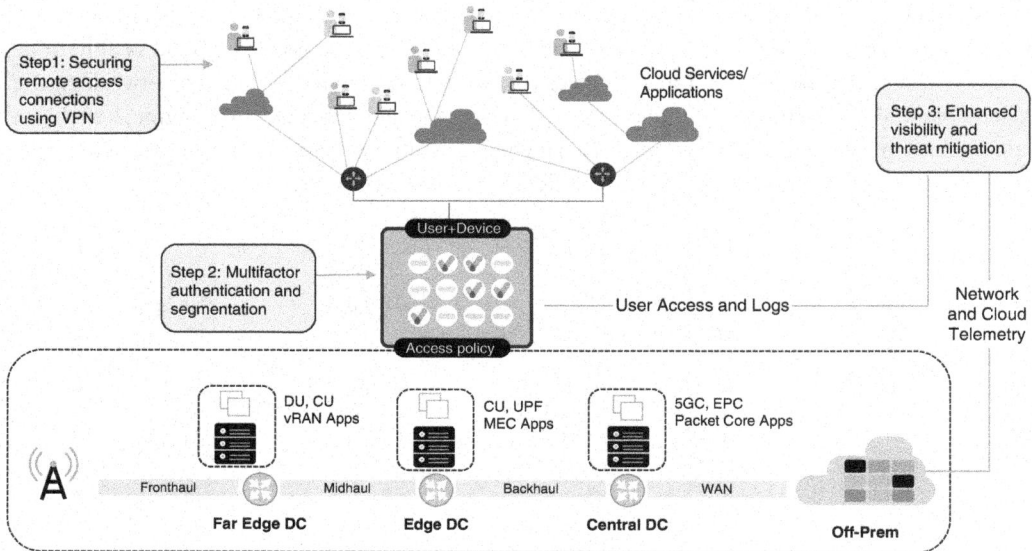

FIGURE 10-6   Applying Enhanced Visibility and Granular Access Controls

The three steps in having a zero-trust access layer for the 5G network are further discussed next:

Step 1.   VPN layer: As the first step, a VPN could be used for any external access to the service provider's DCN. VPNs use the IPsec protocol suite for packet encryption, authentication, and nonrepudiation. VPN includes protocols for establishing mutual authentication between agents at the beginning of a session as well as negotiation of cryptographic keys to use during the session. It can protect data flows between a pair of hosts (host-to-host), between a pair of security gateways (network-to-network), or between a security gateway and a host (network-to-host). Internet Protocol Security (IPsec) uses cryptographic security services to protect communications over IP networks and also supports network-level peer authentication, data-origin authentication, data integrity, data confidentiality (encryption), and replay protection.

Step 2.   MFA and segmentation layer: The next step is to deploy the policy control and enforcement layer along with the multifactor authentication (MFA) layer and segmentation using a Network Access Control (NAC) solution. The user authenticates to NAC, and the device is verified to make sure it is authorized for access to the requested part of the network. The switch or wireless LAN controller (WLC) tags the traffic. As an example, mechanisms such as Security Group Tag (SGT) can be used to tag the traffic. Using an SGT, you will be able to group each vendor's access and control access to network and application resources on a granular level. Then the policy is enforced on the data center firewall with a firewall that allows EPC and 5G core resources (hardware and VNFs) access only to those devices and users that are authorized; all other devices and users are denied access. SGTs are applied to authenticated users to explicitly allow access for authorized users by using security group access control. Using this mechanism, you can provide and enforce access security policies based on user, device, and application risk as well as verify the identity of all users. This integrated solution provides security admins with the ability to enforce consistent user- and device-based policy for VPN access and thereby reduce the risk of data breaches while meeting compliance requirements.

Step 3. Enhanced visibility and threat mitigation layer: The enhanced visibility and anomaly detection layer will require integration with the NAC layer to obtain key information, including MAC address, IP address, last active time, user name, security group, VLAN, domain name, interface device IP, and interface device port ID. These key pieces of information, coupled with network visibility using the network and application telemetry, allow you to provide key insights into endpoint behavior and identity as well as to indicate anomalous behavior. However, not everything unusual is malicious; therefore, enhanced visibility solutions, along with ML and AI integrations, can quickly and with high confidence correlate anomalies to threats such as C&C attacks, ransomware, DDoS attacks, illicit crypto mining, unknown malware, as well as insider threats. In addition to this, the enhanced visibility layer should be able to analyze encrypted traffic without decrypting the encrypted traffic. This is particularly relevant when the level of encrypted traffic is used for API communications between 5G and external networks.

These three steps can help ensure that you authenticate users and continuously monitor and govern their access and privileges. It also provides a method to secure and verify users, user access, and user privileges as they interact with the 5G network, thereby mitigating malicious actions and intent.

Your network will encompass a diverse range of technologies—from legacy signaling protocols to the latest distributed, virtualized environments, including multivendor NFV and multivendor Virtual Network Functions (VNFs) based on virtual machines and containers. This broad attack surface creates gaps in the security posture that can be addressed through the use of trusted platforms, enhanced anomaly detection, granular control on workloads, and limiting communication in line with the zero-trust methodology.

Table 10-1 provides a checklist for granular access control on DCNs mainly used for Day 0 and Day 1 configuration, operations, and maintenance.

TABLE 10-1   Security Control Checklist for Granular Access Control

Recommended Security Controls	Does This Security Control/Capability Exist in Your Network? (Y/N)	Solution/Feature Used
Remote user VPN		
Multifactor authentication		
Device profiling		
Microsegmentation		
Endpoint protection		
Granular visibility and immediate quarantine		

Recommended Security Controls	Does This Security Control/Capability Exist in Your Network? (Y/N)	Solution/Feature Used
DNS security for users and devices		
IoT device authentication and authorization		

Secure Intra/Inter-Network Connectivity

Secure intra/inter-network connectivity caters to securing the following components of 5G:

- Access and aggregation
- DDoS protection
- API security
- Infrastructure security

5G makes it possible for the network functions to be deployed in on-premises, public cloud, and hybrid modes. 5G use cases such as smart factories with ultra-reliable low-latency communication (URLLC) can be realized using various architectures such as Open RAN or Distributed RAN, using multiple 5G network deployment methods such as standalone NPN/private 5G, PNI-NPN, or by reusing the service provider infrastructure using network slicing and deploying NSaaS models. Moreover, all these different deployment modes require integration with an external IT and OT layer, SDN, orchestration layer, and third-party application layer. All this flexibility and openness makes 5G different from its predecessor, but also requires careful planning of security controls to ensure that the network connections are secure.

Although IoT might not be planned in the first year of deployment in service providers, IoT is one of the primary use cases for industry verticals such as smart manufacturing that are planning to use NPN and PNI-NPN 5G deployments. Threat surfaces such as weak built-in security of the network devices and peer-to-peer communication by use cases such as vehicle-to-everything (V2X) bring in new risk vectors.

The fulfillment of NPN and PNI NPN 5G use cases relies heavily on non-3GPP-based communications and traversing insecure public networks, which are prone to man-in-the-middle (MitM) attacks. The required security controls and capabilities would depend on the deployment model being used for 5G and the use case being catered by the deployed 5G network. Figure 10-7 illustrates securing the connectivity of 5G packet core network using inherent methods such as SEPP and SCP introduced by 3GPP and other overlay security mechanisms such as API gateway (GW) and web application firewall (WAF).

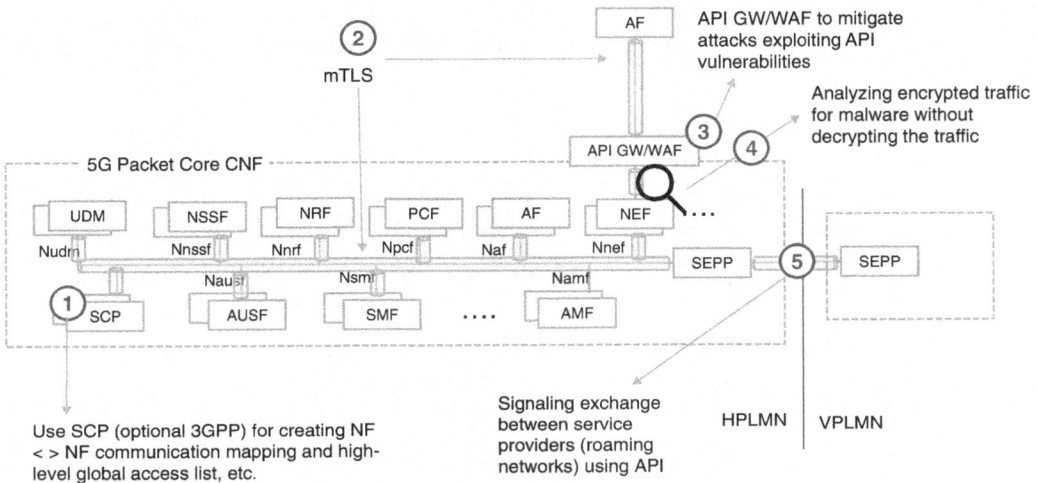

FIGURE 10-7   Securing the API Using Multilayered Security Controls

Figure 10-8 illustrates securing the connectivity of various interfaces in a 5G deployment using SecGW and TLS mechanisms.

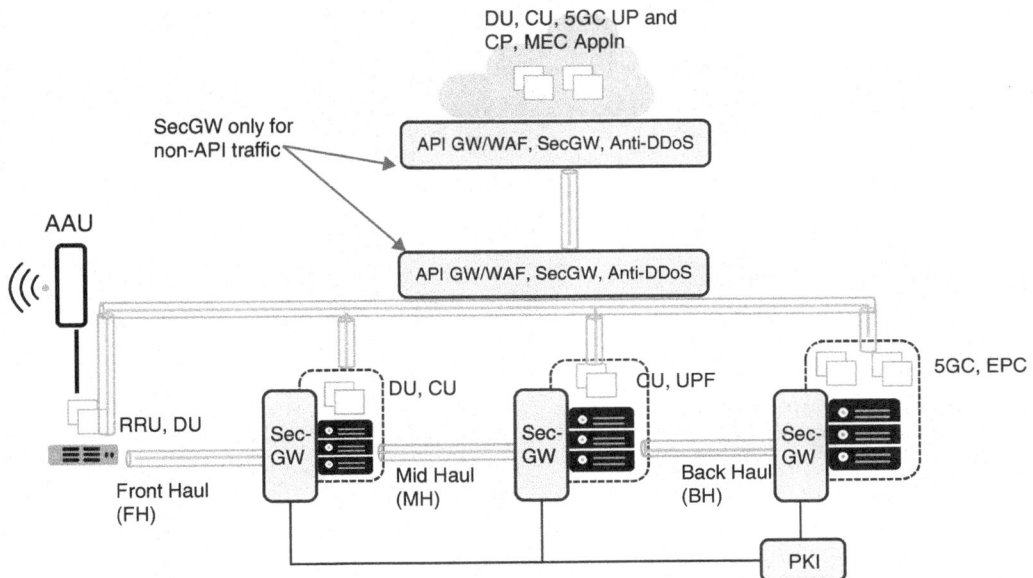

FIGURE 10-8   Securing Intra- and Inter-Network Connectivity

Figure 10-8 shows using a security gateway (SecGW) to provide encryption and authentication of the nodes deployed between the front haul (FH), mid haul (MH), back haul (BH), and the public deployed 5G 3GPP components.

Web application firewalls (WAFs) and API gateways are the two major inline security tools for API protection. While API gateways do usually offer authentication and authorization features, the HTTP and OWASP Top 10 protection offerings are either limited or absent. To protect API-based communications, a WAF can be used to ensure APIs are secured. Any Internet-facing APIs or APIs external to the on-premises 5G packet core components should be secured by using Transport Layer Security (TLS) mechanisms and an API gateway or WAF to mitigate API-related risks.

The internal communication between the 5G NFs should be authenticated and authorized. This could be catered for by using mechanisms like OAuth 2.0, which uses authorization tokens to prove an identity between NFs.

For internetwork connectivity in use cases such as PNI-NPN, the interfaces between the enterprise and the service provider should be secured. Depending on the deployment model, the internetwork communication could be just an API, which can be secured using TLS or the control and user planes and might require both TLS for securing the API and IPsec to secure the non-API requests.

For use cases such as NSaaS, API communication should be secured, and granular user authentication and authorization should be implemented.

Table 10-2 provides a checklist for securing access and aggregation for 5G deployments.

TABLE 10-2   Security Control Checklist for Access and Aggregation

Recommended Security Controls	Does This Security Control/ Capability Exist in Your Network? (Y/N)	Solution/Feature Used
Hardened software and hardware		
Encryption in back haul/xHaul		
Authentication for nodes in back haul/ xHaul		
Visibility on nodes in back haul/xHaul (control plane monitoring)		
DDoS protection (from devices within the network and from devices outside the network)		
Spectrum scanners		
PKI infra		
API security for programmable access and aggregation routers		
Secure Zero-Touch Provisioning server for access and aggregation equipment and devices		

In 5G, new formats of user equipment, including low-cost insecure IoT devices, could introduce new kinds of vulnerabilities that might be exploited to target user data confidentiality and integrity. Abuses on hardware and software implementations on the user equipment (UE) side can stem from installation of malicious components that might compromise confidentiality and the integrity of subscriber profile data. The threat includes the theft of user credentials, brute forcing of user accounts, password cracking, masking the user identity, and impairment of IoT grouping authentication as techniques used by threat actors to abuse the 5G authentication systems. Table 10-3 describes the security controls and capabilities to secure the IoT/IIoT deployment phase.

TABLE 10-3 Security Control Checklist for IoT/IIoT Deployment, API Security, and DDoS Protection

Recommended Security Controls	Does This Security Control/ Capability Exist in Your Network? (Y/N)	Solution/ Feature Used
DNS protection for IoT devices		
PKI config for IoT devices		
API gateway		
API visibility		
Encrypted traffic analysis for control plane		
DDoS protection from IoT devices and external and internal DDoS attacks		
IoT application protection (security policy enforcement per IoT application)		
Web application firewall, including rate limiting		
Authentication and authorization of IoT devices		
Anomaly detection and monitoring of IoT/IIoT network (including OT)		
Segmentation between IT and IIoT/OT network		

Furthermore, you also need security controls to discover network traffic with application-level insight, to classify and manage application sessions such as web browsing, multimedia streaming, and peer-to-peer applications, to monitor application usages and anomalies, and to build reporting for capacity planning and compliance. Table 10-4 provides a checklist for these security controls.

TABLE 10-4 Security Control Checklist for Infrastructure Security

Recommended Security Controls	Does This Security Control/ Capability Exist in Your Network? (Y/N)	Solution/Feature Used
Application-aware firewalling		
Anti-malware using sandboxing for infrastructure users		

Recommended Security Controls	Does This Security Control/ Capability Exist in Your Network? (Y/N)	Solution/Feature Used
SSL decryption and inspection		
Rapid threat containment		
Application visibility and control		
Macro and microsegmentation		

Application-Level Security

5G deployment is based on the CNF deployment model, which is actually a software implementation of a 5G Network Function (NF) using open source software instead of legacy deployments, which traditionally would be performed by a physical device. The CNF deployments could be based on VMs and/or containers that communicate with other 3GPP and non-3GPP NFs via standardized RESTful APIs. CNF deployment models bring in advantages such as being independent of the guest operating system (when deployed as containers) and the ability to scale with a small performance footprint. Service discovery and orchestration add to these advantages by bringing enhanced resiliency to node failures.

While adherence to CNF deployment does bring the aforementioned advantages in 5G deployment, you should allot for new risk vectors such as vulnerabilities from guest hardware and operating systems (OSs) as well as vulnerabilities in the open source software components used to build 5G 3GPP and non-3GPP NFs. 5G cloud-native architecture will bring in the need for industries to adopt a new development framework and move quickly to a cloud-native environment with virtualization, automated deployment, instantiations, and upgrades, by leveraging open source technology and software stacks. This affects not only new 5G services planned to be delivered by service providers, but also represents new ways for enterprise and industry verticals to consume those services. The four pillars of cloud native architecture—DevOps, microservices, containers, and continuous delivery—are all in play as we leverage open source technology. Cloud-native architecture splits the applications and 5G Network Functions (NFs) into individual microservices and focuses on decomposition to allow distributed deployments when it brings value. Some of the key security process, controls, and capabilities are as follows:

- **Application-first security methodology:** Accelerated application development and deployment for 5G networks will require DevOps to bring together software development and operations to shorten development cycles, allow organizations to be agile, and maintain the pace of innovation while taking advantage of 5G cloud-native architecture. However, existing DevOps practices used to develop and deploy software in operational environments can be implemented without consideration of security, leading to reduced efficacy and effectiveness of point-in-time threat modeling of the 5G network. This has shifted the industry mindset from infrastructure-first to application-first security. Processes such as development security and operations (DevSecOps) bake in security at every step of software design. Development, testing, and quality enable the development of secure software with the speed of DevOps. Instead of depending on various inspection algorithms to identify the security gaps within the code, DevSecOps aims to have secure software earlier in the development process. DevSecOps helps ensure that security is addressed as part of all DevOps practices by integrating security practices and automatically generating security and compliance artifacts throughout the process. DevSecOps practice needs to be implemented by 5G vendors and service providers to ensure secure deployment of 5G cloud-native network functions. Both the vendors and service providers will benefit by following the DevSecOps methodology. For vendors, following the DevSecOps methodology ensures reduced vulnerabilities, malicious code, and other security issues in the hardware and software stack being deployed in service providers' 5G critical infrastructure, without slowing code production and releases. For service providers, implementing DevSecOps methodologies will help address security issues such as any vulnerabilities and malicious code before its deployed on the production node, thus improving security efficacy for the 5G network. If you have a software development team to cater for 5G MEC applications, then following the DevSecOps methodology will help you addresses security issues as soon as they emerge and before deployment on the production node, thereby helping you with the vulnerability management and improved security posture of your production network.

■ **Securing 5G CNF images:** You should ensure that the vendor you choose as a part of the 5G deployment has a process in place that ensures all the container images are scanned for any vulnerabilities. They should also have a procedure to track all the Common Vulnerabilities and Exposures (CVEs) in the shipped images and provide resolution if any CVEs are detected. For efficient vulnerability management of 5G CNFs, it is recommended to use risk-based vulnerability management (RBVM) solutions available in the market today, which provides a risk score factoring in the number of instances of each vulnerability in your environment, their potential severity, and the assets that are threatened as a result of each vulnerability. Having such solutions in place will help you better manage the risks in 5G CNF deployments.

■ **Secure registry management:** The vendor should ensure that the images are scanned, approved, and signed before being uploaded to the image registry. This should be followed for both 3GPP and non-3GPP NFs.

■ **Secure software delivery:** The vendor should ensure that they enforce authentication mechanisms to limit distribution to specific customers only based on the SLA and other agreements. As an example, the software from the vendor's software download site should have a stringent authentication mechanism to ensure that only legitimate customers are downloading the software.

■ **Container and resource isolation:** Isolation for CNF deployments can be implemented at different levels. Some isolations are dependent on the vendor, and some require best practices of CNF deployments. An example of isolation provided by the vendor is the container isolation provided to you by using the Kubernetes platform, which enables you to set CPU-relative priority within the CFS scheduler, maximum memory limits, and limits on persistent storage. The main aim of providing granular isolation such as container isolation is to protect your host OS and mitigate against malicious container escape and breakout attempts into other targets hosted on the same host. As an example, the vendor should provide you with options for controlling pod placement. Pods are the rough equivalent of a machine instance (physical or virtual) to a container. Each pod is allocated its own internal IP address, thus owning its entire port space, and containers within pods can share their local storage and networking. In methods such as OpenShift, you are provided with an option for controlling pod placement. In OpenShift, the Pod Node Constraints Admission Controller ensures that pods are deployed onto only specified node hosts using labels, and it prevents users without a specific role from using the nodeSelector field to schedule pods. Using this feature, a cluster administrator can set a policy to prevent application developers with certain roles from targeting specific nodes when scheduling pods. This level of isolation will ensure that rogue application developers working on non-3GPP 5G NFs cannot interfere with your pods with 5G 3GPP NFs.

- **User-to-application mapping:** The vendor you choose as a part of your 5G deployment should allow granular user access to mitigate the risk of unauthorized access to applications in your network. In a container environment, Kubernetes does not implement any kind of user privilege isolation but instead uses RBAC to enable you to configure fine-grained and specific sets of permissions that define how a given user, or group of users, can interact with any Kubernetes object in your cluster, or in a specific namespace of your cluster. Apart from RBAC, it is recommended that you also follow the user access security controls discussed earlier in the section "Securing User and Device Access Using Zero-Trust Principles." This is critical, because once your network evolves to include hybrid deployment models, where you will implement non-3GPP and 3GPP NFs in the public cloud and private cloud, the risks of unauthorized access increase drastically and you will need robust end-to-end granular access controls for users and devices.

- **Microsegmentation:** Apart from the container isolation provided by the vendor, you should implement an end-to-end microsegmentation layer for all the 3GPP and non-3GPP 5G NFs deployed in your network. Microsegmentation is a security best practice that offers a number of advantages over more established approaches like network segmentation and application segmentation. Microsegmentation uses much more information in segmentation policies, such as application-layer information. It enables policies that are more granular and flexible to meet the highly specific needs of an organization or business application. You can also manage and enforce rules on end-to-end connections for just a group of components in a given network segment. The added granularity that microsegmentation offers is essential at a time when many organizations are adopting new container-based environments that make traditional perimeter-based security less relevant. By segmenting the network communication between pods, you can control the flow of traffic, allowing or blocking traffic based on a variety of factors. Segmented networks can reduce the attack surface by containing certain traffic only to portions of the network, and they can prevent unauthorized network traffic or attacks from reaching portions of the network to which they would prefer to prevent access. Microsegmentation also makes the job of monitoring network traffic much easier. Using microsegmentation to implement a zero-trust architecture will involve a paradigm shift and some technical challenges, but it will improve the following aspects of your network security:

- As containers are often deployed as microservices and can be dynamically deployed and scaled across a Kubernetes cluster, segmenting containers should be done on the service level. Because different services can be deployed across shared networks and servers (or VMs and hosts), and each workload or pod has its own network addressable IP address, container segmentation policies can be difficult to create and enforce with the existing network-based solutions.

- You can gain control and visibility over which workloads the open source tools consist of and who they are communicating with.

- Namespaces have been found to be useful for organizing services and to ease the management of services, where each one in a namespace has some attributes in common with others. Creating separate namespaces is an important part of isolation; however, it adds little security and doesn't restrict workload activity. This requires additional microsegmentation strategies to be implemented to apply granular microsegmentation policy to workloads.

- **Service Mesh:** To have a solid cloud-native application security strategy, you should make use of innovations in the cloud-native industry, which provide more granular security controls over applications. One of the key innovations is the service mesh. The use of cloud-native function in 5G deployments creates new security challenges. With microservices being updated much more frequently than they have in the past, service providers who struggled with microsegmentation when dealing with a few VMs for 4G deployments will face greater challenge with 5G NFs, which run dozens of containers that are being updated more frequently. Adoption of service mesh will provide you with a reliable and consistent way to connect, observe, and apply granular microsegmentation to your microservices. It provides you with other key features such as service-to-service authentication, policy creation, load balancing, and traffic routing. It will help to abstract the network topology and routing away from the infrastructure layer and build it into the network function/application itself by adding a sidecar proxy to the application, which abstracts the inter-service communications as well as monitoring and security to the workload. If you are a developer building applications to fulfill 5G customer use cases, a service mesh will allow you to easily link together microservices using a declarative model-based approach that sits on top SDN layers and traditional layers in the underlying infrastructure. At press time, the most popular service mesh is Istio, the open source code started by Google, Lyft, and IBM. There are other service mesh options, such as Consul, Linkerd, Kuma, AWS Mesh, and Open Service Mesh.

Table 10-5 provides a checklist for securing both 3GPP and 3GPP applications—VNFs, CNFs, and third-party applications.

TABLE 10-5   Security Control Checklist for Application Security

Recommended Security Controls	Does This Security Control/Capability Exist in Your Network? (Y/N)	Solution/Feature Used
Hardened software and hardware		
Visibility on VNF and CNF interactions (control plane)		
Visibility on API calls and encrypted traffic		
Secure 5G CNF images		
Secure registry management		
Visibility on multistack public and private cloud interactions (control plane)		
Malware detection on encrypted data (control plane)		
Microsegmentation		
RBAC		
TLS 1.3 encryption		
Authentication between 5GC CNFs (both 3GPP and non-3GPP)		
API gateway		
Web application firewall for API		
Security policy enforcement per VNF/container		
User-to-application mapping (including containers)		
Secondary authentication (for slice-specific authentication)		
Slice-level isolation of CNFs—physical separation of CNFs (design methodology)		
Multifactor user authentication		
Runtime Application Self-Protection (RASP) as part of DevSecOps process		
Static application software testing (SAST) as part of DevSecOps process		
Dynamic application software testing (DAST) as part of DevSecOps process		
Automated continuous integration/continuous delivery (CI/CD)		

Recommended Security Controls	Does This Security Control/Capability Exist in Your Network? (Y/N)	Solution/Feature Used
CNF image scanning		
Vulnerability management (including application development stage)		
Compliance management (including 5G VMs and container deployments)		
Forensics (including 5G VMs and container deployments)		

Vulnerability Management and Forensics

5G NF deployment based on Cloud-Native Functions (CNFs) brings in risk vectors due to open source software being used to build 5G NFs and applications. Because the 5G NFs and applications can run anywhere from on-premises to multicloud and cloud-native microservices, combined with accelerated innovation, the need for an application-led approach to security is paramount. You will need to identify vulnerabilities within the application during production, correlate vulnerabilities and breaches with business impact, and bring together the application and security teams to facilitate speedy remediation. Implementing an application-aware vulnerability detection and forensics layer that can identify vulnerabilities across the physical and virtual infrastructure (including virtual machines and container deployments) is critical. You should follow a DevSecOps model to develop security automation that scales directly alongside your deployment methods, be it an on-premises, cloud, or hybrid model, so that you can ensure security standardization and architectural strength at scale.

Figure 10-9 illustrates an example of implementing vulnerability detection and forensics in 5G deployment, which has both software-defined network (SDN) and non-SDN components. The data and log collection system will collect the non-SDN network and application telemetry in the form of IP network traffic telemetry (for example, NetFlow, S-flow and so on), hypervisor metadata, container sidecar telemetry, and virtual private cloud (VPC) logs. The data and log collection system will also filter the data and apply methods like packet de-duplication. The packet broker (PB) will collect all the network telemetry (similar to the data and log collection system) and optimize the flow collection from the SDN layer.

The collected telemetry from the data and log collection system and packet broker is then fed into the vulnerability detection and forensics solution, which has integrations with threat intelligence (which provides you with a list of threats and threat actors), the Common Vulnerabilities and Exposure (CVE) database (a list of publicly disclosed computer security flaws), your custom policy defined for the network, and machine learning (ML) and artificial intelligence (AI) to detect all the vulnerabilities and list them based on the risk factor. All these integrations with the vulnerability management and forensics solution will help you to detect any vulnerabilities with much higher accuracy and apply forensics investigations when required.

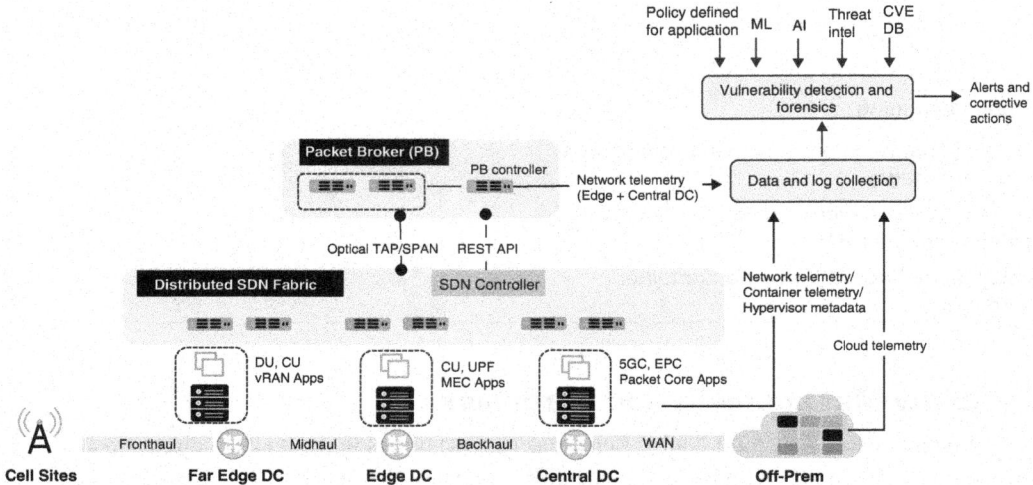

FIGURE 10-9   Vulnerability Management and Forensics

The solution you choose for vulnerability management and forensics should allow you to perform the following key functions:

- Perform continuous monitoring of the end-to-end network components (hardware and software) to detect any anomalous events, behaviors, and malware

- Baseline the behavior or the workloads based on communication activities and processes on the workloads

- Proactively detect anomalous behavior and identify indicators of compromise

- Allow custom definition of critical components

- Prioritize vulnerability and patch management based on criticality of the network component

- View aggregated information about network security risks and vulnerabilities relevant to a specific business application

- Get a baseline of the software inventory and the version information

- Identify any of the package versions that have known vulnerabilities or exposures, along with their severity

- Get an accurate inventory of all the servers that have the vulnerable package and map it to a policy that designates a specific action, such as updating firewall rules or other infrastructure elements to enforce relevant policy elements

- Enable vulnerability alerts to be integrated with your security information and event management (SIEM) systems for further security incident handling

- Allow integration with patch management solutions

Table 10-6 is a checklist for providing vulnerability management and forensics for your end-to-end 5G deployment.

TABLE 10-6 Security Control Checklist for Vulnerability Management and Forensics

Recommended Security Controls	Does This Security Control/Capability Exist in Your Network? (Y/N)	Solution/Feature Used
Vulnerability management of CNFs deployed in multicloud stack		
Vulnerability management of VNFs and CNFs deployed on VMs		
Vulnerability management of VNFs and CNFs deployed on bare metal (BM)		
Forensics capability on VNFs and CNFs		
Automated capability of consolidating, de-duplicating, and correlating vulnerabilities		
Automatically identify and fix vulnerabilities in containers		
Full lifecycle vulnerability scanning (including the build, ship, and CI/CD pipeline and runtime)		

Enhanced Visibility, Monitoring, and Anomaly Detection

One of the primary security controls is end-to-end visibility. Without the visibility of the end-to-end components, it is very hard to detect any threats or any unusual behavior in the infrastructure. If you do not have the visibility of the entire network, it will create blind spots that can cause infiltration of the malicious traffic into your infrastructure and cause serious damage to the network or data exfiltration of sensitive data. Figure 10-10 illustrates the key interfaces that require visibility in your 5G deployment.

By ensuring visibility in the interfaces shown in Figure 10-10, you can continuously monitor and detect advanced threats that have either bypassed existing security controls or originate from within. Monitoring of these network devices and interfaces can be carried out by capturing network telemetry and analyzing the network packet metadata. Network packet metadata can provide useful insights about who is connecting to the organization and what they are up to. Everything touches the network, so these insights can extend from the HQ to the branch, public cloud, and private data centers, roaming users, and even IoT. Analyzing this data can help detect threats that might have found a way to bypass your existing controls, before they are able to have a major impact. Analyzing this data will also help you detect questionable behavior undertaken by hostile insiders. Properly functioning analytics can lessen the burden on your security team and provide them with more opportunity to concentrate on high-probability threats.

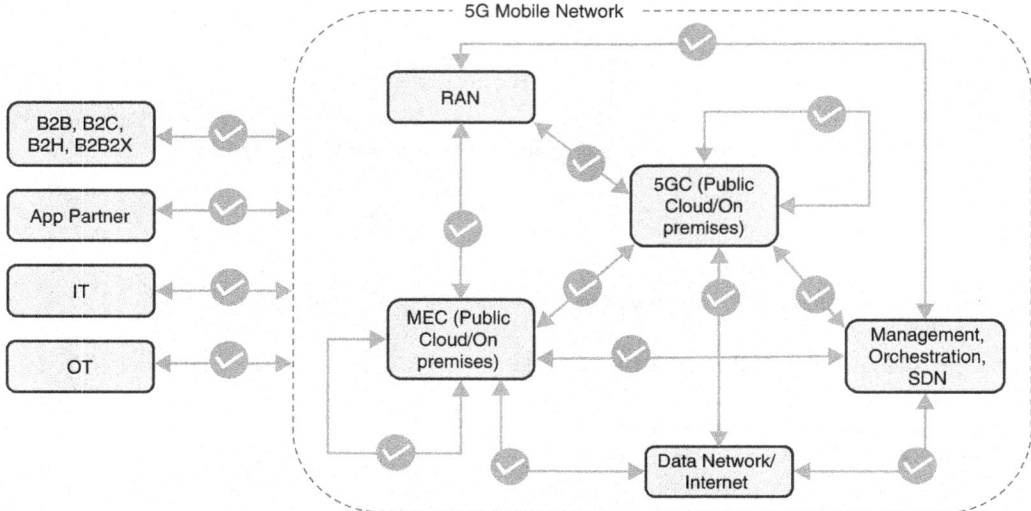

FIGURE 10-10   End-to-End Visibility of 5G Deployment

Figure 10-11 illustrates an example of continuous monitoring for the 4G and 5G packet core. The data and log collection includes flows from various devices, network equipment (both SDN and non-SDN related), and CNF applications used for 5G deployment. Apart from the telemetry from CNFs and network equipment, it is also important to get the system logs (syslog) and network security event logs (NSEL) from the firewalls, web application firewalls, and encrypted traffic analytics (ETA) to identify malware within encrypted traffic without having to decrypt the data traffic. All this collected data is then sent to the enhanced visibility and anomaly detection layer solution. The enhanced visibility and anomaly detection layer solution first creates the baseline of the network by understanding the management, service, and control plane behavior for 2 or 3 weeks, sometimes up to a month if your network is complex and includes multiple layers technologies, such as Wi-Fi, 2G, 3G, 4G, 5G NSA, and 5G SA deployments. Once the baseline is measured and identified, along with the policy defined by you, any variations of traffic against the baseline can be deemed an anomaly. This anomalous traffic is then inspected using ML and AI algorithms to ascertain the legitimacy of and to identify any malware. If any malware is identified, the impacted hosts can then be moved into a quarantine layer to prevent any lateral movement.

The orchestration layer uses a policy-driven approach to automate the network resource allocation to deliver a service or application. Orchestration allows networks to scale as needed, enables network services to be provisioned across multiple devices, and makes it possible to deploy resources as needed, thus making the network more agile and responsive. An API is the most common interface for interactions between the EPC and 5GC to the orchestration layer. A web application firewall or API gateway will provide you with enhanced API visibility and are two major inline security tools for API protection.

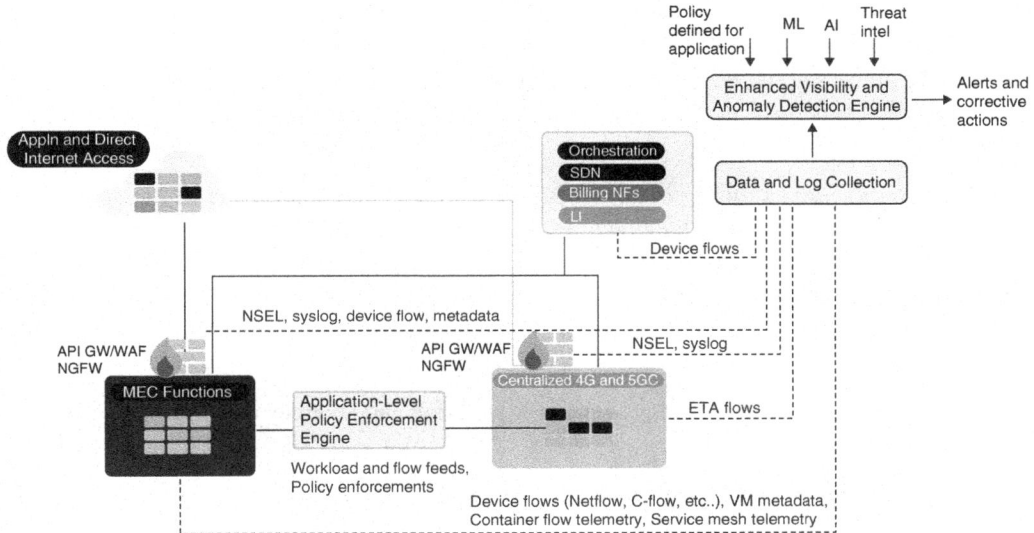

FIGURE 10-11 End-to-End Monitoring of 4G and 5G Packet Core

Figure 10-12 illustrates an example of end-to-end continuous monitoring for ORAN/VRAN deployment.

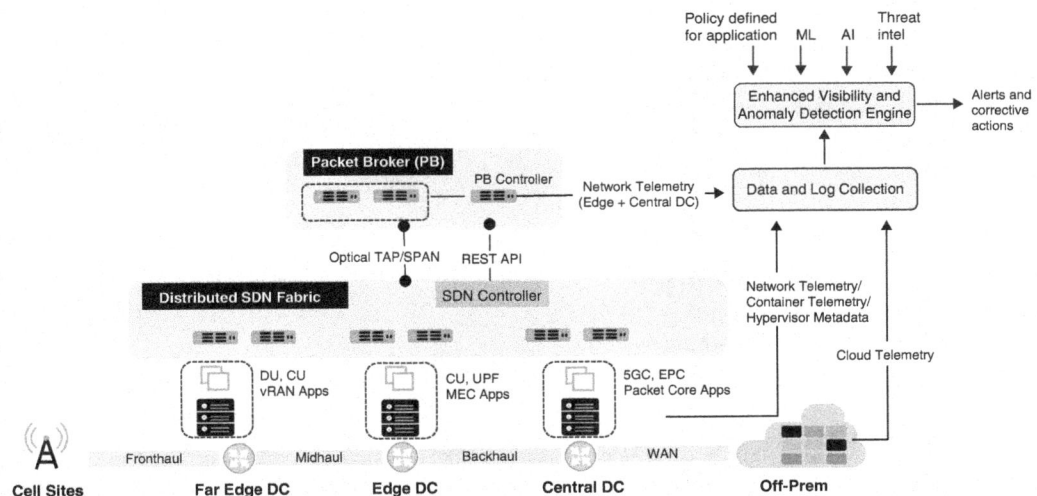

FIGURE 10-12 End-to-End Visibility of 5G ORAN Deployment

As shown in Figure 10-11, the monitoring layer should include both software-defined network (SDN), non-SDN, and legacy network devices. Having a packet broker in your monitoring architecture would highly optimize the flow collection. For example, suppose you need to monitor only the control, service, and management plane; in this case you can drop all the other traffic (for example, user plane) using the customizable filtering rules in the packet broker and then monitor only the control plane traffic. This will optimize the number of flows required to be sent to your SIEM layer, thereby reducing your cost, as the SIEM vendors generally have pricing models based on volume of data ingested by the SIEM solution.

It will also help you to respond quickly and effectively with complete knowledge of threat activity, network audit trails for forensic investigations, and integration with existing security controls. End-to-end monitoring can help you with your Governance, Risk, and Compliance (GRC) requirements, specifically it will help you to detect any regulatory and internal policy violations. You should create policy violation alarms tuned to your required business logic to ensure you capture the relevant anomalies.

Table 10-7 provides a checklist for ensuring granular monitoring and anomaly detection for your end-to-end 5G deployment.

TABLE 10-7 Security Control Checklist for Enhanced Visibility, Monitoring, and Anomaly Detection

Recommended Security Controls	Does This Security Control/Capability Exist in Your Network? (Y/N)	Solution/ Feature Used
Visibility in control plane for inter-VNF, CNF communications		
NFVi visibility		
Anomaly detection		
Detecting anomalies in encrypted traffic without decryption (including encrypted API communication)		
Encrypted traffic malware detection (without need for decryption)		
Detection of rogue/fake eNB, gNB, and MEC gateway		
Monitoring of cloud-deployed 5G CNFs		
Visibility in control plane for inter-VNF, CNF communications		
Storage of collected network telemetry, syslogs, and network security event logs from firewalls		

Slice-Level Security

The openness and flexibility of 5G architecture allows you to peel off certain parts of your network to provide dedicated services to industry verticals and mobile service consumers. It also allows you to integrate with an enterprise or industry vertical that owns and has deployed certain components of the 5G infrastructure. This kind of integration and level of abstraction also allows you to have multiple business models to monetize your 5G deployment.

Models like NSaaS enable network slice consumers (communication service consumers) to manage the parameters for their networks. Another option is for network slice providers (communication service providers) to manage the parameters for the CSC network, depending on the SLA between the CSP and CSC.

Deployment of slicing and the business models introduces a completely different risk vector that you never had to think about in a traditional network deployment. Some of the key risks are as follows:

- Improper management isolation in a multitenant environment, specifically for the management of the service layer in NSaaS use cases

- Improper slice-level isolation (both intra- and inter-slices)

- Improper slice-level authentication and authorization controls

- Improper implementation of APIs

- DDoS attacks at the data plane and control plane

To mitigate the aforementioned key risks, the following key security controls and deployment best practices should be implemented:

- Securing network slices requires you to take a holistic view toward security. You will first need to understand the use case the CSC is aiming to fulfill by using network slice components and then deciding on the security controls and capabilities for the architecture. A similar architecture consisting of RAN, back haul, MEC, and 5G packet core being deployed for different use cases will need different security controls based on the deployment model for the use cases.

    For example, consider a CSP providing parts of 5G network functions such as User Plane Functions (UPFs) and Control Plane Functions like AMF and SMF, while your customer (such as the industry vertical CSC) has deployed its own 5G radio access nodes (gNB) to cater for sensors in a critical infrastructure deployment. In another use case, you are providing your subscribers an eMBB slice to cater for their requirements of broadband usage. Here, although the same network components used are similar, apart from the gNB already deployed by CSC, the security controls are quite different. In the case of the industry vertical CSC, apart from the security considerations that you have planned for the subscriber's user equipment DDoS attacks, you will also need to cater for granular access control of industry vertical CSC users accessing the management in the service layer of the NSaaS.

- Ensuring granular access control on the management within the service layer for NSaaS is very important because any unauthorized access on the management within service layer or insufficient access control could lead to illegitimate users from one CSC having access to slice management of another CSC, thus leading to leakage of sensitive data or slice sabotage, where the NFs of the slices could be deleted or destroyed, leading to loss of service.

■ Another aspect to consider for secure network slicing is the security of the orchestration and SDN components assisting in the deployment and operation of the network slices. The orchestration and SDN layers use API calls in the majority of communications. These APIs are prone to weak security implementation by the developers and vulnerabilities in the API web servers. Best practices for deploying APIs, such as using HTTP over TLS, should be considered to provide secure connectivity. Granular access control should be implemented for the orchestration and SDN layers, as any unauthorized access to these layers could cause serious disruption to the network. An entire slice could be destroyed or deleted by the malicious unauthorized user causing a denial of service.

■ The SDN data plane is a part of the network slice that is prone to DDoS attacks from devices within the network infrastructure and outside the network. This can be mitigated by providing a DDoS detection and protection mechanism at the Internet-facing interfaces and implementing rate limiting at the application level. The SDN layer should also include Control Plane Policing methods to mitigate any attack at the control plane layer.

■ Depending on the use case and the location of deployment of 5G slice NFs, side-channel attacks could be a major concern. This is of particular importance for URLLC use cases, where MEC applications are deployed closer to the user. In industry vertical use cases such as smart manufacturing, these MEC applications could be deployed in an on-premises DC. To mitigate data exfiltration using side-channel attacks, enhanced visibility layer, including anomaly detection in encrypted traffic, should be implemented.

Table 10-8 provides a checklist for securing network slice deployments.

TABLE 10-8   Security Control Checklist for Slice Security

Recommended Security Controls	Does This Security Control/Capability Exist in Your Network? (Y/N)	Solution/Feature Used
Slice-level secondary authentication		
Slice-level isolation of CNF		
Slice-level visibility		
Secure API in orchestration and SDN layers		
DDoS protection for NSaaS offerings		
Slice-level Granular Access Control for devices and applications		

Secure Interoperability

Although this book is aimed at 5G, you should also consider pragmatic deployment models that include integrations with 4G. As an example, the roaming deployment in 5G considers Security Edge Protection Proxy (SEPP), which ensures end-to-end confidentiality and/or integrity between the source and destination network for all 5G interconnect roaming messages. In real deployments, however, there is still some time before we can have widely deployed 5G networks and thus 5G roaming partners. The roaming scenario will be primarily dominated by 2G/3G/4G for a few more years. This means that the GTP, Diameter, and SS7 firewalls should not be discounted and should be planned at the roaming peering points to mitigate risks related to roaming attacks prevalent in legacy technologies.

Table 10-9 provides a checklist for securing interoperability with roaming partners and other public networks.

TABLE 10-9 Security Control Checklist for Interoperability

Recommended Security Controls	Does This Security Control/ Capability Exist in Your Network? (Y/N)	Solution/Feature Used
GTP inspection (if you still have 4G network or roaming with 4G networks)		
SCTP inspection (if you still have 4G network or roaming with 4G networks)		
Diameter inspection (if you still have 4G network or roaming with 4G networks)		
SS7 inspection (if you still have 2G network or roaming with 2G networks)		
Security Edge Protection Proxy (SEPP)		
Service Communication Proxy (SCP)		

Summary

As discussed in this chapter, building a pragmatic security architecture requires you to understand the type of deployment model (on-premises, public cloud, or hybrid), the architecture being deployed (Distributed RAN, ORAN, and so on), and the use cases being implemented.

Furthermore, you also need to consider the legacy technologies already existing in your network. Table 10-10 explains the recommended security controls for multiple radio access technology (multi-RAT) deployments.

TABLE 10-10 Security Control Checklist for Securing Multi-RAT Deployments

Security Control	2G	2G + 3G	2G + 3G + 4G	2G + 3G + 4G + 5G
Centralized security gateway			✓	✓
Distributed security gateway				✓

Security Control	2G	2G + 3G	2G + 3G + 4G	2G + 3G + 4G + 5G
DIAMETER inspection			✓	✓
SCTP inspection			✓	✓
GTP inspection	✓	✓	✓	✓
Carrier-grade NAT	✓	✓	✓	✓
Web application firewall (WAF)				✓
DDoS			✓	✓
Gi-Firewall	✓	✓	✓	✓
SS7 security	✓	✓	✓	✓
Visibility on packet core (CP)	✓	✓	✓	✓
Multifactor Authentication	✓	✓	✓	✓
Network slice security				✓
DNS security endpoints				✓
Application security	✓	✓	✓	✓
API security	✓	✓	✓	✓
Vulnerability management	✓	✓	✓	✓
Management/DCN security	✓	✓	✓	✓
Microsegmentation				✓
Supply chain security	✓	✓	✓	✓

Acronyms Key

The following table expands the key acronyms used in this chapter.

Acronym	Expansion
3GPP	Third-Generation Partnership Project
5GC	5G Core
AAU	Active Antenna Unit
API	Application programming interface
BH	Back haul
BM	Bare metal
C2C	Control and command center
CD	Continuous development
CGNAT	Carrier-Grade Network Address Translation
CI	Continuous integration
CNF	Cloud-Native Function

Acronym	Expansion
CP	Control plane
CSC	Communication service customer
CSMF	Communication Service Management Function
CSP	Communication service provider
CU	Centralized Unit
DCN	Data control network
DDoS	Distributed denial of service
DN	Data network
DNS	Domain Name System
DoS	Denial of service
DU	Distributed Unit
eMBB	Enhanced Mobile Broadband
EPC	Evolved Packet Core
FH	Front haul
GTP	GPRS Tunneling Protocol
HTTP	Hypertext Transfer Protocol
HW	Hardware
IIoT	Industry Internet of Things
IoT	Internet of Things
IR	Incident response
IT	Information technology
M2M	Machine-to-machine
MBB	Mobile broadband
MEC	Multi-access edge compute
MH	Mid haul
MitM	Man in the middle
MNO	Mobile network operator
NAT	Network Address Translation
NF	Network Function
NFVi	Network Functions Virtualization Infrastructure
NOP	Network operator
NPN	Non-public network
NSaaS	Network slice as a service
NSC	Network slice customer
NSMF	Network Slice Management Function
NSP	Network slice provider
NSSMF	Network Slice Subnet Management Function

Acronym	Expansion
NW	Network
ORAN	Open Radio Access Network
OT	Operational technology
PKI	Public key infrastructure
PNI-NPN	Public Network Integrated Non-Public Network
RASP	Runtime Application Self-Protection
RAT	Radio access technology
RBAC	Role-based access control
RRU	Radio Resource Unit
SAST	Static application security testing
SCTP	Stream Control Transmission Protocol
SDN	Software-defined network
SecGW	Security gateway
SLA	Service level agreement
SLS	Service level specification
SOC	Security operations center
SSL	Secure Sockets Layer
TLS	Transport Layer Security
TN	Transport network
UDM	Unified data management
UE	User equipment
UP	User plane
UPF	User Plane Function
URLLC	Ultra-reliable low-latency communication
V2I	Vehicle-to-infrastructure
V2N	Vehicle-to-network
V2P	Vehicle-to-pedestrian
V2S	Vehicle-to-sensor
V2V	Vehicle-to-vehicle
V2X	Vehicle-to-everything
VM	Virtual machine
VNF	Virtual Network Function
VRAN	Virtual Radio Access Network
WAF	Web application firewall

References

Jason Longley and Pramod Nair, "Securing the 5G Core (5GC) and Evolved Packet Core (EPC) with Cisco Security," https://www.cisco.com/c/en/us/solutions/collateral/service-provider/service-provider-security-solutions/white-paper-c11-742166.pdf

Pramod Nair, "Securing 5G and Evolving Architectures," Cisco Knowledge Networks, Cisco Public, September 17, 2020

Pramod Nair, "Securing Your 5G Infrastructure to the Edge and Beyond," Cisco Public, April 1, 2021

3GPP TS 33.501. V17.1.0, "Security architecture and procedures for 5G System"

3GPP TS 28.530 V17.1.0, "Management and orchestration; Concepts, use cases and requirements"

3GPP TS 23.434 V17.1, "Service Enabler Architecture Layer for Verticals (SEAL)"

https://www.cisco.com/c/dam/en_us/solutions/iot/demystifying-5g-industrial-iot.pdf

5G Automotive Association, "C-V2X Use Cases and Service Level Requirements, Volume II," https://5gaa.org

5GAA TR A-200094, https://5gaa.org

https://www.ntia.gov/SBOM

https://www.portshift.io/

https://www.cisco.com/c/en/us/td/docs/solutions/Verticals/Solution_Briefs/IA_Networking_Solution_Brief.html

Chapter | **11**

Prioritizing 5G Security Investments

After reading this chapter, you should have a better understanding of the following topics:

- The methods for creating a prioritized list of security capabilities and controls based on use cases, deployment models, and network deployment stages
- Summary of critical security controls for 5G deployment and their capabilities
- Mapping of security controls and its impact on 5G deployments

While Chapter 10, "Building Pragmatic End-to-End 5G Security Architecture," discussed security controls and capabilities required to secure 5G deployments, this chapter will take you through the considerations for prioritizing investments in securing your 5G network.

This chapter will be of particular interest to the following individuals and teams from enterprise/5G service providers deploying 5GC and cybersecurity vendors planning product developments for 5G security use cases:

- Security leaders within government organizations who need to prioritize investments in security

- Management consultants advising governments and service providers on 5G security investments

- CSO and CTO teams from service providers

- CSO and CTO teams from enterprise teams deploying NPN/private 5G

- Cybersecurity vendor security architects looking to secure 5GC deployments for their customers

- Cybersecurity vendor product managers looking for use cases or features to enhance security products to cater for secure 5G deployments

Normally, investments in service provider domains like Radio Access Node (RAN), Mobile Packet Core, and the like are necessary to improve coverage and capacity, allowing better revenue generation. Unfortunately, in these investment scenarios, security is deprioritized because investments in security do not bring in revenue. Investments in security were usually limited to Internet-facing interfaces and consisted of deploying firewalls for inspection in roaming and interconnect as well as inspecting the user plane in the Gi interface and Carrier Grade Network Address Translation (CGNAT).

5G deployments will require a complete rethink of the way security is perceived due to the openness, disaggregation, and decomposition of the 3GPP and non-3GPP Network Functions (NFs). Furthermore, the use of open source software components to build 5G Cloud-Native Functions (CNFs) introduces higher risks, hence the need for required investments in securing your 5G deployments, be it for deploying a 5G public network or deploying private 5G/5G non-public networks (NPNs) by industry verticals.

This chapter discusses pragmatic 5G security investments, such as the bare-minimum security you need on Day 0 of a 5G deployment and then investing in security in a phase-by-phase manner and using a method to prioritize your security investments. The evolution of security capabilities and controls in your network, listed in Table 11-1, will depend on various criteria:

- What deployment model (on-premises, public cloud, or hybrid) will you implement?

- What are the use cases you plan to fulfill?

- What type of 5G network architecture (Distributed RAN, O-RAN, VRAN, C-RAN) will you choose?

- What is the existing security posture of your network?

Due to so many different variants and dependencies, it is almost impossible to have a single common priority list. The priority list of security investments provided in this chapter is based on the most commonly used deployment methods chosen by my customers. This priority list should help you create your own list, if needed.

Recommended practices dictate that you ensure that user and device security controls such as multi-factor authentication (MFA), endpoint security, and DNS-based security are deployed even before the 5G network NFs and equipment are installed and configured for any of the deployment models. This is to ensure that you have commensurate security controls in your network to identify any malicious activity. If you do not have robust user and device security mechanisms implemented before the 5G network elements are deployed, any amount of security investment will not secure your network, as the malicious software might have already migrated from the user device to your network intentionally or unintentionally.

TABLE 11-1   Summary of Critical 5G Security Controls and Their Capabilities

Security Controls	Primary Security Capabilities Catered for by the Security Control
Supply chain security	Hardware and software supply chain security Secure development lifecycle process Software bill of materials Network access control for users and devices
Granular user and device access control	MFA for employees, vendors, contractors, and sub-contractors Role-based access control (RBAC) for users Device profiling User and device segmentation Securing remote user access Network access control for users and devices Endpoint protection DNS security for users and devices Granular visibility and immediate quarantine
Secure interoperability	Secure inter-public land mobile network (PLMN) roaming Secure roaming between private 5G/NPN 5G networks and public service providers
Enhanced E2E visibility, monitoring, and anomaly detection	Visibility in control plane for inter–Virtual Network Functions (VNFs) and Cloud-Native Network Functions (CNFs) communications NFVi visibility Detecting anomalies in encrypted traffic without decryption, including encrypted API communication Anomaly detection in 5G network configurations Encrypted traffic malware detection (without decryption) Detection of rogue/fake eNB, gNB, MEC gateway Monitoring of cloud-deployed 5G CNFs Backward compatibility with 3G and 4G networks
Slice-level security	Inter-slice security Intra-slice security Slice-based secondary authentication Slice-level visibility Secure orchestration and SDN layers
Application-level security	Secure software delivery Secure 5G CNF images Secure registry management Application-level policy enforcement User-to-application mapping (including containers) Establishing and following DevSecOps methodology Runtime Application Self-Protection (RASP) Static application software testing (SAST) Automated CI/CD Runtime security Container security (image scanning and so on) Application dependency mapping Microsegmentation Vulnerability management

Security Controls	Primary Security Capabilities Catered for by the Security Control
	Compliance management Forensics Secure connectivity between CNFs API gateway
Access and aggregation layer security	IPsec MACsec with 802.1X
Anti-DDoS	Anti-DDoS (volumetric and application centric) Rate-limiting API requests using web application firewall
API Security	API authentication and authorization Encrypting the API using TLS/mTLS/IPsec Token-based authentication between 5G NFs Rate limiting Validation of verbose error messages
Infrastructure security controls	Security zoning Next-generation firewalling Anti-malware using sandboxing SSL decryption and inspection Rapid threat containment Application visibility and control
Vulnerability management and forensics	Vulnerability management of CNF deployed in multicloud stack Vulnerability management of VNFs and CNFs deployed on virtual machines (VMs) Vulnerability management of VNFs and CNFs deployed on bare metal (BM) Full lifecycle vulnerability scanning, including the build, ship, CI/CD pipeline, and runtime Forensics capability on VNFs and CNFs Automated capability of consolidating, de-duplicating, and correlating vulnerabilities Automatically identifying and fixing vulnerabilities in containers

Method of Prioritizing Security Controls

This section will discuss some concrete methods to prioritize the investments in security based on scenarios that consider different deployment methods and network types (public/non-public). The threat surfaces/risks and the priority of security controls to mitigate the risks are discussed using radar graphs. There are basically two key steps in determining the priority of the investments:

Step 1. In this step, we will determine the level of threat surfaces/risks using a radar graph for two scenarios—a public network (service provider) scenario and a non-public network/private 5G (mostly used in industry verticals) scenario. Some of the key points that increase the criticality of the threats are related to possible attacks leading to service disruptions and thus revenue loss. By filling in the radar graph, we will determine the impact of each threat for each scenario and the stages of the scenarios. Here are the threats we will consider:

- Supply chain vulnerabilities

- End-user device-based threats

- End-user-based threats

- Infrastructure user-based threats

- Infrastructure user-device-based threats

- API vulnerabilities

- MEC host vulnerabilities

- Public transport network vulnerabilities

- CNF vulnerabilities

- Threats from customer enterprise networks

- IoT device threats

- C&C-based attacks

- Data exfiltration

Step 2. In this step, we will determine and prioritize the security controls to mitigate the threat surface using a radar graph. By filling in the spider graph, we will determine the priority of each of the security controls to mitigate the threats/risks. Some of the key points that help to improve the priority are related to multiple risk mitigation capabilities, such as the visibility and monitoring layer, which will help you tackle multiple threats with a single security control. Here are the key security controls and threat mitigation techniques detailed in Table 11-1:

- Supply chain security

- Granular user access controls

- Secure interoperability

- Application-level security

- Vulnerability management and forensics

- Slice-level security

- Enhanced visibility and monitoring

- Secure access and aggregation

- Anti-DDoS

- API security

- Infrastructure security

Radar graphs are used throughout this chapter to provide you with a visual representation of the impacts of the threats and security controls. Figure 11-1 illustrates the split of the radar graph into varying levels of impact, from low to critical impact.

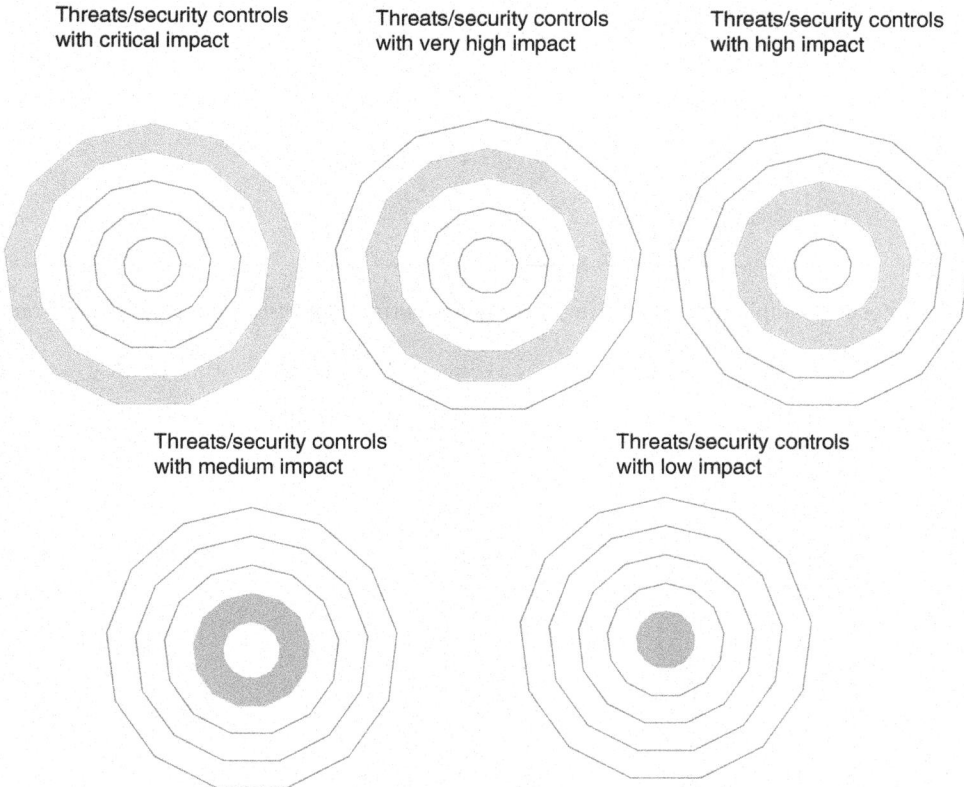

Threats/security controls with critical impact

Threats/security controls with very high impact

Threats/security controls with high impact

Threats/security controls with medium impact

Threats/security controls with low impact

FIGURE 11-1 Key Split of Radar Graph into Varying Levels of Threats and Security Controls

Table 11-2 shows an example of the mapping of threats/security controls and their impacts.

TABLE 11-2 Mapping Threat/Security Controls and Impact

Threats/Security Control	Critical	Very High	High	Medium	Low
A	✓				
B	✓				
C		✓			
D	✓				
E				✓	

Threats/Security Control	Critical	Very High	High	Medium	Low
F		✓			
G			✓		
H	✓				
I					✓
J	✓				
K			✓		
L				✓	
M	✓				

Figure 11-2 illustrates the sample radar plot using the threat/security-control-to-impact mapping from Table 11-2.

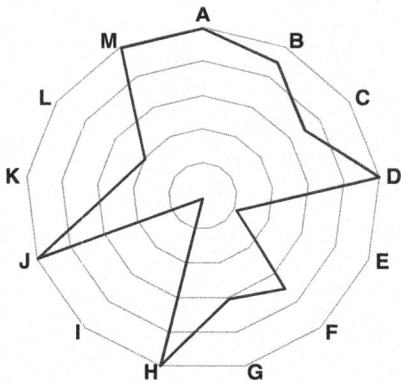

FIGURE 11-2 Sample Radar Graph Plot Using Values from Table 11-2

Figure 11-2 is an example of how the impact of the threats/security controls is plotted on the radar graph, giving you a visual representation to compare different deployment scenarios, such as public/non-public networks, along with the stages of deployment, such as initial/completion stage and so on.

Figure 11-3 illustrates how to interpret the graph.

Figure 11-3 shows you the method used to interpret the radar graphs used in this chapter. It enables you to prioritize mitigating the most critical threats and prioritize investments on the right security controls. As an example, in Figure 11-3, the radar graph under the heading "Critical" has A, B, D, H, J, and M circled to indicate the specific threats/security controls that are critical. Note that the graph for A, B, D, H, J, and M is plotted within the "critical" impact split of the radar graph. Similarly the rest are plotted within respective splits in the radar graph, such as C&F under Very High, K&G under High, L&E under Medium and I under Low.

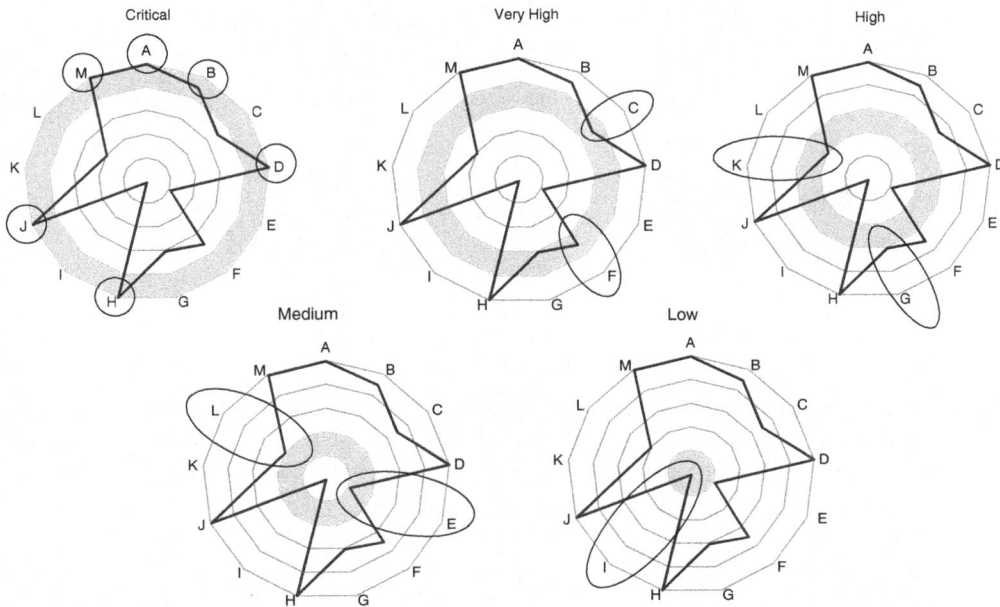

FIGURE 11-3   Interpreting the Graphs Used in This Chapter

The same graphical illustration is used to indicate the priority of the security controls to mitigate the threats and risks in the following sections of this chapter.

Let's look at some of the methods to prioritize the security controls based on various 5G network deployments.

Scenario 1

In this scenario, we will look at a service provider with a hybrid deployment model planning to implement three use cases: eMBB/VoNR, NSaaS, and IoT for consumers and enterprise customers.

The details of Scenario 1 are as follows:

- **Network owner:** Service provider
- **Deployment model:** Hybrid (on-premises and public cloud)
- **5G architecture:** Distributed RAN and Open RAN option 7-2x in some clusters
- **Primary use cases:**
    - **Stage 1:** Enhanced mobile broadband (eMBB) and Voice over New Radio (VoNR) use cases for mobile subscribers

■ **Stage 2:** Network slice as a service (NSaaS) for enterprise customers and industry verticals

■ **Stage 3:** IoT use cases for consumers and enterprise customers

The following steps will help you identify the priorities in investment toward the right security controls:

Step 1.   Identify the threats/threat modeling.

Step 2.   Prioritize the key security controls to mitigate the threats.

The sections that follow discuss Steps 1 and 2 in detail.

Step 1: Identifying the Threats/Threat Modeling

This section will take you through the threat modeling/identification of the key threats for the afore-mentioned three primary use cases, starting with the initial staging, implementation, and configuration onboarding. It then moves to Stage 1, which consists of voice and enhanced mobile broadband use cases (the primary use case many service providers are planning to deploy), followed by network slicing and network slice as a service (NSaaS) in Stage 2, and then moving on to the advanced stage of providing customized IoT use cases for consumers and enterprise customers.

Before we start with the specific use cases, let's have a look at the security controls required during the initial staging and onboarding of the network equipment and applications, which need to be imple-mented before the configuration and activation of the use cases.

Threats and Risks During Initial Staging and Onboarding

Figure 11-4 illustrates the key threat surfaces and risks during initial deployment.

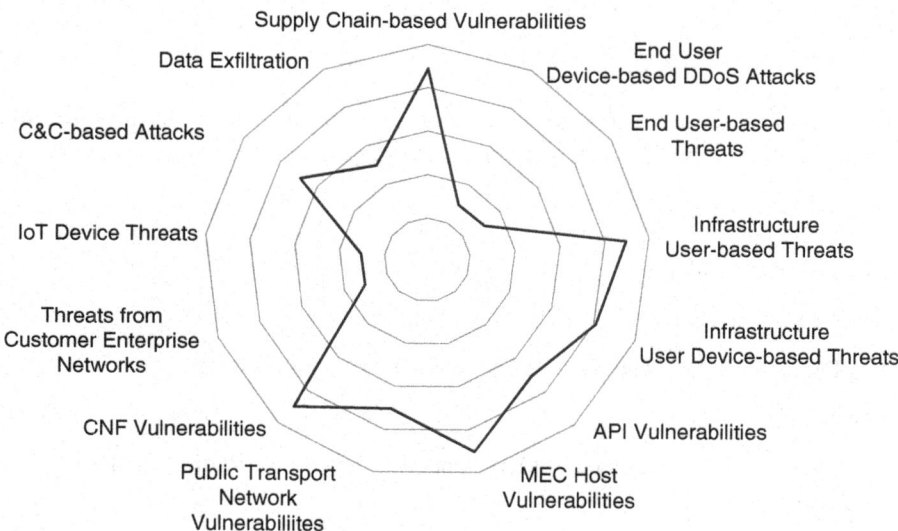

FIGURE 11-4   Key Threat Surfaces and Risks During Initial Deployment Phases

Figure 11-4 illustrates the key threat surfaces and risks during initial staging, onboarding, and implementation of your 5G network, even before any of the use cases are implemented for Scenario 1. The following areas are discussed:

- Hardware and software supply chain

- Risks from devices used by various vendors, subcontractors, and other users of your infrastructure to configure the 5G network

- Risks from intentional and unintentional improper security configurations while configuring various physical and virtual components of the 5G network by users

- Centralized data center and MEC host hardware and software vulnerabilities

- 5G Cloud-Native Function (CNF) image vulnerabilities

Threats and Risks During eMBB and VoNR Use Case Deployment

Figure 11-5 illustrates the key threat surfaces and risks when you implement the eMBB and VoNR use case.

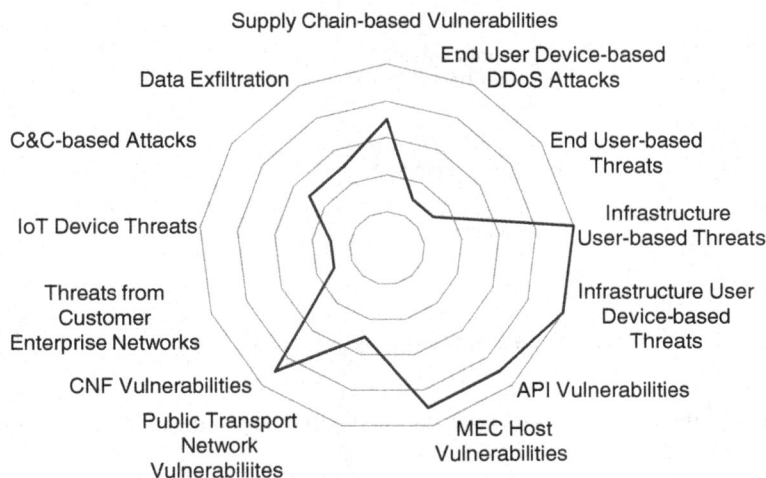

FIGURE 11-5   Key Threat Surfaces and Risks While Deploying eMBB and VoNR

The following threats and risks are critical when you deploy the eMBB and VoNR use case:

- Hardware and software supply chain.

- Risks from devices used by various vendors, subcontractors, and other users of your infrastructure to configure the 5G network.

- Risks from intentional and unintentional improper security configurations while configuring various physical and virtual components of the 5G network by users.

- Risks due to improper configuration of the 3GPP and non-3GPP 5G CNFs for realization of eMBB use cases. The image download of 5G CNFs might be prone to inherent vulnerabilities due to improper checks being carried out by the vendor before uploading it to the container registry.

- Centralized data center and MEC host hardware and software vulnerabilities.

- Risks due to improper API implementation and vulnerabilities of the API web server.

Threats and Risks While You Implement the NSaaS Offering

Figure 11-6 illustrates the key threat surfaces and risks when you implement the NSaaS use case.

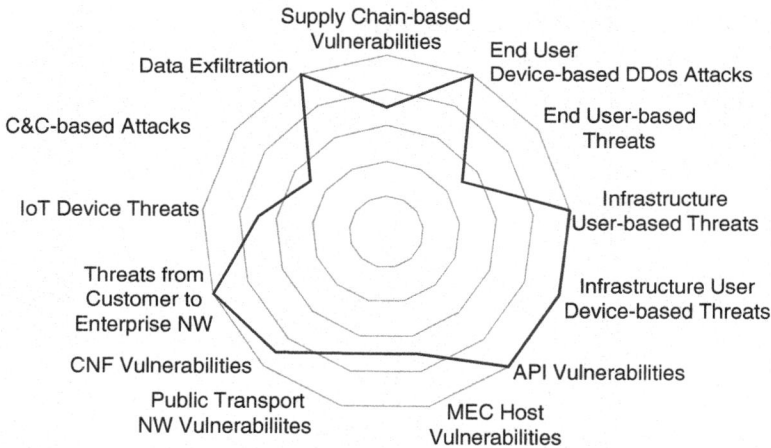

FIGURE 11-6   Key Threat Surfaces and Risks While Deploying the NSaaS Offering

As illustrated in Figure 11-4, the key threat surfaces and risks during deployment of the NSaaS offering are as follows:

- Hardware and software supply chain.

- Risks from intentional and unintentional improper security configurations while configuring various physical and virtual components of the 5G network for NSaaS deployments.

- The image of the downloaded 5G CNFs might be prone to inherent vulnerabilities due to improper checks being carried out by the vendor before uploading it to the container registry.

- Improper API implementation in the management of the service layer in NSaaS deployments.

- DDoS and DoS threats from the devices of the other slices in the NSaaS deployment due to insufficient isolation between slice NFs.

- Improper slice-level authentication for users and devices between NSaaS slices.

Threats and Risks While You Implement IoT Use Cases

Figure 11-7 illustrates the key threat surfaces and risks for when you implement IoT use cases.

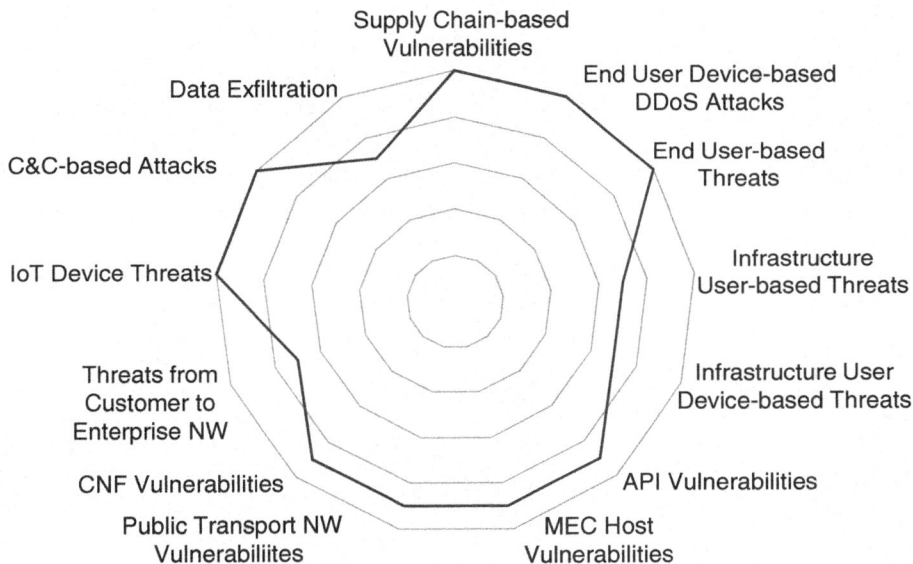

FIGURE 11-7   Key Threat Surfaces and Risks While Deploying IoT

The key threat surfaces and risks during deployment of the massive IoT use case are as follows:

- Hardware and software supply chain

- Risks from command and control (C&C) centers taking over the deployed IoT devices by exploiting their weak security

- Improper API implementation in the management of the IoT devices

- DDoS threats from the IoT devices

- Improper slice-level authentications for users and devices between slices (eMBB, IoT, and other NSaaS slices)

Once we understand the threat model, key threats, and risks for deploying the use cases and creating new offerings, the investments in securing the 5G deployment can be prioritized as discussed in the following sections.

Step 2: Prioritize the Key Security Controls to Mitigate the Threats

Step 1 identified the key threats in each of the three primary use cases. This section will determine the key security controls required to mitigate the threats identified in Step 1.

This section starts by determining the key security controls you require for the initial staging and onboarding stages and then determines the key security controls for each of the use cases related to eMBB, NSaaS, and IoT deployment. The section also includes the priority list of security controls, if you plan to deploy eMBB, VoNR, NSaaS, and IoT all at once in different clusters of your network.

Priority of Security Controls to Secure Initial Staging and Onboarding

Figure 11-8 illustrates the security controls for securing the staging and onboarding phases of 5G deployment.

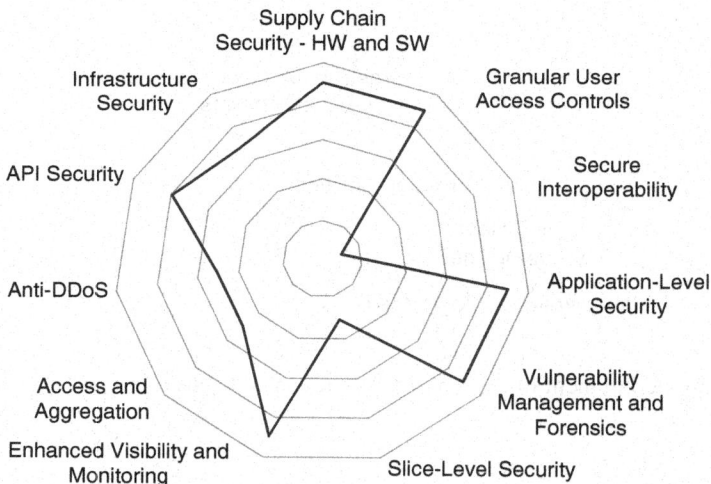

FIGURE 11-8   Priority of Security Controls to Secure Initial Staging and Onboarding

The key threat security controls for secure deployment during initial staging and onboarding are as follows:

- Ensure you follow a robust process for supply chain security, including implementation of Secure Zero-Touch Provisioning (SZTP) procedures for the initial download of software in the access switches and routers.

- Granular access control for contractors, employees, part-time employees, and vendors during the initial configuration of applications and network equipment.

- Application security, including securing inter-NF communications and applying microsegmentation.

- Ensure API communications are secured and have the information of the TLS versions being used in the 5G network. TLS 1.3 is recommended because it is more robust than TLS 1.2.

- Enhanced visibility, monitoring, and anomaly detection on the management and control plane, including encrypted traffic.

- Create network security zones and anti-malware security controls within your data center and network infrastructure.

Priority of Security Controls to Secure the Deployment of eMBB and VoNR Use Cases

Figure 11-9 illustrates the priority security controls for secure deployment of eMBB and VoNR use cases.

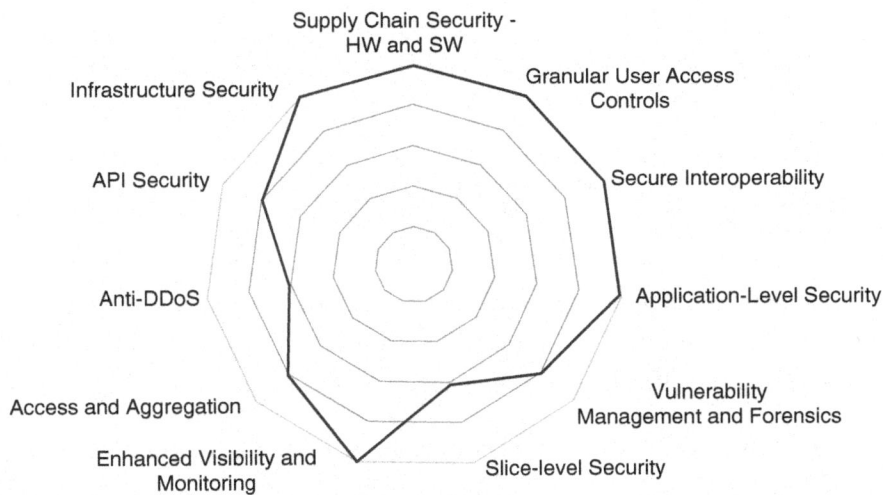

FIGURE 11-9   Priority of Security Controls to Secure eMBB and VoNR Use Cases

The key threat security controls for secure deployment of eMBB and VoNR use cases are as follows:

- Ensure you follow a robust process for supply chain security, including hardware and software.

- Granular access control to prevent unauthorized user access.

- Enhanced visibility, monitoring, and anomaly detection on the management and control plane, including encrypted traffic.

- Application security, including securing inter-NF communications and applying microsegmentation.

- Create network security zones and anti-malware security controls within your data center and network infrastructure.

- Secure xHaul and back-haul transport using IPsec.

Priority of Security Controls to Secure NSaaS Offering

Figure 11-10 illustrates the priority of security controls to secure the NSaaS offering.

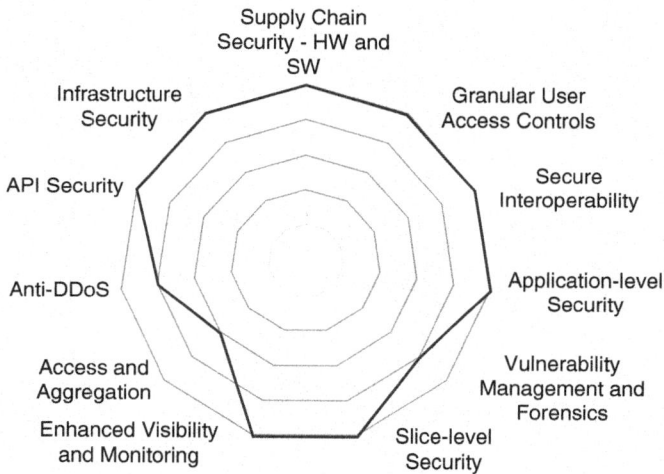

FIGURE 11-10   Priority of Security Controls to Secure NSaaS Offerings

The key security controls for a secure offering of NSaaS use cases are as follows:

- Ensure you follow a robust process for supply chain security, including hardware and software.

- Ensure granular access control in the management of the service layer in NSaaS to prevent unauthorized user access between communication service customers (CSCs).

- Implement enhanced visibility, monitoring, and anomaly detection on the management and control plane, including encrypted traffic.

- Secure interoperability between the NPN and public service providers and between the service provider and roaming partners.

- Implement application security, including securing inter-NF communication and applying microsegmentation.

- Apply slice-level secondary authentication mechanisms.

■ Create network security zones and anti-malware security controls within your data center and network infrastructure.

■ Ensure secure API communication using API gateways or web application firewalls along with rate limiting. API security should also include granular authentication and authorization.

Priority of Security Controls to Secure Deployment of IoT Use Cases

Figure 11-11 illustrates the priority of security controls to secure the IoT use cases.

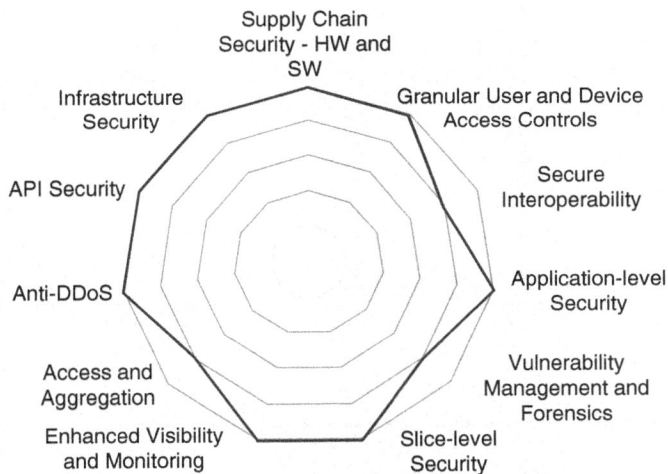

FIGURE 11-11 Priority of Security Controls to Secure IoT Deployments

The key threat security controls for secure IoT use cases are as follows:

■ Ensure you follow robust process for supply chain security, including hardware and software of IoT devices and IoT network gateways.

■ Implement granular access control to prevent unauthorized user access.

■ Implement granular slice-level device authentication.

■ Implement enhanced visibility, monitoring, and anomaly detection on the management and control plane, including encrypted traffic of the IoT network slice.

■ Implement application security, including securing inter-NF communications and applying microsegmentation and application-level security controls.

■ Create network security zones and anti-malware security controls within your data center and network infrastructure.

■ Implement DDoS protection against the IoT devices.

Priority of Security Controls Catering for All Three Use Cases

If you plan to deploy all the three use cases together in the same cluster or different parts of the network, the following process will help you determine the key security controls required to ensure secure rollout of all three use cases—eMBB and VoNR, NSaaS, and IoT—together in your network.

Table 11-3 indicates the priority of the security controls for each of the use cases and then helps you determine the priority of the security controls when all three use cases are deployed together. This table also allows you to understand the highest priority of each of the use cases and then select that priority for the respective security control.

As an example, looking at the Anti-DDoS entry in Table 11-3, you can see it's a "critical" security control for IoT use case and is a "high" and "very high" security control for eMBB/VoNR and NSaaS use cases, respectively. If you plan to deploy all three use cases, the highest security priority for the use case is selected. In this case, the anti-DDoS security control would be a "critical" security control.

If you are interested in only deploying eMBB/VoNR and NSaaS use cases, the priority of the anti-DDoS security control would be "very high" instead of "critical," as "very high" is the highest priority when deploying the eMBB/VoNR and NSaaS use cases.

TABLE 11-3   Security Control Priorities for All Use Cases in Scenario 1

Security Control	eMBB and VoNR	NSaaS	IoT	All Three Use Cases Deployed Together
	Priority/Impact	Priority/Impact	Priority/Impact	Priority/Impact
Supply Chain Security – HW and SW	Critical	Critical	Critical	Critical
Granular User Access Controls	Critical	Critical	Critical	Critical
Secure Interoperability	Critical	Critical	High	Critical
Application-Level Security	Critical	Critical	Critical	Critical
Vulnerability Management and Forensics	Very High	Very High	Very High	Very High
Slice-Level Security	High	Critical	Critical	Critical
Enhanced Visibility and Monitoring	Critical	Critical	Critical	Critical
Access and Aggregation	High	Very High	Very High	Very High
Anti-DDoS	High	Very High	Critical	Critical
API Security	Very High	Critical	Critical	Critical
Infrastructure Security	Critical	Critical	Critical	Critical

Figure 11-12 illustrates the priority of security controls to secure the eMBB, VoNR, NSaaS, and IoT use cases.

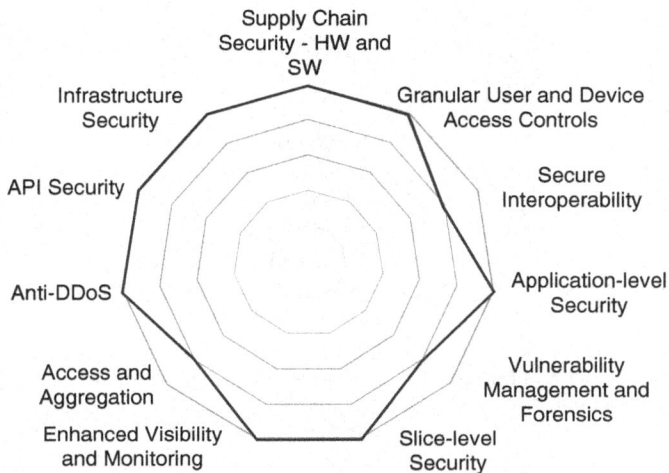

FIGURE 11-12   Priority of Security Controls to Secure Deployments of All Three Use Cases

The key threat security controls for securing all three use cases are as follows:

- Ensure you follow a robust process for supply chain security, including hardware and software.

- Implement granular access control to prevent unauthorized user access.

- Ensure enhanced visibility, monitoring, and anomaly detection on the management and control plane, including encrypted traffic.

- Implement application security, including securing inter-NF communications and applying microsegmentation.

- Secure API communication using API gateways or web application firewalls along with rate limiting. API security should also include granular authentication and authorization. Also, ensure secure API communication between the orchestration and SDN layers.

- Create network security zones and anti-malware security controls within your data center and network infrastructure.

- Implement DDoS protection against the IoT devices.

Summary of Priority Security Controls for Scenario 1

Table 11-4 illustrates the priority of security controls to secure Scenario 1. The Notes column in Table 11-4 explains the reasoning and logic for determining the priority for the security controls catering for all three use cases. The priority will change, as mentioned in the beginning of the section "Priority of Security Controls for All Three Use Cases," based on the number of use cases you plan to deploy in your network.

TABLE 11-4   Summary of Investment Priority for Scenario 1

Security Control	Priority of Investment	Mapping Investment Priority to Deployment Stages	Notes
Supply Chain Security – HW and SW	Critical	Required from initial deployment and staging of equipment and CNFs	Mandatory to be verified and validated during initial deployment and throughout the operation of live networks.
Granular User Access Controls	Critical	Required from initial deployment and staging of equipment and CNFs	Mandatory during initial deployment and throughout the operation of live networks.
Secure Interoperability	Critical	Required before roaming is implemented and for NPN interconnect with public networks	Mandatory once the network is live for roaming use cases and interconnect between NPN and public networks as well as for NSaaS deployments. Not required during initial network implementation stage.
Application-Level Security	Critical	Required from initial deployment and staging of equipment and CNFs	Mandatory to be verified and validated during initial deployment and throughout the operation of live networks.
Vulnerability Management and Forensics	Very High	Required during initial stages but less of a priority than access control, supply chain monitoring, and API security	Mandatory to be deployed for live networks.
Slice-Level Security	Critical	Required before network slicing is implemented	Mandatory to be deployed when slicing is implemented in your network. If your network doesn't use slicing in the initial years of deployment, you can plan it closer to deployment of network slicing.
Enhanced Visibility and Monitoring	Critical	Required from initial deployment and staging of equipment and CNFs	Mandatory during initial deployment and throughout the operation of live networks.
Access and Aggregation	Very High	Required if public network is used in back haul/xHaul and if country regulators have advised using it as a mandatory security control	Mandatory if you are using a public network for back-hauling. If you are using your own back-haul transport network, you can implement these security controls when you have integration points with third-party back-haul transport providers. Some countries have a mandatory requirement of using a security gateway in the back haul/xHaul defined by the country's regulator.

Security Control	Priority of Investment	Mapping Investment Priority to Deployment Stages	Notes
Anti-DDoS	Critical	Required once the network is live	Mandatory once the network is live, primarily in Internet-facing interfaces. Not required during initial network implementation stage.
API Security	Critical	Required from initial deployment and staging of equipment and CNFs	Mandatory during initial deployment and throughout the operation of live networks.
Infrastructure Security	Critical	Required from initial deployment and for providing segmentation of control plane and user plane traffic	Mandatory during initial deployment and throughout the operation of live networks.

Scenario 2

In the previous section, we took the service provider deployment scenario and mainly focused on three use cases and determined the priority of the security controls based on the use cases you plan to deploy.

This section will take you through Scenario 2, which is focused on different stages of non-public network (NPN) deployment, beginning from standalone NPN (SNPN) deployment and then moving toward public network integrated NPN (PNI-NPN). This section also identifies the priority of the security control if you have both SNPN and PNI-NPN deployments in your network. This could be the case if you decide to have part of your industry vertical network air-gapped and choose the SNPN model and decide to have the other part to be more open and integrated to a public network.

By using the radar graph, you can further identify the key threats and determine the key security controls, ensuring that you have the security agility to secure your network from the changing threat landscape.

Here are the details of Scenario 2:

- **Network owner:** Enterprise/industry vertical

- **Industry Vertical type:** Smart manufacturing

- **Deployment model:** Hybrid (on-premises and public cloud)

- **5G Architecture:** Distributed RAN

- **Primary use cases deployed in stages**:

  - **Stage 1:** Standalone non-public network (SNPN) – IoT/IIoT use case for URLLC use case

  - **Stage 2:** PNI-NPN integration with a communication service provider (CSP)

Details of deployment in Stage 1:

- 5G control plane (CP) and user plane (UP) CNFs and equipment deployed within on-premises/private DC of smart manufacturing

- IoT and IIoT devices and sensors to be deployed and integrated with the 5G infrastructure

Details of deployment in Stage 2:

- Integration with public network

- Integration with communication service provider (CSP)/public service provider

The following steps will help you identify the priorities in investment toward the right security controls:

Step 1. Identify the threats/threat modeling

Step 2. Prioritize the key security controls to mitigate the threats

The sections that follow discuss Steps 1 and 2 in detail.

Step 1: Identifying the Threats/Threat Modeling

This section will take you through the threat modeling/identification for Stage 1, starting with the initial staging, implementation, and configuration onboarding. It then moves to Stage 1, which consists of voice and enhanced mobile broadband use cases, which is the primary use case that many of the service providers are planning to deploy, followed by NSaaS in Stage 2, and then moving to the advanced stage of providing customized IoT use cases for consumers and enterprise customers.

Threats and Risks During Initial Staging and Onboarding

Figure 11-13 illustrates the threat surfaces and risks during the initial deployment before Stage 1 goes live.

Here are the key threat surfaces and risks during initial staging, onboarding, and implementation of your NPN 5G network even before any of the use cases are implemented for Scenario 2:

- Hardware and software supply chain, including hardware and software for 5G 3GPP and non-3GPP components

- Risks due to improper segmentation between your IT and OT/IIoT network, leading to threats migrating from IT to the OT and IIoT environment

- Risks from devices used by various vendors, subcontractors, and other users of the infrastructure to configure your NPN 5G network

- Risks from intentional and unintentional improper security configurations while configuring various physical and virtual components of the 5G network by users

- 5G Cloud-Native Function (CNF) image vulnerabilities for 3GPP and non-3GPP NFs
- Centralized data center and MEC host hardware and software vulnerabilities

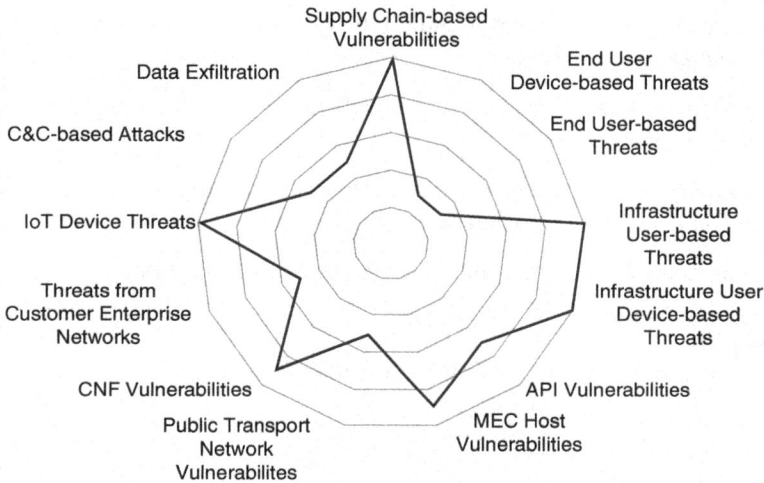

FIGURE 11-13 Key Risks and Threat Surfaces During Initial Deployment Phases

Threats and Risks for Stage 1

Figure 11-14 illustrates the threat surfaces and risks for Stage 1.

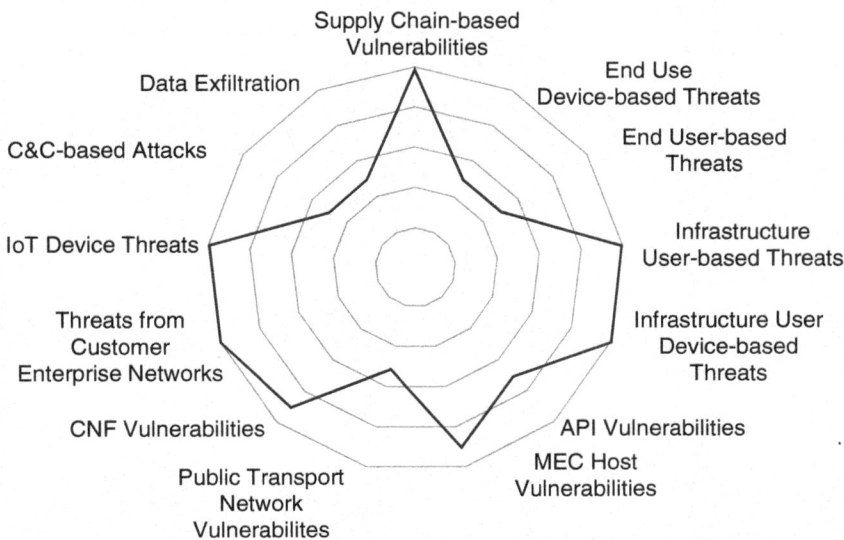

FIGURE 11-14 Key Risks and Threat Surfaces During Stage 1

Here are the key threat surfaces and risks that are critical during deployment of Stage 1:

- Hardware and software supply chain, including hardware and software for 5G 3GPP and non-3GPP components

- Risks due to improper segmentation between your IT and OT/IIoT network, leading to threats migrating from IT to the OT and IIoT environment

- Risks from devices used by various vendors, subcontractors, and other users of the infrastructure to configure your NPN 5G network

- Risks from intentional and unintentional improper security configurations while configuring various physical and virtual components of the 5G network by users

- Risks due to improper implementation of the API and vulnerabilities within the API itself

- 5G CNF image vulnerabilities for 3GPP and non-3GPP NFs

- Centralized data center and MEC host hardware and software vulnerabilities

Threats and Risks for Stage 2

Implementing Stage 2/PNI-NPN will depend on the architecture and use case you plan to implement. Depending on use case, you might not deploy Stage 1/SNPN and might implement only Stage 2/PNI-NPN. In some use cases, you might deploy only SNPN or Stage 1 only. Figure 11-15 illustrates the threat surfaces and risks for Stage 2.

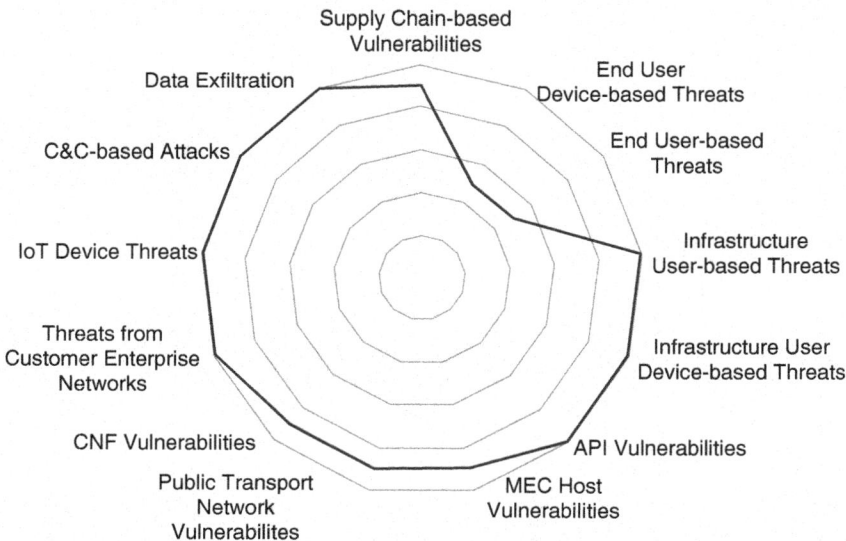

FIGURE 11-15 Key Risks and Threat Surfaces During Stage 2

Here are the key threat surfaces and risks during deployment of Stage 2:

■ Hardware and software supply chain, including hardware and software for 5G 3GPP and non-3GPP components

■ Data exfiltration due to integration with external networks such as public service provider and data networks

■ C&C attacks by exploiting vulnerabilities in an IoT/IIoT device within the smart manufacturing unit

■ Man-in-the-middle (MitM) attacks due to exposure of the NPN network to the public network

■ Risks due to improper segmentation between your IT and OT/IIoT network, leading to threats migrating from IT to the OT and IIoT environment

■ Risks from devices used by various vendors, subcontractors, and other users of the infrastructure to configure your NPN 5G network

■ Risks from intentional and unintentional improper security configurations while configuring various physical and virtual components of the 5G network by users

■ 5G CNF image vulnerabilities for 3GPP and non-3GPP NFs

■ Centralized data center and MEC host hardware and software vulnerabilities

Step 1 so far has identified the key threats and the impacts for each of the threats in Stage 1 and Stage 2. The next step (Step 2) is to determine the security controls required to mitigate the threats.

Step 2: Prioritize the Key Security Controls to Mitigate the Threats

Step 1 identified the key threats for Stage 1 and Stage 2 for NPN deployments. This section will determine the key security controls required to mitigate the threats identified in Step 1.

This section will start by determining the key security controls you require for the initial staging and onboarding stages and then determine the key security controls for Stage 1 and Stage 2. The section also includes the priority list of security controls, if you plan to deploy both Stage 1 and Stage 2 together in your network.

Priority Security Controls During Staging and Onboarding

Figure 11-16 illustrates the priority of security controls to secure the initial staging, onboarding, and implementation phase before Stage 1 goes live.

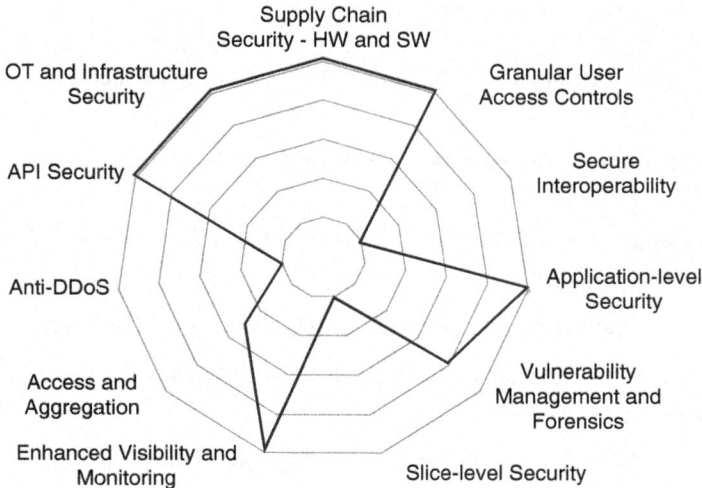

FIGURE 11-16   Priority of Security Controls to Secure Initial Deployment

Here are the key threat security controls for securing the initial staging, onboarding, and implementation phase before Stage 1 goes live:

- Ensure you follow a robust process for supply chain security, including hardware and software of IIoT/IoT devices and IIoT network gateways.

- Implement granular access control to prevent unauthorized access from users of your enterprise IT data center.

- Ensure proper segmentation and isolation between the IT and OT/IIoT network.

- Secure API communication by ensuring encrypted communication between the orchestration and SDN layers.

- Ensure enhanced visibility, monitoring, and anomaly detection on the management and control plane, including encrypted traffic of the IoT network slice.

- Implement application security, including securing inter-NF communications and applying microsegmentation and application-level security controls.

- Create network security zones and anti-malware security controls within your data center and network infrastructure.

Priority of Security Controls to Secure Stage 1

Figure 11-17 illustrates the priority of security controls to secure Stage 1.

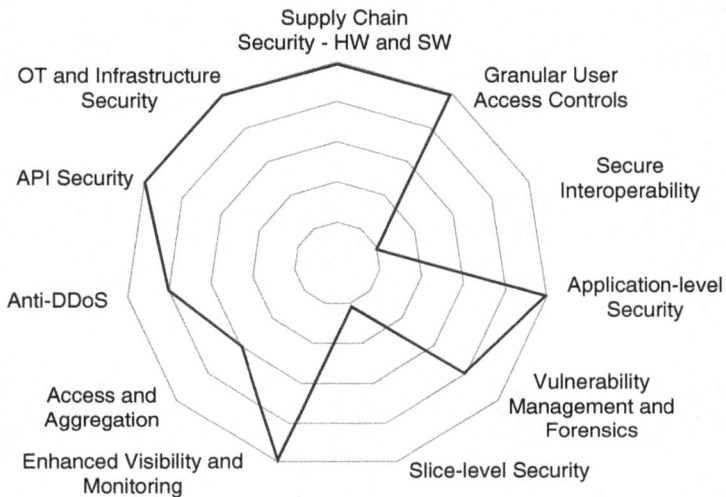

FIGURE 11-17 Priority of Security Controls to Secure Stage 1

The key threat security controls for securing Stage 1 are as follows:

- Ensure you follow a robust process for supply chain security, including hardware and software of IIoT/IoT devices and IIoT network gateways.

- Ensure protection against DDoS and data exfiltration attacks from IoT devices controlled by C&C servers.

- Implement granular access control to prevent unauthorized access from users of your enterprise IT data center.

- Ensure proper segmentation and isolation between the IT and OT/IIoT network.

- Secure API communication by ensuring encrypted communication between the orchestration and SDN layers.

- Ensure enhanced visibility, monitoring, and anomaly detection on the management and control plane, including encrypted traffic of the IoT network slice.

- Implement application security, including securing inter-NF communications and applying microsegmentation and application-level security controls.

- Create network security zones and anti-malware security controls within your data center and network infrastructure.

Priority of Security Controls to Secure Stage 2

Figure 11-18 illustrates the priority of security controls to secure Stage 2.

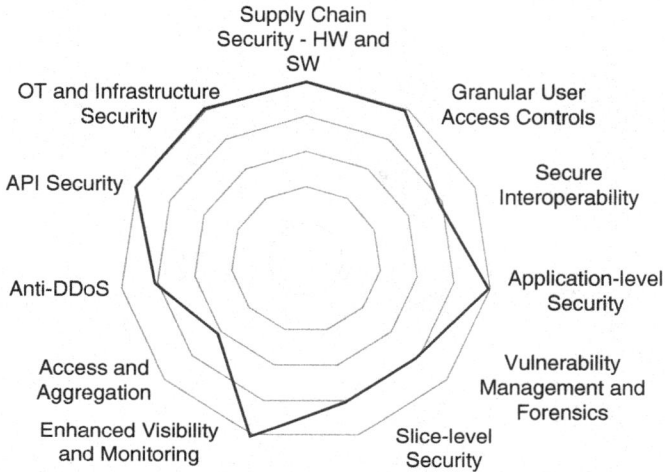

FIGURE 11-18   Priority of Security Controls to Secure Stage 2

The key threat security controls to secure Stage 2 are as follows:

- Ensure you follow a robust process for supply chain security, including hardware and software of IIoT/IoT devices and IIoT network gateways.

- Implement granular access control to prevent unauthorized user access from public service provider–integrated slices or NFs.

- Secure interconnect between your smart manufacturing facility and public service providers.

- Ensure secondary slice-level authentication if the integration with public networks requires implementation of slices.

- Implement protection against DDoS and data exfiltration attacks from IoT devices controlled by C&C servers.

- Ensure proper segmentation and isolation between the IT and OT/IIoT network.

- Secure API communication by ensuring encrypted communication between the orchestration and SDN layers.

- Ensure enhanced visibility, monitoring, and anomaly detection on the management and control plane, including encrypted traffic of the IoT network slice.

- Implement application security, including securing inter-NF communications and applying microsegmentation and application-level security controls.

- Create network security zones and anti-malware security controls within your DC and network infrastructure.

Priority of Security Controls to Secure All Stages

If you plan to deploy both stages together in the same cluster or different parts of the network, the following process will help you determine the key security controls required to ensure a secure rollout of both stages.

Table 11-5 indicates the priority of the security controls for each of the stages and then helps you determine the priority of the security control when both stages of Scenario 2 are used in your NPN network. This table also helps you to understand the highest priority of each of the stages and then select that priority for the respective security control.

As an example, looking at the Slice-Level Security entry in Table 11-5, you see it categorized as a "low" security control for Stage 1 and as a "very high" security control for Stage 2. If you plan to deploy both stages, the highest security priority for the use case should be selected. In this case, slice-level security would be a "very high" security control.

TABLE 11-5   Summary of Priority Security Controls for Scenario 2

	Stage 1	Stage 2	Both Stages Together
Security Control	Priority	Priority	Priority
Supply Chain Security – HW and SW	Critical	Critical	Critical
Granular User Access Controls	Critical	Critical	Critical
Secure Interoperability	Low	Very High	Very High
Application-Level Security	Critical	Critical	Critical
Vulnerability Management and Forensics	Very High	Very High	Very High
Slice-Level Security	Low	Very High	Very High
Enhanced Visibility and Monitoring	Critical	Critical	Critical
Access and Aggregation	High	High	High
Anti-DDoS	Very High	Very High	Very High
API Security	Critical	Critical	Critical
OT and Infrastructure Security	Critical	Critical	Critical

Figure 11-19 illustrates the priority of security controls to secure the network when both Stage 1 and Stage 2 are used in your 5G NPN deployment.

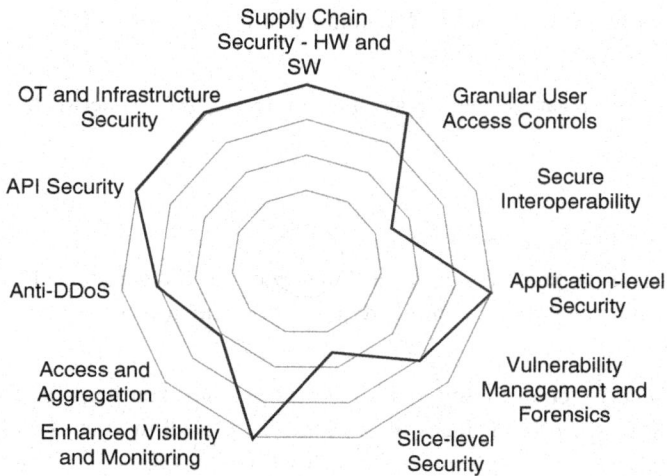

FIGURE 11-19   Priority of Security Controls to Secure Both Stages in Scenario 2

The key threat security controls for securing Stages 1 and 2 are as follows:

■ Ensure you follow a robust process for supply chain security, including hardware and software of IIoT/IoT devices and IIoT network gateways.

■ Implement granular access control to prevent unauthorized user access from public service provider–integrated slices or NFs.

■ Ensure protection against DDoS and data exfiltration attacks from IoT/IIoT devices controlled by C&C servers.

■ Implement proper segmentation and isolation between the IT and OT/IIoT network.

■ Implement API security catering for the Top 10 OWASP threats.

■ Implement enhanced visibility, monitoring, and anomaly detection on the management and control plane, including encrypted API traffic.

■ Implement application security, including securing inter-NF communications and applying microsegmentation and application-level security controls.

■ Create network security zones and anti-malware security controls within your data center and network infrastructure.

Summary of Priority Security Controls for Scenario 2

Table 11-6 illustrates the priority of security controls to secure Scenario 2.

TABLE 11-6 Investment Priority for Scenario 2

Security Control	Priority of Investment	Mapping Investment Priority to Deployment Stages	Notes
Supply Chain Security – HW and SW	Critical	Required from initial staging, onboarding, and implementation of Stage 1	Mandatory to be verified and validated during initial deployment and throughout the operation of live networks.
Granular User Access Controls	Critical	Required from initial staging, onboarding, and implementation of Stage 1 to secure access during staging of equipment and CNFs	Mandatory during the initial deployment and throughout the operation of live networks to mitigate unauthorized access from the IT network.
Secure Interoperability	Stage 1: Low Stage 2: Very High	Required in Stage 2	Mandatory once you move away from standalone NPN deployment and integrate with CSP, data network, and public cloud.
Application-Level Security	Critical	Required from initial staging, onboarding, and implementation of Stage 1 to ensure secure staging of equipment and CNFs	Mandatory to be included in Stage 1, including CI/CD processes by your development team.
Vulnerability Management and Forensics	Very High	Required from Stage 1 but less priority than Access control, Supply Chain, Monitoring, and API security	Mandatory to be included in Stage 1 to detect and mitigate any identified vulnerabilities.
Slice-Level Security	Very High	Required before Network Slicing is implemented	Mandatory to be deployed during Stage 2 when being integrated with CSP.
Enhanced Visibility and Monitoring	Critical	Required from initial staging, onboarding, and implementation of Stage 1	Mandatory during initial deployment and throughout the operation of live networks.
Access and Aggregation	Very High (network dependent)	Required if public network is used in backhaul/xHaul and if country regulators have advised using it as a mandatory security control	Mandatory if you are using a public network for back-hauling. If you are using your own back-haul transport network, you can implement these security controls when you have integration points with third-party back-haul transport providers. Some countries have a mandatory requirement of using a security gateway in the back haul/xHaul defined by the country's regulator.

Security Control	Priority of Investment	Mapping Investment Priority to Deployment Stages	Notes
Anti-DDoS	Very High	Required mainly after network goes live or closer to being live	Mandatory to mitigate DDoS attacks from IoT and IIoT devices connected to CSP, public cloud, or data networks.
API Security	Critical	Required from initial staging, onboarding, and implementation of Stage 1	Mandatory during initial deployment and throughout the operation of live networks.
OT and Infrastructure Security	Critical	Required from initial staging, onboarding, and implementation of Stage 1	Mandatory during initial deployment and throughout the operation of live networks.

Summary

As seen from the aforementioned scenarios, you can prioritize your investments based on the phases of 5G network deployment, the use cases planned to be deployed, and the preferred architecture. As compared to legacy technologies, 5G brings in the requirement to secure network functions at the application level, including user-to-application mapping, application dependency mapping, application-level vulnerability management, and securing the API communications. You should have the following critical security controls at a bare minimum, even before you start any deployment of 5G 3GPP and non-3GPP equipment and NFs. In other words, if you would like to prioritize your investment in securing 5G deployments, the following list should be at the top of your priority list for security:

- Controls related to supply chain security, such as image authentication and vulnerability detection

- Granular access control (users and devices)

- End-to-end visibility and monitoring, including anomaly detection and detection of malware in the encrypted layer (specifically for external API communications)

- Application-level security, including application-level policy enforcement, threat detection, and response

Acronyms Key

The following table expands the key acronyms used in this chapter.

Acronym	Expansion
3GPP	Third-Generation Partnership Project
5GC	5G Core
API	Application programming interface
AUSF	Authentication Server Function
C&C	Control and command center
CD	Continuous development
CI	Continuous integration
CNF	Cloud-Native Function
CSC	Communication service consumer
CSP	Communication service provider
DDoS	Distributed denial of service
DoS	Denial of service
eMBB	Enhanced mobile broadband
HW	Hardware
IoT	Internet of Things
M2M	Machine-to-machine
MBB	Mobile broadband
MEC	Multi-access edge compute
MitM	Man in the middle
NF	Network Function
NPN	Non-public network
NSaaS	Network slice as a service
NW	Network
OT	Operational technology
O-RAN	Open RAN
PNI-NPN	Public network integrated non-public network
RASP	Runtime Application Self-Protection
RAN	Radio access network
RAT	Radio access technology
SSL	Secure Sockets Layer
TLS	Transport Layer Security
UDM	Unified data management
UE	User equipment
UPF	User Plane Function

Acronym	Expansion
VNF	Virtual Network Function
VRAN	Virtual RAN
WAF	Web application firewall

References

Pramod Nair, "Securing Your 5G Infrastructure to the Edge and Beyond Cisco Public," April 1, 2021

Pramod Nair, "Securing 5G and Evolving Architectures," Cisco Knowledge Networks, Cisco Public, September 17, 2020

Jason Longley and Pramod Nair, "Securing the 5G Core (5GC) and Evolved Packet Core (EPC) with Cisco Security," https://www.cisco.com/c/en/us/solutions/collateral/service-provider/service-provider-security-solutions/white-paper-c11-742166.pdf

3GPP TS 33.501. V17.1.0, "Security architecture and procedures for 5G System."

https://www.cisco.com/c/dam/en_us/solutions/iot/demystifying-5g-industrial-iot.pdf

https://www.portshift.io/

https://www.cisco.com/c/en/us/td/docs/solutions/Verticals/Solution_Briefs/IA_Networking_Solution_Brief.html

Chapter | **12**

5G and Beyond

After reading this chapter, you should have a better understanding of the following topics:

- Adoption of 5G standalone technology
- Adaptation of 5G standalone technology on fulfilling new use cases
- Convergence of non-3GPP and 3GPP technologies
- Application of AI and ML in securing 5G and evolving technologies
- Importance of deploying crypto-agile mobile networks

The content in this chapter will continue to evolve, and I want to keep you up to date on developments. I will update the chapter occasionally with relevant new content and insights on the book's website at www.informit.com. To access the chapter, start by establishing a login at www.informit.com and register your book. To do so, simply go to www.informit.com/register and enter the ISBN of the print book: 9780137457939. Answer the challenge question as proof of purchase. Click the "Access Bonus Content" link in the "Registered Products" section of your account page to be taken to the page where your downloadable content is available.

This chapter takes you through the key areas for pragmatic adoption of 5G and evolving technologies, the architecture evolution to cater for 3GPP and non-3GPP convergence, the use of artificial intelligence (AI) and machine learning (ML) to enhance security posture, and the methods to ensure the crypto-agility of your network.

This chapter would be of particular interest to the following individuals and teams from enterprise/5G service providers deploying 5GC and cybersecurity vendors planning product developments for 5G security use cases:

- Security leaders within government organizations who need to prioritize investments in security
- Management consultants advising governments and service providers on 5G security investment

- CSO and CTO teams from service providers

- CSO and CTO teams from enterprise teams deploying non-public networks (NPN)/private 5G

- Cybersecurity vendor security architects looking to secure 5GC deployments for their customers

- Cybersecurity vendor product managers looking for use cases or features to enhance security products to cater for secure 5G deployments

5G Standalone (SA) is the first cellular technology that brings in true openness to 3GPP network deployments. There is industry-wide anticipation in the range of pragmatic use cases that can be enabled through 5G. But if we look at the history of telecommunications, the lifespan the 2G, 3G, 4G technologies is each around 10 years.

Various service providers have announced that 2G technology will be switched off to free the spectrum to deploy newer technologies. Many have already switched off 2G and have performed spectrum re-farming. Although 2G has been superseded by technologies such as 3G and 4G, it is still widely used across Europe, Asia, Africa, and Central and South America, specifically in rural areas and highways. Switching off 2G in many countries will leave the population in rural areas vulnerable and without any coverage. There are also many point-of-sale (PoS) terminals that still use 2G due to low-cost chipsets, leading to lower prices for the terminals. Therefore, the decision to terminate an old technology and adopt a new one is not just based on "getting on the new technology bandwagon" but is primarily use case and customer profile driven.

Adoption and Adaptability of 5G and Evolving Technologies

The timeline of 5G SA adoption will depend on the timeline for deploying your use cases and the availability of 5G-capable handsets. Fully fledged 5G inter-public land mobile network (PLMN) interconnect might take at least 2–3 years, as you will need your roaming partners to be fully 5G compatible as well. The initial 5G specifications came through 3GPP release 15 (Rel-15), which was primarily based on 5G NSA. Rel-16 then brought in 5G SA–related features, which are now being widened to enhance the network slice as a service (NSaaS), voice-to-everything (V2X), integrations with non-3GPP networks, and so on. Figure 12-1 illustrates the timeline of Rel-17 and work package approval for Rel-18.

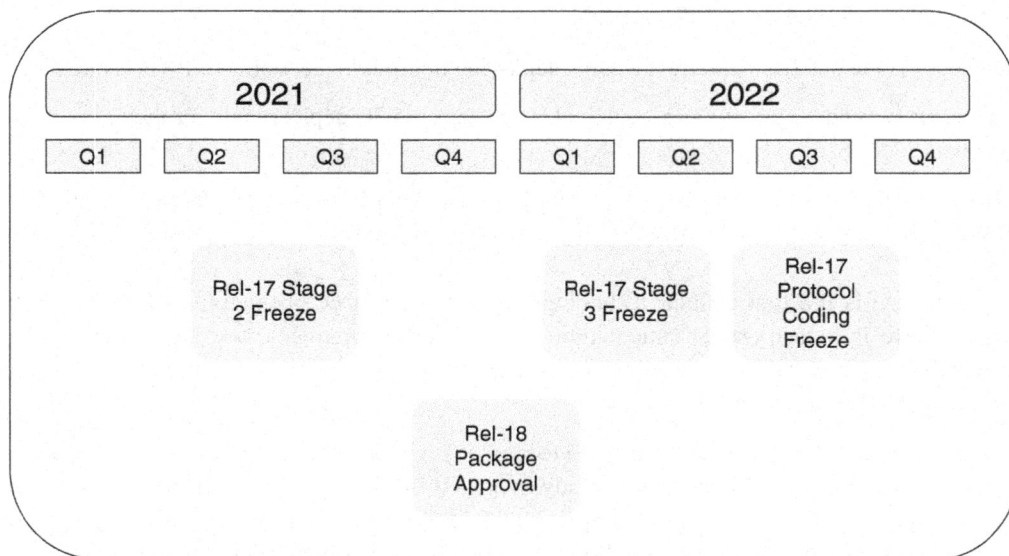

From: www.3GPP.org/specifications/67-releases

FIGURE 12-1   Timeline of Rel-17 and Work Package Approval for Rel-18

At the same time, certain industry verticals such as critical infrastructures as well as military and defense organizations are looking to deploy 5G NPNs to make use of the advantages provided by 5G. Depending on the chosen deployment model, such organizations will require integration with the service provider's 5G Core (5GC) Network Function (NF). You might also see an abundance of 5G-capable phones in the near future, which is a very important ecosystem to consider before you plan to launch 5G.

To provide service to existing subscribers and to enable new 5G use cases, you will require legacy technologies as well as the adoption of 5G technology. Therefore, although investing in 5G security controls is important, you also need to ensure that the security controls for legacy technologies are not ignored. Figure 12-2 illustrates some of the key use cases pinpointed by 5G and evolving technologies.

The range of uses cases targeted for 5G and evolving technologies is much broader than legacy technologies. 5G and evolving networks will transform how you live, work, and travel. To enable this transformation, the standards bodies are working behind the scenes to provide you with a seamless experience. This tight interworking among standards development organizations (SDOs) has never been so important. For example, several SDOs are working together on machine-to-machine standards activity. These SDOs include the Association of Radio Industries and Businesses (ARIB) and the Telecommunication Technology Committee (TTC) of Japan; the Alliance for Telecommunications Industry Solutions (ATIS) and the Telecommunications Industry Association (TIA) of the United States; the China Communications Standards Association (CCSA); the European Telecommunications Standards Institute (ETSI); and the Telecommunications Technology Association (TTA) of Korea.

5G and Beyond

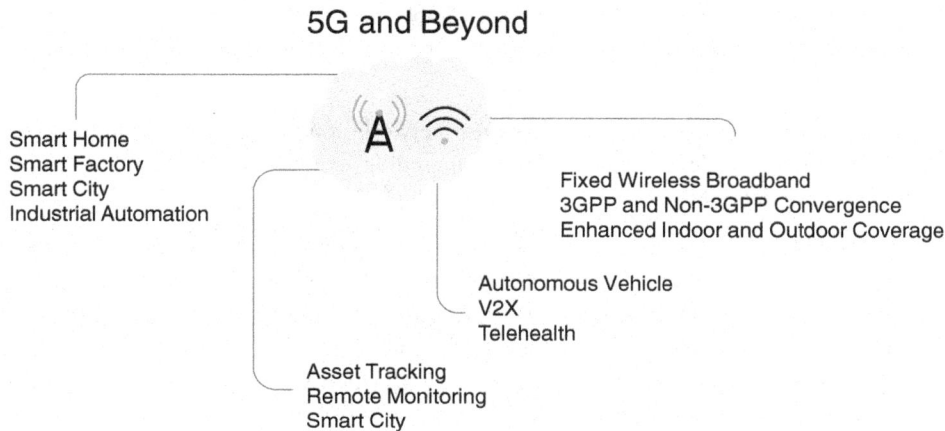

Smart Home
Smart Factory
Smart City
Industrial Automation

Fixed Wireless Broadband
3GPP and Non-3GPP Convergence
Enhanced Indoor and Outdoor Coverage

Autonomous Vehicle
V2X
Telehealth

Asset Tracking
Remote Monitoring
Smart City

FIGURE 12-2   Use Cases Focused on Technologies 5G and Beyond

Once you have deployed 5G SA, you will see a range of use cases that require flexible 5G deployment models. Some deployment models require your infrastructure, such as multi-access edge compute (MEC), to be shared with certain industry verticals, and in some cases you will need to peel off a part of your network and offer it to industry verticals using an NSaaS offering. Such models will expose your network to new risks related to untrusted private and public network integrations. Fulfilling certain use cases might require you to onboard untrusted workloads, leading to risks such as compromised onboarded workloads exposing your entire network to attackers. While it is possible to detect and mitigate such risks manually at a small scale, it is an extremely difficult and nearly impossible task to do manually at a large scale. This requires you to think about scale and feature parity across virtual deployments—be it virtual machines, containers on virtual machines and bare metal, or serverless deployments.

Convergence of Wi-Fi and Evolving Cellular Technologies

ITU-R is an international decision-making body that is in charge of radio communication regulations with 193 member states. It also manages global mobile telecommunication standards and has provided guidance and a roadmap for next-generation communication R&D by defining the vision for next-generation mobile telecommunication, such as 4G and 5G. Starting with completing the 6G Vision by 2023, ITU-R plans to develop technical requirements and recommendations for 6G through industry standards organizations such as 3GPP. Out of those candidate technologies for 6G, the technologies that pass ITU-R's evaluation will be approved as the global standards for 6G around 2030.

ITU-R Working Party 5D (WP 5D) has started to develop a draft new recommendation on IMT Vision for 2030 and beyond, which will include 6G. The 6G Vision Group was newly launched at the ITU-R meeting in 2021. This group is in charge of establishing the 6G Vision, which includes defining the key capabilities, working on technology development, and creating timelines on standardization and commercialization of 6G.

IMT Vision for 2030 and beyond aims at a hybrid dense network consisting of multiple networks (3GPP and non-3GPP), with the desired end result of upgrading existing technologies instead of replacing the technology being used.

Non-3GPP networks like Wi-Fi 6 and 3GPP networks like 5G are built from the same foundation of providing better capacity and throughput, supporting new use cases and applications, and enabling massive IoT. Although it's not the first time the topic of Wi-Fi versus cellular technologies has come up, now the discussions are about Wi-Fi *and* cellular technologies instead of Wi-Fi *versus* cellular technologies. Going forward, any upcoming standard and specification will focus on bringing 3GPP and non-3GPP networks closer.

IEEE 802.11be Extremely High Throughput (EHT), likely to be designated as Wi-Fi 7, is the potential next amendment of the Wi-Fi 802.11 standard, which is currently at 802.11ax (also called Wi-Fi 6). IEEE 802.11be will be built on 802.11ax, focusing on indoor and outdoor operation with stationary and pedestrian speeds in the 2.4, 5, and 6 GHz frequency bands. IEEE 802.11be (Wi-Fi 7) standardization is expected to be ready by 2024. This is around the time when networks will be deploying Rel-17 and initial Rel-18 3GPP 5G networks. The technology evolution in 3GPP cellular networks and Wi-Fi will lead to increased coexistence between them, and service providers might opt for a converged packet core for offering bundled services with Wi-Fi and 5G.

Looking at the overall cost of managing Wi-Fi for large enterprises, network slice as a service (NSaaS) might be an alternative that some might opt for going forward. There was always an option of enterprises choosing 4G instead of Wi-Fi, but the architecture was too constrained, and there was a need for using proprietary solutions to enable a seamless user experience. Looking at the ongoing Rel-17 work items and expected Rel-18 work items, there is a lot of work planned for non-3GPP and 3GPP integrations both at the user device domain and the packet core domain. This might be attractive for large enterprises that might prefer using network slices from service providers, leading to a seamless user experience with improved cost optimization.

Specifically for use cases such as smart cities, the convergence of Wi-Fi and 5G will see increased adoption because many of the free wireless options will still be provided by Wi-Fi (Wi-Fi 7), and critical infrastructure components will use 5G due to enhanced security, support of ultra-reliable low-latency communications (URLLC) for vehicle-to-everything (V2X), and customized use case offerings from service providers or non-public network (NPN) deployment by the industry verticals catering for smart city use cases. New 5G features, such as integration of 5GC Core with untrusted and trusted non-3GPP networks, bring the non-3GPP and 3GPP technologies closer.

3GPP has already started work in having closer integration between 3GPP and non-3GPP networks, starting with Rel-16. Figure 12-3 illustrates a non-roaming architecture for 5GC with untrusted non-3GPP access.

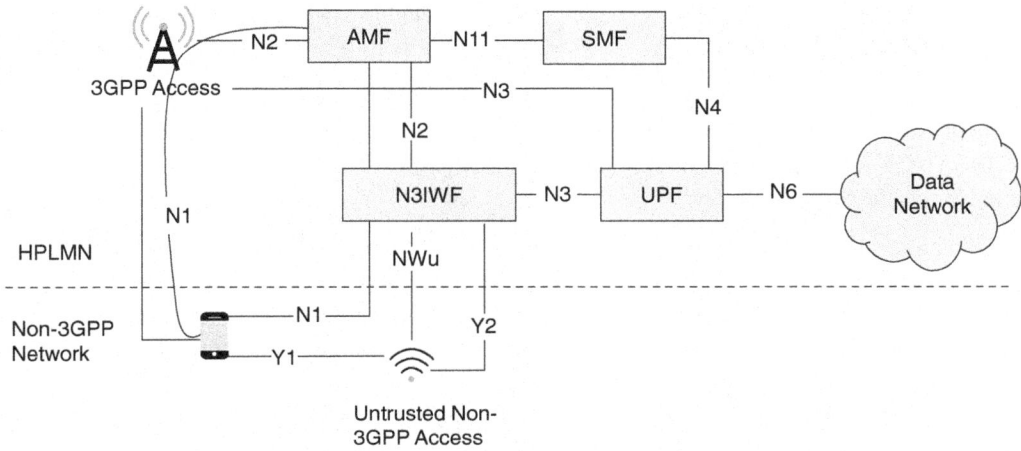

FIGURE 12-3   Non-Roaming Architecture for 5GC with Untrusted Non-3GPP Access

As shown in Figure 12-3, the Non-3GPP Interworking Function (N3IWF) performs routing of messages for the non-3GPP network. N3IWF connects to the User Plane Function (UPF) using the N3 interface and connects the control plane to the Access and Mobility Management Function (AMF) using the N2 interface.

Figure 12-4 illustrates non-roaming architecture for 5GC with trusted non-3GPP access.

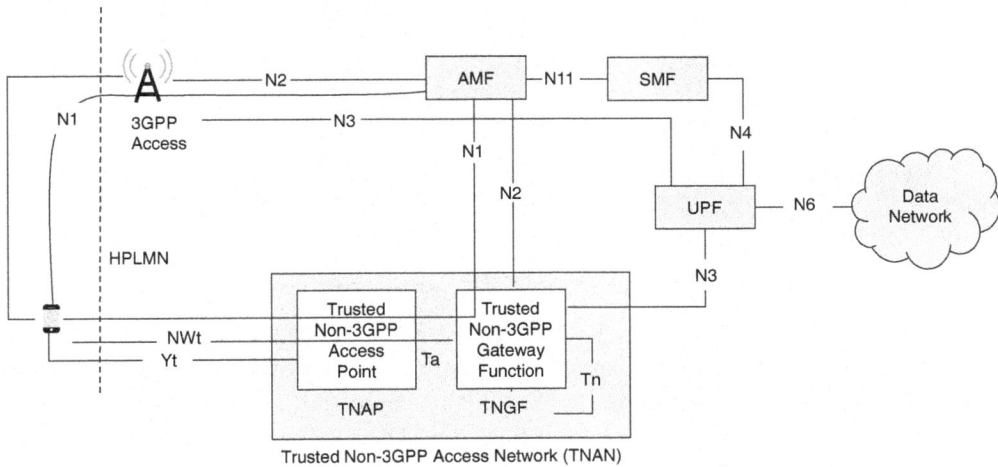

FIGURE 12-4   Non-Roaming Architecture for 5GC with Trusted Non-3GPP Access

The key components of the non-roaming architecture for 5GC with trusted non-3GPP access, as illustrated in Figure 12-4, are discussed next.

The Trusted Non-3GPP Access Network (TNAN) supports WLAN access technology such as Wi-Fi and consists of a Trusted Non-3GPP Access Point (TNAP) and a Trusted Non-3GPP Gateway Function (TNGF). The TNGF integrates with the 5G Core network control plane and user plane functions via the N2 and N3 interfaces, respectively.

3GPP specifications also discuss the integration to allow non-5G-capable devices to access 5GC. The devices that do not support 5GC NAS signaling over WLAN access are referred to as *non-5G-capable over WLAN (N5CW)* devices. N5CW devices are not capable of operating as 5G UE that supports 5GC NAS signaling over a WLAN access network; however, they are capable of operating as 5G UE over NG-RAN, as shown in Figure 12-5.

Figure 12-5 illustrates non-roaming and LBO roaming architecture for supporting 5GC access from N5CW devices.

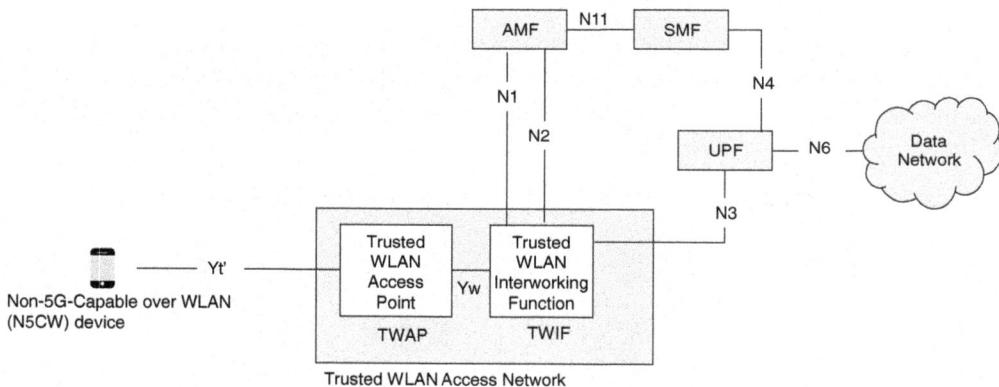

FIGURE 12-5 Non-Roaming and LBO Roaming Architecture for Supporting 5GC Access from N5CW Devices

The key components of the architecture for supporting 5GC access from non-5G-capable devices, as illustrated in Figure 12-5, are discussed next.

A trusted WLAN access network is a particular type of Trusted Non-3GPP Access Network (TNAN) that supports a WLAN access technology such as Wi-Fi. It's composed of a Trusted WLAN Access Point (TWAP) and a Trusted WLAN Interworking Function (TWIF). TWAP is a type of Trusted Non-3GPP Access Point (TNAP) that supports a WLAN access technology such as Wi-Fi. To support 5GC access from N5CW devices, a trusted WLAN access network must support the Trusted WLAN Interworking Function (TWIF). TWIF provides interworking functionality that enables N5CW devices to access 5GC. When an N5CW device performs an EAP-based access authentication procedure to connect to a trusted WLAN access network, the N5CW device may simultaneously be registered to a 5GC of a PLMN. The 5GC registration is performed by the TWIF function in the trusted WLAN access network, on behalf of the N5CW device.

These integrations, with enhanced security being worked on by 3GPP, will enable closer integration of 3GPP and non-3GPP technologies and also allow non-5G-capable devices to access 5GC. However, practical deployments for these architectures are not estimated to be widely available until 2023 or 2024, as there is further work to be done by both UE manufacturers and network equipment vendors to support such deployments. Rel-17, expected to be completed by 2022, and Rel-18, which is in the early study items approval stage at press time, will enhance the interworking and concrete use cases for interworking between 3GPP and non-3GPP networks and provide clarity around the converged core.

Use of AI and ML in Securing 5G and Evolving Networks

Artificial intelligence (AI) and machine learning (ML) generally are used by vendors to showcase how their products are sophisticated and provide you with some "magic" and a cool user interface. Let's dig down a bit deeper and see what exactly AI and ML are and how to use them to secure your 5G and evolving technology deployments.

AI is the intelligence demonstrated by machines. AI was actually founded as an academic discipline in the 1955 and has since evolved to almost mimic human intelligence—of course, without emotions and consciousness.

ML is usually seen as a part of AI. ML is basically an algorithms model based on sampled data, also referred to as training data. ML algorithms make use of this sampled data to make certain decisions or predictions. The larger the set of training data, the more fine-tuned the algorithms become, and theoretical modeling can then be applied by AI to approximate and predict behavior.

In the past, training an ML system was too computationally expensive. The improved availability of the CPUs and GPUs, the emergence of cloud computing, and the widespread use of open source big data technology such as Hadoop has massively reduced the price and encouraged rapid progress in the field of ML and eventually fine-tuned AI systems.

AI and ML are used in variety of applications, such as for research and development (R&D) teams in pharmaceutical industries to provide diagnoses to high-need areas such as cancer research, or in consumer-based areas such as human speech recognition systems in mobile phones and so on. The telecommunications industry will see more AI and ML adoption due to the deployment of 5G and upcoming technologies because of the massive amount of data that can now be collected to enhance various highly dense use cases, such as smart cities, which will see various types of IoT and M2M devices across different verticals, many of which need to coexist. One of the primary areas where AI and ML can be used is to enhance the security posture of your end-to-end networks. They can be used to identify any anomaly in the network and device behavior, decide if that anomalous behavior is malicious, and then mitigate it.

Applying innovative technologies like AI and ML on the collected data from 5G-capable devices, non-5G-capable devices, and network functions to proactively identify issues in the network such as predictive fault detection will be a key catalyst in ensuring full exploitation of the 5G technology within service providers and industry verticals adopting 5G by identifying anomalous behavior and making automated corrective actions.

To reduce human intervention and enhance the security posture of the network, you will require security automation that uses smart ML and AI. Traditionally, security automation is used for high-frequency tasks (in other words, tasks that have low-complexity and high-volume aspects of threat detection and mitigation). But for actual benefit in 5G systems, you need to have ML and AI mechanisms that can proactively predict the threats with high accuracy based on the data collected from different parts of the 5G network.

You can start building up the ML- and AI-based security posture by using service plane, management plane, and control plane information. Discussions with various service providers indicate that data collection on the user plane is not a priority, unless it is mandated by lawful entities based on the country or for specific IoT use cases. IoT use cases will require careful implementation of security controls to ensure that bot-based attacks do not cause a denial of service (DoS) for your legitimate users.

Figure 12-6 illustrates the use of AI and ML to enhance the security posture in your 5G and evolving technology deployments.

As shown in Figure 12-6, AI and ML detect threats by constantly monitoring the behavior of the new and legacy networks, including public cloud–based network function deployments, for any anomalies. Machine learning engines process massive amounts of data in near real time, which will help you discover critical incidents in critical use cases such as critical infrastructure and smart city use cases. These techniques also allow for the detection of insider threats, unknown malware, and policy violations to improve the infrastructure security.

In a constantly changing threat landscape, you need to ensure that the vendors you select use AI and ML to support comprehensive, automated, coordinated responses between various security components.

5G and evolving architectures enable you to provide use cases that define true digital transformation. Use cases such as vehicle-to-everything (V2X), where the vehicle communicates with other cars, pedestrian systems, multi-radio access networks, and sensors to optimize traffic speed will require security capabilities to ensure that attackers don't disrupt the networks and cause chaos. In critical use cases such as the energy and water sectors, gathering data from various sensors and running it through an AI- and ML-based security system will ensure smooth operation and mitigate any threats by detecting and understanding any anomalies in device behavior, thereby enabling proactive, preventive maintenance to eliminate downtime.

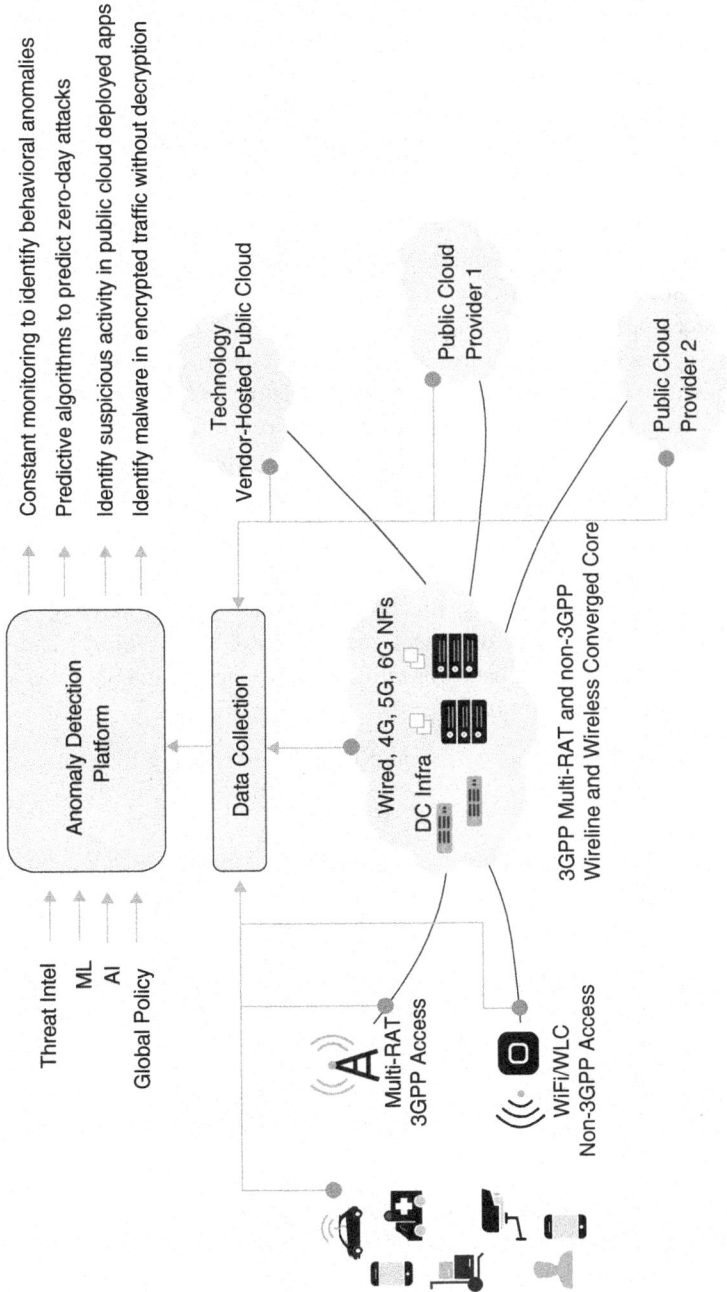

FIGURE 12-6   Example of Using AI and ML to Enhance Your Security Posture

Crypto Agility in 5G and Evolving Technologies

The term *crypto agility* refers to the ability of an organization to swiftly shift from existing cryptographic technology to newer, more robust ones. Crypto agility is very important if you plan to deploy millions of IoT devices, many of which depend on the non-3GPP network to connect to the 3GPP network. With quantum computing becoming more popular, several popular public-key cryptography systems such as RSA could be broken.

The reason existing asymmetric crypto algorithms are vulnerable to quantum computing is due to the fact that they rely on large, universal quantum computers that can compute on a very large scale and solve the integer factorization and discrete log problems used by RSA, Diffie-Hellman (DH), and so on at a much faster rate than usual computers. For symmetric algorithms, the impact is less severe, but the effective key length is at most halved. In other words, this new generation of quantum computers will solve mathematical tasks much faster than existing computers. This means that an attacker with a quantum computer can obtain a private key from a corresponding public key, rendering PKI deployment, which is used in security protocols such as Transport Layer Security (TLS), ineffective with current cryptographic algorithms such as RSA, DH, and so on.

When this will happen, though not clear, is actually just a matter of time. To mitigate such occurrences, you should identify the best options and plan to move your public key systems toward post-quantum-cryptography systems (such as Lattice) at least for the devices in untrusted networks connecting to your 3GPP network. This is important for use cases such as massive IoT (mIoT) and machine-to-machine (M2M), where you might need to cater for millions of devices, as many of these devices use cryptographic algorithms that are not quantum safe.

Here are some key practical considerations you should take before moving to a fully quantum safe cryptographic state (FQSCS):

- Prepare an inventory of the cryptographic algorithms used in your IT and telco environment. This can be done using scanning or monitoring solutions.

- Identify the parts of your network that could be susceptible to attacks, such as critical infrastructures, and ensure that you prioritize those network components for migration toward FQSCS. This can be done using end-to-end enhanced visibility solutions that show you the network topology of your 5G network, helping you determine the most exposed and vulnerable parts of your critical infrastructure.

- Identify and understand the security properties of the quantum-safe cryptographic algorithms and protocols.

- Identify whether the quantum-safe cryptographic algorithm (QSCA) you plan to implement requires being validated by industry efforts like the Cryptographic Module Validation Program (CMVP). CMVP is a joint effort between the National Institute of Standards and Technology, under the Department of Commerce, and the Canadian Centre for Cyber Security. The goal of the CMVP is to promote the use of validated cryptographic modules and provide federal agencies with a security metric to use in procuring equipment containing validated cryptographic modules.

- Validate the compatibility and impact of quantum-safe cryptographic algorithms in your existing architectures. Identify parts of your 5G network and use cases that are dependent on low latency. This can be done by checking the number of network slices configured for your 5G network, which is usually configured in your 5G Core (5GC), and then identifying the slices that are dependent on latency. For example, a vehicle-to-everything (V2X) slice is deemed to be an ultra-reliable low-latency communication (URLLC) use case that is highly dependent on latency. Any increase in latency for such use cases can cause service disruption.

- Consider the compatibility and impact of quantum-safe cryptographic algorithms in security offerings for industry vertical use cases. This can be determined by having discussions with your 5G slice consumers and identifying whether their network devices can implement QSC algorithms.

- Consider the cost of switching to quantum-safe cryptographic algorithms.

The European Telecommunications Standards Institute (ETSI) published a report, TR 103.619, that defines the migration strategies and recommendations for quantum-safe schemes and enhancing cryptography awareness. ETSI recommends three stages for your infrastructure to migrate to FQSCS:

- **Stage 1:** Aimed at compiling all inventories and identifying business process requirements.

- **Stage 2:** Caters for preparation of the migration plan, including the hardware-based security environment, key management, isolation of the sub-systems involved in migration, and business processes to enable execution of the migration.

- **Stage 3:** Caters for execution of the migration.

The National Institute of Standards and Technology (NIST) has also initiated a process to identify, evaluate, and standardize post-quantum cryptography (PQC) algorithms.

Other methods to make 5G and evolving technology deployments crypto-agile include ensuring automation of digital certificates and private key rotations as well as implementing hash algorithms with different lengths of output. If a vulnerability is found in the quantum-safe cryptographic algorithm, it might be necessary to switch to a different quantum-safe algorithm entirely, although in some instances the specific vulnerability can be addressed by restricting modes or by revising the strength of individual parameters. Ensuring cryptographic agility will make this significantly easier.

Ensure that all the members of your security team, from security strategy to the application development engineers, understand the need for crypto-agile implementation; it should be taken into consideration for each and every strategy, design, product, and deployment decision.

You should also focus on some key methods, such as ensuring that your network equipment and device vendors supply you with the most recent cryptographic protocols and algorithms. They should also supply you with software updates and provide you with information on crypto agility of the hardware and software used in their products.

Summary

Migrating to 5G SA and adapting it for various 5G-related use cases requires the right security controls and best practices. Although 3GPP will provide enhancements in Rel-17 and upcoming specifications, it will not be enough for the wider 5G ecosystem to be secure—the main reason being the integration of 3GPP-specified systems with other non-3GPP systems. Although there is ongoing work in these areas, the work is in the early stages and requires further development, and the technology partner ecosystem needs to be widened.

You will also need to use AI and ML systems for your network for variety of security use cases. The security posture of your network can benefit from AI and ML, as they enable near-real-time detection of threats, prediction of possible zero-day attacks, identification of anomalies in the network, and mitigation of threats.

It is also very important to provide crypto agility to your network to ensure that you can adopt newer, more robust cryptographic algorithm systems. This will help better secure your network and devices from advanced persistent threats (APTs) and take you toward the next step of getting your network to a fully quantum-safe cryptographic state (FQSCS).

All these different discussions bring us to the conclusion that we need to look at security from a different angle and secure converging technologies and integrations that allow flexible deployments and an open ecosystem.

As the network evolves, opening up and bringing in the deployment of mobile packet core network functions within the public cloud and closer third-party application integrations, the cybersecurity controls also need to evolve to ensure your infrastructure remains protected to keep the sensitive data of your employees, subscribers, and customers secure.

Acronym Key

The following table expands the key acronyms used in this chapter.

Acronym	Expansion
5GC	5G Core Network
AMF	Access and Mobility Function
API	Application programming interface
AUSF	Authentication Server Function
CNF	Cloud-Native Function

Acronym	Expansion
FQSCS	Fully quantum-safe cryptographic state
IAM	Identity and Access Management
IoT	Internet of Things
M2M	Machine-to-machine
MBB	Mobile broadband
MEC	Multi-access edge compute
N3IWF	Non-3GPP Interworking Function
N5CW	Non-5G-capable over WLAN
N5GC	Non-5G-capable
NEF	Network Exposure Function
NF	Network Function
NSaaS	Network slice as a service
NSSF	Network Slice Selection Function
PKI	Public key infrastructure
QSC	Quantum-safe cryptography
RAT	Radio access technology
SBA	Service-based architecture
SBI	Service-based interface
SEAF	Security Anchor Functionality
SEPP	Security Edge Protection Proxy
SMF	Session Management Function
SUCI	Subscription Concealed Identifier
SUPI	Subscription Permanent Identifier
TLS	Transport Layer Security
TNAN	Trusted Non-3GPP Access Network
TNAP	Trusted Non-3GPP Access Point
TNGF	Trusted Non-3GPP Gateway Function
TNL	Transport Network Layer
TWAP	Trusted WLAN Access Point
TWIF	Trusted WLAN Interworking Function
UDM	Unified Data Management
UPF	User Plane Function
UPF	User Plane Function
VNF	Virtual Network Function
VNFM	Virtual Network Function Manager
W-5GAN	Wireline 5G Access Network

Acronym	Expansion
W-5GBAN	Wireline BBF Access Network
W-5GCAN	Wireline 5G Cable Access Network
W-AGF	Wireline Access Gateway Function
WAF	Web application firewall

References

3GPP TS 23.501, "System Architecture for the 5G System: Stage 2"

3GPP TS 33.501, "Security Architecture and Procedures for 5G System"

3GPP TR 103.619, "Migration strategies and recommendations to Quantum Safe schemes"

Cisco Systems, "Transforming Business with Artificial Intelligence," Cisco Public, 2018

Cisco Systems, "Reimagining End-to-End Network Mobile Network," Cisco Public, 2019

https://www.itu.int/en/ITU-T/focusgroups/net2030/Documents/White_Paper.pdf

https://www.3gpp.org/specifications/67-releases

Index

authentication transforms (ESP), 102–103

authorization, secure API, 353

automation, network slicing and, 305–306

autonomous pedestrian system, 379

AWS Mesh, 487

B

B2B (business-to-business), 40, 472

B2B2X (business-to-business-to-everything), 41, 472

B2C (business-to-consumer), 40, 472

B2H (business-to-home), 472

B2X (business-to-everything), 40–41

back-haul sniffing

mitigation examples, 222–223

real scenario case study, 222–223

bandwidth, 5G, 6

bare metal (BM), CNF as containers on, 240

base band unit (BBU), 39, 117–118

Base Station Controller (BSC), 235

Base Transceiver Station (BTS), 235

BBU (base band unit), 39, 117–118

BGP (Border Gateway Protocol), 312

Bluetooth sniffing, 382

BM (bare metal), CNF as containers on, 240

Border Gateway Protocol (BGP), 312

botnets, 386–391

DNS-based attacks, 390–391

IRC (Internet Relay Chat), 387–388

P2P (Peer to Peer), 388–390

Telnet, 387

Broken Authentication API attacks, 172–174

Broken Functional Level API attacks, 220-C05.9001

Broken Object Level authorization attacks, 172–174, 207–210, 311

BSC (Base Station Controller), 235

BTS (Base Transceiver Station), 235

build, deploy, and run processes, 193–198

built-in device hardening, 5GC virtual environments, 291–292

business-to-business (B2B), 40, 472

business-to-business-to-everything (B2B2X), 41, 472

business-to-consumer (B2C), 40, 472

business-to-everything (B2X), 40–41

business-to-home (B2H), 472

C

C language, 330

C&C (command-and-control) servers, 386–391

DNS-based attacks, 390–391

IRC (Internet Relay Chat), 387–388

P2P (Peer to Peer), 388–390

Telnet, 387

CAG ID (Closed Access Group Identity), 48–49

CAGR (compound annual growth rate), 377

calculated reference time, 43

CAPIF (Common API Framework), 14, 198–199, 263

key features in 3GPP releases, 20

security enhancements in Rel-16, 67–70

smart factory use case, 431

cargo sensors, 379

CAs (Certification Authorities), 132, 182

CCSA (China Communications Standards Association), 464, 538

cellular IoT (C-IoT), 18–19

cellular technology evolution

4G architecture, 4–5

5G enhancements

4G architecture compared to, 4–5

cloud-native technology. See cloud computing

disaggregated architecture, 7–10

flexible architecture, 10–11

key 5G features in 3GPP releases, 18–20

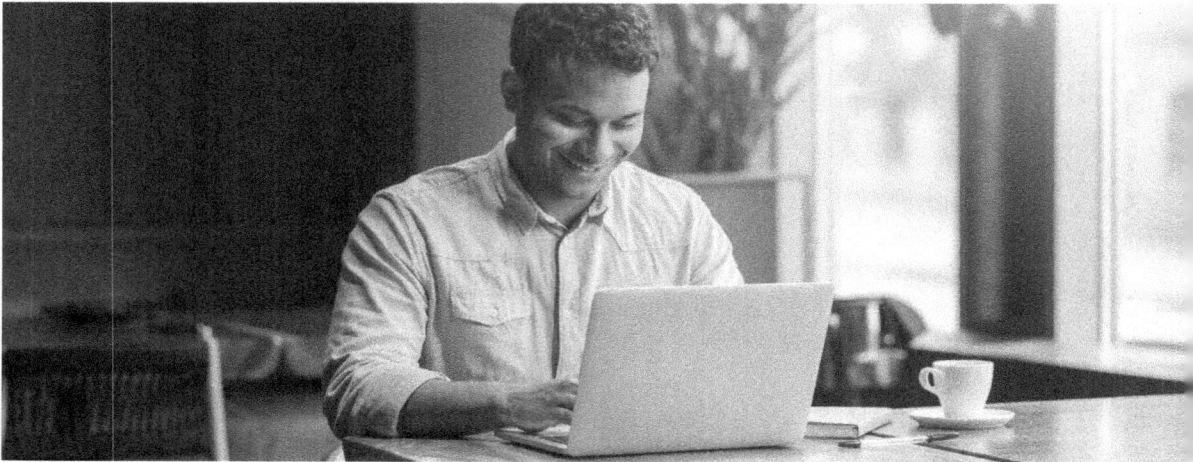

Register Your Product at pearsonITcertification.com/register

Access additional benefits and **save 35%** on your next purchase

- Automatically receive a coupon for 35% off your next purchase, valid for 30 days. Look for your code in your Pearson IT Certification cart or the Manage Codes section of your account page.

- Download available product updates.

- Access bonus material if available.*

- Check the box to hear from us and receive exclusive offers on new editions and related products.

Registration benefits vary by product. Benefits will be listed on your account page under Registered Products.

Learning Solutions for Self-Paced Study, Enterprise, and the Classroom

Pearson IT Certification delivers training materials that address the learning, preparation, and practice needs of a new generation of certification candidates, including the official publishing programs of Adobe Press, Cisco Press, and Microsoft Press. At pearsonITcertification.com, you can:

- Shop our books, eBooks, practice tests, software, and video courses
- Sign up to receive special offers
- Access thousands of free chapters and video lessons

Visit **pearsonITcertification.com/community** to connect with Pearson IT Certification

P Pearson

Addison-Wesley • Adobe Press • Cisco Press • Microsoft Press • Pearson IT Certification • Que • Sams • Peachpit Press